Science and Technology

A FIVE-YEAR OUTLOOK

Science and Technology

A FIVE-YEAR OUTLOOK

Published in collaboration with the National Academy of Sciences by

W. H. FREEMAN AND COMPANY
SAN FRANCISCO

This report was prepared at the request of the National Science Foundation, under contract No. NSF-C310, T.O. 402

Library of Congress Cataloging in Publication Data
National Academy of Sciences, Washington, D.C.
Science and technology.

Includes bibliographies and index.
1. Research—United States. 2. Engineering research—United States. 3. Science—United States. 4. Technology—United States. I. Title.
Q180.U5N29 1979 507'.2073 79-15862
ISBN 0-7167-1140-0
ISBN 0-7167-1141-9 pbk.

Printed in the United States of America

9 8 7 6 5 4 3 2 1

Study Chairman

RALPH E. GOMORY, International Business Machines Corporation

Advisory Board

RICHARD C. ATKINSON, National Science Foundation
PHILIP HANDLER, National Academy of Sciences
FRANK PRESS, Office of Science and Technology Policy

Steering Committee

BERNARD DAVIS, Harvard Medical School
DAVID HAMBURG, Institute of Medicine
JULIUS HARWOOD, Ford Motor Company
JAMES HILLIER, RCA Corporation
THOMAS MALONE, Butler University
DAVID ROSE, Massachusetts Institute of Technology
ROBERT G. SACHS, University of Chicago
JACOB SCHWARTZ, New York University
CONRAD TAEUBER, Georgetown University
GILBERT F. WHITE, University of Colorado
ROBERT M. WHITE, National Academy of Sciences

Contributors

MANUEL AVEN, General Electric Company
DAVID BALTIMORE, Massachusetts Institute of Technology
JOHN BARDEEN, University of Illinois
GORDON BAYM, University of Illinois, Urbana
WILLIAM BRINKMAN, Bell Laboratories
ROBERT BURRIS, University of Wisconsin
ALBERT CLOGSTON, Bell Laboratories
JOHN COCKE, International Business Machines Corporation
MORREL COHEN, University of Chicago
MORRIS COHEN, Massachusetts Institute of Technology
GEORGE E. DIETER, University of Maryland
CHARLES DRAKE, Dartmouth College
RICHARD EASTERLIN, University of Pennsylvania
EDWARD EVARTS, National Institutes of Health
DEBORAH FREEDMAN, University of Michigan
HERBERT FRIEDMAN, Naval Research Laboratory
GERALD T. GARVEY, Argonne National Laboratory

PAUL GLICK, Bureau of the Census
PHILIP HAUSER, University of Chicago
ROGER HEYNS, W. R. Hewlett Foundation
WALTER R. HIBBARD, Virginia Polytechnic Institute and State University
DAVID HUBEL, Harvard Medical School
ROBERT I. JAFFEE, Electric Power Research Institute
JOSEPH KATZ, Argonne National Laboratory
CHARLES KIDD, George Washington University
LEONARD KLEINROCK, University of California, Los Angeles
RICHARD KRAUSE, National Institutes of Health
HOWARD J. LEWIS, National Academy of Sciences
MALCOLM MACFARLANE, Argonne National Laboratory
CLEMENT MARKERT, Yale University
GORDON MOORE, Intel Corporation
NORTON NELSON, New York University
WILLIAM NIERENBERG, Scripps Institution of Oceanography
NILS NILSSON, Stanford Research Institute
ALFRED NISONOFF, Brandeis University
ITHIEL DE SOLA POOL, Massachusetts Institute of Technology
DAVID PRESCOTT, University of Colorado
DAVID PRICE, Argonne National Laboratory
WILLIAM R. PRINDLE, National Academy of Sciences
C. B. RALEIGH, U. S. Geological Survey
EBERHARDT RECHTIN, Aerospace Corporation
JOHN J. SCHANZ, Resources for the Future
JOHN SCHIFFER, Argonne National Laboratory/University of Chicago
SHELDON J. SEGAL, The Rockefeller Foundation
HARRISON SHULL, Indiana University
RICHARD THOMPSON, University of California, Irvine
W. MURRAY TODD, National Academy of Sciences
KING WALTERS, Rice University
MAX WEISS, Aerospace Corporation
HATTEN YODER, Carnegie Institution

Editorial Consultants

DUNCAN BROWN, National Academy of Sciences
MARGARET M. CARTER
ELIZABETH M. FISHER
ELIZABETH W. FISHER
BARBARA FURST
RUTH HAAS, Resources for the Future
SAMUEL IKER

WILLIAM KIRK, Stanford Linear Accelerator Center
MARY LOU LINDQUIST
MICHAEL OLMERT
KENNETH REESE
PHILLIP SAWICKI
LYDIA SCHINDLER
LUCIENNE SKOPEK
ELIZABETH STEPHENS
WALTER SULLIVAN, *New York Times*
WINFIELD SWANSON
DONNA TURNER
PEGGY D. WINSTON
GEORGE W. WOOD, National Academy of Sciences

Production

National Academy of Sciences

BARBARA S. BROWN
ESTELLE H. MILLER
DAVID M. SAVAGE

Support

National Academy of Sciences

LINDA CANNON
EDNA SAUNDERS
CHERYL SULLIVAN
ANN THOMPSON

International Business Machines Corporation

PAULINE THOMSEN

Staff

National Academy of Sciences

JOHN S. COLEMAN
NORMAN METZGER
SUSAN C. PERRY

Contents

PREFACE / xiv

OBSERVATIONS / 1

I SCIENCE

1 PLANET EARTH / 9

Introduction, 9
The Crust in Motion, 11
Earthquakes, 15
Our Mineral Future, 21
Promise of the Oceans, 26
The Changing Climate, 31
View from Space, 40
Afterthoughts, 49
Outlook, 51
References, 56
Bibliography, 57

2 THE LIVING STATE / 59

Introduction, 59
Molecular Genetics, 63
Cell Biology, 75

Immunology, 85
Neuroscience, 95
Biology and Agriculture, 116
Conclusions, 133
Outlook, 135
References, 140
Bibliography, 141

3 THE STRUCTURE OF MATTER / 143

Introduction, 143
Astrophysics and Cosmology, 146
Condensed Matter, 163
Molecular and Atomic Structure, 170
Nuclear Structure, 179
Particle Physics, 190
Conclusion, 204
Outlook, 206
References, 211
Bibliography, 211

II TECHNOLOGY

4 COMPUTERS AND COMMUNICATIONS / 215

Introduction, 215
Current Technology and Systems, 216
Software: Problems and Techniques, 227
Theoretical Computer Science and the Contribution from Mathematics, 231
Computers and Communications, 236
Artificial Intelligence, 246
Outlook, 249
References, 251

5 ENERGY / 253

Introduction, 253
Some Important Time Perspectives, 254
Mid-Term Energy Options, 256
Mid-Term Supply Strategies, 259

Some Outstanding Environmental and Health Effects, 273
Long-Term Energy Options, 275
Some Research Needs, 286
Conclusion, 290
Outlook, 291
Notes and References, 297

6 MATERIALS / 301

Introduction, 301
Developments in Materials, 302
Materials Processing and Manufacturing, 317
Recent Concepts in Materials, 323
Near-Term Issues in Materials, 326
Outlook, 335
References, 340

III SCIENCE AND THE UNITED STATES

7 DEMOGRAPHY / 343

Introduction, 343
Demographic Trends in the United States, 346
Age Structure, 351
Internal Migration, 353
Households and Family Formation, 356
Labor Force, 358
School and College Enrollment, 366
Birthrate Projections, 367
Implications of Geographic Redistribution, 368
Conclusion, 373
Outlook, 376
References, 378

8 HEALTH OF THE AMERICAN PEOPLE / 381

Introduction, 381
Cardiovascular Diseases, 387
Cancer and Related Problems, 396
Cigarettes and Health, 401
Mental Illness and Biobehavioral Sciences, 404

Aging and Health, 412
Genetic Factors in Disease, 417
Innovation in Health Care Delivery, 417
Perspectives on Health, 423
Outlook, 428
References, 433
Bibliography, 435

9 TOXIC SUBSTANCES IN THE ENVIRONMENT / 439

Introduction, 439
Growth of Chemical Technology, 442
Persistence, Transformation, and Movement of Chemicals, 446
Compounds of Special Concern, 451
Adverse Effects on Nonhuman Species, 455
Scientific, Technical, and Policy Responses, 456
Outlook, 461
References, 463

IV INSTITUTIONS

10 ACADEMIC SCIENCE AND GRADUATE EDUCATION / 469

Introduction, 469
Scientific Research and Graduate Education, 471
Enrollment, Degrees, and Jobs, 472
Women and Minorities, 476
Funding of Academic Science, 477
Project Funding, 482
Other Issues, 485
Outlook, 486
References, 488

11 INSTITUTIONS FOR INTERNATIONAL COOPERATION / 491

Introduction, 491
The Rationale for International Cooperation, 492

Some Characteristic International Institutions of Science and Technology, 495
Some Speculation About Future Needs and Prospects, 505
Perspectives: The Near-Term Environment for International Institutional Change, 511
Outlook, 515
Notes and References, 516

REVIEWERS AND ADDITIONAL CONTRIBUTORS / 519

INDEX / 527

Preface

The National Science and Technology Policy, Organization, and Priorities Act of 1976 directed the Office of Science and Technology Policy to prepare periodically a Five-Year Outlook on Science and Technology. Specifically, the Congress asked to be informed of conditions that might warrant special attention within the next five years, involving—

> • Current and emerging problems of national significance that are identified through scientific research, or in which scientific or technical considerations are of major significance; and
>
> • Opportunities for, and constraints on, the use of new and existing scientific and technological capabilities which can make a significant contribution to the resolution of these problems. . . .

A reorganization of the Executive Office of the President in 1977 transferred the responsibility for a five-year outlook to the National Science Foundation. In 1978, the Foundation asked the National Academy of Sciences to assist in the preparation of the Five-Year Outlook by providing a report describing the current state of significant research areas and pointing out issues within those areas that could be of special concern within the five-year period. After preliminary consideration of the intent and content of the report, the Academy began its work under contract in August 1978, with a commitment to deliver a manuscript to the Foundation by the end of March 1979.

With the advice of the President's Science Advisor and the Director of the National Science Foundation, the Academy—acting principally through the National Research Council, in collaboration with the National

Academy of Engineering and the Institute of Medicine—selected a number of areas in science and technology to be included in the first outlook report. Ralph E. Gomory was appointed Study Chairman. In turn, a steering committee was constituted, its membership composed of those principally responsible for the preparation of the chapters of the report.

A two-tier review procedure was also established: a review of initial drafts by members of the National Academy of Sciences, the Institute of Medicine, the National Academy of Engineering, and units of the National Research Council; and a final review, prior to release of the report to the National Science Foundation, by the Governing Board of the National Research Council.

This report was submitted to the National Science Foundation on March 30, 1979. It is anticipated that when the Foundation transmits the formal five-year outlook to the Congress it will include—in addition to this Academy report—a set of papers prepared by selected federal agencies, papers commissioned by the Foundation on selected, policy-related topics, and a summary and analysis. Present plans call for the process to be repeated biennially.

The present report is not a comprehensive statement on the status of all science and technology in the United States. Nor is it a statement of the relative merits or priorities among the large family of disciplines and activities that comprise science and technology. The sweep of the scientific and technological enterprise, as well as the short time allotted for the preparation of this report, compelled a selective approach. Major areas—such as mathematics, economics, synthetic chemistry, and land use and water resources; the technologies of the transportation, buildings, and textile industries; and the general question of industrial technology and innovation, to name but some—are not explicitly discussed.

The chapters that follow do, however, illustrate the present vitality and direction of much of American science and technology. They should enable appreciation of both the substance of science and technology and its interaction with the formulation of public policy. The 11 chapters are divided into 4 units: science, technology, topics specific to the United States, and scientific institutions. The contents and arrangement of the chapters are given in the Contents. An index is also provided.

Each chapter is followed by a short outlook emphasizing five-year trends. These outlook statements are brief; the bases for the trends there noted are to be found in the chapters themselves. A section entitled "Observations" follows the Contents and provides some general thoughts on American science and technology.

The projections for various areas of science and technology derive from assessments of their current status, the work being done, and the evident trends. That approach is tempered by specifics. Since technological products and processes available five years from now should be at the design stage today, we can be surer of the five-year prediction for a given technology, whether it be of new forms of plastics to replace metals in car manufacture or the likelihood that new commercial coal conversion processes will be in place. In basic science, however, the more reasonable approach is to assess current work and its general directions, for example, pointing out the considerable recent progress made in understanding the role of membranes of living cells or the steady addition to the list of precursory phenomena associated with earthquakes.

Observations

Science and technology continue to transform our world. Our view of the universe about us now ranges outward to the galaxies, back in time to the origin of life and of the universe, and inward into the nucleus of the atom and the molecular basis of heredity.

Science today is ripe with discoveries and new explorations. On the grandest scale, that of the cosmos, it is revealing an array of new objects; X-ray bursters, for example, flashing like a million suns for a second and quiet 10 seconds later. The confusions and blind alleys that marked the search for the very heart of matter now seem to be giving way to clarity. Permutations of only a very few particles, the quarks, rather than of hundreds may account for the structure of particles composing the nucleus of the atom. And through the prism of plate tectonics, we now see a planet whose oceans are growing larger or smaller, whose continents shift, and whose seafloors are younger than its continental masses.

The enormous transformation of biology triggered by the discovery of the nature of the genetic material DNA continues. In neurobiology, powerful instruments and techniques now make it possible to examine the electrical activity of single cells in the brain and to understand in molecular detail how impulses are transmitted among nerve cells, by contact and by newly discovered neurohormones.

The new capabilities and understanding from science and technology continue to mesh with the goals and needs of our society. The prevalence of chronic illnesses such as cancer and cardiovascular disease calls for better understanding of their nature and causes. Pressures on energy supplies are driving investigation of the chemistry of different coals, of the relative hazards of different fuels, and of the possible shapes of a more electricity-dependent society. Government programs—from allocation of funds for wastewater treatment plants to estimating the future number of annuitants of the Social Security System—demand continual refinement of demographic techniques and analyses.

Some of the links between science and societal concerns are more subtle. The carbon dioxide issue is instructive. Its science is concerned with long-term measurements at remote stations—at Point Barrow, Alaska; at Mauna Loa, Hawaii; at the South Pole; at American Samoa—with analysis of factors affecting the rate at which the oceans and the biosphere produce, absorb, and release carbon dioxide; and with complex mathematical models demanding the most powerful computers available. The societal issue is whether the increased release of carbon dioxide from burning fossil fuels, which may lead to higher average climatic temperatures throughout the globe, is an acceptable risk.

Science and technology are, of course, done within institutions—universities, industrial and governmental laboratories, and various organizations for international cooperation. The United States does a large part of its basic research in the universities, whose faculties are both researchers and teachers.

This productive intertwining of research and education has given U.S. science its vigor and also some of its problems. For example, with the present system, the continued high quality of U.S. science depends on the ability of the universities to sustain excellence in basic research while accommodating to several changes. These include the sharp decline in the research growth rate of the 1960's to the more gradual level of the 1970's, the decrease in the potential college population as a result of the decline in birth rates from the early 1960's on, and a diminution in available faculty positions.

Developing new technology into commercial products or processes—innovation—is an even more complex matter than assuring the health of basic research in the United States. One reason is the interaction of economic and technical concerns, invalidating simple analyses. The belief that scientific preeminence goes hand in hand with technological leadership is clearly false. While Japan's scientific stature was until recently not comparable to that of the United States, that country is nevertheless challenging us in many areas of high technology. Scientific excellence is important, but not sufficient for the effective support and introduction of new technology.*

*While the complex issue of the relationship of science and technology to industrial innovation is not treated in this report, a very broad analysis by the Department of Commerce of federal policy and industrial innovation is presented in its 1979 report entitled *Domestic Policy Review on Industrial Innovation.*

There is now more emphasis on international collaboration in science, occasioned sometimes by the scale of particular research projects, sometimes by the desire to share the costs of major new instruments, by the need for global observations in some areas of science, and to some extent by the recognition of the scientific maturity of other nations.

Space technology and computers have made possible great multinational programs of observation and research in oceanography, meteorology, and geophysics. A vivid example is the Global Weather Experiment, an effort by 140 countries, the World Meteorological Organization of the United Nations, and the International Council of Scientific Unions to observe the world's weather at one time in unprecedented detail and to analyze the data with unprecedented speed. Its goals include reliable weather forecasts over longer periods and better understanding of the dynamics of seasonal climatic change.

Scientific work on an international scale extends beyond basic research. Efforts to improve the economic status and quality of life in developing nations and to strengthen their scientific and technological bases occupy the time and concern of many scientists. Such efforts include large international centers of agricultural research, employing 8,000 people and having an annual budget of $100 million. High-yielding wheat and rice created at these international centers now comprise a third of the worldwide acreage sown with these grains.

This country's involvement in international science and technology is inevitably affected by the changing position of U.S. science. The United States for three decades clearly has been a world leader in almost all areas of science. This is changing as other countries, building on strong economic bases, mount research efforts that in some areas equal and in some surpass our own.

The changing status of American science vis-à-vis that of other countries will continue to prompt examinations of how the United States can best assure the continued vitality of its basic research—whether its investments are properly allocated and of sufficient amount, whether its system for conducting its research principally through the universities is the optimal one, and so forth. Some of that questioning is evident in this report.

Instrumentation is a recurring theme. Electron beam microscopes open a different world than do optical ones. New accelerators are new windows into the ultimate structure of the nucleus of the atom. Electromagnetic sensors, operating at wavelengths we cannot see, reveal new events in the cosmos. And always more is demanded of these tools as we explore ever more distant celestial objects, chemical reactions on the time scale of their actual mechanisms, more subtle and complex genetic controls, and new clues to the forces transforming the face of the earth.

As the contents of this report have been shaped largely by the instruments available, so will the subject matter of future Outlook reports depend on new tools for scientific research and technological development. Computations that once required hours and now take minutes may be reducible to seconds. The large space telescope, the multiple mirror telescope, and the Very Large Array radio observatory will further probe cosmic mysteries. Tools for reading the properties of surfaces will continue to develop, enabling further understanding of corrosion, catalysis, the growth of crystals, and other surface phenomena. The already exquisite methods of studying very fast, minute events among nerve cells will become even more sophisticated.

As we move farther away from the direct perception of our senses in most areas of science, the tools become more complex, often going beyond what can reasonably be built or used by an individual or the typical small university research team. Of course, the very large, extremely expensive instruments such as particle accelerators are already shared internationally. Arrangements for sharing more modest instruments, growing beyond the reach of individual investigators because of their rising costs and complexity, are also needed to sustain first-class research. Some of these arrangements are already in place, and it is likely that new modes of sharing costly instruments regionally and nationally will evolve as universities seek to adjust from the exuberant growth of the 1960's to the more gradual growth of the 1970's.

Regulation is also an issue pertinent to the nation's scientific effort, regulation both of hazardous actions and of possibly hazardous knowledge. The issue appears in different ways, depending on whether it involves automobile emissions, possible intrusions into privacy and confidentiality through the ubiquitous spread of new computer and communications technology, or the problems of nuclear power. The regulation of actions that may be harmful is a proper and necessary function of society. Quite a different issue is raised, however, by the assertion that the creation of knowledge should also be controlled because knowledge in itself is societally destructive or subject to abuse. Basic knowledge can be used for both good and ill. The possible applications of new knowledge are so varied and unpredictable as to preclude the capacity to foresee net harm. Our historical experience tends to support the view that knowledge is better than ignorance and that it is better to regulate applications than knowledge.

A final observation relates to ignorance and to the limits of science. Scientific knowledge is systematic, enormous in its extent, powerful; but it is slight compared to what is not known. Thus, science has contributed precise knowledge on such seemingly esoteric matters as the electronic structure of atoms, knowledge that has been used to create new technologies and indeed new industries. However, we remain uncertain about seemingly common-sense questions, such as the effects of different air pollutants on human health. These uncertainties simply indicate questions whose answers are not yet part of the core of agreed-on science. That core will expand, but it will always be smaller than needed to answer unambiguously all questions asked by society.

I SCIENCE

(NASA)

1 Planet Earth

INTRODUCTION

We live on a restless earth. The evidence is everywhere. Changing ocean tides and the passage of storms confront our senses as part of our daily existence. We are unaware of the imperceptible movements within the earth except when they are manifested by earthquakes and volcanoes. Movements on and within our dynamic planet take place on many scales of time and space, from circulations within the earth's interior over millions of years, to oceanic motions over centuries, to atmospheric storms over hours or minutes.

Our understanding of the planet has changed in recent years. We now believe that the outer shell of the earth is composed of large plates that move like rafts at sea, but only a few centimeters per year. The cumulative effects of these imperceptible motions over millions of years have shaped our planet's surface. The recognition of these plates and the reconstruction of the history of their motions is one of the profound achievements of modern earth science.

We also better understand the complexity of oceanic motions, the fine structure of the great global currents such as the Gulf Stream, and the importance of the slowly moving deep waters of the world's oceans for marine ecosystems and the chemistry of our environment. We are beginning to comprehend the basis of our changing climate, forces that bring on ice ages, and the droughts and floods that plague us.

Our new understanding results from the application of a rapidly changing technology to scientific inquiry. Space technology has given us a

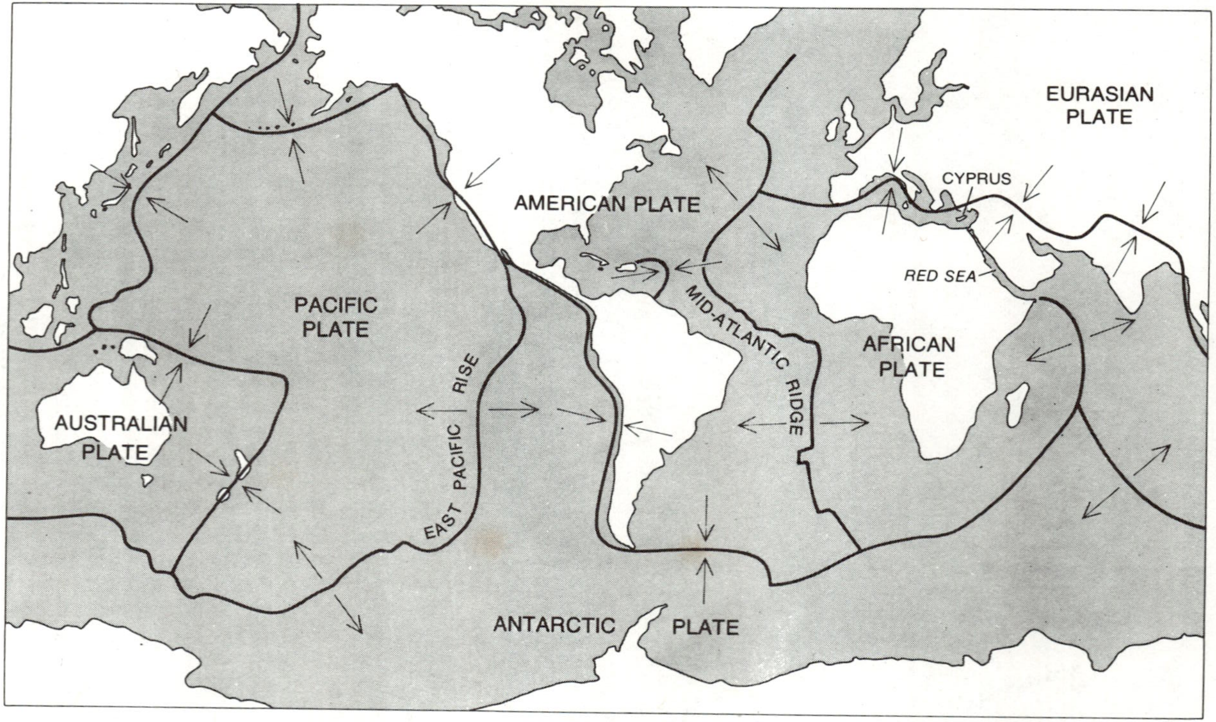

FIGURE 1 The principal plates of the lithosphere, the rigid outer shell of the earth. The paired arrows indicate whether a plate boundary is convergent or divergent. (From "Plate Tectonics and Mineral Resources," by Peter A. Rona, © July 1973, Scientific American, Inc.)

finer view of the planet's surface, its oceans, its enveloping atmosphere, and the atmospheres of other planets. Submarine sediments have been sampled at great depths with modern drilling techniques and the earth's interior has been explored remotely with new arrays of seismic instruments.

Finally, we are learning to interpret the new data with the help of modern communication, display, and computer devices. With these, we can compare observations with hypotheses. We can, for example, now electronically simulate probable conditions within the earth, sea, and air; and we can test concepts in a manner beyond the reach of an army of earlier experimenters. We can simulate the intricate, competing chemical reactions occurring within the upper atmosphere, estimate the effect on climate of more CO_2, assess the effects of the oceans on the weather, or test ideas about motions within the solid earth.

We are witnessing a new age of discovery about the earth, and conceptions of the world about us are being altered.

THE CRUST IN MOTION

THEORY OF PLATE TECTONICS

Not since Copernicus displaced the earth from the center of the universe has there been such a revolution in scientists' concept of the planet as plate tectonics. Plate tectonics touches many of the most critical problems of our time—finding new mineral resources and new sites to search for offshore oil and gas deposits, predicting earthquakes, understanding climate changes, and selecting areas for disposing of nuclear wastes.

The concept of plate tectonics is essentially that the rigid surface of the earth is actually divided into separate plates some 100 km in thickness that, in some places, move relative to one another at a rate of several centimeters a year. A plastic semisolid layer beneath the rigid surface lubricates the movements of the overlying plates and may transport them. Some plates, such as that which forms the northern part of the Pacific Ocean, are vast. Others, like those within the Mediterranean region, are smaller fragments wedged between larger plates such as the gargantuan Eurasian and African plates (Figure 1).

We are directly affected. As the Pacific plate slowly moves northwest, it carries with it the rim of North America from Baja California in Mexico to the seaward edge of San Francisco, causing earthquakes along the San Andreas Fault in California.

The Pacific is thought to be a shrinking ocean, and the Atlantic an expanding one. Much of the rim of the Pacific floor dives beneath chains of islands—the Aleutians, Japan, the Philippines, etc.—causing earthquakes.

Part of the descending seafloor melts a few hundred kilometers down and resurfaces through the overlying crust to feed volcanoes.

The American and European–African sides of the Atlantic are parted by movements beneath the plates that include the ocean's eastern and western halves. New seafloor is constantly created within a rift valley down the ocean's center, a boundary of the two plates. This rift is in the Mid-Atlantic Ridge, which is part of a global ridge system that extends some 80,000 kilometers over much of the earth.

The ocean basins are geologically young. New ocean crust is continually forming along diverging boundaries between plates. Some continental rocks are more than 3 billion years old; but to date, all rocks recovered from the ocean crust by deep-sea drilling indicate an age of no more than about 200 million years—barely older than the earliest mammals.

A record of plate motion for the last 180 million years is preserved in the magnetized rocks of the ocean crust. Plate motions can be tracked back 600 million years, into the dawn of the Paleozoic era, by using older plate materials that are part of today's continental interior and mountain systems. What happened before 600 million years is uncertain. Our best opportunity to draw inferences about the early history of the earth may well come from comparative planetology—studies of the "terrestrial" planets such as Mars, Venus, and the moon, which have had evolutions significantly different from the earth's.

A UNIFYING CONCEPT

The plate tectonics concept, in providing a logical framework for understanding and linking other earth processes, has helped to unify the earth sciences. For example, the widespread flooding of continents during the Cretaceous period, ending about 100 million years ago, may have been part of the breakup of the supercontinent, Pangea, the primordial landmass from which all present continents were formed. This association is of interest to stratigraphers and petroleum geologists concerned with the distribution or character of marine sediments through time, to paleontologists studying the evolution and distribution of marine organisms, to geomorphologists involved with the effects of sea-level change, and to geodesists, who are interested in sea level itself.

Many ore deposits are related to volcanic activity. The plate tectonics model has provided new clues on the role of subduction, melting and assimilation in concentrating metals, giving economic geologists a new tool to use in identifying promising resource areas. The opening of an ocean, from the first rifting to a fully mature ocean basin, produces a predictable sequence of sediments—important to petroleum geologists investigating offshore areas. Changes in the juxtaposition of the continents have major consequences for both oceanic and atmospheric circulation and are important to paleoclimatologists. Also, the global distribution of vertebrate animals is a puzzle until viewed in terms of moving continents. For example, fossil remains of the ancestral horse, Eohippus, are found in North America, but the modern horse evolved outside of North America and did not appear again on this continent until the age of Spanish exploration.

NEW INSIGHTS, NEW PROBLEMS

The plate tectonics model gives us important clues about what is happening, where it may happen, and why it should happen. But its clues are only suggestive. It tells us where earthquakes and volcanic activity are likely, but not when. It indicates several processes that should concentrate mineral deposits, but doesn't tell exactly where. It accounts for the history of a continental margin where hydrocarbons may be concentrated, but doesn't guarantee that they are actually there.

The ability of the model to reveal the nature of the ocean crust, its age, the nature and thickness of ocean sediments, and the relationship of the depth of the ocean to its age is remarkable. In contrast, it says little about the continents; for example, their history or structure; information important for the safe disposal of radioactive and other wastes on land, or for predicting earthquakes far from the plate boundaries. The model does provide a geological structure into which the continents fit, and, as such, may be looked upon as a starting point for learning more about the continental crust, the lithosphere, and how the surface of the earth forms and is changed.

But puzzles remain. Some postulates of plate tectonics theory remain to be tested, especially the rigidity of plates and their continuous motion. We are still unclear about what moves or deforms the plates. We have only poor explanations of the origin, evolution, structure, and dynamics of the continents and their oceanic margins. The study of the continents, and particularly the deeper, older rocks of the continents, will be of critical importance in understanding the dynamics of the earth. We lack adequate understanding of the early history of the earth; there is no record of the first 800 million years. We need to know more about how chemical elements migrate through the crust and how they become concentrated into economically useful mineral deposits. Information is needed about climates of tens of thousands to millions of years ago, information that may be culled and pieced together from the geological record.

IMPLICATIONS FOR THE FUTURE

The development of the theory of plate tectonics is an excellent example of consolidating information from separate scientific endeavors into a new concept of our planet. Observations over the past decade have yielded a new body of intellectual capital that is now available for a wide variety of applications. Our knowledge about the processes of the solid earth, the formation of the continents, and structures of basin sediments and the deep seabed provides a framework for exploring resources and obtaining data about past environments.

The earth does not easily reveal its secrets. Acquiring our present knowledge has required significant national investments in efforts such as deep-sea geological and geophysical research and drilling programs. What we know now is the product of a long, sustained investment of the nation's funds and scientists. We face decisions about the degree and nature of similar commitments in the future.

We probe the earth for many purposes—some unrelated to science—at very great expense. Maximizing the scientific utility of these efforts will be a challenge to our organizational abilities. Drilling programs illustrate the issue. Drilling is one of the important technologies used to study the solid earth. While industry has drilled on continents for many years for mineral and energy resources, its drill-holes are concentrated in areas where resources are likely to be found. More recently, federal agencies started a range of drilling programs costing over a half billion dollars a year. These include holes drilled for scientific and resource assessment purposes by the U.S. Geological Survey; for geothermal prospecting and development of exploitation technology and for the National Uranium Inventory Project by the Department of Energy; and for military purposes by the Department of Defense. Such extensive drilling efforts afford many scientific opportunities and, conversely, can benefit greatly from participation of the scientific community.

Drilling is one approach to increasing our understanding of the solid earth, but one that must be used in conjunction with others. A broad range of surface and subsurface geophysical research on the structure of the earth's crust and the layer beneath it will depend on the availability of new types of seismic and other geophysical data and new forms of remote sensing from space.

Nor is the understanding we seek about the solid earth a task for the United States alone, but rather, a cooperative task for all nations. This exploration has led to remarkable forms of international collaboration: the Deep Sea Drilling Project, for example, which used the drillship *Glomar Challenger* to verify the plate tectonics concept, and the International Geodynamics Project, which is aimed at understanding the dynamics of the earth's interior. The research opportunities of the next decade should foster even greater international collaboration. Over the next few years, we will have to decide how this collaboration is to be accomplished.

EARTHQUAKES

Our relatively recent understanding of plate tectonics explains much that was previously obscure about earthquakes.

The cost of earthquakes can be enormous. In just the past decade, for example, China lost approximately 700,000 people to earthquakes; Guatemala, 25,000; the Philippines, 10,000; Peru, 70,000; and Iran, 20,000. The United States has been fortunate so far; a densely populated, modern city such as Los Angeles has been spared a great earthquake. Although the death rate in such a modern city should be much lower, costs of damage would be astronomical. In 1971, the San Fernando earthquake, releasing only 1/1,000 the energy of the 1964 Alaskan earthquake, caused nearly three quarters of a billion dollars in damage.

However, new results indicate that we may no longer be so helpless. At least three major earthquakes in various countries have been predicted, saving thousands of lives. At the same time, we are gradually learning how to reduce future losses by building earthquake-resistant structures.

CAUSES OF EARTHQUAKES

Most earthquakes occur as plates of the earth's crust slide past or under one another. Built-up stresses are released suddenly. Plates may jump 10 meters, accomplishing in a few minutes what would ordinarily take decades of steady motion. Earthquakes along plate boundaries include those in San Francisco in 1906; Tokyo in 1923; Chile in 1960; Alaska in 1964; Peru in 1966, 1968, and 1970; and Guatemala in 1976. A rupture as long as 1,000 kilometers may occur along opposite sides of a segment of a boundary.

Enormous quantities of mechanical energy are released when plates lurch past one another. Some of that energy is converted to local heating along the fault, and some is propagated as elastic waves that produce intense shaking of the ground for as much as a minute. Motions may be violent enough near the fault to throw cars and other objects into the air. Structures may be heavily damaged. Farther away, the shaking may be less severe but its duration and periods may be longer. The consequence is a potential for more severe damage to tall buildings when they are farther away. The longer the periods of wave motion, the taller the building that may be set into motion by being shaken.

Large earthquakes occur within plates as well as along their boundaries. In 1811 and 1812, earthquakes in southeastern Missouri changed the course of the Mississippi River and were felt on the eastern seaboard. Boston, Massachusetts, and Charleston, South Carolina, have suffered damaging earthquakes; and many of the devastating earthquakes of China are not along recognized plate boundaries. The tentative conclusion is that stresses within the plates are normally high and that large quakes occur because of a weak spot, a fault, in the plate. No evidence suggests that such earthquakes within plates recur on a regular cycle. Our understanding of these kinds of earthquakes needs to be improved.

EARTHQUAKE PREDICTION

A massive failure of the earth's crust that generates great earthquakes is such a momentous event that it seems unlikely that it should occur without warning signals. Yet, often, as in the disastrous earthquake at Tangshan, China, in 1976, no obvious warnings such as foreshocks are observed. Fortunately, observations over the past decade indicate that warning signals sometimes do precede large earthquakes, even though some may be so weak that they can be detected only by instruments. Scientists have begun to systematically gather data to define classes of premonitory events in order to draw a picture of what happens just prior to an earthquake. Research is underway in China, Japan, the Soviet Union, and the United States.

The Chinese Experience

The earliest indications that earthquakes might be predictable came from ancient China, where farmers noted unusual changes in the level or quality of water in their wells prior to large earthquakes. The Chinese began a systematic program of research in 1968, combining observations by peasant volunteers with measurements of more subtle changes by scientists.

Their approach has been largely empirical, involving extensive observational networks that measure many phenomena. Using simple instruments, tens of thousands of volunteers observe water wells, animal behavior, and electrical currents in the ground. The scientists measure deformation of the land surface, detect and analyze microearthquake patterns, and measure the earth's gravitational, electrical, and magnetic fields. The chief success of the Chinese has been in the public prediction of the Haicheng earthquake of magnitude 7.3 about five hours before it occurred in February 1975. The subsequent evacuation of the populace probably saved tens of thousands of lives.

Many early warnings, precursory phenomena, accompany earthquakes of magnitude 5 or greater. The best documented of these involve distortions of the land surface, foreshock activity, and local magnetic and electrical field variations. Variations in the rate of emanation of radon gas in water wells and a slowing of seismic-wave velocities that typically return to normal just before the quake have also been identified. Geodetic measurements of crustal movements are the most widely used techniques for earthquake prediction in China, the Soviet Union, and Japan.

Occasionally, faults slip and surfaces tilt at the same time. Indeed, strains recorded before modern earthquakes are consistent with a fault having begun to slip at great depths days or weeks before the event. The data are still insufficient for scientists to confirm this deep slippage prior to an earthquake as a general phenomenon. Laboratory experiments, field sampling, and on-site measurements of major fault zones are under way to help develop a better physical understanding of the events that cause earthquakes. Without this knowledge, science will progress only slowly beyond the present empirical stage and prediction will remain an uncertain art.

Monitoring Earthquakes

Although all earthquakes do not give the same warnings, there is general agreement on the data needed to predict them. To monitor the stresses and strains along the San Andreas Fault in California, for example, arrays of telemetering sensors measuring strain, seismicity, magnetic field, and perhaps electric field should be installed along active faults. Instrument spacing would depend on the signal-to-noise ratio and on the scale of the precursory phenomena being monitored. Analysis of the collected data would be performed automatically by digital computers. Geodetic measurements made over distances of hundreds of meters, which can determine rate, orientation, and spatial character of deformation fields, are well suited to these purposes. Prototype systems are in place in California. How soon an operational system can be designed and emplaced depends on acquiring more observations on precursors of moderate-to-large earthquakes.

Moderate-to-large earthquakes now are too infrequent in well-instrumented areas of California for sufficient observations. Only rarely have more than one or two monitoring devices been located near the region of recent large earthquakes, even in China and Japan. Until many more such observations are made, possibly by concentrating effort in seismically active parts of the world, the prediction of moderate-to-large earthquakes will be unsystematic and probably unreliable.

EARTHQUAKE HAZARDS

Damages caused by earthquakes are of several types. Ground shaking may break up buildings, bridges, highways, and other structures—the damage depending on the structure, its interaction with the soil beneath it, and the intensity and duration of shaking. Modern design, particularly in California, has minimized the potential for complete structural collapse and consequent loss of life. However, even when obvious damage seems slight and injuries few, the structure's integrity may be vitiated and replacement therefore necessary. In the 1971 San Fernando earthquake, for example, property losses could have been considerably reduced had better structural design practices been applied uniformly. Masonry structures in California and the rest of the United States may be vulnerable because of poor reinforcement, weak mortar, or inadequate rigidity. Possibly as many as 40,000 buildings would collapse or be damaged so badly as to be unsafe in any future magnitude-8 earthquakes striking near Los Angeles.[1]

Also, subsurface materials can act like liquids and lose their frictional bearing strength because of seismic shaking. An artificial fill near San Francisco Bay in the 1906 earthquake caused severe damage because of this phenomenon. It is confined to areas where subsurface layers are poorly packed and water saturated. Such areas are widespread in the low-lying lands near San Francisco Bay, and planning for their best use requires detailed study of construction sites.

Earthquake shaking has triggered landslides, especially, as in 1906, when winter rains saturated the soil. Of the 70,000 lives lost in the Peruvian earthquake in 1970, nearly half were people inundated by gigantic ice and rock avalanches.

Fault displacements in great earthquakes may shift the earth's surface. The San Andreas Fault laterally displaced roads, fences, and culverts by as much as seven meters in the 1906 earthquake. Active faults in some geologic provinces have well-defined surface manifestations and can be avoided as building sites, provided they are adequately mapped.

Great offshore earthquakes can produce tsunamis, the seismic sea waves that caused such devastation on the Hawaiian Islands and in Alaska. The arrival of the wave, which is generated by a sudden vertical displacement of the seafloor near the fault, can be timed with moderate accuracy, but a tsunami's height as it hits shore can be predicted only poorly.

Fires or flooding may be the most costly effects of earthquakes. The potential for serious fire damage is perhaps less now than in 1906, when fire accounted for 80 percent of the losses, but it is still substantial. In some places, for example, fire trucks are kept in garages prone to collapse in great earthquakes. Water supplies may be disrupted by pipeline failure.

Loss of electrical power, nearly certain in a severe earthquake, would shut down 40 percent of Los Angeles' water services and all of its gasoline pumps. Los Angeles' wide streets would probably confine fires, but a general conflagration could occur under certain weather conditions.

Response

Large cities require pipelines for fuel and water, dams for water storage, and, possibly, nuclear reactors for power. The failure of such critical facilities after an earthquake can be devastating. Schools, hospitals, or power reactors, however, can be sited and constructed to minimize the chance of damage. Little flexibility is available for the siting of dams and pipelines. Pipeline failures should be expected where they cross the rupture of a fault.

A special problem is that of older, earthen dams constructed by hydraulic filling. Because of entrapped water, the core of a dam may liquefy and slump during seismic shaking. Had the water level been 4 ft higher when the Van Norman Dam failed in this manner in the 1971 San Fernando earthquake, the overtopping of the dam and subsequent flooding of the valley would have caused large losses of life and property. Since that earthquake, the state of California has accelerated its program of rehabilitating or removing its 25 remaining earthen dams. Modern dams are much less likely to fail because of seismic shaking. Nevertheless, the risk is still high in densely populated areas within a potential floodplain. Federal dam builders are reviewing seismic safety criteria because of this risk.

REDUCING THE RISK

We now know how to construct or rehabilitate structures to make them earthquake resistant, but who will pay the cost? The added cost for new structures is 5–10 percent, which could add about $2 billion per year to construction costs in the more seismic areas of the United States. To rehabilitate or replace possibly unsafe buildings would cost several billions of dollars in Los Angeles alone. We could assess the value if we knew with some certainty that a great earthquake would occur within the next 10 years. In 20 or 30 years, however, attrition would take care of at least half of the problem; within 50 years, very few of the suspect structures would still stand.

Recurrence intervals for damaging earthquakes are now poorly known; for example, the intervals between San Andreas earthquakes in Southern California now range between 275 and 105 years. Since the last great earthquake on the San Andreas Fault near Los Angeles occurred in 1857, we can only say that a major quake could occur there any time during the next century. However, for Charleston, Memphis, Boston, Salt Lake City, and Seattle, where damaging earthquakes have also occurred, there is yet no basis for guessing when, or even if, they will occur again. Faults and fault motions are not observable in some of these areas by any present technique, and an understanding of these earthquakes is needed.

A timely forecast of a large earthquake could save many lives. If clear warning signals were observed a few days in advance, emergency preparations could be made. Unsafe buildings, if identified, could be evacuated. Fire fighting equipment and standby power sources could be augmented; water levels in reservoirs reduced; and nuclear power reactors shut down to protect against fire, floods, and nuclear contamination. People could move to safe places and take precautions against fires with even a few hours' warning.

The effects of earthquake prediction on local economies and social structures are being assessed.[2] It appears that under certain social conditions, predictions issued one or more years prior to an expected earthquake could hurt the local economy, principally by reducing new construction, but a short-term prediction is not likely to provoke panic.

The uncertainty about when an earthquake may strike contributes to lack of preparation for it. Because the cost of preparedness is high, communities may understandably plead that other matters take higher priority. Since the federal government will be expected to compensate for much of any earthquake damage, federal encouragement of adequate preparation to mitigate the effects of a disaster is appropriate.

The amount of preparation, however, must be based on the severity of the risk. Without knowing when the next earthquake will occur or even approximately the interval between earthquakes, the uncertainty in the degree of risk is so large that only limited measures for preparedness can be taken. Studies of earthquake recurrence are, therefore, critical. The Earthquake Hazards Reduction Act of 1977 (PL 95-124) requested that the President prepare a plan for reducing earthquake hazards by community preparedness, land use planning, insurance, building design, and prediction. To do that, we must develop the techniques and knowledge needed to predict the occurrence of earthquakes. National plans to accomplish this objective are being implemented.

GOVERNMENT POLICY ISSUES

Policy issues need to be addressed at the same time that research on prediction proceeds. Improved contingency planning to enable effective community response to earthquakes is needed. Long-range plans to lessen the social and economic effects of an earthquake prediction need to be formulated. Policies for reducing hazards and planning reconstruction immediately after an earthquake should be adopted. In short, we must seek a better understanding of what communities will do if we can predict with precision that an earthquake will occur.

OUR MINERAL FUTURE

In time, the demand for some mineral resources in addition to fossil fuels will far exceed supply, creating shortages throughout the world.[3] Forecasts of supply and demand are uncertain because accurate estimates of potential reserves do not exist. Also, the mining of lower-grade ores will be greatly dependent on economic conditions that are as yet unknown. Nor is it certain that energy will be available at costs needed to make some deposits economically exploitable.

There are several reasons for potential shortages in the United States: Consumption is increasing; there is a geochemical limit to economically accessible and extractable world mineral resources; current markets do not encourage more active exploration and development of domestic sources; and U.S. access to world supplies may be curtailed. (See Table 4 for U.S. imports, p. 328.)

UNDERSTANDING MINERAL FORMATION

The principles of mineral concentration are still poorly understood. Events that concentrate elements into deposits are rare: They result from infrequent and fortuitous combinations of many different chemical, physical, and biological processes that, for the most part, are also poorly understood (Figure 2).

Important concepts of element concentration remain untested in either laboratory or field. For example, one concept is that many ore deposits are associated with ancient molten rock intrusions, yet rarely are they identified with modern volcanoes. Do modern volcanoes have ore deposits at depths equivalent to eroded levels of old magma intrusions? Or do ore deposits develop with time as magma cools, and hot, circulating groundwaters extract and concentrate elements from surrounding rock? Or are additional as yet unrecognized factors responsible for the fact that only a small fraction of all intrusions are mineralized, that is, harbor concentrations of various elements?

Ore deposits appear to be related to hydrothermal waters; yet active hot springs and brines—the Salton Sea and Red Sea are examples—only rarely have commercially exploitable concentrations of metals. Are hot springs the end result of an extensive hot-water plumbing system depleted of its

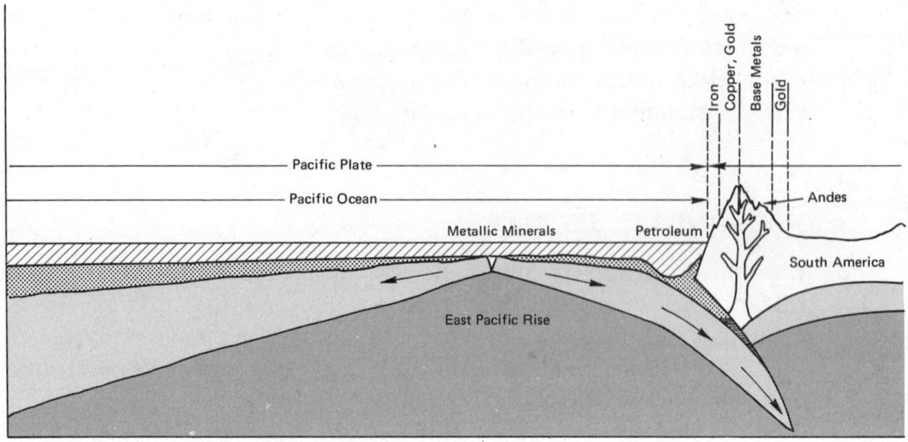

FIGURE 2 Tectonics and mineral formation. The diagram provides one hypothesis of the relationship of plate motion to the formation of mineral deposits. In this view, the spreading of the seafloor outward from the Mid-Atlantic Ridge widens the Atlantic and rafts South America over a trench, or convergent plate boundary. The resulting deformation forms the Andean mountains. Metals also accumulate about the Mid-Atlantic Ridge. Salt originating in the sediments of the continental margins rises to form large, dome-shaped masses that then trap the oil and gas generated from organic matter. (From "Plate Tectonics and Mineral Resources," by Peter A. Rona, © July 1973, Scientific American, Inc.)

minerals by mineral deposition at greater depths? Is there a predictable sequence of elements with increasing temperature and depth? Such concepts of ore zoning have been suggested, but they have not as yet been proven.

Ancient sediments are the most frequent sources of ores. Yet, the source regions and the factors that lead to the concentration of elements are imprecisely defined for these sediments and even for modern sediments. Some investigators believe that the metals dispersed in sediments may be the next major area for resource development.

Biological agents may be responsible for the selectivity and specificity of some concentrations in sediments, yet the relationship of organisms to inorganic mineral deposition is known for only a few living organisms. Can selected organisms be used on a large scale to concentrate specific elements under controlled conditions?

Hypotheses of transport that concentrate elements during rock metamorphism, weathering, groundwater circulations, and evaporation have been formulated. Few have been verified or demonstrated either in nature or in the laboratory.

In short, the major concepts on which efficient exploration must be based have not yet been demonstrated.

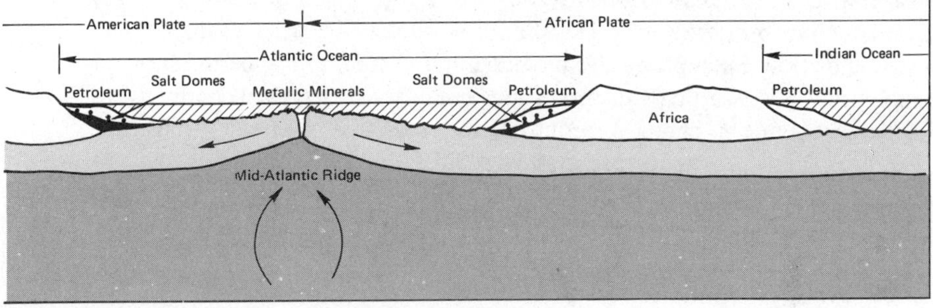

THE STATE OF RESEARCH

The federal government, the mining industry, and universities have only limited programs of basic research for understanding how ores form.[4]

The mining industry has neither the capacity nor the tradition for basic research and there is no cooperative research program for the entire industry similar to those supported in the past by, for example, the steel and cement industries. There may be insufficient proprietary advantage for a single corporation to do the basic research necessary for exploration. Mining companies tend to resolve local problems only within those properties in which they are directly interested.

The U.S. Geological Survey performs regional studies, defines potential resource regions, and makes resource assessments. It also undertakes some limited research on the origins of ore deposits and on exploration methods. The Bureau of Mines is concerned mainly with methods of ore extraction and beneficiation essential for economic evaluation. Substantial improvements in these processes will be made when results are forthcoming from the recently established mining and mineral research institutes that now exist at universities in each of 22 states, as mandated under Title III of the Surface Mining Control and Reclamation Act of 1977.

University research programs are limited in number and scope. A typical effort by a university consists of one or two staff members and their students. The reason is a weak demand for highly trained economic geologists, with universities apparently supplying them at a rate just adequate for present industrial needs.

SOME NEW DEPARTURES

There is a need to determine whether the scope of this research—in universities, industry, and government—is commensurate with the need for new approaches to mineral resource development. For example, many fundamental properties of rocks should be studied in order to understand ore deposition. Such properties include mineral solubilities, permeability of rocks, flow rates of fluids under various conditions, electrical conductivity, and geochemical changes required for selective mineral deposition. Because these properties are closely related, it is essential that they be measured on the same materials. Mathematical modeling of the many factors affecting mineral economics may be a useful approach to evaluating feasibility of mining, yet little is being done.

Research programs are essential to developing unconventional sources of critical elements. For example, aluminum ore, which is almost entirely imported, could be obtained from the large deposits of dawsonite, alunite, kaolinite, or feldspar in the United States, none of which are now mined for their aluminum content. Other examples include gallium from coal and underclays, nickel from residual laterites, rhenium from molybdenite-bearing deposits, iron from nonmagnetic taconites, manganese from silicates and deep-sea nodules, metals from seawater trapped in sedimentary formations, and copper and uranium from black shales. Often, it is not economically attractive for industry to do research on unconventional sources of critical elements, since economic evaluation for such sources requires building pilot and demonstration plants.

THE STATE OF MINERAL EXPLORATION

Of greatest immediate concern is the present state of exploration techniques, many of which are still primitive and developing only slowly. Exploration seems to be going in two directions. First, because the historic economic practice of basing a profitable mine on a single element may no longer be practical, multiple-element yields are sought in relatively shallow but large-volume deposits, even if the element concentrations are low. Some relatively metal-rich but otherwise ordinary rocks, if widely distributed in an area, can be mined for several elements simultaneously on a large scale; but the environmental impacts are potentially large. Second, some exploration is emphasizing higher-grade deposits at great depths under considerable overburden. Some theories of ore zoning predict higher-grade ores with increasing depth. Geophysical techniques will aim at detecting these buried deposits, which are now prohibitively expensive to drill and excavate.

New geochemical techniques for exploration are appearing slowly.

Geophysical techniques are advancing, but only indirectly as the result of other concerns. Thus, the study of rock mechanics has identified properties of rocks that bear on ore deposition, and structural studies outline potential fluid pathways and traps for ore deposition. Radiometric surveys serve to correlate rock units with ore deposits.

Finally, geological mapping is one of the primary methods for finding ore deposits. Only 40–45 percent of the United States has been geologically mapped on a reconnaissance scale (one inch = four miles). Further, increasing portions of the nation's land are not accessible to private exploration. Geological mapping of the United States and its adjacent ocean areas is necessary to discover major ore-producing regions at a rate adequate to meet anticipated demand and to gain a comprehensive inventory of mineral resources. In addition, there is a continuing need for comprehensive regional studies of mining districts and their surroundings.

ECONOMIC AND LEGAL IMPEDIMENTS TO MINERAL DEVELOPMENT

There are many nontechnical restraints on exploration by private enterprise. For example, taxation of proven reserves constrains owners of a mine to verify only those ores that can be mined within a given year. Another constraint is the withdrawal of federal lands, or the reluctance of companies to develop acquired properties for economic reasons. Lack of milling and processing capacity impedes exploration. The potential for adverse environmental impact dampens enthusiasm to explore many potentially economically attractive areas. Finally, capital requirements are high and economic risks are great, tending to exceed the limits of private enterprises involved in developing major new ore deposits.

The extensive lead time required for developing an operating mine is a significant contributing factor to the potential mineral shortages. After discovery through geological exploration and mapping, and before extensive drilling, sampling, and laboratory study can begin, business must obtain land rights to secure tenure. Only if the environmental impact assessment is favorable can access-road and housing construction begin and development work proceed. Only then can mining begin. The total elapsed time for these steps is approximately 15–20 years for an underground mine and 6–10 years for a surface mine.

ENSURING OUR MINERAL FUTURE

A prudent nation should develop its domestic supplies and make provisions for a flexible program of imports from alternative sources and substitutes. Although the question of imports is a matter of how this nation will interact with other nations in assuring sources of essential minerals, the first step is a matter for national action. A systematic program in which both conventional and nonconventional resources in the United States are explored would seem a wise long-range national investment. There is a clear need for fundamental knowledge about ore deposits and their distribution, as well as for the development of technologies necessary to locate and identify them. These investments, however, will pay off only if the policies and regulatory and legal structures within which mining enterprises must operate encourage vigorous exploratory efforts.

PROMISE OF THE OCEANS

The arrangements among nations governing the use of the oceans are changing rapidly. Through the U.N. Law of the Sea Conference, nations are seeking to define their rights to the ocean's use and their obligations for its protection. A new body of domestic law enacted during the past decade frames our national usage. The era of freedom of the seas is becoming an era of the managed ocean. Stakes are high as nations evolve ocean policies for developing ocean minerals, managing fisheries, protecting the oceans against pollution, performing ocean science research, and protecting coastal areas against conflicting economic interests of other nations. Ocean science has much to contribute in providing the scientific basis for evolving national policies.

THE POTENTIAL FOR MINERAL RESOURCES

The oceans are a source of a few important mineral resources, especially petroleum and gas, sulfur, magnesium, and sand and gravel. For example, the United States obtains 20 percent of its total oil and gas supplies from offshore sources. But it is the potential of mineral resources from the oceans that is intriguing. Cores drilled through deep accumulations of seafloor sediments and into underlying basement rocks over the past 10 years by the drillship *Glomar Challenger*, combined with other geophysical information, have yielded new data about the processes of oceanic mineral concentration. Signs of minerals of many different kinds have been found in various parts of the oceans. For example, a few hundred miles off New York City, beneath 240 m of sediment, cores have revealed layers containing copper, iron, and zinc laid down a short distance above basement rock and volcanic debris formed millions of years ago.

Deep dives by U.S. and French research submersibles have collected rocks that are newly formed by volcanic activity in the valleys of the Mid-Atlantic Ridge and that are often heavily coated with manganese-bearing minerals. Interesting prospects for future exploration for petroleum resources may occur in the deep-ocean sediments off the continental shelf of the United States. Such petroleum and other resources are beyond current commercial development, but they signal that on and beneath certain parts of the seafloor may lie deposits of minerals that might one day be accessible.

Exploration and test-mining of the potato-shaped manganese nodules scattered across vast areas in low latitudes of the Pacific Ocean have been under way for many years. Many of these nodules are rich in nickel, copper, manganese, and cobalt. Mining equipment is being tested aboard the *Glomar Explorer* and other vessels and metallurgical problems of nodules are being investigated. How these mineral resources will be developed and under whose aegis is being intensely debated by the U.S. Congress and the U.N. Law of the Sea Conference.

New technological capabilities and a better understanding of basic oceanic and geological processes are needed to explore and develop nonliving ocean resources. Improved knowledge of the forces affecting the ocean's crust and in active and passive continental margins is needed (see p. 10). The technology for doing this is at hand in the form of improved geophysical techniques such as modern multichannel seismic reflection systems, submersibles, and drilling technology that can safely penetrate to greater depths in deeper waters.

MANAGING LIVING RESOURCES

Realizing the promise of the ocean's living resources means solving different kinds of problems. These, unlike nonliving resources, are renewable; they are also fragile. Fishery and marine mammal stocks have been destroyed by overfishing and changes in environments and habitats. The collapse of the New England haddock fishery is associated with massive foreign and domestic overfishing in the 1960's. The collapse of the sardine fishery off California and the anchovy fishery off Peru have been caused partly by overfishing and partly by environmental changes. Uncontrolled commercial hunting of whales has endangered five of the eight species of great whales. The destruction of coastal fishery habitats, as in Chesapeake Bay, has compounded these effects, since a significant fraction of the ocean fish stocks of the United States spend at least part of their life cycles in the waters and marshlands of coastal areas. Many of these habitats have been destroyed by construction and polluted by human activities.

But we have learned. Our improved understanding of ocean ecosystems has provided a basis for establishing new management policies to maintain the health of our fisheries. The passage of the Fishery Management and

Conservation Act of 1976, extending the fisheries jurisdiction of the United States to 200 miles, marked a major turning point. Criteria that provide for a sustained high level of fisheries yield were written into law as a basis for fisheries management, and the law further requires that the best-available scientific information be considered in formulating fisheries management policies and plans. Segments of the fishing industry, in economic straits just a few years ago, are making a comeback. The restoration of depleted marine mammal populations and their protection have been advanced through the passage of the Marine Mammal and Endangered Species Acts, and the International Whaling Commission has introduced new management procedures to protect the world's whales.

Still, wise management of the living resources of the sea has only begun. Many species of fish and crustaceans that abound in the oceans are hardly used to meet human needs; for example, the shrimplike krill of the Antarctic could sustain a harvest of perhaps as much as 50 million tons per year. However, before the development of such resources is begun and other underused fishery resources exploited, it is essential to understand the marine ecosystems that sustain them. The study of marine ecosystems, and biological oceanography in general, should yield practical management tools that will enable us to conserve and maintain our fisheries and other living resources.

UNDERSTANDING OCEANIC PHENOMENA

The influence of oceans extends far beyond their resources. Oceans determine weather and climate. They are the final sink for many wastes. All uses of the seas and their protection against pollution and other insults are advanced by an improved understanding of oceanic phenomena.

TREND TO BIGGER SCIENCE

While field oceanographic research has traditionally required the deployment of ships and other expensive facilities, during the past decade it has become evident that many of the vital problems requiring investigation are so large and require such extensive facilities that they can be carried out only as multi-institutional efforts. To acquire observations of processes involved in ocean–atmosphere interactions, changes in major ocean current systems, circulation of the southern oceans, and coastal upwelling, it has become necessary to mount comprehensively planned, long-term investigations using the ships and facilities of many institutions.

For example, in the North Pacific Experiment (NORPAX), scientists have attempted to examine the nature of the seasonal changes in average temperatures at the sea surface. Large patches of warm surface water covering several hundred kilometers and persisting for many months, and sometimes for years, have been investigated. Scientists believe that these patches, penetrating about 100 meters below the surface, are related to seasonal fluctuations in global weather patterns. Other projects, such as the Mid-Ocean Dynamics Experiment (MODE) in the Atlantic and its U.S./U.S.S.R. extension known as POLYMODE, are examining the nature of the Gulf Stream. These investigations have revealed a complex structure composed of continuously forming and dissipating smaller oceanic eddies. These eddies have been found to resemble in some ways atmospheric disturbances that form along the jet streams of the prevailing westerlies.

Oceanic Research and New Technology

New technology may provide new kinds of data on the oceans. Global synoptic observations of surface ocean conditions are now possible by remote sensing from satellites. These observations will also be useful in providing global information on surface currents and temperatures, wave conditions, surface pollutants, and biological productivity. Another new source of data will be satellite altimeters, which may measure changes in sea level with time and, in association with gravimetric surveys, will indicate changes in the mean sea level.

Although the study of what occurs beneath the surface of the oceans is naturally more difficult, it is critical to understanding three-dimensional movements and the nature of vertical mixing among ocean layers—information important to assessing ocean pollution and ocean-climate relationships. With ocean buoys that sink to set depths, we can obtain information about three-dimensional flows within the oceans. One of the promising approaches is the wide deployment of instrumented buoys that can be interrogated via earth-orbiting satellites.

We need to proceed in new ways, however, and, possibly, developments in theoretical oceanography will eventually help us interpret the internal structure of the oceans in terms of what we see from the surface. Thus, internal waves at moderate depths in the oceans are visible in satellite optical photography. Backscatter radar can detect surface shears. It also should be possible to estimate internal heat flow from direct measurements of temperature gradients near the surface of the oceans.

An additional remote-sensing technology is available using low-frequency, long-distance acoustic ranging. Considerable information about the propagation of sound waves in the ocean—and the technology to generate, detect, and interpret them—has come from the nation's need for more information about subocean acoustics for national defense purposes. By using our knowledge of acoustic propagation in the oceans, it might be possible, with appropriate installations, to map remotely the variability of ocean properties. Such measurements might immediately provide information about the ocean thermocline, of probable value to ocean and weather forecasting. These acoustic techniques have major potential.

OCEANS AND CLIMATE

The oceans affect climate in many different ways. A basic uncertainty in estimating the effects of increased CO_2 in the atmosphere (see pp. 34 and 273) is the fraction of CO_2 that enters the oceans in various forms. This problem will be an important concern of ocean science in the years ahead.

To make reasonable estimates of the amount of CO_2 that can and actually would be absorbed by the oceans, we need better knowledge of the extent of oceanic carbonates and bicarbonates, oceanic alkalinity, and related physical and chemical factors. The extent to which the oceans are a reservoir for carbon depends on the rate at which surface and deeper waters mix. The mixing rate is not well known, but efforts such as the Geochemical Sections Program (GEOSECS) should determine that. The mixing rate may be determined by using radioactive tracers such as tritium from nuclear-weapons tests.

Carbon levels cannot now be measured in the oceans with the precision achieved in the atmosphere, and this technology must be improved. Particularly serious is the lack of information concerning the nature and extent of organic compounds in the oceans. The total oceanic living matter is only about 0.2 percent of the living matter on land, and, logically, has no significant role in the carbon cycle. However, the amount of dissolved organic carbon is comparable to the carbon existing on land. Therefore, dissolved carbon—its origin, turnover rate, and fate—becomes a major factor in estimating the ocean's role as a reservoir for CO_2.

Carbon is but one substance whose fate in the oceans is of concern. Growing pollution of coastal areas, including oil spills, has spurred new studies of the impacts of pollution on marine ecosystems—in the New York Bight and Puget Sound, for example. Legislation has authorized major research efforts on ocean pollution. Protection of the seas includes the wise management of coastal zones and adjacent waters; and knowledge of the fate and effects of many different substances in the oceans becomes essential.

IMPROVING OCEAN MANAGEMENT

Problems of conflicting use are inherent to the oceans. What happens in one part of the ocean affects conditions in another. Some fish migrate from one part of the ocean to another, and habitat destruction in one place, therefore, can damage fisheries in another. Development of oil and gas resources may conflict with access to important fishing grounds. Refinery construction or routes of oil tankers may threaten the oceanic environment.

The United States has established a set of ocean-management policies over the past several years. Laws have been enacted for managing fisheries, developing oil and gas resources at sea, siting deep-water ports, managing coastal zones, and conserving marine mammals. If such policies are to work, they will require an improved base of scientific knowledge about the oceans to illuminate the consequences of various management options.

Finally, a major scientific, political, and international issue of great concern to ocean scientists is that access to ocean areas for research is being increasingly restricted as nations begin to exercise jurisdiction over the oceans within 200 miles from their shores.[5] This issue, being debated in the U.N. Law of the Sea Conference, poses major questions for the Congress of the United States and international bodies. For the continued vigor of the national ocean research effort, we need to ensure maximum freedom for ocean science research.

THE CHANGING CLIMATE

Like the oceans, climate has its own rhythms—rhythms that influence the ebb and flow of civilization. During the warm climates of the Middle Ages, between 900 and 1350, Arctic sea ice diminished, the Norse colonized Greenland and cultivated grains to 60°N latitude in Norway. During the little ice age, from about 1450 to 1750, the climate became colder and more severe, at least in Northern Europe: glaciers expanded, Arctic sea ice returned, and human settlements in Greenland collapsed.

A fluctuating climate has been evident in our own century as a general warming early in the century seems to have shifted toward cooling from the 1940's to the late 1960's. From a geological perspective, however, the climate has been unusually warm for the past 10,000 years. The peak of the last major ice age occurred about 18,000 years ago, when average global temperatures were about 6°C colder and the oceans 100 meters lower.

Evidence about ancient climate comes from deep-sea sediments, fossil records, and other sources. From time to time over geological history, our planet has been subject to major shifts from ice age to warm, largely ice-free periods, such as the present, and back to ice age. Over the last million years, these shifts have taken place at intervals of tens of thousands of years.

CLIMATE AND THE ENERGY CYCLE

Climate is the cumulative product of more familiar weather phenomena—patterns of rainfall, cloud cover, daily and weekly temperature changes, and so forth. Scientists generally consider climate to be the aggregate of weather phenomena for time periods in excess of those for which daily weather prediction is possible—between one and three weeks. As a practical matter, the statistics of weather conditions for periods of a month or greater can be considered as climate. Its character is determined by the interplay between parts of the planet, including the atmosphere, oceans, lakes and rivers, snow and ice, the solid crust of the earth, and plant and animal life.

A global energy cycle links these elements to other parts of the planet and to the sun. Through this cycle, energy can be delivered, extracted, and transformed in the atmosphere. The cycle starts with the sun's energy. This radiated energy creates ozone from molecular oxygen in the high stratosphere and the ozone then absorbs almost all the ultraviolet radiation. Most of the remaining solar energy passes through the atmosphere where it can be absorbed by water vapor or scattered or absorbed by clouds and dust. About 40 percent of the sun's energy is absorbed at the earth's surface, mainly in the oceans. About 30 percent is reflected back into the atmosphere and space.

The energy cycle has only begun. The oceans and the atmosphere become working elements of a global energy transfer system. Ocean currents move energy to other parts of the planet. Energy is injected into the atmosphere when water vapor is evaporated from oceans, lakes, and rivers. The vapor is transported by winds and released in other parts of the atmosphere when it condenses as clouds and rain. Storms in the atmosphere transform potential energy from heating and cooling to the kinetic energy of the winds, which in turn transfers energy from the tropics to polar latitudes.

Anything interfering with or modifying the channels by which energy is moved and changed may in turn change the climate. There are many possibilities. Dust from plowed land, cleared forests, or volcanic explosions can significantly affect the transmission of solar energy. Gases produced by industry or by natural biological and physical processes selectively absorb transmitted solar or earth radiation. Changes in ocean temperatures or varying snow and ice cover affect the global distribution of energy.

We need to find out how these and other perturbations interact and what their effects are on the global circulations that determine climate. Fortunately, science and technology are developing ways to measure and understand the consequences of these interactions.

GROWING VULNERABILITY TO CLIMATE CHANGES

Natural fluctuations of climate frequently and, often, severely disrupt society. Moreover, pressures of growing population, demand for greater food supplies, and deterioration of land intensify the impact of climatic fluctuations. Recent examples include the drought in the Sahelian countries of Africa in the late 1960's and early 1970's, which caused famine and death on a continental scale; the 1972 and 1975 droughts in the Soviet Union, which affected the world grain markets; the failure of the Indian monsoons in 1974, which caused a food crisis in that subcontinent; and the abnormally cold winters of 1977 and 1978 in the eastern and midwestern United States, which closed schools and industries and caused widespread temporary unemployment. Of course, changing climates can benefit society: Bountiful grain harvests during the past two years have produced the grain surplus the world now enjoys.

Human Impact on Climate

We have growing evidence that not only nature but also humans may alter the climate; that industrial and agricultural practices may be causing climatic fluctuations not only on local but also on global scales.[6] We have inadvertently learned to alter the basic energy processes that control the climate in ways that are subtle but may have large effects. Humanity is not only vulnerable to climate, but climate now appears to be vulnerable to humanity. We must confront the fundamental question: Are we changing the climate of the earth?

How can we change it? A straightforward way is to dump great amounts of heat into the atmosphere. We already have done this on an urban scale. Cities are concentrated heat islands; in some cases, the amount of heat energy released in the city is equal to or greater than the average amount of energy absorbed from the sun. But since only within cities does the heat released approach a level comparable to that received from the sun, we can discount direct heat addition as a global concern, at least for the next several centuries.

There are, however, more subtle influences (see p. 447). Gaseous wastes from industry and agriculture selectively absorb different wavelengths of radiated energy. In some important instances, these gases are transparent to incoming shortwave solar energy but opaque to longwave infrared radiation from the earth's surface. The wastes thus trap energy much like greenhouses do, hence the "greenhouse effect." The net result, in the absence of other processes, is a warming of the lower atmosphere.

The CO_2 Problem

Carbon dioxide, which absorbs infrared radiation, is released to the atmosphere when we burn fossil fuels (see p. 273). More is added when we clear forests and either burn the debris or allow it or the soil humus to decay. The CO_2 content of the atmosphere has been recorded at a network of observatories extending from Point Barrow, Alaska, through Mauna Loa, Hawaii, and American Samoa to the South Pole. Everywhere, the record is similar: The CO_2 content in the atmosphere indicates a systematic increase over the globe.[7]

The amount of CO_2 in the atmosphere seems to have increased about 10 percent in the past 50 years, or from about 300–335 parts per million. Projections of fossil fuel use suggest that the amount of CO_2 will reach about 380–390 parts per million by the end of the century. By the middle of the next century, the present level of CO_2 would be double, and, by 2100, quadruple current levels.

Mathematical models of climate indicate that global surface temperatures will increase as CO_2 levels increase. For each doubling of CO_2, the models project a 2–3°C average global temperature rise. The inference is that a global surface temperature increase of 4–6°C would occur by the end of the next century. Such temperature differences are characteristic of major climatic shifts to and from ice ages.

It is important to distinguish clearly between observed facts and projections based upon calculations. That CO_2 is increasing is a fact established by measurements. All else is calculation and projection. There are major uncertainties. For example, the observed increase in CO_2 is only about 50 percent of the amount estimated to have been introduced into the atmosphere during the first half of this century by burning fossil fuels. While it is likely that much of the remaining expected CO_2 has entered the world's oceans, we have no verification; and we are uncertain about the role of plant and animal life in the CO_2 balance.

Other Problems

Although CO_2 levels have been increasing since the beginning of the industrial age, there are more recent insults. The chlorofluoromethanes (CFM's) used as propellants in spray cans and as refrigerants absorb long-wave infrared radiation from the earth, as does CO_2, and are transparent to shortwave solar radiation. Over the next several decades, a continuation of past production rates of CFM's could result in a global surface temperature rise of up to 1°C as early as the year 2000. This projected temperature increase would be in addition to that caused by CO_2. There is also nitrous oxide, another infrared-absorbing gas released when green plants convert nitrogen to their use. The atmosphere has long since adjusted to natural releases of this gas; but increasing use of commercial nitrogen fertilizers may significantly increase the amount of nitrous oxide released into the atmosphere. A doubling of the use of nitrogen fertilizer would significantly add to the warming caused by other gases.

Other human practices may pose problems for climate, though none so far reaching as those that add infrared-absorbing gases to the atmosphere. Anyone who has observed smog in Los Angeles, the dust raised by a plow, or the smoke plumes of factories can appreciate that particles can obscure sunlight. Particles both scatter and absorb sunlight; and scientists are still not sure whether the net effect of aerosols is to increase the temperature of the lower atmosphere because they absorb sunlight or decrease it because they scatter sunlight back to space.

Changing patterns of human habitation, and more significantly, agriculture, may also affect climate. Generally, cleared areas reflect more sunlight; and when trees are replaced by crops and grassland, the amount of sunlight reflected from that land increases, especially when covered with snow. The consequences of this for the energy balance of the atmosphere are not well known.

Acid rains are of increasing concern. While not directly a climate problem, the location of acid rains and their duration is controlled by the climatically determined wind and rain patterns in relation to the source regions of the industrial effluents that cause them. These rains, extremely dilute acids, form from the entry into solution of sulfur oxides and other industrial pollutants. They are deleterious to crops, forests, and natural ecosystems. Fish populations in lakes in the eastern United States and Canada and in Scandinavia have been affected.

The Stratosphere Problem

There recently has been increasing concern about the effects of contaminants within the stratosphere.

Temperatures tend to drop with increasing elevation in the lowest layer of the atmosphere, the troposphere, leading to vertical air movements that mix air masses. And tropospheric precipitation tends to wash out pollutants. In the stratosphere, however, temperature increases with height, suppressing vertical mixing. There is no precipitation. Substances injected into the stratosphere remain there for a long time. Therefore, that region is particularly sensitive to the cumulative effects of small amounts of contaminants.

The stratosphere contains small concentrations of the gas ozone, maintained by photochemical processes. Ozone, while a minor constituent of the stratosphere, nevertheless is important to life on earth, for it absorbs those wavelengths of solar ultraviolet radiation that are destructive to life and that can, for example, induce skin cancer. Any long-term change in average abundance of ozone will also affect the temperature structure of the stratosphere. Because of the interaction between the stratosphere and troposphere, such a temperature change could affect the circulation of the lower atmosphere and, therefore, the climate.

The balance between ozone formation and destruction can be shifted by the addition of chemicals such as oxides of chlorine and nitrogen. The oxides of chlorine come mainly from chlorofluoromethanes and the oxides of nitrogen mainly from nitrous oxide flowing from metabolism in plants and from the effluents of stratospheric aircraft or rockets.

These chemicals can catalytically reduce the amount of ozone in the stratosphere. If one assumes 1973 rates of production of CFM's, calculations indicate a possible future reduction of from 7–15 percent in total ozone. Observations do not yet show such a reduction since the normal variability of ozone is large enough to obscure it. The effects of an ozone reduction upon global surface temperature would be slight, but those on the stratosphere, human health, and global ecology would be significant.

The history of predicting the effects of nitrogen oxides on the ozone layer illustrates the need for caution in advising the government. It was suggested in 1970 and 1971 that nitrogen oxide might reduce ozone.[8,9] Calculations indicated that reductions up to 23 percent might result from operating a fleet of supersonic transports. The Department of Transportation, at the request of Congress in 1971, initiated its Climatic Impact Assessment Program[10] to determine the effect of SST operations on ozone. In 1975, after several years of investigation, that study and a parallel one by the National Academy of Sciences, supported such projections and indicated that ozone reductions of about 10 percent were likely from a large fleet of supersonic transports.[11]

Deep concern about these effects led to expanded investigations of the critical photochemical reaction rates in the stratosphere. The original assumptions about chemical rate constants in the stratosphere were modified by improved laboratory and field measurements. The new evidence, although not yet definitive, now suggests that nitrous oxide additions to the lower stratosphere by SST's might increase the amount of ozone, reversing earlier projections. But in the case of chlorofluoromethanes, the effects have not only been confirmed but also found to be twice as destructive of ozone.

If, as it now appears, we will have to address the climatic impact of CO_2, the oxides of nitrogen, CFM's, and as yet unknown substances in the future, we will need a more systematic approach to understanding the processes that control their atmospheric concentrations.[12] This means understanding how they are cycled among the ocean, atmosphere, biosphere, and solid earth. The emphasis now is on the biogeochemical cycles of carbon, nitrogen, and chlorine. We are most likely to influence climatic processes through these cycles.

IMPACT OF NATURAL CLIMATE FLUCTUATIONS

Our vulnerability to natural climatic changes will continue and intensify. Natural seasonal and interannual fluctuations of climate will continue both to plague and to benefit us. Because of increasing world demand for food and energy in the face of limited supplies, the impact of natural climatic changes will become of greater concern to all nations, including our own.

Since the 1930's, we have been fortunate to have experienced generally good growing weather in the major grain-producing regions of the United States with only occasional, short-lived dry periods. In some years, as in 1972, the world depended on U.S. grain to make up shortages. At other times, 1977 and 1978, for example, the world as a whole experienced good growing conditions. Fluctuations between world shortages and surpluses probably will be more frequent as world food supplies tighten in the future.

Improved warning systems and pre- and postdisaster planning and assistance can help us deal with the impacts of climatic change. Food reserves, water storage facilities, and postdisaster financial aid are prudent options for governments until predictions are improved.

The very long-term cycles of climatic change marked by the beginning and end of ice ages are now widely thought to result from changes in solar energy reaching the earth due to changing orbital relationships between the sun and earth. There is evidence that the long-term (thousands of years) trend is a return to a colder climate. Some scientists have suggested that the projected warming of the global climate from increasing CO_2 is a short-term counter to this trend. The long-term changes in climate due to the variations in the amount of solar energy reaching the earth will probably be gradual, although there is much uncertainty about the suddenness of the termination of an interglacial period and the beginning of a new ice age.

IMPACT OF CLIMATIC CHANGES CAUSED BY HUMAN ACTIVITIES

The most difficult policy choices arise from climatic fluctuations that occur over decades to centuries. Evidence already recounted on the effects of infrared-absorbing gases suggests a warming of the earth's atmosphere over the next several centuries. But, to repeat, this is a very uncertain projection.

What would a warmer or cooler earth be like? What changes in storm tracks may occur? How will patterns of rainfall and the variability of temperature be affected? What changes might occur in the planetary circulations that determine climatic belts, or the location of deserts and rain forests? How will such changes affect polar ice caps and sea ice? We can now only speculate about such questions; mathematical models have given us some insight. But if we were wise enough to read nature's record more exactly, we might be able to say more, because nature has provided past examples of climates warmer or cooler than the present one. Global temperature increases of nearly the same magnitude now projected, due to more CO_2, actually occurred 4,000–8,000 years ago. Paleoclimatologists and geologists have attempted to portray past climates from ocean sediments, pollen and lake sediments, histories of lake conditions, records of mountain glaciers, and other sources. Our knowledge is sketchy; but evidence is strong that several thousand years ago present subtropical deserts were wetter than they are today. North Africa was probably more favorable for agriculture; parts of Europe and western North America were wetter, and eastern North America drier.

The lesson of the past and our knowledge of climatic processes suggest that the largest effects on agriculture will be felt in the arid and semiarid regions, where rainfall is impeded by descending air motions that are associated with semipermanent high-pressure circulations. Such belts of descending air motion would likely move poleward and perhaps widen. On the other hand, with more CO_2 in the air, there is experimental evidence that crops, grasslands, and forests may grow at a faster rate.

The effects of a warmer atmosphere might be most noticeable in the polar regions of the world, possibly through a reduction in the amount of sea ice. The projected warming may eventually cause sea ice to melt, leaving an ice-free Arctic Ocean. Many scientists consider the West Antarctic ice sheet to be unstable to modest changes in temperature; there are some signs that this ice sheet is already retreating. During a significant warming, it might retreat much faster, leading to global rises in sea level of several meters.[13] In general, however, the effects on sea level are not expected to be large over the next century.

TASK FOR SCIENCE AND TECHNOLOGY

Given the unprecedented stakes, the task for science and technology during the next several decades is to reduce the uncertainties in our knowledge of climatic change. National and international scientific and technological efforts are under way. The first task is to monitor climatic variations and their causes. The technology for many aspects of this monitoring is available: for example, the global monitoring platforms provided by space technology. Earth-monitoring satellites—the weather satellites, Landsats, and Seasat—let us monitor the earth's radiation budget, trace gases, changes in the biosphere, the extent of snow, ice, and cloud cover, sea-surface conditions, and other important processes and features. More conventional land- and sea-based observations will also be needed, such as those already made at stations that monitor climatic change and CO_2 levels. There is no longer a technological bar to making many necessary observations.

Mathematical Models and Climate Predictions

The central problem is to understand the dynamics of climate: the forces that shape it. The mathematical model, which, with computers, can simulate the climate system, is one key to understanding these dynamics. With such models, we can study the probable causes of climate variations and the sensitivity of climate to human activities. Mathematical models have already been used successfully to simulate the first-order effects of CO_2, volcanic eruptions, and stratospheric changes. However, many potentially interactive processes were neglected in those simulations, and there is currently no model that fully represents the interactions among the atmosphere, oceans, and cryosphere. Only with such complex models can we understand and simulate important feedback mechanisms among the various parts of the climate system.

There is a more fundamental question. Is climate predictable? We know that our ability to forecast the daily progression of storms and other weather events decays with time to a point where forecasts are no better than guesses. The reason is that we can neither precisely specify the initial conditions of the earth–atmosphere system as a whole nor precisely describe all of the physical processes involved in the day-to-day evolution of weather. Small errors due to our lack of knowledge of the state of the atmosphere or in approximating physical processes inevitably become magnified until they limit our ability to forecast weather. The theoretical limitation to our ability to predict daily weather events is somewhere between one and three weeks. We need to know the extent to which climate is predictable. This question remains unresolved.

SOME SOCIOECONOMIC ISSUES

Societal issues raised by the prospects of human influences on climate could not be more pervasive. Energy policy is an example. Best estimates today indicate that demand for oil will outrun supply (see p. 263), and we will turn to coal as one of our basic fuels until new and renewable sources of energy can take over. The primary limiting factor in energy production from fossil fuels in the next few centuries may turn out to be the climatic effects of the release of CO_2. How will we balance our energy needs, usage, and allocations against possible impacts of a changed climate?

Food and agriculture policy issues are equally grave. The specific effects of climate variability or changes on agricultural productivity and a knowledge of where specific shifts of climate will take place are uncertain; but it is not too early to apply existing knowledge about variability and to begin thinking about adjustments that might be necessary in food supply systems. What changes will be needed in water conservation, storage, and irrigation methods to accommodate a different climate? Should we step up development of drought-resistant crop strains? The time when the amount of CO_2 will have doubled, about a century from now, is also the time when a tripling of world food production will be needed to meet the demand of increased population. Since the world's best agricultural land is already under cultivation, special attention will have to be given to raising productivity on existing cultivated lands and reducing the vulnerability of more marginal lands to climate. How will the United States assist in doing this?

Many aspects of the problem have been recognized. The U.S. Climate Program Act of 1978 provides for the necessary study of climate, its changes, and its impacts. The World Meteorological Organization (WMO), other U.N. agencies, and the International Council of Scientific Unions have joined to mount a World Climate Program.[14,15] Steps already taken in planning this program are the first toward a common basis of understanding climate processes and their possible consequences. Such a common base is needed for any international measures to mitigate the adverse impacts of climatic variability or changes. Our concerns can stimulate a new appreciation among nations that climate is a common resource essential to all and its protection a common responsibility.

VIEW FROM SPACE

The advent of the space age two decades ago put the earth's landmass, oceans, and atmosphere in a new perspective. What earth-borne sensors were piecing together as a puzzle, space-borne sensors began to integrate as a whole. The view from space, whether of the earth or of other planets, has contributed to our understanding of the physical processes of our planet, our philosophical concept of the earth as our habitat, and is beginning to help us to monitor the state of certain earth resources and the earth's environment.

The technology for remote sensing of the earth and other planets from space is evolving rapidly. During the next decade, the Space Shuttle will help to advance this technology; and new earth-application satellites will expand the range, diversity, and utility of data from space. Science missions to other planets will offer additional prospects for new understanding of the earth.

LOOKING AT THE EARTH

The events and features of the earth we wish to monitor occur on many scales of time and space (see p. 237). Tornadoes and squalls have lifetimes of minutes and hours, and continuous observation is necessary. Large-scale storms can be monitored with observations taken only once or twice a day; ocean currents and temperatures, once a week; crops, once every two or three weeks; and some geological processes, from months to decades.

Designing a space-monitoring system to detect specific events is a difficult technological and scientific task. The system includes sensors ranging from a simple camera to advanced electro-optical scanning devices or radar systems to collect data; satellite platforms with support functions such as power and thermal control and communication links to send data to ground-receiving terminals; data-processing systems, either on the satellite or on the ground or both, to convert sensed information into usable form; and data-dissemination devices to provide useful information to specific users. Each system is a chain from observation to use of information; and it is essential that all links be planned and implemented in concert if the observations are to be of practical use.

Types of Satellites

Frequency of observation from satellites depends on orbital conditions such as altitude, eccentricity of orbit, inclination to the equator, and the local times at which the satellites pass over a point on the earth. The orbit depends on the specific mission. For example, geosynchronous satellites are enormously valuable for tracking global weather patterns continuously. These satellites, when placed at an altitude of approximately 35,000 kilometers over the equator, appear to hover at a fixed point above the earth. Such a satellite can continuously monitor about one-fourth of the earth's surface but cannot adequately view high latitudes. Four geosynchronous satellites can continuously cover most of the earth but, because of their high altitude, may not provide the detail required for many applications.

Low-altitude satellites can reveal small-scale features, but cover a particular area on earth less frequently.[16,17] For example, Landsat has a circular, near-polar orbit at an altitude of approximately 915 kilometers. It completes 14 orbits per day, and observes a given area at the same local time every 18 days (Figure 3). For some uses, such as forest and range

FIGURE 3 Cape Cod region seen by LANDSAT in orbit some 900 kilometers above the earth's surface. (USGS EROS Data Center)

inventories, observations every 18 days may be adequate. More frequent observations may be necessary to assess storm damage or episodes affecting yields of food crops.

Remote measurements of our planet from space record electromagnetic radiation reflected, scattered, or emitted by the earth and its atmosphere.[18] Visible light represents only a small fraction of the electromagnetic spectrum that is important for remotely sensing the earth and its atmosphere. Instruments are available that can sense across a wide range of the electromagnetic spectrum, from the very short ultraviolet wavelengths to long radio wavelengths. Interpreting these observations requires an understanding of the characteristics of solar radiation at different wavelengths and the variations with wavelength of the reflection, absorption, and emission properties of the land, its vegetation, the seas, human objects, and the atmosphere.

Using television cameras, scanning devices, microwave radiometers, and other sensing devices, we have been able to detect a wide range of natural and human phenomena from space. Volcanoes, movements along geological faults, coastal erosion, flooding, storms and hurricanes, and forest fires all have been monitored by satellite imagery or data-collection systems.

Earth Applications

Events of the past two decades have demonstrated practical applications of information from space.

Landsat satellites have provided information useful to agriculture. Experiments in forecasting yields of wheat in the United States, the Soviet Union, and Canada using satellite observations have been encouraging. The conditions of crops and forests have been assessed.

Information from Landsat and meteorological satellites, such as areas of water impoundment and snow cover, has helped in managing the country's water resources.[19] Snow mapping by satellite can improve predictions of runoff from melting snows. Operational environmental satellite data are used routinely to monitor snow cover in scores of river basins in the mountainous western United States.

Landsat data represent a new source of information for geologists looking for minerals, petroleum, and geothermal energy sources.

Weather services depend in many ways on satellite data for essential weather and severe storm warnings. Geostationary weather satellites monitor hurricanes and tornadoes and provide the basis for improved national and international severe-storm-warning systems. Polar-orbiting weather and environmental satellites provide data about global cloud cover, temperature, and moisture, enabling major improvements in worldwide computer forecasting.[20]

EARTH SCIENCES

The earth sciences are among the major beneficiaries of earth-viewing satellites and other space technology. New Landsat systems will carry a new generation of instruments for geological studies. These instruments include multispectral remote sensing for discrimination among rock types, active and passive microwave systems, and stereoscopic imagery for the third-dimension information necessary to make structural interpretations of space-sensed data.

The solid earth will be studied as a global entity, its geological aspects with the new Landsat, its geophysics with special satellites to measure the earth's gravitational and magnetic fields, and its geodynamics with two other satellite systems that have been under development over the past decade. Both systems are capable of measuring positions on the surface of the earth within a few centimeters, the accuracy needed to measure plate motions. They will provide laser ranging to natural and artificial satellites and very long baseline microwave interferometry. These measurements are important in understanding the nature of earthquake occurrence and, hence, in the design of practical methods of earthquake prediction.

Ocean sciences have benefited from many earth-viewing satellites. Geodetic satellites, such as the Geodynamic Satellite-3, yield measurements of sea level and the earth's gravitational field. Meteorological satellites and the Seasat satellite provide a synoptic view of global ocean conditions.

The meteorological sciences are midstream in the international Global Atmospheric Research Program in which satellites play a central role. The objective is to examine the predictability of daily weather events and provide data to examine the dynamics of seasonal and annual variations in climate. The nations of the world are conducting the Global Weather Experiment during 1979. This experiment involves the deployment of five geosynchronous satellites launched by the United States, Japan, and a group of European nations; polar-orbiting satellites launched by the United States and the Soviet Union; and fleets of aircraft, surface ships, and unattended buoys. Only earth-orbiting satellites could have made such programs possible.

EARTH AND THE OTHER PLANETS

Additional insight into atmospheric and solid planetary processes is being obtained by studying our sister planets. Evidence of major past climatic changes has been discovered on Mars in the form of the channels on the surface and the extensive layered deposits that surround its north polar region. The knowledge that such changes have taken place on other planets may help us to understand causes of climatic change on earth. As other planetary systems are examined, we should gain further helpful knowledge. For example, scientists believe that the very high surface temperatures on Venus are the result of a runaway greenhouse effect. More information on that planet and its atmosphere could help us understand the potential hazards of similarly increasing atmospheric CO_2 on earth. Satellite studies of Venus are just beginning.

By studying the atmospheric circulation on Mars and Venus, each with its unique atmosphere, rotation rate, inclination, and distance from the sun, and by comparing what we learn with the general circulation of the earth's atmosphere, we may be able to construct computer models that can predict the atmospheric circulation of any given planet.

Insight into planetary volcanism and plate tectonics has been important to the geological sciences (Figure 4). Basaltic volcanism is common to terrestrial planets and the moon. Planetary basaltic lavas from the mantle make their way to the surface by the same mechanisms as on earth—from fissure and central-vent eruptions. Basaltic volcanism on the moon was confined to a restricted period of lunar history between 3 and 4 billion years ago. If it is assumed that the Martian lithosphere is so thick and rigid that movement is difficult, it is possible to deduce that basaltic volcanism lasted much longer on Mars, a bigger planet, with eruptions from single central vents persisting much longer than on Earth.

Our knowledge of terrestrial processes is augmented as information on other planets accumulates. We are also beginning to gain further knowledge of such diverse topics as early planetary evolution, early crust formation, earthquake mechanisms in nonplate-tectonics settings, tectonic processes, and many others. Terrestrial application of planetary discoveries is just beginning. Those benefits can be expected to continue as we visit and revisit the planets. The remarkable and unexpected results from the Voyager mission to Jupiter and the recent measurements on Venus are tantalizing indications of what lies ahead.

SPACE SHUTTLE

The introduction of the Space Shuttle[21] in the next decade will mark a major operational change in the U.S. space program. Its advent as a relatively inexpensive transportation system will open an era of more ambitious operations in space. The shuttle should reduce the cost of transporting satellites into earth orbit. With its 60-foot by 15-foot diameter compartment, larger and heavier payloads can be lifted into space. This large volume may be essential for earth observations that require large antennas. The shuttle will return payloads to earth for refurbishment and reuse.

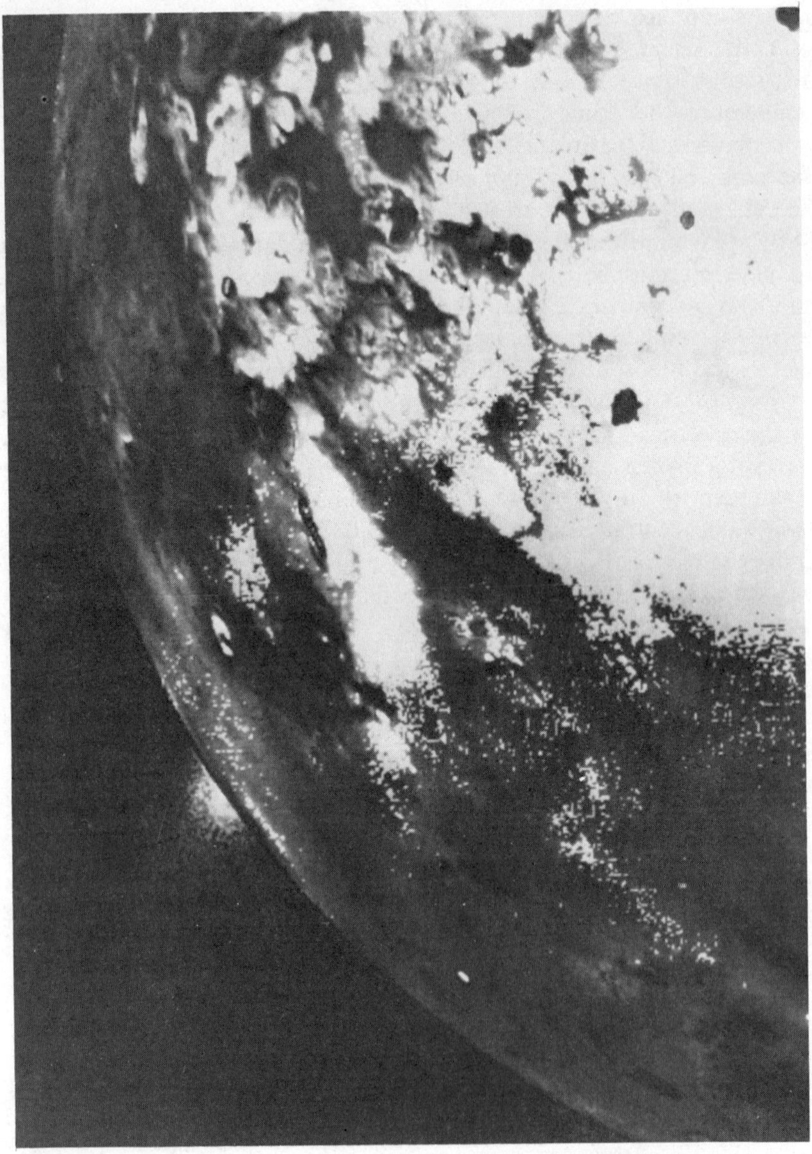

FIGURE 4 Volcanic eruption on Io, a satellite of Jupiter. This plumelike eruption rising almost 100 kilometers above Io's surface indicates that the satellite may have the most active surface in the solar system. The volcanism is very explosive with velocities above 3,000 kilometers an hour, or faster than any observed on terrestrial volcanoes, including Etna, Vesuvius, or Krakatoa. (NASA)

An important attribute of the shuttle, which is to be flown in the 1980's, is its potential for reducing the cost and time required to test new remote-sensing systems. Development tests of such systems now require expensive, self-contained satellites that must include power, propulsion, pointing, command, and telemetry subsystems. With the shuttle and its manned onboard module, the Spacelab, development prototypes can be flown to investigate how useful certain measurements are, and to optimize the design characteristics of future flight sensors.

USING DATA FROM SPACE

Users face formidable problems in applying satellite data. A single frame of the Landsat Multispectral Scanner is made up of 32 million picture elements with six bits of information each; 192 million bits of information for each frame. The weather satellite of the National Oceanic and Atmospheric Administration yields about 100 billion points of new data every day. Computer processing is indispensable.

Computers can construct images to closely resemble the original scene.[22] Geometric distortions introduced by the sensor or by satellite and earth motions can be computer-corrected. Computers can calibrate and compensate for certain data errors. They can enhance images, modifying the original data to emphasize features of special interest. For many purposes, however, digital information rather than images is required, and computers process these directly.

By comparing measurements in different spectral bands at different times, the computer can differentiate between corn and wheat, determine, in conjunction with other data, the health of plants, and help predict crop yields. Further research on crop "signatures," models of growth and yield, and pattern-recognition processes is leading to automatic extraction of specific information that agricultural users need. Automated systems already are used to produce atmospheric and surface-temperature analyses and displays of upper-air wind fields.

Solid-state devices are advancing remote-sensing technology. Arrays of silicon charge-coupled devices have made possible solid-state, visible-light cameras of very high sensitivity and resolution for use on the ground. When flight modules are developed for these cameras, reduced cost, weight, power, and volume can be expected. Similar developments coming soon are solid-state array sensors for various regions of the infrared spectrum to replace mechanical scanners currently in use. In addition to enhancing sensitivity and resolution, an important added advantage of solid-state array sensors will be that signal-processing circuitry will be combined with sensors within the array to select the information to be transmitted.

Current developments in microprocessors, charge-coupled devices, memory devices, and very large-scale integrated circuits may permit the development of spaceborne instruments that can process the data and transmit to earth only the information needed by specific users. One can conceive of a spacecraft that could locate a forest fire in its early stages, sound an alarm, and transmit an image of the fire and the surrounding terrain, including the roads to be used by fire-fighting teams.

MANAGING NEW SPACE ENTERPRISES

The technological directions for the next five years appear to be well defined, although much research and development still needs to be done. Transferring the results of this rapidly advancing technology into actual practice poses some troublesome issues. Although there are basic questions about costs and benefits and, hence, differing views on the desirability of moving to an operational system, one major problem is institutional. Space missions are costly, and effective institutional arrangements are essential to ensure that their output is used and that the benefits are commensurate with costs. The operational weather-satellite system was introduced with relative ease because of the existence of a working domestic and international infrastructure—the government weather services and WMO—that could fund and coordinate the operation of the system. Also, the diverse users of weather data were already aware of the value of such data and thus receptive to data from new sources.

Such a well-developed infrastructure does not exist for earth resources or other environmental services and must be established if the transition from research and development to an operational system is to be made. An important factor delaying this transition has been the mixed population that could benefit from satellite information. This community includes scientists of many disciplines, farming groups, government agencies, fuel and mineral industries, land-use planners, and water-resource managers in this and other countries. There is a wide disparity in the needs, technical sophistication, and financial resources of these different users. In addition, the paucity of institutional arrangements that permit them to agree on information needs makes it difficult to resolve questions about what should constitute an operational system, who should operate it, and how the costs should be shared.

INTERNATIONAL ISSUES

Other nations have strong interests in many space applications. Several successful models for international participation in space activities and for making the benefits accessible to other nations have been developed (see p. 498). In meteorology and communications, two quite different but successful models of institutional frameworks have been devised. In satellite communications, Intelsat Corp. is the model of an independent international corporation responsible for developing and managing a single global satellite system. The meteorological model is different. A coordinated, internationally agreed-upon regional system of geosynchronous satellites has evolved, with individual or regional groupings of countries committing themselves to the development, management, and operation of individual weather satellites as part of an international system.

Applications of other satellite data are more difficult. Developing countries have meager data on which to base their resource and land-use planning and may find Landsat data particularly useful.[23] However, they have expressed deep concern about how such data will be used. In some of these cases, effective use of the data is limited by domestic political considerations and shortages of qualified personnel. The problem is further complicated because, unlike the weather and communications cases, some countries are concerned about the dissemination of Landsat satellite data to third parties. Such countries can obtain processed data from the United States or directly through their own or a neighboring country's Landsat ground stations. Canada, Italy, and Brazil operate substations. Memoranda of understanding for establishing stations also have been signed with Iran, Argentina, India, Chile, and Zaire and are being negotiated with several other countries.

The way in which the United States should proceed over the long stretch is not clear. Involving other nations in planning and structuring a global resources information system or any other earth-sensing satellite system is a sensitive and difficult problem that will need to be overcome. Remote sensing can generate questions of national security and national sovereignty. These issues have been and are continuing to be discussed in the Outer Space Committee of the United Nations.

AFTERTHOUGHTS

It is tempting to bewail the complexities of modern society. In earlier days, we located mineral resources by walking the land, we collected oil from natural seeps, and we drew water from abundant sparkling streams or shallow wells. Weather and natural hazards were in God's hands and their ill effects were to be borne with grace. Population was kept at modest levels by plague, pestilence, famine, and war. Society is no longer so simple nor so limited in size. Our real or perceived need for resources and our concern about the global environment call for more imaginative ways of finding resources and for more sophisticated knowledge of the earth.

Both future and past discoveries depend on the imaginative weaving together of ideas and tools; of concepts such as plate tectonics, with its unique view of how the earth's crust moves; and new technologies that enable us to drill in the deep ocean, see the earth from the vantage of space, and detect and measure the faintest tremors and movements in the earth's crust. We have achieved better understanding of the nature of climatic change by acquiring observations from space and from deep in the earth, and by using theoretical models of the climate with ever more powerful computers to simulate climatic conditions.

Moreover, the classical boundaries of the earth sciences—geology, meteorology, oceanography, and so on—are being eroded and replaced by a planetary multidisciplinary view. For example, portraying and understanding the long-term evolution of climate depends on understanding the movement of crustal plates and the interpretation of deep-sea cores and sediment samples. Understanding and predicting the shorter period changes in climate depends on knowledge of the oceans, their temperature, currents, ability to act as a reservoir, and their role in the global energy cycle.

A central theme is that the new knowledge gained by the vigor of earth sciences and by pertinent technology is now vital to the wise management of our planet. Plate tectonics is essential to the effort to understand and predict earthquakes and to improved reconnaissance for new mineral deposits. Atmospheric chemistry enables us to make a reasoned estimate of the likely future effects of trace amounts of chlorofluoromethanes on trace amounts of ozone in the stratosphere. Basic work in marine biology and ecology is indispensable to structuring effective policies for managing the living resources of the seas. Research on the chemistry of ocean water will enable us to fix more precisely the role of the oceans as a reservoir for CO_2, helping to yield, in time, precise estimates of the climatic effects of CO_2 and a more rational base on which to plan the future use of fossil fuels.

Our appraisal of recent trends in the earth sciences is dominated by the role of technology and the approach to planetary problems through organized and collaborative efforts of institutions and scientists—big science. There is a current question about big science and its relation to the science of individual investigators. It should be noted that the big science efforts described here grew from little science—the ideas of individuals—and provide to individual scientists data that could be obtained in no other way.

While it is true that further advancements will continue to be dependent upon new technology and massive efforts to acquire observations of natural phenomena, we must recognize that we would accomplish little without the creativity of individual scientists who integrate and synthesize information about these phenomena and who develop the concepts that ultimately spur advance in any field.

New concepts sometimes require many years before they are verified and accepted. Examples in the earth sciences serve as reminders of the need to support the inquisitive leanings of individual scientists. The theory of continental drift and, indirectly, plate tectonics, was first proposed by a German meteorologist, Alfred Wegener, in 1922. Its proof waited for half a century to pass and for the technology of modern geophysical research and deep-ocean drilling. The theory that long-term climate changes resulted from the earth's orbital variations was first advanced by the Serbian astronomer, Miltvin Milankovitch, in 1930. Again, it required close to half a century and the advent of the technology of ocean drilling and paleoclimatic dating to document this as a real possibility. Finally, the massive Global Weather Experiment is the culmination of proposals made by Jule Charney and a few others in the early 1960's.

Our experiences lead us to the view that neither must the support of individual scientists be neglected nor the claims of big science. To date, neither have been. We have diversity. It is necessary that this continue.

OUTLOOK

The following outlook section on the planet earth is based on information extracted from the chapter and covers trends anticipated in the near future, approximately five years.

Humanity is dependent on the resources and environment of the earth for its sustenance. Their wise use depends on the understanding of the planet that the earth sciences provide. The achievements of these sciences in the recent past have had a broad influence on our daily lives. Only a few aspects of the earth sciences are treated in the chapter—the structure of the earth, earthquakes, minerals, climate, the oceans, and the impact of space technology. The earth sciences also encompass vital activities such as water resources and hydrology, weather forecasting, the geology of the continents, the high atmosphere and near space, and other subjects not treated in this volume. They will be examined in subsequent reports.

WHERE TECHNOLOGY IS LEADING

Progress in the earth sciences depends upon the acquisition and interpretation of observations of the earth, its atmosphere, and its oceans. Observational information becomes available only as small samples of natural phenomena, some of which vary rapidly with time and some of which may have global dimensions. The plates that make up the ocean floor may cover half an ocean, planetary wind systems girdle the globe, and oceanic currents extend over thousands of kilometers. While laboratory experiments are important for the earth sciences, data from the natural world are essential. Much of the modern technology that can acquire these observations requires long lead times for development. For this reason, the technologies that will influence the course of earth-science research during the next five years are reasonably well known.

Because the earth itself is the laboratory, acquiring data is difficult and costly. Much of the technology used by marine geologists and geophysicists was developed by industry or government for other purposes. For example, the search for oil and national interest in the nuclear test ban stimulated the development of the technology for geophysical surveying of the earth's interior with multichannel seismic arrays. The direct recovery of rock cores from the seabed by deep-sea drilling draws much from the industrial technology of offshore drilling for oil.

Space technology provided the first truly global platforms for observing the earth. The practical achievements and the scientific data about the earth obtained from them during the past two decades have influenced many fields, such as weather forecasting and environmental and resource monitoring, and have added greatly to our knowledge of the earth. Building on these achievements, spacecraft being planned for the next decade should expand the diversity of observations available for earth applications and for the earth sciences.

New generations of satellites will yield new data for more precise mapping of the earth, improved assessment of the state of crops and forests, more precise measurements of sea level and sea state, and measures of the imperceptible motions of the solid earth. New satellites will enable us to make global observations hitherto unobtainable in any other way such as those important to an understanding of climate—trace gases and the earth's radiation budget.

Remote sensing technology is developing rapidly, both from space satellites and from earth and airborne platforms. It will yield an observational capability that will enable us to probe phenomena deeper in the earth, higher in the atmosphere, and at greater distances at sea. Multichannel seismic reflection systems will enable us to penetrate deeply into the earth's interior, acoustic ranging offers hope of oceanic measurements at depth, and remote electromagnetic probing with radars, lasers, and other devices will enable us to sound the atmosphere and measure the sea state at great distances.

We will also probe the earth directly with more powerful tools. We will have the capability to penetrate the deep-ocean sediments to great depths in any part of the oceans, sample the lower atmosphere with specially designed aircraft and reach the high atmosphere with giant balloons that stay aloft for a long time.

OUTLOOK FOR SCIENCE

Trends in science may be molded by events and socioeconomic needs, often expressed in new legislation, or by pressures due to international concerns; examples include the intense interest in climatic change stimulated by the passage of the U.S. Climate Program Act in 1978 and the search for alternative energy options stimulated by the 1973 oil embargo and subsequent price rises. Such trends are usually clear. What we cannot foresee are the directions that will be taken as a result of discoveries yet to be made.

In marine geology and geophysics, one major emphasis will be to explore continental margins. Both the drive for basic understanding of plate tectonics as well as the societal need to know more about the mineral potential of seabeds will channel these sciences.

More intense studies of earthquakes, their possible prediction, and their socioeconomic impacts will be undertaken. A much stepped-up effort will be made to detect and understand earthquake precursors using the new seismic networks and geodetic measurements along faults. These studies will be parallelled by new investigations of ways to reduce the hazards from earthquakes.

Two major national oceanic research efforts of the past decade are now drawing to a close: the International Decade of Ocean Exploration (IDOE) and the Deep Sea Drilling Program. Future trends in ocean science will depend largely on the allocation of resources among competing needs of the science. Now competing for investment in ocean research are the re-equipment and replacement of ships that belong to an aging research fleet, the continuation of IDOE investigations that have examined such major problems as the major ocean current systems, study of the interaction between ocean and atmosphere, and the processes of upwelling of ocean waters, and the continuation of a program of geological and geophysical research based on the use of newer geophysical techniques and more effective drilling capabilities. Marine biological and ecological research will make an important claim to this investment based on the needs of improved fisheries and coastal zone management.

The U.S. Climate Program Act of 1978[22] has set the course of research on climate. The principal focus of this effort will be on developing more precise estimates of the consequences of human interference with climate, with special emphasis on carbon dioxide. Considerable effort will be directed at improving our ability to make climatic predictions for periods of months and seasons. Two areas of climate research that need special attention concern ocean/climate relationships and biogeochemical cycles.

SOCIAL AND ECONOMIC ISSUES

Our new knowledge of the earth frames an interesting set of social and economic issues. What is evident in these issues is the interweaving of science with wider concerns: of climatic change with future use of energy and evolving patterns of agriculture; of new space-borne sensors with more sensible land use; of analyses of tectonic processes with better protection against earthquakes and the discovery of new mineral resources. Thus, while the viewpoint in this chapter is that of scientists, the implications are universal.

Response to Natural Hazards

Social response to natural hazards and warnings of them will continue to pose serious problems. As we learn more about the nature of earthquakes, droughts, and other hazards, we must also learn more about possible responses, for example, what to do when an earthquake is predicted. Policies aimed at mitigating social and economic effects of hazards will demand a base of sound scientific data.

Ocean Management

Managing ocean and coastal resources will be a continuing and increasingly difficult issue. We will have to draw from our bank of scientific knowledge about the oceans as we turn to our continental shelves for oil and gas, increasingly assume for ourselves the right to coastal fishery resources, consider the oceans for waste disposal, and seek to protect the habitats of marine species. Over the next five years, much of the legislation enacted during the past decade will be reviewed and modified. Science can, as a minimum, assist in outlining the consequences of various ocean-management policy options.

Energy and Mineral Policy

The questions of energy and mineral policy should attract major interest during the next five years. Our knowledge of earth sciences will contribute importantly to framing these policies. What is the potential for new oil and gas resources in the deep ocean? What are the possibilities for radioactive waste disposal at sea? What are the climatic implications of scenarios of future fossil-fuel use?

The United States, whose economy depends on ready access to nonfuel minerals, has an inadequate domestic supply of certain strategic minerals. There exists no integrated U.S. program of basic research for understanding ore-forming processes. If we are to stimulate mineral research, exploration, and development in the United States, we will also need to examine the legal and regulatory structures that govern mining activities. The policy questions are numerous and their resolution will depend on the scientific and technical information that can be provided.

Agricultural Policy

No area of national policy is so sensitive to climatic variations as food policy. Worldwide growing conditions determine the nature of the global food market. Agricultural export policy is involved as are decisions on acreage restrictions. International disaster relief and our participation in a global food reserve system are all dependent on climate. Improved use of climatic data in such policy decisions will become increasingly important.

Environmental Policy

We will need to base environmental policies on what is known about the impact of human activity upon the environment and vice versa. Basic to such policies is the fundamental knowledge of the capacity of the atmosphere, the hydrosphere, the lithosphere, and the biosphere to disperse or concentrate pollutants.

USING HIGH TECHNOLOGY

The pursuit of information to further knowledge drives the development and use of high technology in the sciences. But different considerations apply when that same high technology finds general uses. Illustrative is the current problem in transferring the knowledge and technology of the earth-orbiting Landsats from research and development to operational use. Introduction of an operational Landsat satellite is a complex matter that depends upon costs and benefits and how those will be shared among users. It raises the question of how institutional mechanisms can effectively relate the interests of the federal government, private concerns, and state and local governments. How this will be done will depend upon the utility of the data acquired from space, the distribution of the users, their willingness to pay, and the needs and responsibilities of the federal government. The Landsat case is likely to be the first of several, and how it is resolved will affect how we approach analogous problems arising in the use of ocean or geodetic satellites.

International Affairs

Developments in the earth sciences have special implications for international affairs. Much research in the earth sciences requires observations from large areas of the globe and, in some cases, from the entire planet. The acquisition of some kinds of data has been made easier by remote-sensing devices such as earth-orbiting satellites. However, satellites cannot provide all data needed for earth-science investigations. Experience has shown that satellite data need to be supplemented by ground data because of ambiguous "signatures," i.e., several different objects giving the same remote-sensing results. The study of earthquakes requires a worldwide seismic network; that of climate, a network of global observation stations; and that of oceans, the facilities of many nations. For these reasons, international cooperation in the earth sciences is already a long-standing tradition.

The mechanisms for conducting international earth science activities are diverse, and include bilateral arrangements between governments, informal arrangements between institutions, and arrangements within specialized agencies of the United Nations and within scientific organizations outside the U.N. system. Present trends in the earth sciences indicate that such cooperative activities will not only need to be continued but intensified over the years ahead. The resources required to study many of the problems are too great for any single nation to provide. Furthermore, in many cases, access to territories under national jurisdiction for scientific observation is indispensable. Of special importance during the next five years will be the issue of scientific research at sea. Developments in the U.N. Law of the Sea Conference foreshadow serious restrictions of freedom for oceanic research.

The earth sciences are unique in their planetary nature. Information about the earth and its global environment is essential to all countries of the world if they are to meet their own social and economic objectives. For this reason, the earth sciences offer a channel for cooperation among many nations of the world, both developed and developing. Because of the increasing importance of scientific information to the development objectives of other countries, the outlook is for greater interest and, in many countries, for closer collaboration in earth-sciences research. How this is done and whether it is desirable represents an area of policy and programmatic decision that will need considerable attention during the coming years.

REFERENCES

1. *Earthquake Prediction and Public Policy* (NRC Advisory Committee on Emergency Planning). Washington, D.C.: National Academy of Sciences, 1975, p. 38.

2. *A Program of Studies on the Socioeconomic Effects of Earthquake Predictions* (NRC Committee on the Socioeconomic Effects of Earthquake Prediction). Washington, D.C.: National Academy of Sciences, 1978.

3. *Review of National Mineral Resource Issues and Problems* (NRC Board on Mineral and Energy Resources). Washington, D.C.: National Academy of Sciences, 1978.

4. Rose, A.W., *et al. Research Frontiers in Exploration for Nonrenewable Resources.* College Park, Pa.: Pennsylvania State University, 1977.

5. Ocean Policy Committee. The Marine Scientific Research Issue in the Law of the Sea Negotiations. *Science* 197(4300):230–233, 1977.

6. Study on Man's Impact on Climate. *Inadvertent Climate Modification.* Cambridge, Mass: MIT Press, 1971.

7. *Energy and Climate.* (NRC Geophysics Study Committee). Washington, D.C.: National Academy of Sciences, 1977.

8. Johnston, H. Reduction of Stratospheric Ozone by Nitrogen Oxide Catalysts from Supersonic Transport Exhaust. *Science* 173(3996):517–522, 1971.

9. Crutzen, P.J. The Influences of Nitrogen Oxides on the Atmospheric Ozone Content. *Quarterly Journal of the Royal Meteorological Society* 96:320–325, 1971.

10. *Final Report of the Climatic Impact Assessment Program.* Washington, D.C.: U.S. Department of Transportation, 1974.

11. *Environmental Impact of Stratospheric Flight: Biological and Climatic Effects of Aircraft Emissions in the Stratopshere* (NRC Climatic Impact Committee). Washington, D.C.: National Academy of Sciences, 1975.

12. *Effects of Human Activities on Global Climate.* World Meteorological Organization, Technical Note 156, WMO No. 486, 1977.

13. Whillans, I.M. Inland Ice Sheet Thinning Due to Holocene Warmth. *Science* 201(4360):1014–1016, 1978.

14. White, R.M. Organizing a World Climate Program. *Bulletin of the American Meteorological Society* 59(7):817–821, 1978.

15. *Toward a U.S. Climate Program* (NRC Climate Research Board). Washington, D.C.: National Academy of Sciences, 1979, Appendix B.

16. Lintz, J., and D.S. Sinnot. *Remote Sensing of Environment.* Reading, Mass.: Addison-Wesley, 1976, p. 326.

17. Bishop, B.C. Landsat Looks at Hometown Earth. *National Geographic* 150:140–147, 1976.

18. Barrett, E.C., and L.F. Curtis. *Introduction to Environmental Remote Sensing.* London: Chapman and Hall, 1976.

19. *Practical Applications of Space Systems: Inland Water Resources Supporting Paper No. 5* (NRC Space Applications Board). Washington, D.C.: National Academy of Sciences, 1975.

20. *Weather and Climate* (NRC Space Applications Board). Washington, D.C.: National Academy of Sciences, 1975.

21. *Space Shuttle* (National Aeronautics and Space Administration Information Office). Washington, D.C.: National Aeronautics and Space Administration, 1976.

22. National Climate Program Act. Report No. 95-1489, House of Representatives. August 14, 1978.

23. *Resource Sensing from Space: Prospects for Developing Countries* (Ad Hoc Committee on Remote Sensing for Development). Washington, D.C.: National Academy of Sciences, 1977.

BIBLIOGRAPHY

Brobst, D.A., and W.P. Pratt. *United States Mineral Resources.* U.S. Geological Survey Professional Paper 820. Washington, D.C.: U.S. Government Printing Office, 1973.

Cox, Allan (ed.). *Plate Tectonics and Geomagnetic Reversals.* San Francisco: W.H. Freeman, 1973.

Tilton, J.E. *The Future of Nonfuel Minerals.* Washington, D.C.: The Brookings Institution, 1977.

Uyeda, S. *The New View of the Earth,* O. Masako trans. San Francisco: W.H. Freeman, 1978.

Van Rensburg, W.C.J., and D.A. Pretorius. *South Africa's Strategic Minerals.* Johannesburg: Valiant Publishers, 1977.

Wylie, P. *The Way the Earth Works.* New York: John Wiley and Sons, 1978.

Government and the Nation's Resources (National Commission on Supplies and Shortages). Washington, D.C.: U.S. Government Printing Office, 1976.

Mineral Resource Perspectives. U.S. Geological Survey Professional Paper 940. Washington, D.C.: U.S. Government Printing Office, 1975.

Problems of U.S. Uranium Resources and Supply to the Year 2010 Supporting Paper 1, (NRC Committee on Nuclear and Alternative Energy Systems). Washington, D.C.: National Academy of Sciences, 1978.

Special Report: Critical Imported Materials (Council on International Economic Policy). Washington, D.C.: U.S. Government Printing Office, 1974.

Technological Innovation and Forces for Change in the Mineral Industry (NRC Committee on Mineral Technology). Washington, D.C.: National Academy of Sciences, 1978.

(William Reaves, EPA—Documerica)

2 The Living State

INTRODUCTION

In recent decades biological science has grown explosively, not only in scale but in depth. Initially a descriptive field, concerned with recognizing the structures and the behavior of whole organisms and their organs, biology has become increasingly analytical as it has probed ever finer levels of organization. We now know a great deal about the mechanisms responsible for the formation, function, and regulation of cells and their components.

These advances have depended in part on the development of elaborate instruments: electron microscopes, which extend a thousandfold the dimensions that can be visualized; the scanning electron microscope, which provides three-dimensional images; ultracentrifuges, which separate cellular particles and molecules for many sciences; X-ray crystallography, nuclear magnetic resonance, and other physical probes into the intimate three-dimensional structure of molecules; a great variety of radioactively labeled compounds whose fate in living organisms can readily be followed; and electronic equipment for recording the activity of nerve cells. In addition, hundreds of biochemical compounds, which investigators used to have to make themselves at the cost of much valuable time, are now available from an industry that sprang up as research expanded. Other industrial advances with polymers made possible simple and ingenious chromatographic methods, in which a single passage through a gel can separate a complex mixture of hundreds of substances, in any size range, into its components. Finally, antibiotics have made it easy to culture animal or plant cells free of contamination by much faster-growing bacteria; genetic and regulatory properties of these cells, and of infecting viruses, can then be studied in ways that would be much slower, more expensive, and often impossible in the whole animal.

With these developments, the barriers between areas within biology have broken down, and fruitful marriages have taken place. Thus the field of cell biology has emerged with the advent of techniques for exploring a range of dimensions that used to be too large for the biochemist and too small for the investigator of cell structure. Similarly, at the molecular, mechanistic level, genes are linked to origins in the past and to functions in the present; the fusion of genetics with biochemistry has brought together those biologists concerned with how organisms arose in evolution and in embryonic development and those concerned with how organisms work.

Because research in biology spans a great range of biological systems and experimental approaches, we shall review in this chapter only a few selected areas, mostly at the molecular level. We shall also add a few comments on evolutionary studies, those concerned with whole organisms and populations. Other large areas, such as population biology and ecology, which are of great importance for agriculture and for conservation of resources, will not be covered.

MOLECULAR STUDIES

The groundwork for the growth of the molecular studies that now pervade most of biology was laid in the biochemistry of several past decades. Research in this field identified the chemical components of living cells, worked out the sequences of reactions that synthesize these components from foodstuffs and that provide the energy for these syntheses, and disclosed a good deal about the enzymes that catalyze the individual reactions. By now we know the complete pathways, involving several hundred different intermediates, for constructing all the known building blocks that are common to all living organisms. Such biomedical studies continue to be extended fruitfully to other small molecules with specialized roles in various organisms, such as hormones, as will be illustrated particularly in the sections on neurobiology, and biology and agriculture.

The search for a coherent conceptual framework in biology took an enormous step forward with the emergence of the field of molecular genetics. With the discovery in 1944 that the genetic material is DNA (deoxyribonucleic acid), and the finding in 1953 that this long-chain molecule is composed of two precisely complementary strands, so that the sequence of bases of either specifies that of the other, the key to gene duplication was evident. In the subsequent quarter-century, intense investigation has revealed a great deal more about how genes work: i.e., how they serve as blueprints for fashioning the cellular machinery, and how they are regulated, mutate, repair errors, recombine in sexual reproduction to yield endless diversity, and expand in number as evolution creates increasingly complex organisms. Furthermore, we can now describe the metabolic activities of a cell in a thoroughly logical, comprehensive way, proceeding from the genes to the enzymes that they code for, to the small molecules made by some of those enzymes, to the linking of building blocks into long chains by other enzymes, and to the regulatory feedback effects of various substances on the activity of specific genes and enzymes. Genetics, like biochemistry, now pervades all of biology. This point will be particularly illustrated in the sections on cell biology and immunology.

In addition to this wealth of insights into subtle mechanisms, molecular genetics has yielded a profound, germinal concept: that of molecular information transfer. Previously, biochemists viewed compounds only in terms of structure, chemical reactions, and energy relations. The most important feature of DNA, however, is that the sequence of units in its chain is a store of information, quite analogous to that in a magnetic tape or a line of type. This information specifies the chemical structures that a cell can build, and it also specifies the program of successive reactions that, over time, channel the development of a fertilized egg into a higher organism. Eventually it was realized that all specific interactions between molecules in biological systems involve information: for example, the recognition of a specific receptor on a cell by other cells, or by hormones, or by antibodies; or the feedback response of a gene or an enzyme to the concentration of a specific substance in its environment, as a means of regulating the amount of that substance to be synthesized.

In higher organisms the reception, storage, and transfer of information in the nervous system also involves molecular changes. Accordingly, we now see a continuity between two kinds of information in biological systems: the many bits inherited in the genes, and the many others acquired from the environment. These two stores of information interact to make organisms what they are.

Along with deepening understanding of how the information in genes is translated into protein sequences has come equally satisfying understanding of how the three-dimensional structures of proteins endow them with highly specific surfaces, which allow the proteins to function as the working machinery of the cell—catalytic, regulatory, and structural. Moreover, we now know a good deal about how enzymes fit other protein molecules and induce specific chemical changes in them; and how the activities of specific enzymes are regulated by interactions with specific surrounding molecules, which convey information about the biochemical needs of the organism at each moment. We also know, in principle, that the specific affinities of proteins on cell surfaces guide the cells to find their place during the growth of an organized tissue. Detailed understanding of these structures is one of the major challenges of biology.

With the recognition of all these mechanisms, it is not unreasonable to say that the secret of life has been discovered—or better, that many secrets have been discovered. Each involves known physical and chemical forces. These results make it very unlikely that any vital forces remain to be discovered in the major mechanisms still buried in neurobiology or developmental biology. However, this triumph of the mechanistic approach to biology in no way diminishes the unique qualities and the marvel of the living world and the human spirit. Rather, we can only stand in awe that life, evolving from inorganic matter and working with inorganic forces, could develop these qualities—and could develop an organism with the capacity to understand its own origins in remarkable detail. Although there is no unique vital force in living organisms, there is a unique molecular basis for their organization which is not found in the inorganic world: the molecular storage of information programming the development and function of the organism.

CHANGING ORGANIZATION OF BIOMEDICAL SCIENCES

The large recent changes in the level and the focus of biological research have affected its organization in various ways. One is the shift, in many branches of biology, from simple to complex instruments. It may cost in excess of $100,000 to equip a laboratory for a new independent investigator, and inevitably the costs will continue to mount.

Another change has been a fading of the boundaries between many branches of biology. As a result, in many universities resources and teaching responsibilities are shifting from medical schools, where many new lines of basic research originated, to basic science departments in faculties of arts and sciences. Although this trend is logical, there would be serious losses if medical schools were divested of their interest in fundamental science, and reverted to being trade schools limited to applied research.

A third trend is worrisome. Although the current generation of medical students is intrinsically better qualified than were their predecessors, fewer are pursuing careers in medical research. The percentage of M.D.'s among those applying to the National Institutes of Health (NIH) for research support declined from 41 percent in 1966 to 28 percent in 1976, while the percentage of Ph.D.'s was rising from 48 percent to 63 percent. If this trend continues, it could undermine the transfer of basic research findings to clinical practice, while research itself might be denied the special insights arising from human pathology.

PROGRAMMING OF BIOMEDICAL RESEARCH

Since the great medical advances of the past resulted in the prevention or the cure of a remarkable number of infections and nutritional or endocrine disorders, it has been only natural to expect similar rapid applications from the intellectual achievements of the present. The expectations include cures for cancer, vascular diseases and arthritis, and healing of the mentally disturbed or retarded. But the realization of these expectations may be far off, because these major medical problems today arise to a considerable extent from the inherent weaknesses, imperfections, and aging of human organisms; and these processes are not so readily changed. On the other hand, one may be more sanguine concerning the prospects for preventive measures to reduce the incidence, for example, of cancer and atherosclerosis.

Impatience and unfulfilled expectations, as well as exaggerated promises from some scientists, have led to disappointments and to complaints that scientists are more concerned with satisfying their own curiosity than with investigating matters of greatest significance to their public sponsors. But until the requisite understanding is at hand, large-scale targeting of research is all too likely to engender expensive but unproductive attempts to apply the inapplicable.

We must therefore try to advance our understanding of cell and organ function on all possible fronts, using as study objects those natural biological systems that offer special research advantages. Our interest in the life and times of *Escherichia coli*, the neurobiology of the squid, the aging of the rotifer, or the alarm reaction of the clam derives largely from the fact that each serves as an easily studied model for some process highly relevant to man.

Moreover, we must recognize why support of apparently esoteric research is especially important in biology. Breakthroughs have frequently depended on fortunate accidents and unexpected observations that revealed isolated components in the incredibly complex network of events in the living process. A battalion of eager scientists is waiting to transfer the results to medicine.

However well motivated the pressures to channel research to societally desirable targets, if the time is not right, such channeling is more likely to retard than to advance the desired outcome.

MOLECULAR GENETICS

Molecular genetics stands at the threshold of major socially important advances. The basic mechanisms elucidated in bacteria and viruses are rapidly being extended to higher organisms, where novel features are also being discovered. Since several of the main health problems in the United States today are diseases of cellular malfunction, and since these malfunctions often involve defects in gene function, fruition of research in this area can reasonably be expected. In addition, modern genetics promises new capabilities for agriculture and pharmaceutical and other industries.

TRANSLATION OF GENE INFORMATION

The previously formal concept of the gene was given physical reality in 1944, when Avery and his colleagues showed that genes are located in long-chain molecules called DNA. Each chain is composed of only four different kinds of molecules called bases, which are denoted by the letters A, G, T, and C. The sequences of these units store the information that makes up our inheritance. The mechanism became clear when Watson and Crick discovered that DNA is composed of two complementary strands, i.e., A in either strand is always matched with T in the paired strand, and G is always matched with C. Thus, when DNA duplicates, either strand can automatically specify the sequence of the other.

A chromosome consists of an extremely long chain of DNA in which specific sequences function as genes. Most of the information in DNA is translated into the sequence of another class of long-chain molecules—proteins—which constitute most of the working machinery of the cell. Proteins are composed of 20 different kinds of amino acids. The formation of proteins begins with the process called transcription, in which one strand of a gene directs the synthesis of a complementary sequence in a similar long-chain (but single-stranded) molecule called RNA (ribonucleic acid). The sequence of bases in RNA is ordered by essentially the same base pairing as in DNA. The RNA serves as a messenger between DNA and protein; in the next step its sequence is translated, inside a complex subcellular particle called a ribosome, into a corresponding sequence of the amino acids of a protein chain.

Ten years ago, in a major triumph, the genetic code that translates the language of DNA and RNA into the language of proteins was deciphered. In messenger RNA, successive three-letter words—made up of different combinations of three of the four base letters—specify which amino acid is to be linked next onto the growing protein chain. All of the three-letter words possible with a four-letter alphabet are actually utilized. These words (including those used as the signals for start and stop) have all been identified. Moreover, the code, and the basic machinery of protein synthesis, is universal. All proteins are made in the same way in all living cells.

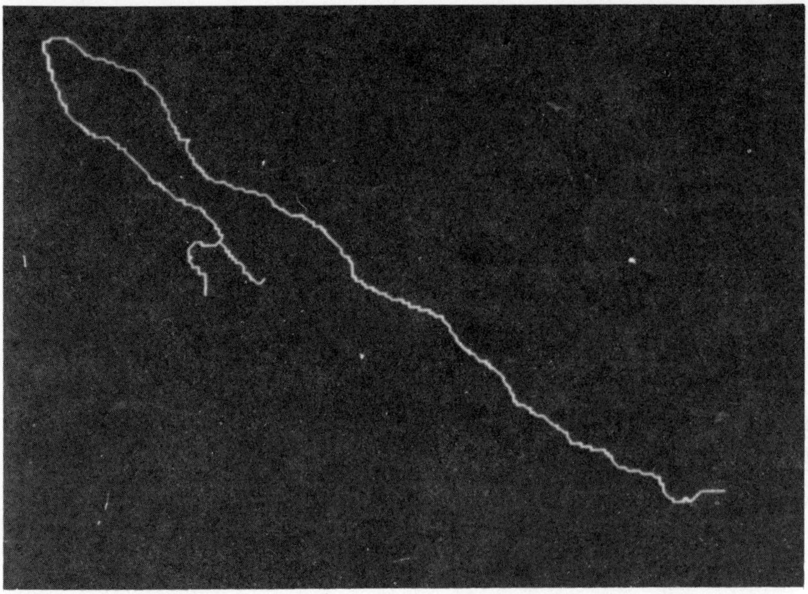

Skeleton of a single-stranded SV40 DNA molecule computer-generated from an electron microscope image. (National Institutes of Health: B. A. Shapiro, L. E. Lipkin, P. F. Lemkin, E. M. Smith, M. Schultz, National Cancer Institute; J. Maizel and M. Sullivan, National Institute of Child Health and Human Development; N. Salzman and M. Thoren, National Institute of Allergy and Infectious Diseases)

Gene Regulation

DNA not only codes for the structure of messenger RNA and thereby of proteins; it also contains specific regulatory sequences, which determine how much of each messenger is made. By turning specific genes on or off, these regulatory mechanisms adjust the composition of cells, and their output, to their circumstances. Even the most primitive organisms, the bacteria, have evolved elaborate regulatory mechanisms that promote the efficiency and speed of their growth by preventing the synthesis of components, such as amino acids, in excess of their requirements.

For example, *E. coli* makes the amino acid tryptophan only when it is lacking in the culture medium and its intracellular concentration is very low. This is made possible by the fact that, in the *E. coli* DNA, there are consecutive genes that code for five different enzymes needed during the synthesis of tryptophan. A separate protein called the tryptophan repressor (coded for elsewhere on the DNA) can bind to a region of DNA (called the operator) immediately preceding the five genes; when so bound, the repressor prevents transcription of all five structural genes. But to attach to the operator region of DNA, the repressor must have tryptophan bound to it; such binding occurs readily when the concentration of tryptophan is about that normal for *E. coli* but not at lesser concentrations. Thus the concentration of tryptophan governs tryptophan synthesis, mediated by a regulatory protein which senses the concentration of end product and transmits this information from the surroundings to the genes. The molecular basis of these interactions and the mechanisms of several other, more intricate regulatory mechanisms in bacteria are now understood in considerable detail.

APPROACHES FOR STUDYING HIGHER CELLS: RECOMBINANT DNA

Much less is known about regulatory mechanisms in animal and plant cells than in bacteria. We do know, however, that all the different kinds of cells in a higher organism contain essentially the same set of genes, of which only a fraction is functional in a given cell type. Accordingly, the regulatory mechanisms discovered in bacteria provide a simple analogy to the mechanism, for example, that restricts hemoglobin synthesis in higher organisms to cells destined to become red cells. However, there is a major difference. In bacteria, regulatory changes are instantly reversible, while in the differentiated cells of higher organisms they are stable and self-perpetuating, through yet unknown mechanisms.

The striking success of studies of regulation in bacteria has depended on the ability to perform genetic and molecular studies on specific regions of DNA in these simple cells, which have about 5,000 genes each and can double every 20 minutes. In higher systems, with 1,000 times as much DNA per cell and with generation times of years, these methods are ineffective. It is therefore extremely encouraging that in the past few years the experimental barriers have fallen, owing largely to the development of what has come to be known as recombinant DNA methodology.

In this approach, the DNA from any source can be cut into small pieces by a family of so-called restriction enzymes, each of which recognizes and excises a specific short sequence in the DNA chain. The fragments can then be spliced into bacterial DNA, where their behavior can be studied as precisely as that of bacterial genes. In addition, the inserted DNA, grown in unlimited amounts in the bacteria, can easily be recovered and purified. Its base sequence can be determined quite simply by recently developed methods. With viruses, which contain much shorter molecules of DNA than do cells, the restriction enzymes can be applied directly to that DNA, and the total base sequence easily determined from the fragments. Finally, viruses and DNA fragments can be subjected to agents that produce persistent changes, or mutations, in known loci rather than in random loci, as in cells.

At one time, scientists considered it prudent and desirable to avert any potential danger that might be associated with research on recombinant DNA. The NIH guidelines were a reasonable response to public anxiety, given the limited knowledge that was available at the time. Research during the intervening years has shown that the earlier concerns were without basis, and has essentially dissolved the apprehension that originally led scientists to bring the problem to public attention. As a result, the restrictions have been somewhat relaxed.

Worst-case analysis with a virus highly infectious to mice indicates that it is utterly devoid of infectious or other noxious properties when joined to other DNA inside *E. coli.* No precautions other than the conventional safety procedures of routine microbiology now seem necessary or appropriate to research with recombinant DNA.[1,2]

SEGMENTED GENES IN HIGHER ORGANISMS

The magnitude of the breakthrough provided by recombinant DNA is hard to overestimate. One example of its impact is the recent revolution in the concept of a gene.

Classical genetics defined a gene both as a unit of function (which determines a particular trait, such as the color of a flower) and as a unit of structure (located at a particular position in a chromosome). However, with the advance of molecular genetics, it became clear that each gene is not a molecule but rather a particular sequence of bases in the long DNA helix. It has no beginning or end, except as signals to start and stop transcription are encoded in its sequence. In bacteria the whole sequence is transcribed, one letter after another, into messenger RNA, which is similarly translated into the corresponding protein.

In higher cells, however, the very first applications of recombinant DNA methodology yielded a most unexpected result: The DNA sequence for a given protein (for example, hemoglobin) is not found as a single contiguous series of bases. Instead, several regions, each coding for part of the protein, are separated by long intervening sequences which are not translated. When a gene is expressed, the entire long DNA sequence is transcribed into RNA; then special enzymes cut out and splice together the several segments that together constitute the messenger. This RNA molecule is then translated continuously, as in bacteria. This discovery has led to optimistic speculation that the intervening sequences, and the joining mechanism, may play a central role in the complex gene regulation of higher organisms. Many tests of this hypothesis are in progress.

The fragmentation of genes goes even further in the synthesis of antibodies. A single antibody molecule is encoded in three libraries of DNA fragments, which are found in separate places on a chromosome. In making an antibody messenger, first one piece of DNA from each of the three libraries is physically translocated to form a single sequence, then that DNA is transcribed into RNA, and finally parts of that transcript are spliced together to make the messenger RNA.

Duplication of DNA

One of the major challenges of molecular genetics is to understand how DNA is precisely duplicated, with an accuracy that allows a sequence of a million bases to be transmitted intact from one generation of cells to the next. In this area, as in gene structure and regulation, a great deal is known about bacteria and their viruses and very little about higher cells, except that the latter have new levels of complexity. When Watson and Crick first grasped how DNA is organized, they were gratified to realize that the molecular mechanism of its duplication was inherent in its structure: Either strand can be a template for synthesizing the other. But many questions remain to be solved. How does the double helix unwind to open up the paired regions? How is synthesis regulated to assure one complete round of DNA synthesis each time a cell doubles? How is the very high fidelity of duplication maintained? At how many sites along the strand is DNA synthesis initiated?

Central elements of the answers to these and many other questions are lacking. We have learned, though, that DNA can take many physical forms and that it is duplicated in various ways. In some viruses, for instance, DNA is linear and in others it is circular, with neither beginning nor end. Some linear viral DNA have a unique sequence (e.g., A–Z), while others occur as a family of sequences derived by cutting a circle of DNA at different places (e.g., A–Z, D–C, Q–P). Both in bacteria and in higher cells not quite all DNA is in the chromosomes. There are independent, smaller DNA pieces—called plasmids—that find expression and replicate autonomously like chromosomal or viral DNA but are not released like viruses. The ability of bacteria to form toxins or to resist various antibiotics is usually due to information in such DNA pieces, which can be transmitted from one organism to another by special mechanisms. Another complement of autonomous DNA is found in mitochondria, small intracellular structures that inhabit all cells, as well as in the chloroplasts of plant cells.

The synthesis of chromosomal DNA in higher cells differs in a crucial way from that in bacteria. Bacteria generally start duplication at one point on their circular chromosome, and a wave of synthesis passes from that site around the circle. In the much larger chromosomes of higher cells, duplication starts at many sites, and the resulting segments are then joined. The mechanisms coordinating this process are yet to be studied.

In summary, although the elementary act of DNA duplication is evident in the Watson-Crick model, many individual aspects vary with the specific DNA molecule and the cell type. And the picture is far from complete.

Chromosome Structure

Great progress has recently attended the study of chromosome structure in higher organisms. Small strands of the DNA double helix wind around balls of a protein called histone. These balls then clump together to form the compact chromosome.

This mode of packaging for DNA raises many questions. Does this arrangement break apart to allow the DNA to duplicate, or can the DNA replicate within this chromosomal organization? How does this organization change in different stages of the cycle of cell division? How can the DNA specify RNA sequences when it is so compactly packaged? Does this package mechanism help to regulate DNA expression, or is it simply a good way to pack a lot of DNA into a small nucleus? Investigation of this field has just begun.

GENETIC RECOMBINATION

Another goal of molecular genetics is to understand how the DNA molecule, in all forms of life, can recombine as it does during reproduction of sperm and egg cells, breaking at identical points in a pair of homologous chromosomes and then resealing with fragments exchanged. Genetic recombination is pivotal to evolution, for while mutations are the ultimate source of genetic variation, their accumulation within a succession of progeny would be a very slow source of the variation on which natural selection acts. Genetic recombination, by reshuffling the mutations accumulated in different members of a species, enormously increases the number of variations.

In higher cells, recombination seems quite complex, and even in bacteria, where recombination is essentially the same, the details are not clear. Because of the simplicity of viruses, their recombination mechanisms are better understood. Recently it has become possible, using viruses, to carry out recombination in solution, outside cells. These new systems promise to provide models through which recombination in cells may be understood.

Mobile DNA

One important new discovery is the presence in bacterial DNA of sequences that "hop" from one region to another rather frequently, through high-frequency recombinations at specific short sequences near their ends. How they do this is still debated, but the process clearly plays an important role in bacterial evolution and in certain regulatory processes. This finding should help us to understand similar obscure processes in the cells of plants and animals. For example, as mentioned earlier, the production of antibodies requires pieces of DNA to move within the chromosomes.

MUTATIONS AND RADIATION DAMAGE

One area of genetic research with great practical significance is the study of mutations. Errors are incorporated into DNA, either by mistakes made during copying or by chemical alterations in existing DNA. Mutations can have either positive or negative consequences for organisms. The negative aspect is obvious; most changes in DNA impair normal functioning by altering or inactivating a protein or by changing the control of its synthesis. But mutations are also the ground substance of evolution, for occasionally they improve the adaptation of an organism.

Spontaneous mutations cannot be totally eliminated, because occasional errors are inherent in the copying process, despite its remarkable accuracy. These errors can occur at random in any gene in an organism. Errors would be much more numerous if it were not for repair processes, which eliminate most, but not all, of the mismatches in base pairs. The mutation rate per unit length of DNA, which varies widely among species, depends on two mechanisms—the precision of the replicating enzyme and the effectiveness of various repair systems.

In addition to spontaneous mutations, others can be induced by agents that increase the mutation rate in various ways. These agents include ultraviolet and high-energy radiation. Many chemicals, both naturally occurring and man-made, are known to cause mutations in intact living cells; only a few have, to date, been seen to so affect DNA *in vitro*. With the very sensitive bacterial systems now available, it is possible to detect slight mutagenic effects of certain normal constituents of food, such as caffeine, or even of moderately elevated temperature.

Closely related to the study of mutations are the lethal effects, on cells and on organisms, of damage to DNA by radiation and by some chemicals. Many kinds of damage cause errors in replication (mutation); but others, which prevent further replication, are lethal to the cell. Cells have developed systems for repairing such damages just as for repairing mutations. The effects of radiation on organisms, long studied at a largely descriptive level, have now been translated into specific chemical terms in some detail.

While mutations in germ cells are passed on to offspring, it appears that some mutations in normal body cells (somatic mutations), passed on to daughter cells (but not to the offspring of the organism), can result in cancer. Knowledge of the mechanisms of mutation and repair is therefore fundamental to understanding how cells become cancer cells and how we may prevent this process.

Because of the apparently close links between mutation and cancer, inexpensive and rapid tests for mutations in bacteria are used to screen for potentially cancer-causing substances. The Ames test was the prototype of this approach. These tests, now widely used in industry, are a landmark in the application of molecular genetics to public health problems.

VIRUSES

Until the advent of molecular genetics, viruses could be studied only descriptively, as submicroscopic agents of disease. Virology has now become an intensely active, sophisticated field, as modern methods, gradually extended from bacterial viruses to those of animals and plants, have made it possible to analyze the properties of various viruses in great detail. Viruses are not cells smaller than bacteria; they are autonomously replicating blocks of DNA (or in some cases RNA) encased in a protein coat. The protein, by binding to a specific receptor in a host cell membrane, may facilitate its entry into that cell.

Like plasmids, viruses multiply within cells. The virus sheds its protein coat and its nucleic acid enters the cell, where it appropriates the cellular machinery it needs to duplicate itself and to synthesize its protein components. Then the nucleic acid and its coat are assembled and exit from the cell as complete virus particles.

Because of the extreme simplicity of viruses (some have only three genes), virology has contributed much to molecular genetics, as well as to the development of improved vaccines and the beginnings of effective chemotherapy of viral infections. The present stage of the field is one of dynamic progress.

Some viruses have chromosomes made of RNA instead of DNA. These viruses generally induce in the cell a novel system for copying (transcribing) RNA from RNA. But in one group called retroviruses the virus codes for an enzyme that makes a DNA copy of the viral RNA; this DNA copy can be spliced into the DNA of cellular chromosomes. These viruses can thus alternate between being independent entities and being cellular genes. This phenomenon is somehow crucial to the induction of tumors by viruses of this class.

PROTEIN SYNTHESIS

The ultimate role of DNA is to direct protein production. The conditions required for making protein outside cells, in extracts, were worked out about 20 years ago, and with this development the components of the system could be separated and identified. Progress was rapid. Proteins are now known to be synthesized in complex particles called ribosomes, with the help of several dozen enzymes as well as several dozen RNA molecules called transfer RNA, all of which are found in the cytoplasm. Each transfer RNA provides a link between a three-letter word in messenger RNA and the specific amino acid that it codes for.

The ribosome is an extraordinarily complex structure, consisting of 54 different proteins and 3 different RNA molecules, fitted together in a specific way. These components provide many different binding sites, and go through an orderly cycle of complexing with other components of the system in the course of adding each amino acid as the ribosome reads the RNA message. A major triumph has been the reassembly of active ribosomes in the test tube from their dissolved components.

Observing the functions blocked by various antibiotics that act on bacterial ribosomes (e.g., streptomycin), has made it possible to work out the mechanisms of action of these drugs, and to clarify aspects of ribosomal functions. Bacterial mutants resistant to streptomycin contain altered ribosomes, and by reassembling ribosomal components from sensitive and resistant cells, researchers have also identified the basis for resistance.

Protein Structure

The specific surface of each protein enables it to play its particular enzymatic or structural role by closely fitting to the surface of some other molecule. Accordingly, the amino acid chain translated from the one-dimensional information in DNA must fold into a three-dimensional structure with a highly specific surface. The mechanism has turned out to be both remarkably simple and elegantly subtle. Not only the information but the forces for the correct folding are built into the protein chain, through the specific affinities of the 20 different amino acids in that chain for each other. The resulting internal binding causes each protein chain to fold up on itself spontaneously in a unique way, without requiring an external template or enzyme. On the resulting surface, unique to each protein, will be found regions with affinities for various other types of chemical structure such as substrates (for enzymes) or antigens (for antibodies). Moreover, the spontaneous aggregation of specific structural proteins with each other and with other cellular components initiates a cascade of steps, in an increasingly complex assembly, that accounts for the formation of membranes, cells, and even organs and organisms.

GENETICS AND HUMAN DISEASE

Folklore has long suggested that some diseases are transmitted from generation to generation. Current understanding began in 1908 with the description by Garrod of half a dozen "inborn errors of metabolism." The subsequent development of biochemistry and genetics continued to expand that understanding (see p. 417); over 2,000 hereditary disorders are now known, and the list continues to grow. Most are seriously disabling, and many are fatal in early life.

To date, the specific defects in more than 150 specific hereditary metabolic abnormalities have been learned. Examples include cystic fibrosis, Tay-Sachs disease, sickle-cell anemia, various hemophilias, and phenylketonuria. In each a specific protein—an enzyme or other functional category—is synthesized either in a defective form or not at all. Such single-gene diseases are inherited in the manner predicted by classical Mendelian genetics. In sum, they constitute a massive burden of illness for which cures in the ordinary sense are generally unlikely.

At the same time, evidence has mounted that other disorders are genetic in a more complex sense, as in diabetes, for instance, where the defective biological process is affected by a multiplicity of genes. Thus, while juvenile diabetes is surely the expression of an individual's genetic constitution, the genetics are subtle and complex and differ in different individuals with much the same disorder. This may be the case for many endocrine disturbances.

More recently, the essentially genetic character of yet another broad set of diseases has begun to be understood. These include ankylosing spondylitis, rheumatic fever, rheumatoid arthritis, lupus erythematosus, thromboembolic purpura, and Reiter's disease. The individuals who acquire these diseases usually (with some diseases, always) bear certain genetic markers, a specific one of the dozens of antigens on the surfaces of red blood cells, of lymphocytes, or of tissue cells generally. The converse does not hold true: Only a fraction of those who bear a given tissue antigen ultimately develop the associated disease. The relationship between the specific antigen and the manifestations of the specific associated disease is unknown. In some cases, occurrence of the disease also requires previous exposure to some environmental agent, for example, streptococci for rheumatic fever, dysentery-causing organisms like Shigella for Reiter's disease. Yet it is clear that individuals so affected are genetically predisposed to these diseases.

Lymphocyte and tissue typing of humans has barely begun. It is clear that as an increasing variety of tissue antigens is typed in larger numbers of individuals, additional correlations of this character will surely be found. They perhaps will predict which of us are most likely to develop cancer when exposed to an environmental carcinogen, which are most likely to develop atherosclerosis, multiple sclerosis, and so on. In fine, from this standpoint, all disease has a genetic component. But that should not be surprising, since that is true of all significant traits. What is new and hopeful is that it may become possible to ascertain, in early life, the special vulnerabilities of any individual human and, hence, to minimize exposure to the particular environmental factors that might trigger disease.

FUTURE CHALLENGES

Molecular genetics, born 25 years ago, has made great progress in its short lifetime. But in contrast to earlier developments in biochemistry, which identified small molecules, such as vitamins and hormones, that could immediately be put to use in medical practice, discoveries in molecular genetics have not been so readily converted into applications. Rather, this field has dealt with the giant molecules that lie at the heart of the cell, and these were often best studied best in bacteria and viruses. However, molecular genetics is now analyzing problems of health and disease in human cells, and also the influence of small molecules such as antibiotics on DNA synthesis and function. As this understanding progresses, so will our understanding of such diseases as cancer, autoimmune disorders, thalassemia, and formation of abnormal hemoglobins. Other rapidly growing applications include the ability to determine the presence or absence of various fetal defects by amniocentesis (sampling of intrauterine fluid), and to detect mutagenic compounds.

Even more predictable than contributions to health will be the contributions to our understanding of our own physiology. We can confidently expect that in a few decades we will understand in some detail how the growth of a specific organ system and the harmonious growth of an entire organism are regulated. Moreover, though the particularly interesting behavioral and physical traits in human beings must involve huge numbers of genes, it may not be too long before we begin to unravel the genetic components of much of human individuality.

Beyond the new understanding of the human body in health and disease, other valuable consequences can be foreseen. One is the harnessing of biological processes for human welfare. Recombinant DNA methods are already being applied to convert bacteria into factories for synthesizing specific products, such as a brain hormone (somatostatin) and insulin. Synthesis of the powerful natural antiviral compound, interferon, is a goal in a number of laboratories (see p. 94). Synthesis of a wide range of pharmaceuticals is expected.

Other possibilities have yet to be examined. Will recombinant DNA methods make it possible to design and synthesize new enzymes to be used in place of chemical synthesis of many organic compounds? Are there other transformations of biological systems that can generate new energy sources, such as improved biological fixation of solar energy? Can other food plants be rendered, like legumes, independent of the need for nitrogen fertilizer? Industries may come to rely on genetic technologies.

Another consequence might be gene therapy—the correction of genetic diseases by replacing defective genes with normal ones. The possibility of success is hard to evaluate. We will soon be able to isolate or synthesize any gene we wish, but the obstacles to inserting them into body cells in a useful way are large. It seems likely that this step may become possible before too many years for the defective precursor cells of blood cells, since these are located in the bone marrow in a way that allows them to be replaced from the bloodstream. With more highly organized organs, however, the problems seem insuperable.

This type of genetic engineering would benefit the afflicted individual; it is much less likely that a gene replacement can be carried out in sex cells, as would be required to eliminate further inheritance of a genetic disorder.

CELL BIOLOGY

SCOPE AND DEFINITION

It has been known for 140 years that all higher organisms, regardless of size or complexity, are constructed of cells of about the same size (about 1/100 millimeters diameter). Only in the past two decades has the electron microscope permitted studies of visible structure to penetrate to ever finer dimensions. Meanwhile, biochemistry has moved up from small molecules to the detailed structure of giant molecules, such as proteins and nucleic acids, and indeed to the isolation and analysis of even larger components of the cell. Biochemists and morphologists have thus met in what was formerly a no-man's land between their domains.

The dynamic, functional properties of the cell have been advanced by new biophysical techniques, such as study of the electrical potential across cell membranes, and by the use of mutations in microorganisms to produce a variety of sharply defined changes in specific cell components. We are beginning to understand how many components of cells fit together, and even beginning to carry out such assembly in the test tube.

UNIFYING PRINCIPLES OF CELL FUNCTION AND STRUCTURE

In the 3 billion years since life arose on this planet, millions of species have evolved, from relatively simple bacteria to highly complex plants and animals. Contemporary cells thus vary greatly in their properties; the structural organization, functions, and life styles of, for example, a bacterial cell, an amoeba, and a human liver cell might seem to have little in common. But despite the glaring differences, these cells share profound similarities in the principles underlying their structure and function. All cells store their genetic information in DNA, and they use the same genetic code and much the same kind of machinery to transfer this information into proteins. Every cell has an outer membrane that determines which materials may pass in and out of it. And though various cells derive energy from very different foodstuffs—for example, some bacteria oxidize sulfur to sulfuric acid—this energy is utilized, by common final pathways, to synthesize ATP (adenosine triphosphate), which provides the energy needed to drive reactions within all living cells.

Although it is, thus, justifiable to speak of "the cell," it is also often useful to distinguish two groups with major differences in their complexity: the prokaryotes (bacteria) have only one chromosome and no enclosed cell nucleus; the more complex eukaryotes (all species of animals, plants, and higher microorganisms) have multiple chromosomes and a nuclear membrane.

A bacterial cell is about 1/1,000 as large as most animal or plant cells. It is a good deal simpler (containing perhaps 5,000 genes, whereas the DNA in a human cell is equivalent to 1 million genes), and it has been much easier to study genetically. Therefore, the mechanisms that regulate and integrate the activities of the genes and their products are much more completely understood in bacteria than in the cells of higher organisms. However, with the knowledge derived from these simpler cells, and with the recently developed possibility of carrying out genetic studies in cultures of eukaryotic cells just as in cultures of bacteria (e.g., isolation of mutants; genetic recombination between different mutants), knowledge of the cell biology of higher organisms is just beginning to reach a similar level of sophistication.

Cell Adaptation and Evolution

The cells of unicellular organisms, such as bacteria, can to some degree adapt to changed environments by turning appropriate genes on and off. There is a counterpart of this process in multicellular organisms. Liver cells, for example, can respond to the chronic presence of drugs, alcohol, and other toxic substances by increasing their synthesis of a limited group of enzymes that destroy such substances. Muscle cells respond to their own exercise (contraction) by synthesizing more contractile proteins, and immune cells that synthesize a given antibody multiply faster in response to stimulation by the corresponding antigen.

In addition to such reversible adaptations, cells of unicellular and multicellular organisms can undergo changes in genes that are inherited by the descendants of those cells. A particularly striking case of such cellular evolution occurs among tumor cells. We have long known that these cells may undergo a genetic change as evidenced by visible changes in their chromosomes. We have also learned that tumor cells exposed to a chemotherapeutic drug sometimes become resistant to the drug, in some cases because of the emergence of genetically altered tumor cells endowed with a means of resistance. For example, the drug methotrexate blocks an enzyme necessary for cell duplication. Recent research has shown that cultured cells can become resistant to methotrexate by overduplicating the gene that codes for that enzyme. A normal cell has one or a few copies of that gene, but a resistant cell has several hundred copies.

This development of resistance to methotrexate provides a guide for investigating the evolution of resistance to other drugs; it also reveals an unexpected instability of the genes—a problem of the broadest biological significance. Since the methotrexate has simply selected for those occasional spontaneous mutants that have duplications of a particular gene, similar duplications must be appearing for all kinds of genes as cells multiply, but the changes are perpetuated only when they are useful.

Cell Reproduction and Its Regulation

Every organism requires cell reproduction for its long-term existence, and every cell arises from a preexisting cell by division. A human, for example, starts as a single fertilized egg cell and grows to 100 trillion cells in adulthood. Cell reproduction continues throughout life. Although brain and muscle cells are long lived, certain kinds of white blood cells are replaced every day.

The division of a cell is preceded by a period of growth, during which all of the structural components increase in size. A striking feature of this growth is its balance: Each part increases in proportion to all other parts. The kinds of intricate regulatory mechanisms that integrate thousands of gene and enzyme activities to achieve this balance in prokaryotic cells are reasonably well understood, but a great deal of research will be needed to reveal the full nature of the regulatory interactions of our own cells.

Another major preparation for cell division is the duplication of the chromosomes, so that a full set of genes can be distributed to each of the two daughter cells at cell division. But the molecular control of the switch that regulates the initiation of DNA replication remain unknown.

These are critical areas, not only because of their importance for normal cell reproduction, but also because the conversion of a normal cell into a cancerous one involves loss of its ability to control its reproduction.

Tightly coupled to this loss is the cell's failure to continue to perform its normal, specialized functions. Thus, cancer cells are characterized by uncontrolled reproduction and by functional defects. Both of these changes are considered to derive from one proximate cause, still unknown.

Carcinogens and Mutagens

A paramount question is how certain agents, in whole animals or in cell cultures, convert a normal cell into a cancer cell. Particularly puzzling is the extraordinary diversity of chemical structures active in this regard. It is difficult to imagine some common aspect among the small and large molecules, which can be water-soluble, nonpolar, acidic, basic, electrophilic, or nucleophilic. Since the change is passed on through cell division to all of the descendants of a single original cancer cell, in the absence of the original causative agent, the original cell must have undergone a change in gene structure or a persistent change in gene regulation. Indeed, radiation and many cancer-producing chemicals can cause mutations, and tumor viruses (oncogenic viruses) bring additional genes into cells. We may therefore reasonably hypothesize that cancer is caused by persistent changes (mutations) in a limited number of genes that affect normal cell reproduction.

According to this hypothesis, radiation and chemicals are carcinogens largely because they are mutagens. The case for radiation is strongly documented: All forms of ionizing radiation and ultraviolet light radiation are mutagenic and all can cause cancer. Admittedly, some evidence suggests that radiation may also cause cancer in animals by activating a latent oncogenic virus already in the cell or by activating a chemical carcinogen

A particularly puzzling aspect of carcinogenesis is the long lag—sometimes decades—that usually intervenes between initial exposure to the carcinogen and the appearance of the cancer. Probably, this delay will be understood only when the origin of cancer cells and the nature of the biological defenses against such cells are known in greater molecular detail.

A major question about radiation-induced cancer is whether there is a lower limit below which radiation does not cause either irreversible cell damage or cancer; that is, is there a safe level of radiation? Some evidence suggests a threshold below which radiation does not permanently damage or change cells, perhaps because the cellular DNA repair mechanisms can keep up with DNA damage when it is infrequent enough. Other evidence suggests that the mutagenic and carcinogenic effects are proportional to the lowest levels tested. The question remains open because human exposures, to both radiation and chemicals, are generally to dosage levels far below those tested in the laboratory. This problem is difficult and it continues to be the subject of research.

The proposition that chemical carcinogens are mutagens is currently under extensive study. Results of the Ames test, which measures the capacity of a substance to cause certain mutations in bacteria have been positive for about 90 percent of the hundreds of chemicals that are known to be carcinogenic (see p. 449).

Because of this striking evidence, any chemical that is mutagenic in the bacterial test must tentatively be considered a possible carcinogen. Yet there is doubt that mutagenesis may be equated to carcinogenesis. Considerable effort is being made to substitute mammalian, including human, cells in culture for bacteria in order to develop a more directly applicable test that is still fast and inexpensive.

While recognition of the role of environmental carcinogens has been a major advance, it would be a serious mistake to think of these agents as the unique cause of cancer. Just as in the simpler case of mutations, there is undoubtedly a background rate of carcinogenesis that arises through the inherent production of errors in the process of gene duplication, and by background radiation. Mutagens/carcinogens increase the frequency of such errors.

Tumor viruses also cause cancer in various animals by altering a cell's genetic makeup, but in this case the change is not mutational. Rather, part or all of the viral chromosome becomes integrated into a chromosome of the cell. One or more of the integrated viral genes functions somehow to override the cellular mechanism that regulates cell reproduction.

Studies of the viral induction of cancers have thus made major contributions to understanding virus–cell interactions. It may also be easier to trace the mechanism of deregulated cell growth when it is caused by one or very few added viral genes, rather than by the action of a mutagen that can change any gene in the cell. Moreover, there is strong evidence that some human cancers are of viral origin—for example, Burkitt's lymphoma, which is prevalent in parts of Africa.

The concept that cancer is the result of a genetic change in a cell bears on the question of a hereditary disposition to some kinds of cancer. Cancers of the eye (retinoblastoma) and of the kidney in children involve inheritance of a specific defective gene from one parent or the other. The basis for the less rigid but seemingly real inheritance of susceptibility to some other kinds of cancer, such as breast cancer, remains to be uncovered.

In a sense, any research that increases knowledge of the normal and abnormal behavior of cells may have the potential to increase our understanding of the cancer cell. But in particular we must continue the large effort to elucidate the molecular mechanisms that underlie normal cell reproduction, regulation of that reproduction, and cell differentiation, as well as the transformation of a normal cell to a cancerous one.

Cell Membrane and Transport

The various interactions of cells with their environments intimately involve the cell's outer membrane. This extremely thin, continuous barrier prevents cell components—even very small molecules—from leaking out of the cell, and it similarly prevents substances from entering the cell except by special uptake mechanisms.

The membrane is essentially a continuous layer of lipid (fatty) molecules, much like a soap bubble but stronger. These molecules are closely packed, creating the impenetrable barrier that separates the cell from its environment. Special protein molecules embedded in the membrane between the lipid molecules enable selective entry and exit (transport) of appropriate substances.

Some of these transport systems act as pumps, actively concentrating in the cell interior substances that are present in the environment only in low concentration. As a result, bacteria growing in very dilute external environments, such as pond water, have concentrations of substances inside their cells just as high as bacteria grown in a rich culture medium in the laboratory.

In the past two decades, a number of proteins that bind specific small molecules and transport them across the membranes has been isolated. More recently, some insights have been developed into the mechanism that converts metabolic energy into the work of active transport by such systems; however, the details of this conversion remain a great challenge.

Membrane pumps are responsible for the remarkably constant concentration of ions (e.g., potassium, magnesium) in all cells of the mammalian body. Other molecular pumps in kidney cells, selectively excreting ions and waste products, ensure that the fluid surrounding body cells (and in the circulation) enjoys an equally constant, though very different, composition. In addition, the mechanism responsible for the electrical impulses characteristic of nerve cell activity exploit transient alterations in membrane permeability (see p. 96).

Nutrient and ion transport is only one aspect of research into the interactions of cells with their environments. We are only beginning to understand chemotaxis, the process by which motile cells sense and respond to the presence of substances that cause them to change direction of their motion. In bacteria, for example—in a striking breakthrough—some molecular components of this process have been recognized.

In multicellular organisms, such directed movements of cells, far more complex than in bacteria, are extraordinarily important and still very poorly understood. White blood cells, for example, find their way toward sites of infection, where they engulf the invading organisms and destroy them. Similarly, during embryonic development a myriad of directed cell movements occurs as tissues and organs are being formed, guided by cell-to-cell interactions. Regulated cell movements also occur in tissue repair after damage and in normal tissue maintenance in an adult organism. The response of the immune system (including the chemotaxis of white blood cells just mentioned) similarly depends on a complex set of specific surface interactions among the various kinds of cells that make up the immune system (see p. 85). The detailed molecular mechanisms of signal and response are a major challenge.

Membrane Receptors

Until recently, only molecules present in soluble form in cells or extracts could be purified and identified by biochemical procedures; organized microscopic structures that remained after cells were broken up (membranes, for the most part) were sometimes labeled "cell debris"; their components could not be separated and manipulated by available techniques. Recently developed methods, however, can isolate cell membranes, separate their components, and identify specific protein molecules that have the properties of receptors, *viz.*, they specifically bind to some compound.

This extension of biochemical analysis has revolutionized several fields. In endocrinology, which is concerned with the formation of various hormones in certain cells, the molecular basis for hormonal actions on the development and the function of various target cells is being scrutinized. For example, when a steroid hormone such as an estrogen reaches its destined target cells, it binds to specific membrane receptor molecules and enters the cell as a complex. This complex migrates to the nucleus where its binding then turns some genes off and others on, resulting in major changes in the cell's metabolic activity. Similarly, several powerful bacterial toxins, such as those of diphtheria or of cholera, are now known to complex with specific membrane receptors, enter the cell, and alter the activity of specific enzymes.

Finally, the focus on cell surface receptors has also led to the discovery of a number of novel, hormonelike transmitters in the brain, which stimulate or inhibit nearby target cells (see p. 104). Awareness of membrane receptors and the capability to study them is but a decade old; a wealth of understanding must surely lie immediately ahead.

Chromosomes

Chromosomes, whose DNA constitutes the genes, play a most important role in determining the structure and function of the cell. In any mammalian cell only a small fraction of the genes is active. An important advance of the last few years has been the discovery that the DNA is coiled in a highly regular pattern (see p. 69). Whether this coiling pattern is related to the activity (expression) of genes in the DNA, and what role other chromosomal proteins play in chromosome function, are central questions now receiving major attention.

In ordinary cell division by mitosis all the chromosomes in the cell nucleus duplicate. The resulting pairs (still connected) condense into tightly packed bodies, and then the two products of each duplication are pulled apart to form two groups of daughter chromosomes. In the next step of this process these two groups are drawn to opposite sides of the cell, which then splits into two halves. An identical group of chromosomes is distributed to each daughter cell.

The essentials of this process, visible in the light microscope, have been known to cytologists since the beginning of the century. Cell biologists are now trying to clarify the underlying molecular mechanisms: What makes the duplicated chromosomes condense? How and by what are they moved to opposite ends of the cell? How does the cell split into two daughters? What is the motile mechanism so busily engaged in the division process? (It begins to appear that this last has features in common with other aspects of cell motility.)

The Cytoskeleton and Cell Motility

The most obvious form of cell motility is the contraction of muscle cells to perform mechanical work. However, cell motility takes a variety of other forms, such as the beating of flagella (as in sperm, flagellated protozoa, algae), the waves of movement of cilia (which, for example, sweep mucus along the lining of the bronchial tubes of the lung), the streaming of cytoplasm in a cyclic manner (as in nerve cells, many plant cells, and algae), and the migration by protrusion of regions of the cell (as in amoebae, white blood cells, some types of cancer cells, and all animal cells in culture). All these forms of motility appear to be based on specialized intracellular protein molecules that are capable of sliding along one another to produce contractions, but only in muscle is the role of proteins in the contraction understood in any detail. It will be necessary to unravel the various mechanisms of cell motility if we are to understand such phenomena as cell migration in development, migratory movement of white blood cells, migration of cancer cells, etc.

Cell motility involves the internal cytoskeleton of a cell. This structural component, scarcely known to exist a decade ago, is made up of a system of extremely thin fibers that determine cell shape and give the cell its mechanical rigidity. It is not a skeleton in the usual sense, since the cell constantly assembles and disassembles the fibrous components, changing its shape from minute to minute.

Cytoskeletons are built of microtubules and microfilaments. Microtubules are rigid fibers that are constantly growing longer in one part of a cell and disappearing in another, as their specific component protein molecules are being assembled or disassembled. Microfilaments are built of a protein, actin, that participates in muscle contraction, and also creates a gelatinous texture in parts of virtually all eukaryotic cells. They are often assembled into fiber bundles that function as miniature muscles, providing the force, against the microtubular network, that changes the cell shape and causes movement.

The plasticity of the cytoskeleton is dramatically demonstrated by changes that occur during cell division. Immediately prior to division, the whole cytoskeletal system is disassembled, and the molecular parts are recruited to build the apparatus that moves chromosomes and splits the cell in two. When division is complete, the division machinery is disassembled, and the molecular parts reassemble into new cytoskeletons in the daughter cells.

Finally, the cytoskeleton is also modulated by environmental signals that induce the cell to change directions. Study of the molecular basis of the cytoskeleton's dynamic nature is enjoying rapid progress.

Cell Aging

Normal animal cells in culture undergo aging (see p. 413); they can continue to reproduce for only a finite period, usually a few months. In contrast, cancer cells grown in culture are very nearly immortal, reproducing without limit. The prime example is the human cervical carcinoma cell known as HeLa, which was removed from a patient in the early 1950's and is still reproducing as vigorously as ever.

Normal cells taken from an elderly individual have shorter life-spans in culture than those derived from a young individual. This finding suggests that the limited life span of normal cells in culture reflects the normal aging process of cells in an animal. Much current research is aimed at relating the possible role of the change observed in cultured cells to the processes of senescence of animals and at identifying the molecular bases of the immortality of cancer cells.

Cell Differentiation

In the development of a multicellular organism from a fertilized egg, cells become differentiated, that is, specialized to perform particular functions. Since all these cells contain an apparently full set of genes (though recent developments suggest that there may be fine differences), the great differences in their function and structure are due to differences not in genetic makeup but rather in the ensembles of genes that are expressed and their degree of expression. Thus muscle cells produce large amounts of contractile proteins, endocrine cells are specialized to synthesize hormones, certain white blood cells produce antibodies, etc. A key element in differentiation is therefore the molecular events that regulate gene expression differently in different cells, and that perpetuate the differentiated state. In contrast to gene regulation in bacteria, we know virtually nothing about this process, or about the mechanisms by which differentiation prescribes different lifetimes for different animal and plant cell types. For many years to come, the problems of cell differentiation will present a virtually endless frontier.

CELL FUNCTION AND STRUCTURE IN HEALTH AND DISEASE

All diseases in plants and animals are expressions of the malfunctioning of cells of one kind or another, or of a tissue or organ made of specific cells. For example, circulatory diseases, which currently account for 48 percent of deaths in the United States, are mostly caused by the behavior of certain cells in the walls of arteries. As a result, patches of deposited lipids, called plaques, form on the inner surfaces of arteries (atherosclerosis), block the flow of blood, and serve as foci for formation of blood clots. Recent research has suggested a resemblance between this process and cancer; plaques appear to start as tiny overgrowths of smooth muscle cells in the arterial wall—something like a benign tumor.[3] The early stages of this process are visible in a high proportion of the aortae of young Americans who die of trauma.

Cancer is a particularly clear example of the cellular basis of disease: Its essential feature is a persistent loss of some, as yet unidentified, normal genetic regulatory mechanism, resulting in the overproduction of cells that are functionally defective. Aging, not a disease process in the usual sense, is a result of poorly understood losses in the functions of cells.

To increase our understanding of disease processes (and aging) we must continue the current, broad-based research on the normal structure and function of cells; continue to investigate the molecular basis of degenerative changes or defects in structure and function of diseased cells; and identify the specific causes of such degenerative changes or defects.

Collectively, these activities will continue to account for a major thrust of research in cell biology for the next decade or more.

IMMUNOLOGY

Immunology is concerned with a broad range of phenomena relating to human disease, some beneficial and some harmful. Best known is resistance to infectious disease, induced either by natural infection or by artificial immunization. The latter practice has virtually eliminated diphtheria, whooping cough, poliomyelitis, and tetanus in the United States, and smallpox worldwide. On the other hand, the immune system can also occasion harmful reactions, including allergies, rejection of transplants, and various autoimmune diseases in which the body reacts against its own components. The major medical purpose of immunological research is to enable manipulation of the immune system so as to augment the beneficial effects and curtail the harmful ones.

After more than half a century of descriptive observations, the past two decades have witnessed the development of deep insights into the cellular, genetic, and molecular mechanisms of the immune system. Immunology has also become of great interest to other biological sciences, more broadly taken. It provides a model system for studying the challenging problems of selective gene activation during development of the fertilized egg, and it has provided extremely sensitive, highly specific analytical tools for quantifying many compounds.

NATURE OF THE IMMUNE SYSTEM

The main elements in the immune system are lymphocytes, small white cells found in the bone marrow, lymph nodes, spleen, and circulating blood. One class of lymphocytes secretes protein molecules called antibodies or immunoglobulins (Ig). Each individual lymphocyte makes many copies of one specific antibody, each molecule of which can bind very tightly to two molecules of a given foreign substance (antigen). As a consequence of binding to antigens, antibodies can neutralize toxic products of bacteria, or promote the uptake of bacteria by scavenger cells that destroy them, or prevent the penetration of viruses into cells.

Another class of lymphocytes also bears antibodylike molecules on their cell surface, but does not secrete them; instead, by means of these molecules they can attach to, and kill, cells that carry the corresponding specific antigen on their surface: for instance, infecting microbes, foreign (transplanted) animal cells, and host cells infected with a virus.

Antibody Molecules

The binding sites of antibodies are relatively small regions, each of which fits the combining site of the corresponding antigen as precisely as a lock matches a key. The repertoire of antibodies in an organism is enormous: There are perhaps a million.

The total amino acid sequences of several antibodies are known and their general three-dimensional structures are apparent (Figure 5). Antibodies of different specificity hold large regions in common but differ at their binding sites.

Several distinct classes of human antibodies exist, each with a unique structure and biological function, and a set of binding sites for the full range of antigens. For example, the class called IgE, constituting less than 0.1 percent of the total antibody population, is responsible for allergies. Contact with allergens such as ragweed pollen or penicillin may provoke formation of several classes of antibodies, but only the IgE antibodies cause allergic reactions.

Lymphocytes

A number of different classes of lymphocytes, with major differences in function, have been recognized in the past two decades. This knowledge should lead, perhaps in a very few years, to beneficial control over formation of these alternative responses to an antigen.

B cells, arising in the bone marrow, are the lymphocytes that produce antibodies. A mammal contains hundreds of thousands of different B cells, each specific for a particular antigen. Each kind of cell is normally present in very small numbers, but when such a cell comes into contact with the corresponding antigen it divides rapidly, and all of its descendants secrete the same antibody molecules.

T cells are also lymphocytes of bone marrow origin, but they are processed further, in some unknown way, in the thymus gland—whose function was unsuspected 20 years ago. As a result of recent research, three major subclasses of T cells are now recognized, each with distinct biological functions; within each subclass, different T cells have specificity for different antigens.

The *killer T cell* is one of our principal defense mechanisms; it can attack and destroy tissue cells that have been invaded by certain viruses. Killer T cells can attack tumor cells, too, at least in experimental animals; and they are largely responsible for the rejection of skin or organs transplanted from one individual to another. The success rate in kidney transplantation has been increased markedly through the development of drugs that prevent the multiplication of killer T cells, as well as through more careful matching of recipients and donors so that the antigens

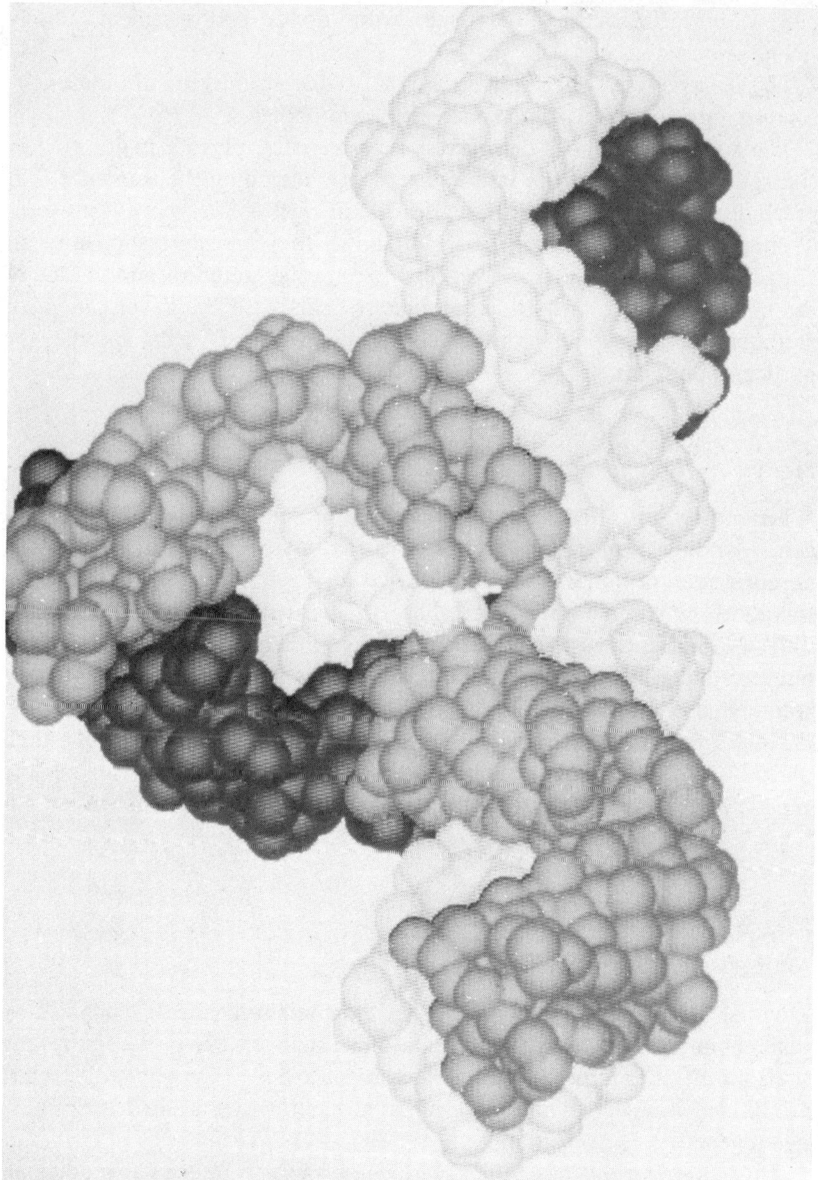

FIGURE 5 Computer-generated model of intact human antibody molecule. The two heavy polypeptide chains appear dark gray and almost black, both light chains are lighter gray. The almost-white circles near the center make up a complex carbohydrate. (M. A. Navia. Coordinates from E. W. Silverton, M. A. Navia, and D. R. Davies, *Proceedings of the National Academy of Sciences* 74:5140–44, 1977.)

present on cell surfaces of tissues from donor and recipient closely resemble one another.

The *helper T cell*, once activated by a particular antigen, stimulates the corresponding B cell to divide and release antibodies.

The *suppressor T cell*, discovered very recently, plays a major role in controlling the level of the immune response, including the tolerance that precludes us from forming antibodies to our own constituents. An excess of suppressor T cell activity can shut down the immune response to an antigen, whereas a deficiency can lead to an exaggerated response. One of the most promising areas in immunology deals with the development of methods for controlling the level of suppressor T cells specific for particular antigens.

Molecular Basis for Antibody Diversity

When molecular genetics showed that each protein chain is coded for by a corresponding gene, immunology faced a major problem: Could the chromosomes of a mammal carry a separate gene for each of hundreds of thousands of specific antibodies? There is increasing evidence for a quite different mechanism, in which the individual inherits a much more limited number of genes for antibodies, and mutations in these genes, or recombinations between them, occur at an exceptionally high rate in the precursor cells, thereby giving rise to the great variety of final, differentiated, antibody-producing cells. This unusual behavior of the genetic material in these special precursor cells is under intense investigation.

Cell Fusion; Development of Permanent Cell Lines Producing Antibodies

Most large antigens possess many different combining sites, recognized by different antibodies. Moreover, even a single such site can be recognized by several different antibodies of overlapping specificity. Consequently, when an antigen is injected into an animal, the antibodies evoked are highly heterogeneous.

This heterogeneity has been a serious obstacle to molecular studies on antibody formation, structure, and function. However, an elegant solution has recently been developed. If an antibody-producing cell (which has a life span of only a few days) is fused with a cell from a certain tumor line (which can be cultivated in the test tube indefinitely), the resulting hybrid product—which encloses the chromosomes of both in a single membrane—can multiply and secrete the desired single antibody indefinitely.

These homogeneous antibodies promise immense variety of applications, in research and in practice.

APPLICATIONS

Vaccines

Since the nineteenth-century discovery that infectious agents killed in certain ways retain the capacity to elicit antibodies, such preparations have been used as vaccines with great success in preventing a number of diseases. Live (attenuated) vaccines have also been developed against viruses (and the tubercle bacillus), through the use of harmless mutants. Immunization against viruses (e.g., polio) became much more feasible after methods were developed for growing the virus inexpensively and in large quantities in cultured animal cells, rather than in infected animals.

Some killed organisms have proved too toxic for use in immunization. However, the range of effective vaccines is now being extended by purifying the surface antigens of bacterial cells (which are the components attacked by protective antibodies) and eliminating the more toxic, deeper constituents.

New vaccines are contributing to the control of a variety of disorders. Pneumococcal (lobar) pneumonia still kills 25,000–50,000 Americans annually (especially older or debilitated persons), and a vaccine against the 14 commonest types was licensed in 1978. About 20,000 Americans contract bacterial meningitis annually, with around 3,000 deaths, and antigens against two of the three meningococcus types (A and C) have been licensed and are now given routinely to members of the armed forces. German measles (rubella) is a mild disease, but in pregnant women the virus is often lethal or damaging for the fetus. Following the 1964 epidemic, some 40,000 infants were born with mental or other defects. The virus was first isolated in 1962, and an attenuated vaccine has now been licensed.

Viral hepatitis A (infectious hepatitis), with about 30,000 cases reported annually in this country, is usually transmitted by fecal contamination. We have not learned how to cultivate the virus, and so it has not been possible to produce a vaccine against this debilitating disease. However, with the similar hepatitis B (serum hepatitis), an antigen from the virus, designated as HBs, was accidentally discovered in the blood of Australian aborigines, in the course of a search for new blood types (for which it was at first mistaken). The incidence of this disease, which is usually transmitted by blood transfusions, has been markedly decreased by screening all blood being used for transfusions for this antigen. Moreover, HBs antigen, isolated from the blood of carriers and then inactivated, appears promising as a vaccine.

A warning note is in order. When a sufficient fraction of all children is immunized against a disease such as pertussis or measles, the incidence of the disease markedly declines. One result is that nonimmunized children are very much less likely to contract the disease and develop active immunity. When they, as adults, travel to countries where the disease is prevalent, they are highly susceptible to an infection that is far more serious in adults than in children. Similarly, if the very success of a vaccination campaign engenders subsequent carelessness, with a decreasing fraction of children being immunized, the stage can be set for a future epidemic in the adult population.

Parasitic Infections

Parasitic infections of man constitute the main health hazard in tropical regions, where they affect hundreds of millions of individuals. In Africa alone an estimated million children die annually from malaria. A number of drugs are effective, to varying degrees, against some of these infections, and vaccines have been developed against several parasitic infections of domestic animals. So far, however, vaccines against parasites in man have not been successful.

Current work on malaria is very promising. The parasite that causes most human malaria has recently been cultivated in the test tube. This development provides a rapid test for antimalarial drugs, and it should also make many other kinds of research on this organism more effective and less expensive. The prospect for a vaccine against malaria in man now seems bright.

Rh Incompatibility (Hemolytic Disease of the Newborn)

All red blood cells have numerous antigens on their surfaces, including those of the Rh series. If a fetus inherits from its father antigen Rh-D, but the mother lacks that antigen, she may form antibodies to red cells that leak from the fetus into her circulation. These antibodies, in turn, can reach the fetus and destroy its red blood cells. The risk of this disease increases with each pregnancy: Fetal cells are particularly likely to invade the mother's circulation during childbirth, and a second antigenic stimulus elicits a larger response than the first. Prior to 1968, an estimated 10,000 infants died of this disease each year in the United States, and about 10 times as many were affected. However, the disease can now be almost completely prevented: Anti-Rh-D antibodies, injected into the mother immediately after each delivery, prevent her from forming antibodies to this antigen.

Organ Transplantation

Immune responses to foreign tissues are the main limitation to effective organ transplantation. Kidney transplants are the most common, but some progress is also being made in transplantation of bone marrow, heart, and liver.

Approximately 40,000 Americans with kidney failure are at present receiving regular dialysis treatments, at a total cost of over $500 million annually. The potential economic benefit of transplantation is thus very large.

Most kidney transplants come from recently deceased individuals, but only about 50 percent of these grafts have survived for four years, compared with 75 percent of those from properly matched related donors. However, major developments in the past three years promise to greatly increase the survival rate of cadaver grafts. Two sets of antigens that are present on all tissues, designated HLA-A and HLA-B, have generally been used for matching. In one study in which tissues were matched as well to a third set (DR), which is present on lymphocytes, 90 percent of cadaver grafts survived for at least two years.

There has also been progress in suppressing the immune system of patients receiving transplants, by using drugs, irradiation, or antibodies to killer cells.

Inherited Diseases of the Immune System

In several inherited diseases, the production of B and/or T cells is abnormal, and the patients often succumb to infections early in life. Some patients can now be treated with transplants of bone marrow, thymus tissue, or fetal liver.

Another serious abnormality of the immune system, hereditary angioedema (a type of giant hives or swelling), has been traced to alteration in the function of a component of the complement system (a set of proteins that interacts with antibodies to increase resistance to infection). This disorder can now be successfully managed with a synthetic steroid, which usually returns complement function to normal.

Allergies

The most frequent disorder of the immune system is allergy. A serious form, asthma, afflicts about 9 million persons and causes over 2,000 reported deaths a year.

Allergic reactions are caused by antibodies of the IgE class, which thus far are known to have only deleterious effects. Certain white blood cells (basophils) have receptors that can bind complexes of IgE with corresponding allergens. Binding triggers the release of substances, such as histamine, that then act on various other tissues, causing such symptoms as sneezing, rash, or hives. Identification of these agents can lead to the development of specific drugs, such as the antihistamines, that minimize their formation or interfere with their action on target tissues. Moreover, the receptors for IgE on basophils have been identified in the past two or three years, and this breakthrough may permit the design of chemical agents that will interfere with their interaction with IgE.

In another recent advance, certain suppressor T cells and helper T cells have been found to act specifically, in opposite directions, on the IgE response. Since the balance of the two cell types can be influenced (e.g., by the dosage and the form of antigen), this finding offers promise of a rational, guided approach to desensitization. Another new approach seeks to block the IgE response, on the basis of the finding that some antigens induce tolerance, instead of antibody formation, if they are linked to certain nonantigenic large molecules.

Purified specific components of many pollens that cause allergy are now being used to improve diagnosis. Moreover, these antigens can be used to measure the amount of corresponding IgE antibody in the blood, a procedure that has some advantages over the common skin test for allergy. These purified antigens are also being used in immunotherapy (also known as hyposensitization, or allergy shots), which presumably works by shifting the balance of immune responses.

Similarly, purified venoms of bees or hornets are proving much more effective in hyposensitization than the previously used extracts of whole insects. Unfortunately, because the market is limited, and extensive tests are now required before a new pharmaceutical product can be licensed, this advance is unlikely to be made generally available through the normal marketing mechanisms.

Autoimmune Diseases

Under ordinary circumstances one does not develop an immune response to constituents of one's own body tissues. However, occasionally regulatory mechanisms break down and an individual produces antibodies or active lymphocytes that react with specific tissues of his body. Such autoimmunity has been implicated in recent years in an increasing number of important diseases, including lupus erythematosus, rheumatoid arthritis, myasthenia gravis, hemolytic anemia, thrombocytopenic purpura, Addison's disease, hyperthyroidism, multiple sclerosis, Guillain-Barré syndrome, chronic glomerulonephritis, and possibly pernicious anemia and certain forms of diabetes mellitus. Autoimmunity may also appear after tissue damage or an infection. For example, after a myocardial infarction, antibodies to heart tissue can be detected, and in rheumatic fever or glomerulonephritis, which may follow a streptococcal infection, antibodies to certain streptococcal antigens cross-react with antigens on the specific tissues.

The mechanisms that normally prevent autoimmune responses are referred to as immunologic tolerance; that is, one is normally tolerant to one's own tissues or body constituents. The recently discovered suppressor T cells play an important role in mediating tolerance, and a major advance in understanding is promised by the finding that the level of specific suppressor cells and of corresponding antigens varies with the activity of certain diseases, such as lupus erythematosus and juvenile arthritis. In addition, it has been found that killer T cells, which may be involved in autoimmune processes, can be inactivated by antibodies directed against specific receptors on them.

This newly acquired and rapidly increasing understanding of the mechanism of tolerance offers promise of leading, relatively soon, to improved methods for identifying and controlling autoimmune processes. For example, many patients with myasthenia gravis have shown improvement after alteration of immune function by removal of the thymus gland, or by treatment with steroid drugs.

Immunology and Cancer

It has been possible to transplant tumors, without rejection, by using highly inbred, genetically almost identical mice, which no longer react immunologically to each other's normal tissues. Certain tumors were thus found to have characteristic antigens that elicit immune responses in the recipient host. These responses are detected because they result in rejection of a second transplant from the same donor. The occasional unexpected spontaneous regression of a malignant tumor in man may well be due to an immune response. It has been hoped that immunotherapy would completely eliminate any cancer cells remaining after chemotherapy, just as the two procedures complement each other in the therapy of infections. So far, however, results have been disappointing.

Immunological Analytical Techniques

The extreme specificity of antibodies, combined with radioactive labeling, has recently led to their widespread use in radioimmunoassay as analytical tools to measure very low concentrations of many important biological substances, including hormones, drugs (such as digoxin, gentamicin, or penicillin), and metabolic intermediates. A modern clinical laboratory may run 30 different tests of this type routinely, and hundreds are used in research. This approach has revealed, for example, that many patients with diabetes produce adequate quantities of insulin but inactivate it with their antibodies.

A powerful derivative tool has been the attachment of fluorescent compounds to antibodies. These reagents can single out a specific bacterial cell in a population of other bacteria, or identify organisms in tissues, making possible much more rapid diagnosis than that dependent on cultures. (In Legionnaire's disease, diagnosis depends entirely on immunofluorescence.) In addition, immunofluorescent staining of normal cell constituents has become a major tool in studying the organization and dynamics of animal cells. For decades antibodies have been used as diagnostic tools to identify infectious microbes: Known antibodies are used to identify an organism recovered from a patient, and known organisms are used to detect a rise in antibodies in a patient.

Interferon

Interferon is a powerful antiviral protein manufactured in tiny amounts by the body's own cells when they are infected by a virus. When this substance is released, it renders nearby cells more resistant to infection by any virus. Interferon has proved therapeutically effective in preliminary tests. However, both research and application are limited by the extremely small quantities available. Recombinant DNA technology may soon alleviate this problem.

FUTURE DIRECTIONS

The past 20 years have witnessed a revolution in research in immunology, with improved tools for study at a cellular and a molecular level. The results have revealed a cascade of reactions, and a balance between antagonistic processes, which offer many opportunities for therapeutic intervention. In addition, the recombinant DNA technology should soon make available unlimited quantities of any specific component of the system. The next years therefore offer strong promise of marked increase

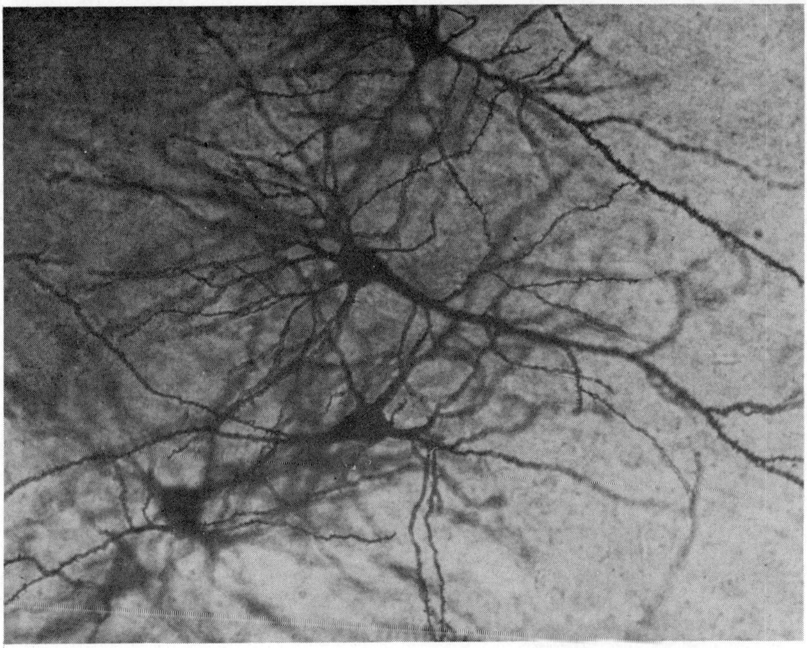

Photomicrograph of nerve cells. (Fritz Goro, courtesy of Polaroid Corporation)

in our ability to control both desirable and untoward immune reactions, which are relevant for a large fraction of all human disease.

NEUROSCIENCE

Neuroscience is the comprehensive research endeavor that seeks to understand the biological bases of behavior and experience. The human brain is the most remarkable and complex structure in the known universe. Composed of more than 50 billion nerve cells, each connected with as many as a thousand other nerve cells, a single human brain has a greater number of possible connections among its nerve cells than the total number of atomic particles in the entire universe. This three pounds of tissue is the structure that has allowed the human species to dominate its environment, write poetry, compose symphonies, make war, and undertake scientific understanding of the cosmos and of itself.

Whatever the mind is, it is in the brain. Application of physical, biological, and behavioral techniques to the study of the nervous system has just begun to reveal the way individual nerves function, the functional organization of the brain, the nature of sensations and perceptions, the way language is organized, and the physical basis of consciousness. Indeed, by the twenty-first century, neuroscience may well become the dominant science.

Study of the brain and nervous system developed as a special field within each of a number of traditional disciplines; anatomists worked on the structure of the brain, chemists focused on its chemistry, physiologists studied the electrical activity of nerve cells, psychologists analyzed the behavioral functions of various parts of the brain, and so on. Only recently, these efforts began to merge as the single enterprise called neuroscience.

The goal of neuroscience is a comprehensive scientific understanding of the nervous system and brain, particularly the human brain, extending to the final product of the brain's activity: behavior and experience. With this understanding should come treatment for many mental and neurological disorders; psychoses such as schizophrenia and depression, mental retardation, learning disabilities, stroke, blindness, Parkinson's disease, multiple sclerosis, head and spine injuries, drug addiction, muscular dystrophy, myasthenia and even, if we are very lucky, more effective approaches to the prevention of crime and war. These are, of course, long-term goals. The young field of neuroscience has made impressive advances in the past few decades and will accomplish much more in the next few years. This report is focused on the immediate past and future.

THE NERVOUS SYSTEM—A BRIEF OVERVIEW

To facilitate discussion of recent achievements and prospects, it may help to begin with a brief summary of the subject matter of neuroscience: the nervous system. In vertebrates, the brain and spinal cord are a continuous structure that is termed the central nervous system (CNS). The peripheral nervous system comprises the various nerves that lie outside the CNS, mostly sensory nerves that convey information from receptors to the CNS and motor nerves that convey commands from the CNS to muscles.

Although many invertebrates have rather differently organized nervous systems, some of them relatively quite simple, the basic plan of the human nervous system can be traced back at least to the worm, with its tubular nerve cord. Even in the worm, the head end of the nerve cord is enlarged to accommodate the specialization of the front end of the worm for sensing and feeding. The human CNS maintains this basic tubular organization from the spinal cord up to about the middle of the brain. However, the tube's front end, the forebrain, is enormously expanded and laid back over the core tube to form most of the matter we call the "brain."

The Neuron

During the nineteenth century the gross structures and subdivisions of the brain were described by anatomists. However, the nature of these structures was not at all clear. The Spanish neuroanatomist, Ramon y Cajal, mapped out the detailed microscopic structure of the subdivisions of the brain and established the neuron doctrine, *viz.*, the nervous system is composed of a great many discrete cells, the nerve cells or neurons, that influence one another by specific connections in a highly patterned manner. These connections are termed "synapses." The electron microscope revealed that a synapse is not a direct or continuous connection but rather a very close approximation (about 20 millionths of a millimeter) between a terminal of one neuron and the cell membrane of another neuron. On a typical neuron in the brain are hundreds of synaptic endings from other neurons.

The characteristic neuron has a central cell body, containing the nucleus and much of the biochemical machinery of these specialized cells. From the cell body extend fibers as long as 3 feet in man and over 60 feet in the whale. Most chemical substances in the neuron are synthesized in the cell body and then transported out to the fibers. The principal fiber, the axon, conducts information to an end-organ, for instance, muscle, or, in the brain, the end of the axon subdivides into numbers of small fibers that end in synaptic terminals on other neurons. Other fibers, dendrites, are specialized to receive information from synapses with other neurons, as is the membrane of the cell body itself.

The brain, then, is essentially a vast network of interconnecting neurons. Each neuron receives information from other neurons or from sensory receptors and transmits information to still other neurons or to muscles or glands to produce behavior. All stimuli impinging upon us, all sensations, thoughts, actions, and feelings, must be "coded" into the languages of the neuron. A particular neuron is affected at all times by many other neurons, some of which activate, some inhibit, in a graded-decision process. When the sum of these inputs activates sufficiently, the nerve cell transmits this information along its axon fiber by propagating an electrical action potential (the all-or-none digital language of the nerve fiber), which reaches to all the distal presynaptic endings of the axon. The basic processes of neuronal activity thus are synaptic input, the action potential and synaptic transmission.

Research in the 1940's and 1950's, primarily in England and the United States, led to an understanding of the nerve action potential. This electrochemical process, a wave of altered potential across a localized region of cell membrane, travels down the axon at a speed of 10–50 meters per second to all the synaptic terminals of that neuron. Once started, the action potential runs its course and is not normally influenced by other processes in the nervous system.

When the action potential arrives at a presynaptic terminal—the end of the axon that forms the synaptic connection to another neuron or muscle fiber—the presynaptic terminal releases a small amount of a chemical substance. This substance, called a transmitter, diffuses across the narrow synaptic space and binds to specific receptors, protein molecules lodged in the membrane of the second or postsynaptic neuron. This binding causes a small change in the electrical potential across the postsynaptic membrane, most prominently in the immediate vicinity of the synapse. For the entire neuron so affected, the change is also graded; the more synapses that are thus active at any moment, the greater the total change.

Several different compounds are now known to serve as synaptic transmitters. They are all relatively small molecules (molecular weights 150–3,000) that are either normal constituents of most cells, e.g., amino acids, such as glycine and glutamic acid, or are made from such, e.g., noradrenaline, acetylcholine, peptides, γ-aminobutyric acid (GABA). A given synapse uses only one chemical transmitter, and a given transmitter affects the receiving neuron in only one of two modes: excitation or inhibition. Excitation stimulates the neuron, ultimately to the point where it generates an action potential that travels down the axon to influence other neurons. Inhibition makes the neuron less likely to generate an action potential in response to excitation from other synapses. Importantly, at all synapses there is a mechanism that rapidly removes the transmitter, terminating its effect within a few milliseconds after it has been released.

With the exception of "electrical" synapses in certain invertebrates and in a very few locations in the mammalian CNS, information is transferred from the axon terminal of one neuron to the cell membrane of another only by a chemical transmitter. This fact, which has been clearly established only for the past 25 years, has extremely important implications. A chemical synapse is, by its very nature, modifiable; other chemicals, hormones, drugs, the local environment of the neuron, and perhaps even its past experience, can influence the synapse.

The chemical synapse has become a unifying concept in neuroscience. Most drug actions on the nervous system occur at synapses. Different types of drugs can have different types and modes of action at the same synapse; one drug might block the release of chemical transmitter, another might occupy the binding site on the receptor in place of the normal chemical transmitter substance, and so on. In a similar vein, various hormones appear to exert their influences on the brain by attaching to specific hormone-binding receptors on the surfaces of certain sets of neurons, particularly in the brain structure termed the hypothalamus, which is closely connected to the pituitary gland.

Important behavioral phenomena like motivation appear to rest on such mechanisms. The experience of "thirst" is due to the action of a "thirst" hormone, activated by the kidney under appropriate conditions, which attaches to and activates certain sets of neurons in the brain. Although adequate evidence is lacking, it seems likely that learning and memory, the bases for behavioral phenomena in humans, rest on persistent changes in chemical synapses in certain regions and systems of the brain.

General Structure of the Brain

The brain is traditionally described as consisting of several major levels. The lowest level—the brain stem—is, evolutionarily, the oldest part of the brain, shared relatively unchanged with primitive vertebrates. It is essentially an upward elaboration of the spinal cord, important to certain specialized senses and to such vital regulatory and reflex functions as breathing, cardiovascular activity, and digestion.

The next major step in brain evolution was the development of the limbic system, about 150 million years ago. It controlled primitive behaviors—anger and attack, fear and flight, hunger, sexual activity—and has been the highest region of the brain for reptiles like the crocodile. The limbic system remains well developed in the human brain, although it is in some ways much changed and may serve rather different functions. However, it is overshadowed by the cerebral cortex, even in lower mammals like the rat. Over the course of evolution of the mammals there has been a steady increase in the relative size of the cerebral cortex. The cortex is a complex neuronal structure about two millimeters thick, that overlies most of the brain, much like the rind on an orange. In higher mammals it becomes increasingly folded and convoluted, so that its total area and extent are much increased. It is the dominant brain structure in primates.

The evolutionary development of the cerebral cortex that separates human from ape began only about 2 million years ago and proceeded at a rate unprecedented in the earlier evolutionary history of the brain. The human cerebral cortex has existed in its present form for only about 100,000 years. And the rational and abstract thought and subtle feelings of the cortex do not always dominate the surges of cruder feelings from the reptilian core, the limbic system, within.

Another way of looking at the organization of the brain is in terms of systems that have been identified by observing the effects of experimental lesions in animals or accidental lesions in humans. The sensory systems convey information to the brain from the various receptors; the motor systems control the muscles (and glands) and hence generate behavior.

For humans, the most important sensory systems are the visual and auditory. To take the visual system as an example, light entering the eye activates receptors in the retina at the back of the eye. This visual information undergoes considerable processing by succeeding layers of neurons in the retina. This processed information is then transmitted to the brain by the optic nerves. Within the brain, the information receives further processing in a portion of a region called the thalamus, which then transmits the information to the visual region of the cerebral cortex, where still further processing is done. Thus, sensory systems extend from the receptors, through various brain levels, to the cerebral cortex.

Motor systems of the brain are those sets of neurons and pathways most directly involved in the control of muscles. Most behaviors that we observe in others are muscle movements, sequences of contractions and relaxations controlled by the motor regions of the brain. Motor neurons, with cell bodies in the spinal cord and brain stem, send their axons out of the CNS to innervate muscles of the body and head. Higher order systems and structures in the brain—ranging from reflex pathways in the spinal cord and brain stem to elaborate, specialized structures like the cerebellum, the basal ganglia, and the motor region of the cerebral cortex—play on these motor neurons. The microstructure of the cerebellum is surprisingly constant from reptile to human. It plays an important but not fully understood role in regulation and coordination of movement. The basal ganglia, large structures lying mostly under the cerebral cortex in the forebrain, are somehow involved in the control of movement—Parkinsonism is a disorder of the basal ganglia—but their functions are very poorly understood.

The limbic system, noted above, is generally believed to be concerned with the motivational and emotional aspects of behavior. It consists of several large and ancient structures—hippocampus, amygdala, septum—closely interconnected with each other and with the hypothalamus, a very ancient structure at the upper end of the brain stem consisting of a number of small groups of nerve cells that exert rather direct and powerful control over hunger and eating, thirst and drinking, sexual behavior, pleasure and pain. The hypothalamus is adjacent to and closely connected with the pituitary gland, the master control gland of the endocrine system.

Another very important brain system is the reticular formation, a collection of groups of nerve cells and fibers that ascends in the core of the brain from the spinal cord up to the level of the thalamus and has close conections to the hypothalamus. The reticular formation appears to play a critical role in the control of sleep and waking. Also, those neurons that utilize noradrenaline and serotonin as transmitters have their cell bodies only in certain regions of the reticular formation.

It must be emphasized that the various brain "systems" are in part useful abstractions. They are categorized as systems largely on the basis of what appear to be close anatomical interconnections or apparent functional similarities. So little is yet known about many regions and structures of the brain that current descriptions of brain systems must be viewed with reservations. For example, the hippocampus is a large structure in the limbic system, whose functions are believed to be motivational and emotional. Yet recent evidence indicates that the hippocampus plays a critically important role in learning and memory in humans and other mammals. Much of the excitement in neuroscience today stems from the fact that so little is yet known—it is a field of science that has only begun.

Some Aspects of Neuroanatomy

Neuroanatomy is in the midst of a revolution, due largely to techniques developed in the past 5–10 years for the tracing of "pathways." Methods for tracing the connections of neurons were difficult, unreliable, and time consuming until the development, in the early 1950's, of a technique based on the fact that, if a cell body or axon in the brain is destroyed, the axon terminals that connect to other neurons will degenerate. At a certain phase of this degeneration, the terminals selectively absorb silver salts, which can then be visualized with a microscope, revealing terminals far removed from the cell body, thus permitting tracing of remote connections.

Recent tracing techniques, which work with undamaged brain, make use of the fact that neurons transport various substances along their axons. Thus, radiolabeled amino acids, injected in the vicinity of cell bodies, are absorbed and transported to the axon terminals, where they can be visualized by appropriate techniques. Conversely, the enzyme peroxidase is taken up by nerve terminals and transported back to the cell bodies, where it can be stained and visualized. Many laboratories are now using these methods to trace out the wiring diagrams of the brain. Not only is new circuitry being plotted; some of the classical connections in the brain are being shown not to exist. More has been learned about the circuitry of the brain in the past 10 years than in all previous history.

A recent technique of great promise depends on the fact that neurons, like all other cells, use glucose as their fuel. A radiolabeled substance (deoxyglucose) that is chemically similar to glucose but not metabolized by neurons is injected. Neurons take up this substance and accumulate it in proportion to their levels of metabolic activity. It is thus possible to determine the effects of various conditions, such as sensory stimulation, on the activity of specific regions and neurons in the brain. Moreover, if the isotope used for labeling the deoxyglucose decays by positron or γ-ray emission, localization in the living, intact brain can be achieved from without.

A method that is just being developed utilizes antibodies to label certain classes of neurons in the brain. As an example, γ-aminobutyric acid, a transmitter operative as an inhibitor at certain synapses, is synthesized by the action of the enzyme glutamic acid decarboxylase (GAD). Application of radioisotopically labeled antibodies specific for GAD permits identification of all neurons that use this transmitter. In principle, analogous procedures can be developed for most of the known transmitters.

Another important approach to analysis of the structure of the nervous system is electron microscopy. The ultrastructure of neurons and synapses is being catalogued at a rapid rate. New methods such as freeze fracture permit ever better visualization of fine structure, which, one day, will enable better understanding of the functional properties of neurons.

FUNCTIONING OF INDIVIDUAL NERVE CELLS

Axonal Conduction

The achievements of neurophysiology in the past 30 years have been monumental. As indicated above, considerable understanding has been gathered concerning the two basic processes of the neuron: axonal conduction and synaptic transmission. These advances depended upon the development of electronic apparatus for amplification, display, and analysis of the tiny electrical signals generated by individual neurons, and upon the use of the microelectrode—a very slender wire that can be placed close to or inside a neuron.

The action potential generated by a neuron can be recorded by inserting a microelectrode so that its conducting tip is very close to a specific neuron. When the neuron fires, the action potential is recorded as a brief (1/1,000 second) wave, a "spike," thus permitting examination of the pattern of activity of that neuron under various circumstances. The method has been applied in studies of the various neuronal sources of aspects of behavior, such as, sensory processes, motor systems, and learning, and will be discussed later in these contexts.

Microelectrodes inserted inside a single neuron allow measurement of the potential across the nerve cell membrane. The normal resting potential across the membrane is about 70 millivolts, with the inside negative. When synaptic excitation occurs, the transmembrane potential of the affected region of the postsynaptic membrane becomes less negative (depolarized) until it reaches the spike discharge threshold required to generate an action potential that sweeps down the axon. In synaptic inhibition the local potential becomes more negative than at rest (hyperpolarized). Hence it is possible to measure the synaptic effects of other neurons on a given neuron.

A major area of current research is concerned with the molecular basis for the events registered as the action potential. Much of current knowledge rests on studies of the giant axon of the squid. The immediate consequence of excitation is a sudden, brief opening of ion channels ("gating") in the membrane so that sodium ions (Na^+)—which otherwise are blocked from diffusing into the neuron—rush in; this is followed by the exit of a lesser number of potassium ions (K^+) through a separate set of channels. This is the mechanism of depolarization. An ion channel appears to be a micropassage through a specific protein molecule that traverses the membrane, opened or closed by a subtle change in the three-dimensional conformation of the protein. It is the transiently lowered potential across one region of membrane that causes opening of the "gates" in the adjacent region. The toxic principles of the puffer fish (tetrodotoxin), of *Gonyaulax*, the dinoflagellate that causes "red tides" (saxitoxin), and of the scorpion, all of which affect the nervous system by binding to the gating protein on the outside surface of the membrane, have been invaluable in the study of this process. Shortly after the influx of Na^+, a special enzyme, close by in the membrane, utilizes the energy of ATP to pump these ions out, through channels in the pump itself, until the resting potential is restored. At each locus on the axon the entire cycle is complete in about two milliseconds. This pair of phenomena occurs consecutively down the full length of the axon, accounting for the sweep of the action potential. The ATP-dependent sodium pump-protein binds and is inhibited by cardiac glycosides, such as ouabain, accounting for their effects on the conducting system of the heart and elsewhere.

Synaptic Transmission

The immediate events leading to the release of chemical transmitter at the presynaptic axon terminals when the action potential arrives at the terminals are also becoming clearer, as are the events that occur when released transmitter arrives at the postsynaptic membrane receptors and excites or inhibits the next neuron.

The enormous growth of general biochemical understanding in the last generation has had a marked effect on neuroscience. Thirty years ago neurochemistry was largely preoccupied with the structures of unusual lipids that occur in the nervous system. Unraveling the detailed structure of the insulating myelin sheath wrapped around the larger, more rapidly conducting, nerves was a great triumph; multiple sclerosis is the consequence of demyelination of certain nerves. A number of disorders of the nervous system, such as Tay-Sachs disease, involve derangements in the metabolic pathways of these special compounds, and work in this area is continuing. For the past 10 years, however, the main preoccupations of many neurochemists have been the chemistry of synaptic transmission and nerve membranes (Figure 6). The rapid expansion of understanding of these areas has been rewarding; already this has led to major improvements in the treatment of some of the commonest of human ailments, and the prospects for rapid progress are bright.

By the 1950's only two chemical substances had been positively identified as neurotransmitters: acetylcholine and norepinephrine. It now appears that there are probably over 20 different compounds used as transmitters in various parts of the nervous system.

Since chemical transmission was first established by Loewi in 1921, there has been general awareness of a similarity between the actions of hormones and of nerve transmitters. Each system involves the secretion of a special chemical by a specific type of cell, and this compound acts by binding and modifying the properties of a special protein molecule, the receptor, lodged in the membrane of a second, recipient, cell. The main difference between hormone action and synaptic transmission lies in the directness of the action. Most hormones are released into the circulation and travel a great distance before finding a receptor, whereas only a miniscule fraction of a millimeter separates the release site of a transmitter from its receptor; the effect is virtually instantaneous. There is little chance, then, that the transmitter can affect receptors remote from the junction where it is released.

The most thoroughly studied synapses are those employing acetylcholine, thanks to their abundance and easy access in the electric organs of the sting ray and the electric eel. These have revealed that in the presynaptic terminal are collected miniature packets of transmitters bound to a protein and wrapped in a thin membrane (vesicles); entire vesicles are discharged into the synaptic space when the action potential arrives. The vesicles burst, acetylcholine binds to a receptor protein on the postsynaptic membrane and this, somehow, occasions the opening of Na^+ channels, exciting the postsynaptic cell. The effect is fleeting since, in the presynaptic membrane, there is an arrangement that rapidly withdraws acetylcholine back to be repackaged while in the postsynaptic membrane there is an enzyme (acetyl cholinesterase) that can degrade (hydrolyze) all of the acetylcholine in the synaptic space within one–two milliseconds. The Na^+ channels close, excess Na^+ is expelled by the "pump," and the way is then clear for another burst of transmitter.

The entire synaptic arrangement is accessible to chemical agents from the outside. Thus, botulinus toxin specifically prevents the release of transmitter vesicles to the synaptic space; nicotine mimics the action of acetylcholine on the receptor; the toxins of certain snakes such as the

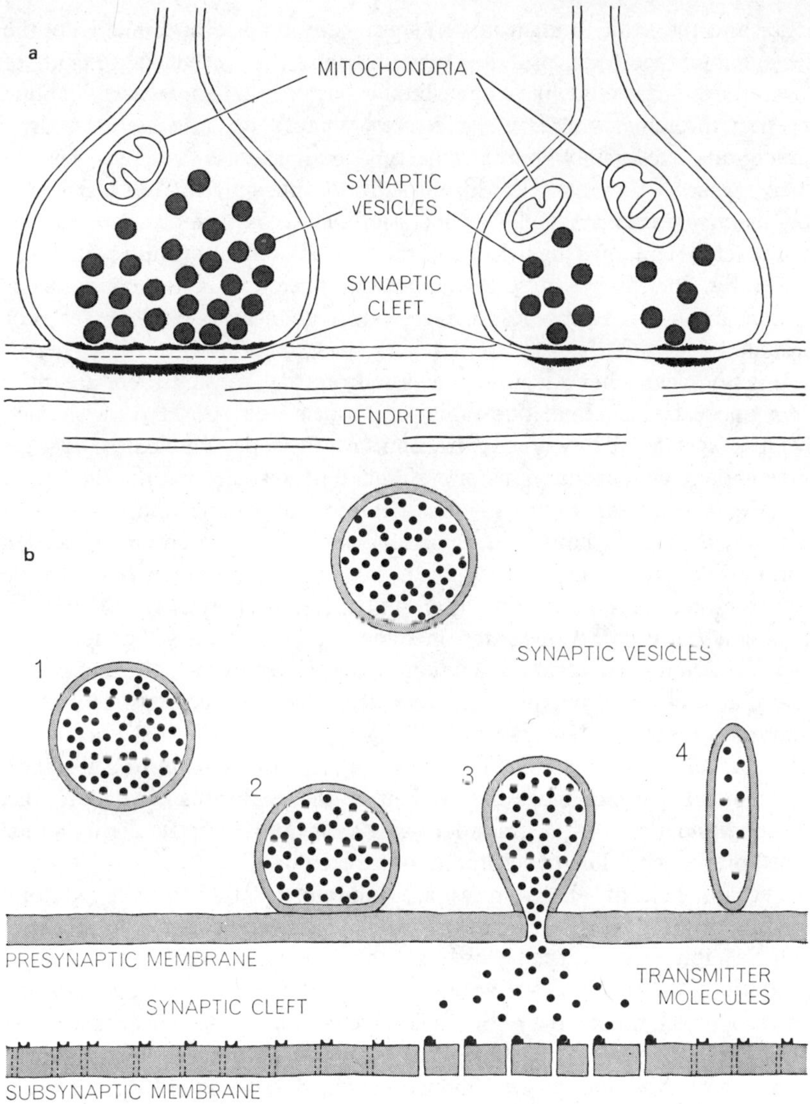

FIGURE 6 (a) Nerve impulses and chemical transmitters. Cued by a nerve impulse, synaptic knobs deliver short bursts of a chemical transmitter into the synaptic cleft. Mitochondria supply the cells with energy. (b) The transmitters, which are stored in synaptic vesicles, require only a few microseconds to diffuse across the synaptic cleft and to attach to specific receptor sites on the surface of the adjacent nerve cell. The probable pattern of movement, as shown here, is that synaptic vesicles move to the cleft, discharge their contents, and then return to the interior for recharging. (From "The Synapse," by Sir John Eccles, © January 1965, Scientific American, Inc.)

cobra and the krait (bungarotoxin) specifically bind to and inactivate the postsynaptic receptor protein; tubocurarine, the active ingredient of curare, acts somewhat like bungarotoxin but only on those acetylcholine receptor proteins where motor nerves synapse on muscles; the drug prostigmine reversibly inhibits acetyl cholinesterase whereas certain "nerve gases," for example, diisopropylfluorophosphate (DIFP), irreversibly inactivate this enzyme. This accessibility offers enormous potential for the development of drugs that can modify or control synaptic function. Ironically, one disease, myasthenia gravis, appears to be the consequence of formation of antibodies against one's own acetylcholine receptor protein.

It is not clear what advantages accrue from the fact that there are more than one excitatory and one inhibitory transmitter substance with each synapse specific for only one transmitter. Perhaps it is but a relic of independent evolutionary history of different systems within the brain.

At least three transmitters—norepinephrine, dopamine, and serotonin—function at their synapses in a manner similar to the action of certain hormones elsewhere in the body, in that their binding to their postsynaptic receptors activates an enzyme that synthesizes a substance called cyclic adenylate (CAMP) and the latter, in some manner, initiates the membrane depolarization that starts conduction of the impulse down the axon. And again, at each such synapse there is a very efficient mechanism for rapid removal of the specific transmitter. Agents are known that specifically affect various components of these synaptic mechanisms, for instance, cocaine and reserpine block the mechanism that reabsorbs norepinephrine, theophylline prevents degradation of CAMP, and lysergic diethylamide (LSD) prevents binding of serotonin to its receptor.

Perhaps half of all synapses are inhibitory rather than excitatory; binding of their transmitters to the receptor appears to occasion inflow of chloride ion (Cl^-) rather than Na^+, thereby hyperpolarizing the postsynaptic membrane. γ-Aminobutyrate was the first such transmitter to be recognized; the simple amino acid glycine appears to so serve at most inhibitory synapses in the brain. The convulsion-inducing drugs picrotoxin and strychnine, as well as bee venom, block binding of γ-aminobutyrate and glycine to their respective receptors.

Polypeptides of the Brain

A huge new chapter in understanding the brain has barely been opened. It has been known for some years that a series of "releasing factors" flows from the hypothalamus to the pituitary, each specifically signalling the latter to release one of its specific hormones. The short connecting vein may be regarded, strategically, as an exaggerated synapse conveying transmitters. The latter, the releasing factors, have been identified as a set of small polypeptides. Surprisingly, these polypeptides, such as somatostatin and thyrotropin releasing factor, have been found widely distributed in the higher levels of the brain, including at certain synapses. So too is a family of other polypeptides, each of which was also first recognized in some other connection, for example, the hormones gastrin and cholecystokinin, and other less familiar peptides. In some areas, they seem to serve as synaptic transmitters, while in other regions, a given polypeptide may merely be released by one or more nerve cells, free to diffuse and slowly influence dozens or hundreds of cells in their surroundings—a mechanism appropriate to mood rather than perception or action. Thus, information transfer in the brain occurs only in part through tightly wired connections. It seems likely that there are many more functional polypeptides than have yet been recognized, evoking an image of the brain as a unique endocrine organ controlled by its own secretions.

Thus the still incomplete but greatly improved understanding of the chemistry of synapses has had profound effects on pharmacology. As we have seen, many drugs, old and new, have effects at one or another stage in these sequences of events involving nerve transmission. Some prevent release by the first nerve's terminals, some mimic the transmitter, some actually *are* transmitters, others prevent the removal or destruction of transmitter, still others are involved in the metabolism of cAMP reactions, or block receptor actions. Knowing roughly how nerve-to-nerve junctions work, even when the knowledge is incomplete, makes it much easier to invent new drugs and predict their actions, and to use older drugs more effectively. Even the unwanted side effects of new drugs have occasionally had a great impact on understanding of disease.

The current treatment for Parkinson's disease offers a good example. Parkinsonism, the cause of which is still in doubt, and which is characterized by tremors and stiffness of the limbs and difficulty in getting movements started, originates in the brain structures termed the basal ganglia. A few years ago, high concentrations of the transmitter dopamine were found in normal basal ganglia. But, in Parkinson's disease, in which these structures are visibly abnormal, dopamine levels were much below normal, suggesting that administration of dopamine to Parkinson patients might relieve their symptoms. Whereas dopamine does not cross certain barrier membranes that separate the brain from the blood stream, its metabolic precursor, dioxyphenylalanine (L-dopa), can cross these barriers. When L-dopa was given to Parkinsonian patients it was found to be amazingly effective in relieving the symptoms of many, an elegant example of the use of basic scientific knowledge in designing rational therapy.

NEURONAL BASES OF BEHAVIOR

The central concern of this branch of neuroscience is to understand what happens in the brain and spinal cord when an organism perceives, thinks, learns, remembers, feels emotion, speaks, is motivated, and acts. Most neuroscientists take it for granted that to understand the mind means to understand the brain, and that it is within the capabilities of humans to do so, one day. Progress in the field in the last 25 years has certainly not discouraged this view.

Work on the structure, functioning, and chemistry of neurons and synapses has provided a powerful armamentarium of methods and techniques to tackle the neuronal mechanisms underlying basic aspects of behavior and experience. To take only one example, by recording the action potentials of a neuron in the visual part of the brain, it is possible to determine the manner in which nerve cells represent or "code" various aspects of visual stimuli—the neuronal basis of sensation and perceptual experience. Recording the action potentials of individual neurons is being used successfully in the analysis of motor systems of the brain, and of neuronal activity associated with motivation, sleep, learning, and memory.

Genetic and Developmental Roots

The extent to which behavior and even brain structure is due to genes or to environment (the "nature–nurture" issue) is no longer viewed as either-or but rather as a continual process of interaction. To understand how the brain develops into its adult form is a stupendous problem. Brain development can usefully be separated into development before birth, which is largely programmed genetically, and after birth, when both genetic and environmental influences are at work. Regrowth of the nervous system after injury involves similar problems.

A central problem in development is the way in which the genome is translated into an organism. For the nervous system, this involves such matters as the cell-to-cell connections within various structures, mutual spatial relationships among cells and structures, and their connections with distant structures. For example, consider the connections between the eye and the brain. The optic nerve contains about 1 million fibers that originate from the retina; each fiber is connected to a tiny part of the retina and any one part of the retina is represented by a small number of fibers. These all connect to a platelike layer of brain cells in a systematic way. Each region of retina connects to a particular region on this plate; the retina can be regarded as being "mapped" onto the plate. During development the optic nerve fibers grow out of the retina, reach the plate, and distribute themselves with absolute topographic precision. How each fiber finds the right destination is not known.

Other specifically wired sets of cablelike connections between topographically mapped areas are common in the nervous system. How such precise wiring is laid down remains a great unsolved problem that is the focus of intense research activity, frequently involving experiments in which a nerve is cut and allowed to regrow. One can rotate the target tissue, such as an eye, remove half of it, or remove half the source of the growing nerve cable, and see whether the fibers regrow properly. In amphibia this occurs with high precision. But in a number of comparable clinical circumstances, nerve regrowth is chaotic. For example, if the nerve growing to a muscle in the hand is severed, regrown fibers rarely find their proper targets, and the result may be almost as bad as paralysis.

The prospects are reasonably good that in a decade or so we will have learned how nerves seek their targets. More pointedly, there might be gained the insight so badly needed to give direction to attempts to stimulate the regrowth and reconnection of a severed peripheral nerve trunk or spinal cord.

An important area of studies concerns the development of nerve–nerve and nerve–muscle junctions. When a developing nerve reaches a muscle, it induces changes in the muscle membrane it innervates. Synapses appear and the transmitter substance at these junctions sometimes changes as the structure matures.

A promising tool for studying development is the use of genetic mutations. Mutant mice have already improved understanding of a few human neurological diseases, including retinal dystrophy (a cause of blindness) and various cerebellar diseases. Work with mutant strains of fruit flies offers great promise of illuminating the genetic control of neural development, and studies of nerve tumor cells in culture may shed light on the basic aspects of neuron development.

The potential importance of developmental studies is great, not only for what they reveal of how the brain works, but because many neurological conditions seem likely to be developmental. These include most birth defects, Down's syndrome, muscular dystrophy, common epilepsies, and a number of rarer diseases.

Sensory Processes—The Neural Basis of Sensation and Perception

Recent progress in the analysis of neuronal aspects of sensation—vision, hearing, touch, pain, smell, and taste—has been rapid. Sensation is probably the most active and highly populated area of neuroscience. This is due, in part, to the fact that one can study sensory systems in anesthetized animals. The main strategy has been to record from single cells using microelectrodes in the appropriate part of the brain while stimulating with sound, light, or other stimuli. The specific problem with respect to the visual part of the brain is how the brain interprets the messages it receives from the light receptive cells of the retina. A fairly satisfactory knowledge has been gathered of the molecular events whereby absorption of light by the pigment in the retinal cells engenders an action potential or "message." These messages have been followed into the brain for some six or seven stages, including the part of the cerebral cortex concerned with vision. The retina is now one of the best understood structures in the nervous system. Incidentally, color blindness, an inherited condition that affects a significant percentage of the male population, results from absence of one of the three proteins used to make the light-sensitive pigments in the cells of the cones of the retina.

A typical cell in the visual cortex responds actively (generating action potentials in rapid succession) when light shines on the retina, but only when the pattern is exactly the right one for the cell. For example, a cell might respond to a tiny line shaped out of light shining in just the right part of the retina and tilted in just the appropriate way. Such a cell would not react to a line anywhere else in the retina or with a different tilt. Since there is a huge area of the visual world where a line may fall, and tilts include every angle around the circle, this obviously requires that there be a huge number of cells in the visual cortex.

This brings us only a little way toward knowing how we see, but it approaches that goal more closely than could have been imagined 20 years ago. At least one region of cerebral cortex—the size of a few postage stamps as compared with roughly 1 1/2 square feet of human cortex—can be understood and in rather simple terms. Of course there is far more to be learned about the visual cortex, and other areas may not be as easy to study. Nevertheless, in principle, the cerebral cortex is capable of being understood, and these results greatly encourage attempts to understand other areas.

An unexpected by-product of this new knowledge about the visual cortex was the possibility of investigating what goes wrong when a young child becomes cross-eyed or develops cataract. Experiments in animals showed that, in these cases, the visual cortex soon loses its correct connections. As a result, such children now have corrective surgery as early as possible.

The properties of the individual neural elements of the visual cortex have yet to explain sensation, and the higher-order influences of the visual neurons on "association" areas of the cortex, processes that may underlie perception, remain unknown. Hopefully, as understanding of sensory systems increases—and it seems likely that the auditory system will come to be understood as well in the next 5–10 years as the visual system is now—approaches to effective prostheses for the devastating disorders of blindness and deafness may emerge. Indeed, there are promising indications that, in certain forms of deafness, differential electrical stimulation of auditory nerve fibers may produce "hearing," even understanding of speech.

Motor Systems—The Neural Basis of Movement

That a particular region of the cerebral cortex produces movement when stimulated electrically was demonstrated just over a hundred years ago. Long before that, however, physicians knew that injury to the brain produced paralysis or difficulties in movement. In recent years, the neural correlates of movement have been widely studied, using a variety of anatomical, physiological, and chemical techniques.

From these investigations has emerged the concept of a motor system or systems that control an organism's movements. Among the structures of the motor system, the ventral region of the spinal cord, the cerebellum, basal ganglia, and motor regions of the cerebral cortex are the most important. Although precise knowledge of how specific components of the systems function together to produce a movement or series of movements is not yet available, new procedures for studying electrical and chemical activities of individual nerve cells in awake and moving trained animals show promise for solving this problem. As an example, it has been found that certain classes of neurons in the motor area of the cerebral cortex "code" the force of a movement rather than its extent.

In the next few years, understanding of how the different components of the motor systems function together to generate and control movements, as well as their failure in certain neurological diseases, should slowly develop.

Sleep

Why do we sleep? The most promising theories rest on new knowledge about brain structures and neurotransmitters. In the 1950's it was discovered that there are two distinct kinds of sleep: "slow wave" sleep and "rapid-eye-movement" (REM) sleep. A person (and all higher animals) goes from one to the other of these states of sleep four or five times a night, spending about 20 minutes to 1 hour in each. There is considerable, although disputed, evidence that REM sleep is associated with dreaming. Much recent research has sought to identify regions of the brain specifically involved in these two phases of sleep. Two small clusters of nerve cell bodies in the brain stem, the locus coeruleus (which utilizes norepinephrine) and the raphe nuclei (which utilizes serotonin), are the current focus of such studies.

Some years ago it seemed that why we sleep would be known by now. The simplest hypothesis—that a sleep "substance" exists—has been investigated intensively. None has yet been found. Discovery of a "sleep-inducing factor" might account for how the brain falls asleep or awakens; but it might still not clarify the physiological "purpose" of the sleeping state. Progress in this field is likely to be slow, while the many brain systems that show changes during the two phases of sleep are analyzed in detail.

Motivation and Emotion

Knowledge of the biological bases of such elemental motives as thirst, hunger, and sex has grown rapidly. In the early 1950's the hypothalamus was implicated as the most importantly involved brain structure. To take hunger as an example, damage to one region of the hypothalamus eliminates eating behavior in animals; they starve to death if not force-fed. Damage to an adjacent region produces a voracious animal that becomes obese. Such hypothalamically damaged obese animals, like obese humans, are much influenced by taste of food, and will not overeat if they have to work hard to get food.

It has been known for some time, of course, that sexual behavior is strongly dependent on the endocrine system. Hormones from the ovaries and testes are reciprocally controlled by hormones secreted by the pituitary gland, itself controlled in part by the hypothalamus. The gonadal hormones act directly on neurons in the hypothalamus and other brain regions that, in turn, control further the actions of the pituitary gland. It is difficult to overstate the importance of the various sex hormones in the biological development of organisms and in sexual behavior. Study of such neural–hormonal interactions is growing at a rapid pace.

Another example of hormonal influence on behavior is thirst. When cellular dehydration occurs, the kidney releases an enzyme, renin. Renin cleaves a piece from a blood protein, ultimately to yield the polypeptide angiotensin II, which is, in effect, the thirst hormone. It appears to have direct actions on the hypothalamus and other brain regions that result in an intense feeling of thirst.

The discovery in 1953 of reward centers in the brain was one of the most important findings yet made concerning behavior. When an electrode was implanted in a region of the hypothalamus and connected to a lever so that a rat could deliver weak electrical stimuli to its own hypothalamus, the animal would press endlessly. The same phenomenon occurs in other species, including humans. Some starving rats, given a choice between such electrical self-stimulation and food, will self-stimulate until they die of hunger. This behavior is reminiscent of severe drug addiction. The brain reward region overlaps the areas of the hypothalamus concerned with drinking, eating, and sex, but seems to be more general.

Other regions of the upper brain stem, some closely adjacent to the reward area, appear to yield pain when stimulated. Strongly addicting drugs like morphine, which seem to exert their actions primarily on these regions of the brain, both alleviate pain and produce strong feelings of pleasure. When it was found that many neurons in the pain (midbrain and thalamus) and pleasure (amygdala) regions of the brain have receptors that tightly bind morphine and related substances, the question arose as to the nature of the material that normally binds at these opiate receptors. The answer has proved to be that regions of the brain itself, particularly the hypothalamus, make and release peptides that bind to the opiate receptors even more effectively than morphine. These substances, called endorphins, very effectively relieve pain, induce a sensation of pleasure, and are strongly addicting. They are the brain's own natural opiates. The possible implications of this discovery, both for relieving pain and for understanding and combatting drug addiction, are considerable. Whether endorphins function at synapses or on other cell surface receptors is not clear.

This area, in which several different fields are just coming together, is one of the most rapidly moving and promising fields of science. It seems quite possible that the underlying causes of a range of human disorders from obesity to drug addiction will, at the least, be much better understood in the next 5–10 years.

Learning and Memory

Language, science, and society endure only because each individual learns anew. There is a general relationship between the evolutionary status of animals and their ability to learn. *Homo sapiens* differs from other primates because of what humans can learn. Apes learn much more than rats, which, in turn, learn a great deal more than flies.

The problem of analyzing the neuronal bases of learning and memory is formidable. It has been surmised that a well-educated adult human has more bits of learned information stored in memory than there are neurons in the brain. This fact alone argues that memory must somehow be coded in the synaptic interconnections among neurons, where the possible patterns are virtually limitless.

Fortunately, basic processes of learning seem to be quite similar over a wide range of animals. Habituation, a simple decrease in the strength of a reflex response as a result of repeated stimulation, has essentially the same properties in primitive invertebrates—reflexes of the neurally isolated spinal cord—and in intact humans. Pavlovian conditioning and simple instrumental learning seem to be similar in mammals from rat to human. Consequently, animal models may be used to study these simpler basic processes of learning.

Considerable progress has been made in the analysis of the neural mechanisms underlying habituation. In simple reflexes of certain invertebrates and in the isolated spinal cord, where it has been possible to obtain habituation across a single set of synapses, it seems due to a decrease in the probability of transmitter release in response to a given stimulus. While a very simple form of behavioral plasticity or learning, habituation has adaptive value. It has been suggested that habituation is the primary mechanism whereby humans adapt to unpleasant but unavoidable situations like urban stress.

More important in human behavior is associative learning, ranging from simple conditioning to language. Learning is thought to occur in two phases in both animals and humans. When information is first learned it is fragile; it can be disrupted, for example, by electroconvulsive shock. However, after it has been well-learned for a period of minutes to hours it is stable and not subject to disruption. Short disruptable neuronal processes are converted to more permanent, long-term storage mechanisms. In animals the two processes can be prevented by quite different drugs, one of which blocks protein synthesis in the ribosome, while the other prevents RNA synthesis on the DNA template.

The neural bases of associative learning are being sought, using animal models, by recording the action potentials of individual neurons in various brain structures during the learning and retrieval (i.e., memory) of information. This search for the "engram" (a term used to refer to the neuronal coding of memory) entails identification of the neurons or systems of neurons where activity changes as a result of learning, and the underlying synaptic mechanisms of storage.

A structure of the limbic system, the hippocampus, seems to play an important role in the memory system in both humans and infrahuman animals. Certain neurons in the hippocampus begin to increase their rates of activity from the very beginning of training in animals. Their activity continues to grow and is projected to other brain structures. The hippocampus is not, per se, the locus of long-term memory storage—humans with severe damage to the hippocampus have intact memory for information learned before damage but have great difficulty in placing new information into storage or retrieving it.

There are tantalizing hints from neuroanatomical studies that either new synapses may form or existing synapses become enlarged in the hippocampus and cerebral cortex as a result of learning experiences. One neurotransmitter, norepinephrine, seems most important in the learning process. Animals learn best when optimal levels of this neurotransmitter are in the brain. ACTH, the stress hormone, exerts a modulatory action on learning. Certain drugs can actually facilitate learning and memory storage. The relations of these effects to synaptic transmission is under current investigation.

Cognition—Language and Consciousness

The neuronal bases of higher mental processes in humans is a field for the future. However, some significant advances have been made in the past 15 years. Information has come primarily from study of patients with brain damage due either to injury or as a result of necessary neurosurgical procedures. It has been known for some time that language is represented in the left hemisphere of the brain in virtually all right-handed individuals. Two regions of the cerebral cortex seem to be specialized for language—an anterior area involved more in the expression of language and a posterior area involved in the understanding of language.

Studies in the past few years indicate that consciousness is also differentially represented in the two hemispheres. Because of the way the eyes project to the brain, visual information can be addressed separately to each of the two hemispheres in persons in whom the large bands of fibers interconnecting the two hemispheres had been surgically severed to prevent epileptic seizures. In essence, left hemisphere consciousness is verbal and right hemisphere consciousness is nonverbal. Both seem to exist more or less independently in these patients.

This is a very young field, and unless expected new technological developments occur, progress is likely to be slow. Thus, methods are needed to record the activity of single nerve cells and to safely stimulate small regions of the brain without drilling holes in the skull. Positron emission techniques can now give a picture of brain activity with about centimeter resolution. Much will be learned about human brain function as this method is applied, even though it will not permit recording the activity of single neurons.

Abnormalities of Behavior and Experience—The Psychoses

Extraordinary progress has been made in the past 30 years in treatment of the psychoses (see p. 404). Most neuroscientists and psychiatrists agree that schizophrenia, the most severe and widespread psychosis, reflects some fundamentally abnormal brain function. This is consonant with the fact that the closer a person's hereditary relationship to a schizophrenic, the greater the likelihood that person will develop the disorder. At this time, the primary defect in schizophrenia is not known.

Research on schizophrenia in the past few years has strongly implicated one neurotransmitter—dopamine. L-dopa, used for the treatment of Parkinsonism (see p. 107), elicits symptoms rather like schizophrenia; chlorpromazine and other antipsychotic drugs that relieve schizophrenia elicit the symptoms of Parkinsonism. These observations are explained by the fact that the antipsychotic drugs block the action of dopamine at dopamine receptors—suggesting that schizophrenia results from overactivity of some set of nerves that utilize dopamine as transmitter.

Whatever the ultimate cause of schizophrenia, it is, thus, unlikely that the schizophrenic brain is, in the usual sense, physically entirely normal. It is hard to exaggerate the importance of accepting that concept, if there is ever to be a real cure or prevention.

Changes in scientists' attitudes toward the psychoses, partly as a result of genetic studies and partly because of the advances in neurochemistry and pharmacology, will have important effects on psychiatry and psychology, even on attitudes toward the neuroses, the lesser behavioral disorders.

BIOLOGY AND AGRICULTURE

In the past two decades advances in plant science have marched in step with the progress of the other sciences. Molecular genetics has conferred a more penetrating understanding of plant variation, and promises new approaches to plant breeding; cell biology has demonstrated how nutrients move across membranes; photochemistry has clarified the primary events of photosynthesis; biochemistry has elucidated the subsequent events of photosynthesis, the process of atmospheric nitrogen fixation and various factors that regulate plant growth. Application of this newly discovered fundamental knowledge holds exciting prospects for the future.

PHOTOSYNTHESIS

Life is sustained on this planet by photosynthesis, through which green plants capture the sun's radiant energy. In ordinary combustion, oxygen from the air reacts with organic compounds, e.g., a carbohydrate such as cellulose, to form carbon dioxide (CO_2) and water (H_2O)—releasing energy as heat, as in the flames of a wood fire.

$$\text{(A)}\quad C_6H_{12}O_6 + 6O_2 \longrightarrow 6\,CO_2 + 6\,H_2O + \text{energy}$$

In living cells the controlled oxidation of glucose by oxygen releases the same amount of energy, but in a chemical form suitable to power the processes of life.

Photosynthesis is, in overall effect, the reverse of this process. Plants convert CO_2 + H_2O into carbohydrate, obtaining the necessary energy from sunlight, and liberating oxygen in the process. This occurs within a specialized subcellular organelle in leaf cells called the chloroplast.

In fact, there is no direct reaction between CO_2 and H_2O. Rather, the overall process can be viewed as a set of partial reactions. To begin with, in the specific process that utilizes solar energy, electrons are withdrawn from H_2O and transferred to a carrier substance, the special nucleotide denoted as NADP:

$$(B_1)\ 2\,H_2O + 2\,NADP \longrightarrow O_2 + 2\,NADPH_2$$

Reaction B_1 should spontaneously flow from right to left; however, within a chloroplast light energy is used to drive the system to the right. In subsequent processes, the $NADPH_2$ is utilized to convert CO_2 to glucose:

$$(B_2)\ 6\,CO_2 + 12\,NADPH_2 \longrightarrow C_6H_{12}O_6 + 12\,NADP$$

Let us first consider how light makes possible reaction B_1.

Chlorophyll Photocenters

Absorption of light energy by chlorophyll is the critical event in photosynthesis. But how is that energy transduced into a form of chemical energy that the plant cell can utilize? Energy from absorbed light migrates to two types of photochemical centers. At each photocenter the functional chlorophyll molecule exists as a sandwich between two other molecules. Chemically they are quite different, and current research is focused on identifying them and understanding the precise three-dimensional arrangements between them. For the present purpose, however, it suffices to denote them A and B, respectively, at one type of photocenter and C and D, respectively, at the second type.

When the chlorophyll absorbs a photon (usually of red light, which is why plants appear green), the electrons in its resonating structure are activated to such an extent that one is ejected, leaving behind a positively charged chlorophyll molecule (Chl^+, a free radical). The ejected energetic electron is instantly accepted by the adjacent molecule of B°. But A then donates an electron to the chlorophyll radical, which returns to its original state:

$$A^\circ Chl^\circ B^\circ \longrightarrow A^\circ Chl^+ B^- \longrightarrow A^+ Chl^\circ B^-$$

If A^+ and B^- were in contact, they would react to reform A° and B°. But the large, greasy molecule of chlorophyll keeps them apart; hence the process is termed "charge separation." This process is complete in a few picoseconds (10^{-12} seconds). Analogous events occur at the second photocenter.

There follows then the process, summarized above, wherein, by drawing electrons from H_2O, O_2 is formed and A^+ is restored to A°. At the same time the excess electron in B flows to the second photocenter and then on to NADP to form $NADPH_2$. Schematically the flow of electrons in the entire arrangement can be depicted as:

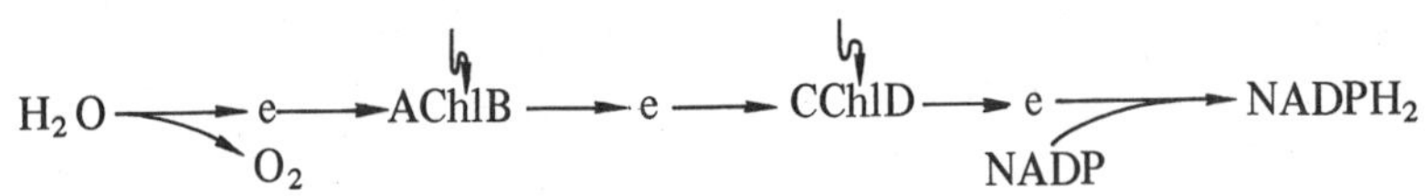

so that the tiny electrical current flows only when the chlorophyll at both photocenters is being illuminated.

The $NADPH_2$ that is produced is used in the conversion of CO_2 to carbohydrate, but the reducing power (chemical potential) of $NADPH_2$, alone, is insufficient to the task. A second form of chemical energy, that in the structure of adenosine triphospate (ATP), is also required. Like $NADPH_2$, ATP is made by the chloroplast photochemical apparatus, from adenosine diphosphate (ADP) and inorganic phosphate (P).

$$\text{(C)}\quad ADP + P \xrightarrow{\text{light energy}} ATP$$

Knowledge of how this happens has been hard gained. The photochemical center chlorophyll molecules and all of the other intermediate electron carriers are imbedded in the chloroplast membrane in fixed physical relationships. As electrons move via consecutive carriers between the two chlorophyll centers, protons (H^+) are discharged across a membrane to the interior, building up a strong electrochemical potential. This drives the protons back through a channel in a complex enzyme, lodged in the same membrane, providing the immediate energy for reaction (C).

Formation of $NADPH_2$ and of ATP is the sole role of the light-using apparatus. All other events in photosynthesis can happen in the dark, given a supply of $NADPH_2$ and ATP.

The overall accomplishment of the photochemical apparatus of the chloroplast may usefully be summarized as:

$$\text{(D)}\ 12\ H_2O + 18\ ADP + 18\ P + 12\ NADP \xrightarrow{\text{light energy}} 6\ O_2 + 18\ ATP + 12\ NADPH_2$$

In the subsequent "dark reactions" the reducing power of NADPH₂, together with energy suplied by ATP, is used to reduce CO_2 to carbohydrate:

$$\text{(E)}\ 6\ CO_2 + 18\ ATP + 12\ NADPH_2 \longrightarrow C_6H_{12}O_6\ \text{(sugar)} + 12\ NADP + 18\ ADP + 18\ P + 6\ H_2O$$

CO_2 Fixation

The conversion of CO_2 into carbohydrate (process E) is accomplished by a series of reactions requiring more than a dozen distinct enzymes. But CO_2 itself participates only in one chemical reaction, that in which a five-carbon compound, ribulose bisphosphate, reacts with CO_2 and then splits into two molecules of the three-carbon compound, 3-phosphoglyceric acid.

Using the supply of ATP and $NADPH_2$ from the photochemical process, the system operates in such fashion that, overall

$$6\ \text{Ribulose bisphosphate} + 6\ CO_2 \longrightarrow 12\ \text{3-phosphoglyceric acid} \longrightarrow 6\ \text{Ribulose bisphosphate} + C_6H_{12}O_6\ \text{(glucose)}$$

Glucose thus accumulates, and ribulose bisphosphate is regenerated, in a combination of reactions sometimes referred to as the C_3 cycle (Figure 7a).

Improving the Efficiency of Photosynthesis

Considerable research is aimed at increasing the efficiency of photosynthesis in the intact plant.

The simplest approach involves altering plant structure. Usually full sunlight provides more light than necessary to saturate the photosynthetic systems in the upper leaves, but not those below. By selecting plant strains with a more upright alignment of the leaves, more light penetrates to the light-limited lower leaves. This approach has yielded a number of hybrid maize varieties that are widely used in the United States. A similar strategy should be applied in the future to other plants.

Another approach is to breed plants for photosynthetic units that operate more efficiently, that is, convert more of their incident light into

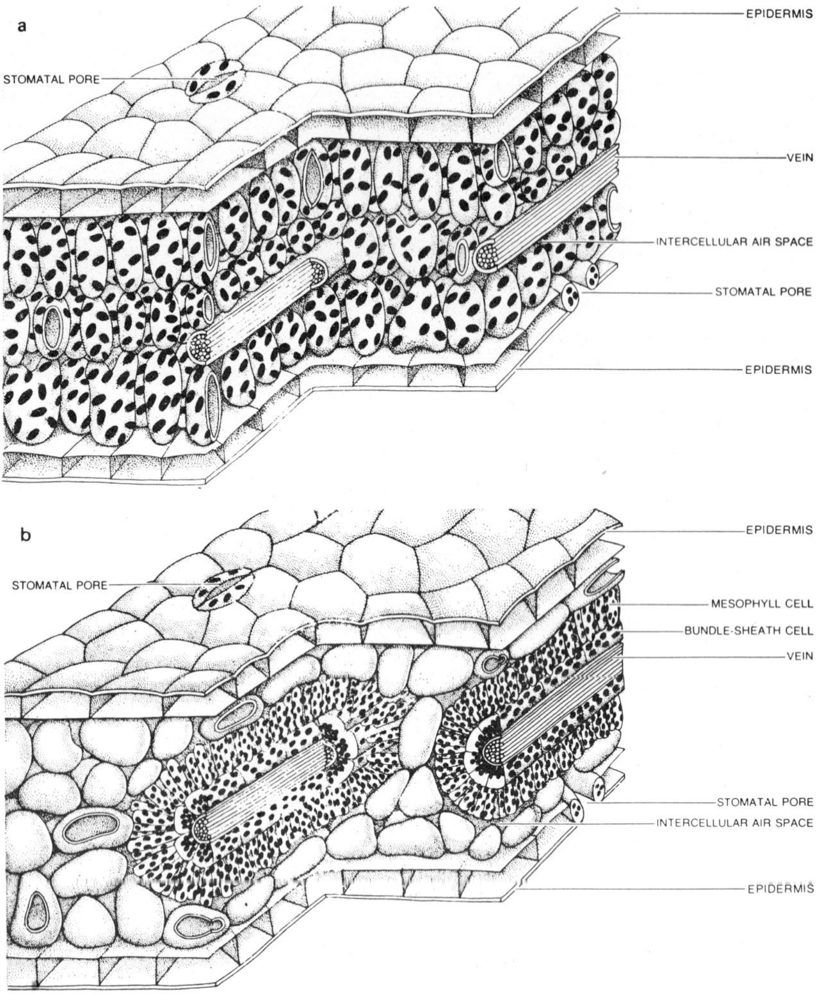

FIGURE 7 Different leaf structures for plants using different photosynthetic pathways. Both plants are varieties of the saltbush, but one uses a three-carbon pathway, the other a four-carbon one. In the three-carbon variety, *Atriplex patula* (a), chlorophyll-containing chloroplasts, shown as black dots, are distributed throughout the plant; but in the four-carbon variety, *Atriplex rosea* (b), they are concentrated in two types of cells that form concentric layers around the fine veins of the leaf. The structure of the latter plant is designed to provide a net transport of carbon dioxide from the outer cell array to the inner one, producing a higher concentration of carbon dioxide at the inner cells and enabling a faster diffusion of carbon dioxide into the leaf, and therefore a higher rate of photosynthesis. The result is that *A. rosea* grows better than *A. patula* in dry, hot weather. (From "High-Efficiency Photosynthesis," by Olle Björkman and Joseph Berry, © October 1973, Scientific American, Inc.)

chemical energy. The most efficient crop plants, such as maize, have an efficiency of only about 5 percent during their period of most rapid growth, when their leaf canopy is full and light capture is nearly complete. In contrast, some algae have an efficiency of 25 percent under optimal conditions in the laboratory. Although there have been some hints that the photosynthetic efficiency of plants can be improved, it should be appreciated that the light reactions in plant photosynthesis are already unusually efficient when compared with most other photochemical reactions.

Another research aim is to improve energy utilization in the dark reactions of photosynthesis. The key enzyme, ribulose bisphosphate carboxylase, is not very efficient at its job compared with other enzymes; it has both a low affinity for CO_2 and a slow turnover time. The plant manages to maintain a reasonable rate of photosynthesis only because it contains large quantities of the enzyme. (In fact, this is the most abundant enzyme known; it may constitute 50 percent of the soluble protein in chloroplasts.)

The main problem, however, is that this enzyme not only fixes CO_2 but is also quick to catalyze a wasteful reaction, called photorespiration, between ribulose bisphosphate and oxygen.

$$\text{Ribulose bisphosphate} + O_2 \longrightarrow \text{3-phosphoglyceric acid} + \text{phosphoglycolic acid}$$

Since the phosphoglycolic acid is then further oxidized through photorespiration to CO_2 and H_2O in an uncoupled reaction that generates no ATP, photorespiration dissipates energy and depletes the supply of ribulose bisphosphate needed for photosynthesis.

There is a small chance that one might be able to enhance the affinity of ribulose bisphosphate carboxylase for carbon dioxide, or otherwise avert the waste of photorespiration. To do so, one must better understand the enzyme's intimate three-dimensional structure. However, the protein is large and complex, presenting a formidable task for crystallographic analysis. The challenge remains, nonetheless, and undoubtedly there will be attempts to alter and improve this critical enzyme, however small the chances of success. The carboxylase from photosynthetic bacteria, which has a lower molecular weight, may be more amenable to X-ray analysis.

C_4 Cycle

To carry on photosynthesis, a plant needs CO_2 from the atmosphere. But the low concentration of CO_2 (about 0.03 percent in air) limits the rate of photosynthesis. A special arrangement that mitigates this problem is present in such plants as sugarcane, maize, sorghum, and crabgrass. In certain leaf chloroplasts of these plants, an appropriate set of enzymes catalyzes the addition of CO_2 to a three-carbon compound to form a four-carbon acid. This four-carbon acid then migrates to other chloroplasts deeper in the interior of the leaf, where it releases CO_2 for use in the C_3 cycle.

By furnishing the C_3 system with additional CO_2, the C_4 system favors CO_2 as it competes with O_2 for the enzyme ribulose bisphosphate carboxylase, and thus minimizes the waste of energy through photorespiration (Figure 7b). Although the $C_4 + C_3$ system requires more ATP than the C_3 system alone, the double advantages of operating effectively at a relatively low concentration of CO_2 and of decreasing the losses from photorespiration more than compensate the C_4 plants for this cost.

Accordingly, there have been attempts to convert other plants to C_4 type metabolism. This seems a reasonable goal, for both systems occur in nature, and intermediate forms between the pure C_3 plants and $C_4 + C_3$ plants have been discovered.

In addition, specific inhibitors that reduce the rate of photorespiration could, potentially, increase photosynthetic efficiency and utilize available energy more efficiently. To date, such attempts have not been successful, but they will be continued because of the tremendous importance of operating the photosynthetic process at the highest possible efficiency.

NITROGEN FIXATION

The importance of biological nitrogen fixation to contemporary agriculture may be second only to photosynthesis. All forms of life need nitrogen for proteins and nucleic acid, and the nitrogen supply is probably the commonest factor limiting plant growth. Although nitrogen gas (N_2) is abundant in the atmosphere, it can be used by biological systems only if it has been "fixed": the bond between the two atoms of nitrogen is broken, and each of the two atoms binds three hydrogen atoms to form ammonia (NH_3), which is then used for diverse metabolic purposes.

Nitrogen can be fixed either chemically or biologically. The industrial Haber-Bosch process utilizes high temperatures and high pressures to make ammonia from gaseous hydrogen and nitrogen. Chemical nitrogen fixation is a major industry, annually producing nearly 40 million metric tons of fixed nitrogen to be used as fertilizer. Unfortunately, this process requires a great amount of energy, most of which now comes from natural gas. It is possible, though perhaps not now economically feasible, to substitute coal or other fuels for the natural gas. The Tennessee Valley Authority is installing a coal–gas pilot plant that will begin to test this approach in the 1980's. As supplies of fossil fuels dwindle, the price of chemically fixed nitrogen will rise. These concerns have spurred a recent surge of interest in biological nitrogen fixation.

In nature, N_2 fixation can be accomplished only by microorganisms: certain bacteria that live freely in the soil, blue-green algae, and the rhizobia—bacteria that infiltrate, and form nodules on, the roots of leguminous crops. Here, too, the product is ammonia, and here, too, the energy cost is great. In contrast to the industrial process, however, the biological system operates through the nitrogenase enzyme system, at normal temperatures and with the pressure of nitrogen less than one atmosphere.

The nitrogenase system contains two proteins, dinitrogenase (a molybdenum–iron protein) and dinitrogenase reductase (an iron protein). These proteins are readily inactivated by oxygen, from which they are protected by special mechanisms in N_2-fixing microorganisms.

Driving the nitrogenase reaction requires much ATP (about 12 ATP per N_2) as well as a strong reducing agent such as ferredoxin (an iron–protein complex).

$$N_2 + 6 \text{ electrons} + 12 \text{ ATP} + 6 \text{ H}^+ \longrightarrow 2 \text{ NH}_3 + 12 \text{ ADP} + 12 \text{ P}$$

The ferredoxin, in turn, must be reduced by the metabolism of the cell. Reduced ferredoxin transfers electrons to the dinitrogenase reductase. The latter uses energy from ATP to make it a strong enough reductant to reduce the dinitrogenase which, in turn, reduces N_2 to NH_3. Even in the presence of N_2, however, the system wastefully reduces H^+ to H_2 gas.

Nitrogenase can reduce not only nitrogen, but also protons, acetylene, and several other substances. The reduction of acetylene to ethylene furnishes an easy and sensitive test for nitrogenase, widely used in the laboratory and in field testing nitrogen fixation.

Structure of Nitrogenase

There is keen interest in the structures of the complex components of the nitrogenase system. A promising new development is the isolation from dinitrogenase of a substance referred to as FeMo-co (Fe = iron, Mo = molybdenum, co = cofactor). When this compound is added to a nitrogenase-deficient mutant of the bacterium *Azotobacter vinelandii*, it restores nitrogenase activity. Because this relatively simple compound carries the active site of the enzyme, it becomes particularly attractive to attempt to crystallize it and to obtain its detailed three-dimensional structure by X-ray crystallography. It might provide a model for developing improved catalysts for chemical nitrogen fixation.

Improving Nitrogen Fixation

Because improved systems for biological nitrogen fixation could translate into better nutrition for millions of people with less drain on energy reserves, scientists are working to develop more effective bacteria, more responsive plants, and more productive associations between the two.

Genetic and Biological Aspects of Nitrogen Fixation Several types of free-living nitrogen-fixing bacteria already exist in nature; the task that confronts scientists is to improve their effectiveness. In the past, scientists working to improve nitrogen fixation have simply selected the most effective naturally occurring strains. Today, a more fruitful approach may be genetic manipulation.

The gene that controls the nitrogenase system is referred to as the *nif* gene. In a recent major advance, the *nif* gene has been transferred from nitrogen-fixing *Klebsiella pneumoniae* to the non-nitrogen-fixing organism, *E. coli*,[4] prompting speculation that it might be possible to transfer the gene to non-nitrogen-fixing higher plants, such as corn or wheat. However, merely transferring the *nif* gene does not establish a useful nitrogen-fixing system. Unless the recipient—*E. coli*, corn, wheat, or whatever—can be protected from oxygen, no nitrogen will be fixed. Moreover, as we have noted, nitrogen fixation requires a large supply of energy in the form of ATP, as well as an ample supply of ferredoxin. Because this energy would come from energy normally used for other plant processes, the crop yield would be decreased unless overall photosynthesis could be increased. Developing auxiliary systems for supplying more energy and for protecting against oxygen may well prove more formidable than actually transferring the *nif* gene. Despite these complications, genetic techniques, including those employing recombinant DNA, may open ways to develop effective nitrogenase systems; and gene-transfer studies should be pursued vigorously.

Nitrogen-Fixing Legumes Another approach to improved agricultural production resides in more widespread use of nitrogen-fixing legumes—peas, beans, clover, alfalfa, and soybeans. (There are over 10,000 known species of leguminous plants.) Various species of *Rhizobium* invade the roots of various legumes where they produce nodules that consist of plant cells packed with nitrogen-fixing bacteria (Figure 8). In this symbiotic relationship the plant supplies carbohydrate to the bacteria, and the bacteria supply the plant with fixed nitrogen. Commercial preparations (inoculants) of the rhizobia, added to the soil, assure an adequate infection of legume roots. Farmers have long exploited the ability of such

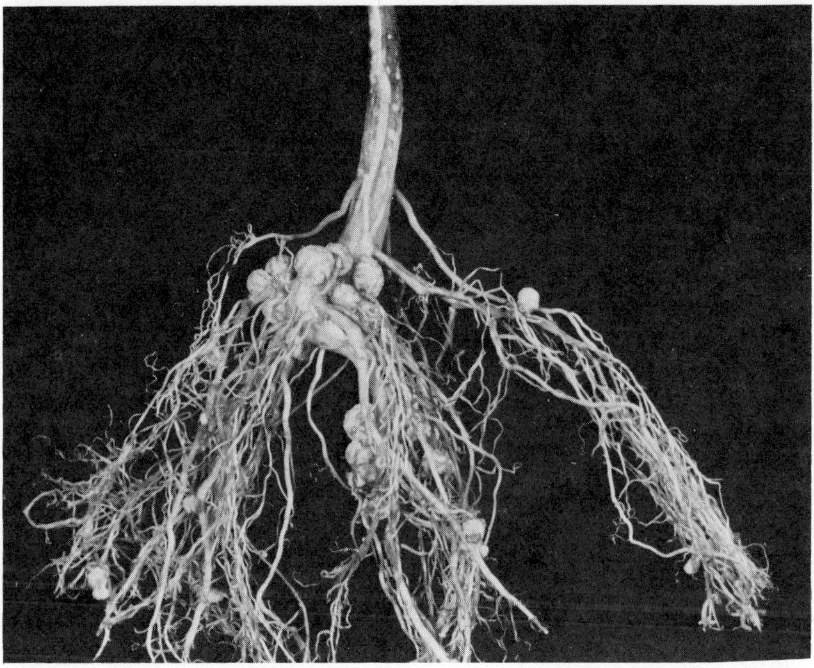

FIGURE 8 Nodules formed by the nitrogen-fixing bacterium *Rhizobium japonicum* on a young soybean root. (Harold J. Evans, Oregon State University)

leguminous crops to revitalize the soil, by rotating them with crops that do not develop such nodules.

There is an extensive program for selecting native bacterial strains for improved nitrogen fixation; this program could be broadened to include testing of artificially produced mutants. Studies must include tests of the leguminous plant as well as bacterial strains, because effective N_2 fixation depends upon proper interaction between the specific plants and bacteria. Moreover, added strains of the root-nodule bacteria must compete effectively with native strains that are present in the soil: It is of little use to select a strain that is effective under greenhouse conditions if it does not compete well in the field. Hence, there is great interest in establishing the basis of this characteristic so that it can be incorporated in inoculants for leguminous seeds. Unfortunately, testing is slow and laborious, because the index of success is the increment in crop production.

Recycling Hydrogen To the extent that the nitrogenase system readily diverts electrons to H^+ rather than to N_2, the system is wasteful. And there is no known way to limit H_2 production by nitrogenase without also impairing nitrogen fixation. But certain root-nodule bacteria contain a hydrogenase, an enzyme that activates H_2 and permits its reoxidation; this yields ATP that can be used for nitrogen fixation. Field studies support the idea that rhizobia selected for their hydrogenase activity fix nitrogen more efficiently than bacteria lacking hydrogenase. Although this approach has been limited to soybeans, it should also be applicable to other species of legumes.

Delaying Senescence In the legume–rhizobia symbiosis it would be useful if plant aging could be delayed, for during the most active period the rates of growth and nitrogen fixation of the plant are essentially exponential. Extending the growing season by even a week could thus markedly increase crop yield.

The leguminous plant, however, operates in a self-destructive manner: At the time it sets seed, it senesces, utilizing all of its nitrogen resources, including nodules, to form seeds. By selecting plants that age more slowly or by using chemical agents, it may be possible to control senescence. Such leads need to be followed vigorously to convert them to practical agricultural techniques.

Other Approaches Legumes, as C_3 plants, are less efficient photosynthetically than C_4 plants. There would be great reward if legumes could be converted to a $C_4 + C_3$-type metabolism (see p. 121). This has not yet been accomplished, but efforts are continuing.

Increasing the concentration of CO_2 available to plants dramatically increases their nitrogen fixation. It is seldom practical to increase CO_2 concentrations under field conditions, but this phenomenon suggests that the full potential of the nitrogen-fixing system has not yet been reached. It may be possible to alter control mechanisms so that organisms will fix nitrogen more effectively at normal levels of CO_2.

Associative Symbioses In the associative symbioses, which are much less structured than the rigid symbiosis of leguminous plants and the rhizobia, bacteria capable of fixing nitrogen by themselves grow on or inside the roots of a host plant without producing a structured nodule. The bacteria obtain photosynthetic products from the roots and use them as a source of energy for fixing nitrogen, which is then available to the plant.

Under tropical conditions, *Azospirillum* associated with the roots of grasses actively fixes nitrogen. Tests have demonstrated that nitrogen can be fixed in such associations with maize, but the fixation characteristically is only 1 to 10 percent of that in better associations of leguminous plants and their root-nodule bacteria.

Because so much of the world's population depends primarily upon cereals for food, it would be of great importance if associative symbioses could be developed to supply a substantial percentage of the nitrogen needed for cereal crops. The possibilities in this field have been examined only superficially; an exhaustive examination of both bacterial species and plant cultivars may reveal associations that will fix a high percentage of the needed nitrogen. Field trials to date have not been encouraging, but the great potential of effective associative symbioses for cereal culture makes it imperative that the studies continue.

Blue-Green Algae The numerous blue-green algae that can fix nitrogen offer a fine example of nature's ingenuity. These organisms grow in chains of cells; about every twentieth cell differentiates into a special thick-walled cell known as a heterocyst. Unlike the normal vegetative cells in the chain, which photosynthesize and produce oxygen, the heterocysts lack the oxygen-producing phase of photosynthesis. The vegetative cells produce sugars photosynthetically and transfer them to the heterocysts. The heterocysts, in turn, fix nitrogen in their oxygen-deficient environment—an arrangement that keeps the heterocyst nitrogenase from being destroyed by oxygen—and, in exchange, they transfer fixed nitrogen compounds to the vegetative cells. These self-sufficient organisms are widespread, from deserts to lakes.

Nitrogen fixation by blue-green algae is crucial to maintaining rice production in Asia. The algae grow in abundance in paddies during flooding; as the water retreats, they decompose at the soil surface, leaving their fixed nitrogen to nourish the developing rice plants. This explains why rice can be cultivated year after year in paddies without adding nitrogenous fertilizer.

It is important that modern scientific methods focus on the nitrogenase system of blue-green algae to improve its effectiveness in rice culture and in other applications as well. Strain selection, genetic manipulation, control mechanisms, nutrient requirements, and conditions of culture all offer avenues toward improving fixation by these organisms.

Control of Growth and Plant Stress

Plants show tremendous diversity in size, morphology, metabolism, composition, and response to environmental conditions. Nevertheless, they are much alike in their cell structures, ability to photosynthesize, need for water and mineral nutrients, and responses to growth substances. In nature, plants are subjected to extremes of temperature, water and nutrient supply, solar radiation, wind velocity, disease, and insect attack. Surviving plants have adapted to these stresses, but crop plants bred for high productivity may lose some of their resistance to stress. Fortunately man can control some stresses and minimize the effects of others.

Plant Growth Substances

Plant growth substances are products that spur or stunt plant growth. The first such substance to be isolated and characterized, about 50 years ago, was the auxin, indoleacetic acid. During World War II a variety of compounds with profound growth effects were synthesized, and some were adopted as selective weed control agents shortly thereafter. Other agents, developed to provide broad spectrum and selective weed control, were eagerly accepted in agricultural practice.

Growth substances control a variety of plant responses. The auxins and gibberellins control cell enlargement, whereas the cytokinins control cell proliferation and leaf expansion. Other agents control responses to light, cell differentiation, internodal elongation, germination, rooting, bud development, and so on.

These substances have found important practical applications. An early use was to stimulate the rooting of cuttings. Another was to delay development of buds on fruit trees until danger of frost damage had passed. Growth substances can minimize the sprouting of potatoes in storage, break the dormancy of seeds, and control the senescence of maturing plants.

The precise biochemical locus of action of herbicides is well defined in only a few cases. Most herbicides have been developed empirically; when an active compound has been synthesized or isolated from a natural source, a family of similar compounds has been synthesized and tested for activity. A more rational approach, based on knowledge of plant metabolism, may lead to new herbicides with high specificity and low toxicity to animals.

The largest application of growth substances, used in combination or succession, is as selective herbicides in field crops. Pre-emergence herbicides can be applied at the time of planting, effecting early weed control with a saving in labor. Moreover, herbicides are important for minimum tillage, a procedure that saves energy, reduces compaction of the soil, and minimizes erosion and runoff. Perhaps 10 percent of the cropland in the United States is now being worked on this basis, and there is pressure to increase this percentage. Development of herbicides specifically designed for minimum tillage in different areas and for different crops is a challenge to agricultural industry.

Continuing research on plant growth substances will surely uncover new and improved agents and will suggest new applications.

Although the use of herbicides has engendered concern, it is doubtful that there has been a well-documented case of serious injury in humans from the proper use of herbicides on crop plants in the United States. Nevertheless, in the development of new agents for the protection or control of growth in plants, it will be prudent to place emphasis on the production of nonpersistent compounds that are not toxic to animals.

Plant Tissue and Cell Culture

Just as with animals, it has become easier to study many plant responses under highly controlled conditions with tissue cultures instead of intact plants. Growth control processes, for example, can be studied very effectively with plant tissue cultured in an appropriate medium. Small pieces of tissue or even single cells from certain plants are totipotent, i.e., with proper manipulation a single cell can be grown into an entire plant.

In the usual reproduction of plants, by seeds or vegetatively, viruses and other pathogens often are carried along. By manipulating plant tissue cultures, it is possible to develop and maintain virus-free and bacteria-free plants. Culturing specialized plant cells by fermentation techniques may prove more effective for generating useful products than growing whole plants. Genetically altering cells in tissue cultures may also speed the search for plants resistant to disease and other stresses. The culture of cells and meristems has become a common horticultural practice with prospects for wide agricultural application.

Effects of Light and Photoperiod

Among the most spectacular responses of plants to light and photoperiod (regularly recurring changes of light and darkness) are those generated through the action of phytochrome, a blue-green protein that serves as a light-activated biological switch. The form of the phytochrome molecule can be changed in different ways by different wave-lengths of light, and the response of the plant depends on the light to which the phytochrome was last exposed. Phytochrome exposed to far red light, for example, elicits changes in the bioelectrical potential, root tip adhesion, coleoptile elongation, leaflet closure, plastid movement, ATP levels, promotion of seed germination, and flowering of plants. Practical work on the phytochromes could be emphasized with benefit.

Plant growth and development is generally dependent upon the length of the light and dark periods. Certain plants will stay in the vegetative stage during long days and enter a reproductive stage and flower only when the photoperiod is shortened. In contrast, some familiar garden plants stay vegetative during the short spring days but exhibit reproductive growth and set seed when the days lengthen. Most photoperiodic responses can be effected at low light intensity. However, the control of growth may be influenced by the intensity as well as the quality of the light, and also by plant growth substances. Although control over light is seldom practical in the field (shade-grown tobacco being one exception), it is used in greenhouse operations to bring plants into flower at desired times. Chemical modification of light responses offers potential for field applications.

Water Supply

Frequently, the water supply limits plant growth. By controlling the opening or closing of the stomata—the pores on leaves—it is possible to control the rate at which water is lost. Plants in dry habitats frequently restrict water loss by closing their stomata during the day and opening them at night. However, chemical control of stomatal opening will have rather limited practical application until agents are developed that cause little damage to the plant while they restrict its water loss. A natural growth factor that regulates stomata, abscisic acid, may prove useful. Attempts also have been made to breed plants that assert greater stomatal control, or that have a reduced number of stomata. However, these approaches are limited by the need for open pores for uptake of CO_2 commensurate with the growth rate.

The method of supplying water also warrants attention. Overhead sprinkling distributes water uniformly, but it is subject to heavy losses through evaporation on hot and windy days. Open trench irrigation incurs much less evaporation, but it can leach nutrients from the root zone and transfer them to areas not penetrated by the roots. Trickle irrigation from porous or perforated tubing is attractive, because the water can be placed directly in the area of use, supplied under close control, and losses avoided of nutrients from the root zone. Improved porous tubing should increase the popularity of this practice for specialty crops.

Frost Resistance

Some plants show remarkable winter hardiness, whereas others show damage even above the freezing point. The speed of freezing and thawing, the physiological condition of the plant at the time of exposure, the adjustment of the plant to gradually lowered versus suddenly lowered temperature, and the state of water in the plant all influence the plant's response. Certain diseased plants exhibit an increased cold hardiness, suggesting that subtle changes in a plant may influence its response to cold. Breeding has improved the cold hardiness of a number of crop plants, but more knowledge of the mechanisms of cold resistance and susceptibility would be helpful.

Plant Stress from Diseases and Pests

A wide spectrum of chemical agents has been developed to control plant diseases and pests. Some are effective but costly; some are only slowly degradable. Agents that are not biodegradable can build up in the soil; if assimilated into the plants, they can be carried into the human or animal food supply with opportunity for toxicity. Breeding plants for disease or pest resistance avoids this problem, and indeed it is one of the more spectacular successes of agricultural research. Research costs are modest relative to benefits; inbred resistance eliminates investment in special field labor, and it avoids use of toxic compounds. However, breeding for resistance is a never-ending task, because as resistant plants are developed, new strains of pathogens may appear.

Crown gall is a classical bacterial disease that attacks a variety of plants. The growth of a tumorlike gall does not require the continued presence of the intact bacterium, but only of a large plasmid (an autonomously replicating piece of DNA) carried in the bacterium. Bacteria lose their virulence if these plasmids are removed, and regain it if the plasmids are restored. This development suggests possibilities for establishing a more rational method for control of the crown gall.

Insect pests present a different problem. Although there has been some success in building insect resistance into plants, this has proved more difficult than incorporating resistance to microbial infection. The usual rationale is to breed plants with an increased level of compounds that are repellant, distasteful, or toxic to the insect. Whether plant- or man-made, agents toxic to insects must be monitored with great caution to assure that they are innocuous in the food supply. It may be that agents that disturb the life cycle of the insect (e.g., juvenile hormone) or which counteract pheromones (insect attractants) may be incorporated into plants for their protection.

Mineral Nutrition of Plants

The availability of an essential mineral from the soil depends upon the acidity of the medium, the presence or absence of other compounds or mineral ions, the binding properties of the soil, and the feeding effectiveness of the plant. Plants may be bred for high feeding capacity. Nutrient elements not used by the plants may be leached from the soil, or the soil itself may be eroded and carry away nutrients on the soil particles.

The usual procedure for improving crop yields in nutrient-deficient soils has been to adjust nutrient levels through fertilizer and lime applications. Recent studies indicate that by applying insights from plant physiology and plant breeding it will be possible to develop economic crops much more efficient in nutrient utilization. Adaptation of the plants to specific nutrient deficiencies is a promising alternative to the traditional fertilization and liming of nutrient-deficient soils. Maize strains that feed more efficiently on both added fertilizer and soil nutrients, particularly potassium, have recently been developed. Continued research on this problem promises excellent returns.

NUTRITIONAL QUALITY OF PLANTS

Alteration of plant cultural practices has little influence on the nutritive quality of plants. However, it is possible to breed plants with increased abundance of vitamins and altered balance among their carbohydrates, proteins, and lipids. After the biochemical nature of a genetic alteration has been established, breeding may proceed on a more rational basis. Plant breeders try to achieve high nutritional quality together with high yield, disease resistance, and other desirable qualities. A striking case in point is the development of a strain of maize relatively abundant in lysine, an amino acid essential in the human diet; classically the low lysine content of maize has limited the nutritional value of its protein. As yet, the yield of this strain is low, but this limitation will surely be overcome by further breeding.

In addition to improving traditional crops, new crops such as the winged bean are being examined for their nutritional quality, in the hope of finding highly productive plants that will be useful in producing more balanced diets.

Leaves, which contain a limited amount of high-quality protein, are usually discarded or returned to the soil in agricultural practice. Leaf protein can be separated from structural materials by simple techniques. Protein from alfalfa leaves has been used successfully in poultry feed, but it has been used for human consumption only on an experimental basis. Nothing precludes such use, and the yield of leaf protein per acre is impressive. Residues left after extruding soluble proteins from leaves can be converted to silage for dairy cattle.

Nowhere in this presentation have we forecast an early, major breakthrough. None is in sight. Rather, we observe the need to pursue fundamental studies of plant physiology and the applied research necessary to take advantage of myriad opportunities to achieve modest increases in productivity, reduction in cost, and conservation of resources. Their sum could well be of huge economic benefit and provide greater assurance of a continuing food supply. Emphasis has been given to those opportunities relevant to domestic agriculture. Research opportunities for agriculture in developing nations, and the arrangements that might facilitate such research, have been described in greater detail in the recent NRC report *World Food and Nutrition*.[5]

CONCLUSIONS

UNITY AND DIVERSITY

The selections in this chapter have illustrated the dynamism of current molecular and cellular studies in various areas of biology, and the remarkable unity of the living world at this level of organization. The source of this unity is the evolutionary continuity of the living world: Organisms today perpetuate many successful biochemical structures and mechanisms that emerged in the earliest living cells 3 billion years ago.

However, our survey of biology would be unbalanced if we did not also note the other side of evolution: diversity. For evolution arises from the continual production of novel combinations of genes, together with the preferential survival, by natural selection, of those organisms that are better adapted to their environment. This process has yielded literally millions of species, of gradually increasing complexity and adaptability; and it has culminated in our species—one whose unique capacity for abstract thought, deep feeling, communication of complex information, construction of efficient tools, and creativity have added a new chapter: cultural evolution.

For some years the dramatic advances in molecular studies rather eclipsed studies concerned with the properties of whole organisms and populations. However, concern for conservation of the environment, and of the rich legacy of evolution, has aroused wide interest in ecology. Furthermore, a bridge now links evolutionary and molecular studies, as the latter have made it possible to quantify precisely the amount of genetic difference between species (and between individuals within a species), by measuring the degree of similarity in the sequences of their DNA or of its protein products.

Nevertheless, evolutionary studies have a unique quality: The physical sciences, and their applications to biology, deal primarily with uniform entities (i.e., the identical molecules of a pure substance); but population biology deals with groups in which almost every individual is genetically unique. So the population distribution of genetically determined traits (such as blood groups), and of genetically determined potential (physical and behavioral), must be characterized in statistical terms. The result has been the replacement of stereotypes with a true appreciation, in general terms, of the scope of human diversity: Within every group individuals differ genetically over a broad range.

An important recent development has been the emergence of the discipline of sociobiology, which aims at understanding the biological foundations of social behavior in animals, and in man in particular. This approach aims to throw light on the biological constraints that evolution has imposed on our species. Within these constraints cultural forces produce an enormous, but not limitless, range of social behavior.

CONCERNS WITH FUTURE IMPACTS OF BIOLOGY

With the increasing depth of our advances in biology some have come to fear that certain areas of research, especially in molecular and human genetics, threaten public welfare and thus need to be regulated. The recent intense debate over recombinant DNA research (see p. 65) is a case in point. The potential dangers perceived involved both those inherent in the research itself (Could a novel virulent organism escape from the laboratory to annihilate whole populations?) and those latent in new knowledge that could confer potentially dangerous powers (Might genetic engineering be used to manipulate our personalities?).

The anxiety over recombinant DNA research has abated considerably, for several reasons. Among these is the failure of five years of experience to produce evidence of any illness or other harm. In addition, sober professional analyses have gradually displaced earlier unrealistic demands for absolute protection against hypothetical risks. Accordingly, the NIH guidelines are gradually being relaxed; and the Congress has determined that the guidelines are adequate for handling the problem without legislation. Nevertheless, the costs of both the debate and the resulting restrictions, in money, time, and morale, have been large.

When the hazards of research are well defined, as with toxic or radioactive substances, or when experimentation on human subjects involves real risk, the problem is straightforward, though sometimes complex. However, when available facts do not clearly define or quantify the risk, it is more difficult for society to reach the best judgment. The experience with recombinant DNA suggests that in similar discussions in the future it is important to try to separate the technical phase of assessing the risks and benefits, on the basis of expert and informed judgment, from the political phase of legislative and public participation in formulating the required value judgments and reaching policy decisions.

In contrast to risks inherent in carrying out research, there is the risk that the knowledge produced by research may be used for harm as well as good. Since virtually any knowledge is double-edged, and since its consequences cannot be predicted in detail, our society has proceeded wisely, in our view, in the belief that *on balance* knowledge is less likely to be dangerous than is ignorance. Our society is trying to assess and to regulate harmful technological applications earlier before damage occurs, but has declined to regulate research on the basis of speculations about possibly harmful applications.

Still other types of new biological understanding raise fears that deeper insights into human nature might imperil a just and decent society. The example most often cited is study of human behavioral genetics, which would sharpen our perception of inborn individual differences in various intellectual capacities, talents, drives, and learning patterns. Here, too, knowledge is two-edged, with possible misuses and valuable uses.

The problem will not remain confined to genetics. Eventually, advances in sociobiology, neurobiology, and the behavioral sciences are also likely to conflict with treasured preconceptions, widely held to be indispensable foundations for public morality. But human curiosity cannot be permanently extinguished, nor can the scientific method be unlearned. Someone will learn, somewhere, sometime. Moreover, the realities will be there, whether or not scientists are permitted to find them; and if we build social policies on false assumptions, which contradict reality, we will be building on a crumbling foundation. A democratic and open society, therefore, has no choice but to defend freedom of inquiry, just as it defends freedom of expression.

OUTLOOK

The following outlook section on the living state is based on information extracted from the chapter and covers trends anticipated in the near future, approximately five years.

A remarkable burst of advances is propelling biology into a new era. Powerful new tools are generating profound insights into the secrets of life, and scientists are now able to move with assurance among realities that they could only contemplate 10 years ago. There is no reason for this momentum to falter. Rather, biologists look for a flowering of their science in the years ahead. Extrapolating new knowledge and exploiting new technologies, they are eager to discern ever more clearly the molecular organization and functioning of living cells, tissues, and organisms.

MOLECULAR GENETICS

In molecular genetics, basic mechanisms that have been elucidated in bacteria and viruses are now being studied in higher organisms. Since major health problems almost invariably involve cellular malfunction, frequently reflecting intrinsic or imposed defects in gene structure or regulation, molecular genetics is in a position to exert a major impact.

The potential contribution of recombinant DNA technology is hard to exaggerate. By splicing fragments of DNA from higher organisms into simple bacteria, one can purify the DNA segments, obtain them in quantity, determine their base sequences, and study the behavior of their genes. Early experiments using recombinant DNA methodology in higher cells are showing the types of surprises that can be in store: The structural genes of higher organisms, which code for the synthesis of specific proteins, are not simple continuous sequences of DNA, as in bacteria and viruses. Instead, each is a set of DNA base sequences interrupted on the DNA helix by several long, unexpressed sections. Scientists are investigating the hypothesis that the intervening, nonactive sequences may somehow function in determining which genes in a cell are expressed at a given time. Other studies are exploring the precise mechanisms by which DNA is duplicated when cells reproduce. Still other studies are investigating the mysteries of the structure of chromosomes, where DNA resides.

Molecular biology is also facilitating studies of mutations in genes, both those in germ cells, which may cause inherited disorders, and those in body cells, which may lead to the development of diseases such as cancer. It may take 10–20 more years before we adequately understand diseases of cellular malfunction such as cancer or autoimmune disorders. Other diseases, however, such as defective hemoglobin synthesis, will probably be understood sooner. In the years just ahead we can also expect improved ability to detect, via amniocentesis, various genetic defects in fetuses. There will be improved procedures to assay the mutagenic properties of chemical compounds and, hence, their potential for carcinogenesis.

Advances in understanding the subcellular structure called the ribosome have revealed how a number of antibiotics work, and how cells develop resistance to them.

Another prospect involves using recombinant DNA methods to manipulate microorganisms so that they will synthesize specific gene products. Human hormones will become more readily available. It may become possible to design and synthesize new enzymes or to manufacture known enzymes cheaply enough to substitute them in certain processes. Microorganisms may yield useful sources of energy, or food plants may be freed of their dependence on commercial fertilizer as a source of nitrogen.

A relatively distant advance is the possibility of treating genetic diseases by replacing defective genes with normal ones. The first such gene therapy to become a reality will probably be replacement of defective blood cells, which are produced from a stem cell line in the bone marrow in a way that allows them to be replaced from the blood stream. With more highly organized organs, however, the problem seems insuperable. It is even less likely that a gene replacement can be carried out in sex cells, as would be required to eliminate further inheritance of a genetic disorder.

CELL BIOLOGY

With knowledge derived from simpler systems, and using DNA technology, the study of the cell biology of higher organisms will move toward a new level of sophistication. Research efforts will be trained on the intricate mechanisms that integrate the complex interactions of genes and enzymes. Such knowledge is pivotal for understanding not only normal cell reproduction and differentiation, and presumably the conversion of a normal cell into a cancerous one, but also the process by which cells age.

The fascinating structure and workings of cell membranes, which maintain the vital balance between fluids inside and outside a cell, and which are the site of interactions of a cell with hormones, viruses, bacterial toxins, and drugs, are the object of intense study. Such investigations will have importance for a great variety of health-related concerns. Critical to new understanding are studies of the newly discovered specific receptor proteins in cell membranes, which are the specific sites of attachment of hormones, neurotransmitters, etc.

Studies are probing the molecular basis for the cytoskeleton's dynamic nature, seeking to understand the motility that allows white blood cells to mobilize to fight infection, or cancer cells to invade nearby organs, or, within the cell itself, for subcellular components to move about with precision and efficiency.

IMMUNOLOGY

Enriched by the new tools and understanding of biochemistry, molecular genetics, and cell biology, immunological research has gained new sophistication and become a powerful source of new insights. Accordingly, one may confidently expect a marked increase in the capability to manipulate both desirable and untoward immune reactions.

New vaccines are in the offing for several infectious diseases, including serum hepatitis and malaria. Interferon, a powerful antiviral protein manufactured by the body, is promising to be therapeutically effective; its current scarcity may be remedied through the use of recombinant DNA technology in a few years. Kidney transplants can expect to enjoy much better survival rates in the years ahead, mostly as the result of improved matching of donors and recipients. Graft survival will also benefit from new therapies with drugs, irradiation, and antibodies against destructive white cells.

In allergy research, new discoveries about the types of white blood cells that influence allergic reactions promise to lead to a rational, guided approach to desensitization therapy. Moreover, the recent discovery of cell receptors for the antibodies that provoke allergic symptoms may permit the design of chemicals to block these receptors and thus abort the symptoms. Chemical studies on the purified components of the allergens will bring improvements to hyposensitization therapy.

Individuals with defects in their immune systems that destroy their resistance to infection may soon benefit from transplants of appropriate tissue such as bone marrow or thymus. New discoveries regarding mechanisms of tolerance promise to lead, relatively soon, to improved methods for controlling several autoimmune diseases in which the immune system attacks a normal tissue; already patients with myasthenia gravis are showing improvement after surgical or drug therapy to alter immune function.

There is hope that advances in tumor immunology will eventually make it possible to eliminate cancer cells that remain after chemotherapy. So far, however, efforts at cancer immunotherapy have been disappointing.

NEUROSCIENCE

New insights will continue to illuminate the remarkable workings of the human brain. Recent research has revealed the nature of the action potential and of synaptic transmission; the synapse, the basic mode of communication in the brain, has become the unifying concept in neuroscience.

Thanks to the development of new techniques, analysis of the structural organization of the nervous system—the wiring diagram—is proceeding at a rapid rate, and a detailed topography of the functional anatomy of the brain is imminent. Spectacular progress in the identification and study of the modes of action of synaptic neurotransmitters promises to enhance markedly the understanding of the biological basis of behavior ranging from neurological disorders to motivation, sleep, motor systems, memory, and psychoses. Most importantly, to the extent that behavioral disorders have their bases in altered neurochemical functioning, they are subject to pharmacologic remedies.

The study of hormones and their actions, both those produced in the brain and elsewhere, is also advancing rapidly. Hormones, now believed to act on the brain in a manner analogous to synaptic transmitters, underlie such important phenomena as sexual development and basic motivations like thirst and response to stress, and may modulate higher order processes such as learning and memory.

That the brain also produces its own hormones, most of them polypeptides, has just been recognized. Some were first known to exist as the releasing substances of the hypothalamus that occasion synthesis and release of pituitary hormones; others as hormones functioning in the gastrointestinal tract. Most recent are the endorphins, the brain's own opiates, with high affinity for opiate receptors. How many such brain hormones there are, what are the physiological and psychological consequences of their activity, and whether they act at synapses or other receptors, constitute a chapter that has just opened and will be pursued apace.

It seems likely that in a decade or so we will know how developing nerves seek their targets and we may even be able to improve regrowth of damaged neural tissue. In view of the severity of derangements of neurological development—birth defects, muscular dystrophy, epilepsies—these advances cannot fail to make a tremendous clinical impact. However, to understand how the gene program is translated into a nervous system remains one of the most gigantic challenges that neuroscience offers.

Much new information will be gained in the next 5–10 years about how the brain codes and integrates sensory information. New treatments for sensory disorders will emerge and prostheses for certain sensory defects may be developed. The most rapid advance in analysis of motor systems will probably relate to certain of the synaptic chemical transmitter systems and may result in improved treatment of certain abnormalities of movement.

In the next decade or so, much will be learned about the neuronal substrates of basic motivations, and more effective treatments may become available for such important and intractable behavioral disorders as drug addiction and obesity. Systematic research will probably elucidate the circuits in the brain that underlie simpler forms of learning and memory. Since memory probably involves persistent physical changes of some kind at nerve-to-nerve junctions, insights into the basic mechanisms for learning and memory are surely forthcoming, although precise localization of the memory trace for a specific bit of information may remain an elusive goal. It is possible that treatments may be developed to improve certain forms of learning disabilities. Finally, with rapid growth in our knowledge of synaptic transmissions, we seem on the threshold of major improvements in the understanding and treatment of schizophrenia, the most severe and widespread form of mental illness.

In most neurological conditions that lead to substantial brain destruction, the prospects for a real cure are indeed poor, for such a cure would involve the removal of scars and debris and the regrowth and rewiring of countless nerves. However, one can look for significant advances in mitigating the effects of such conditions. Better drugs will be developed against epilepsy, psychoses, Parkinson's disease, spasticity, etc. The outlook for devastating diseases such as multiple sclerosis and Alzheimer's disease is somewhat different. The hope for a cure seems to be reasonably good—especially if damage to the nervous system has not progressed very far. However, the cures will probably come from immunology or virology, not neurobiology.

While neuroscience is not promising immediate cures for the wide range of neural and behavioral disorders that afflict humanity, the current rapid growth of knowledge about the brain and nervous system makes it seem highly probable that improved treatments for many disorders will become available in the next few years. Perhaps even more importantly, as we learn more about the brain we will approach closer to a genuine understanding of the human condition.

BIOLOGY AND AGRICULTURE

Biological research in agriculture, stimulated by advances in so many areas, is striving to increase productivity while reducing dependence on chemical fertilizers and pesticides.

More detailed understanding of the remarkable process of photosynthesis is engendering efforts designed to enhance its efficiency. Research will be directed to develop crop plants that utilize light more efficiently, while other work will focus on altering key enzymes so that they are less easily diverted to unproductive activity during the dark reactions of the photosynthetic process.

Other research will attempt to convert the ordinary C_3 plant to the more efficient type of metabolism seen in such C_4 plants as maize.

Because improvements in biological nitrogen fixation could translate into better nutrition for millions, with fewer demands on energy supplies, scientists will be working to develop more effective bacteria, more responsive plants, and more productive associations between the two. Free-living nitrogen-fixing bacteria already exist in nature; scientists will be employing genetic techniques, including those of recombinant DNA, to make them more effective. Cereals feed much of the world's population, and there will be continuing efforts to develop plant-bacteria symbioses that could fix nitrogen for cereal plants. Similarly, a variety of approaches is being explored to improve the natural fixation that blue-green algae provide in rice cultivation.

In the area of plant growth and stress, there will be continuing efforts to discover new and improved plant growth substances, and to foster new applications. Plant tissue cultures offer many advantages. Not only do they make it possible to develop and maintain plants free of infection with viruses or bacteria, they may provide ready access to useful plant products, and they may speed the search for plants resistant to disease and other stresses.

Chemical substances, too, open many avenues to progress. In addition to their role in controlling diseases and pests, they will be useful for manipulating a plant's responses to periods of light and dark, or for regulating the opening and closing of its stomata.

Breeding plants for specific characteristics is yet another biological approach to improving agricultural productivity. The breeding of plants for resistance to disease and pests has been one of the more spectacular successes of agricultural research. Current work indicates that there are excellent prospects for developing economic crops that utilize nutrients much more efficiently. At the same time, plants are being bred for better nutritional value, with an increase in vitamins and an altered balance among carbohydrates, proteins, and lipids. New, highly nutritious crops will be developed, and previously untapped sources of nutrition such as leaves, which are usually discarded but which contain a limited amount of high quality protein, will be exploited.

Although no major breakthrough is in sight, basic studies of plant physiology and applied research promise to yield many modest advances. Taken together, these promise high economic benefit as well as a greater assurance of continuing food supply.

REFERENCES

1. Chan, H.W., *et al.* Molecular Cloning of Polyoma Virus DNA in *Escherichia coli*: Lambda Phage Vector System. *Science* 203(4383):887–892, 1979.

2. Israel, M.A., *et al.* Molecular Cloning of Polyoma Virus DNA in *Escherichia coli*: Plasmid Vector Systems. *Science* 203(4383):883–887, 1979.

3. Benditt, E.P. Implications of the Monoclonal Character of Human Atherosclerotic Plaques. *Annals of the N. Y. Academy of Sciences* 275:96–100, 1976.

4. Dixon, R.A., and J.R. Postgate. Genetic Transfer of Nitrogen Fixation from *Klebsiella pneumoniae* to *Escherichia coli.* Nature 237:102–103, 1972.

5. *World Food and Nutrition Study: The Potential Contributions of Research* (NRC Steering Committee Study on World Food and Nutrition). Washington, D.C.: National Academy of Sciences, 1977.

BIBLIOGRAPHY

Benditt, E.P. Origin of Atherosclerosis. *Scientific American* 236(2):74–85, 1977.

Brill, W.J. Biological Nitrogen Fixation. *Scientific American* 236(3):68–81, 1977.

Cairns, J. *Cancer: Science and Society.* San Francisco: W. H. Freeman, 1978.

Cooper, J.R., F.E. Bloom, and R.H. Roth (eds.) *The Biochemical Basis of Neuropharmacology,* 3rd ed. Oxford University Press: New York, 1978.

Davis, B.D. The Recombinant DNA Scenarios: Andromeda Strain, Chimera, and Golem. *American Scientist* 65(5):547–555, 1977.

Eccles, J.C. *The Understanding of the Brain,* 2nd ed. McGraw Hill: New York, 1977.

Eisen, H.N. *Immunology.* Harper and Row: New York, 1974.

Kuffler, S.W., and J.G. Nicholls. *From Neuron to Brain: A Cellular Approach to the Function of the Nervous System.* Sinauer Assoc.: Sunderland, Mass., 1976.

Mazia, D. The Cell Cycle. *Scientific American* 230(16):54–64, 1974.

Watson, J.D. *Molecular Biology of the Gene,* 3rd ed. Benjamin-Cummings: Menlo Park, Calif., 1976.

(Hale Observatories)

3 The Structure of Matter

INTRODUCTION

The effort to understand matter has challenged the human mind throughout recorded history. It has been motivated both by the desire to manipulate and control matter for human uses and by simple curiosity. As our understanding of matter has deepened, so has the range of study to include the less tangible phenomena of visible light and other forms of electromagnetic radiation.

Understanding matter includes knowledge of its forms and behavior under all imaginable conditions, the transformations that can occur among various forms of matter, and the natural laws that underlie these phenomena. Such knowledge makes it possible to predict the behavior of matter under new circumstances, to establish how it is fashioned and how new forms may be created, and to control its forms and uses.

In view of the great diversity and complexity of matter, it must be an act of faith to search for an underlying simplicity. While the modern physical scientist continues the pursuit of a goal that began in ancient Greece, he is under the severe constraint that his explanations must account consistently for reproducible phenomena and must unfailingly predict the behavior of matter. There is no greater miracle than the success of this enterprise. Indeed, the successes of the past justify faith in the future.

The structure of matter may be viewed on a scale that ranges from the very large to the very small, from the macroscopic to the microscopic (see Figure 9). Cosmology is concerned with the structure of matter on the largest scale, with the grand structure of the universe: How did it

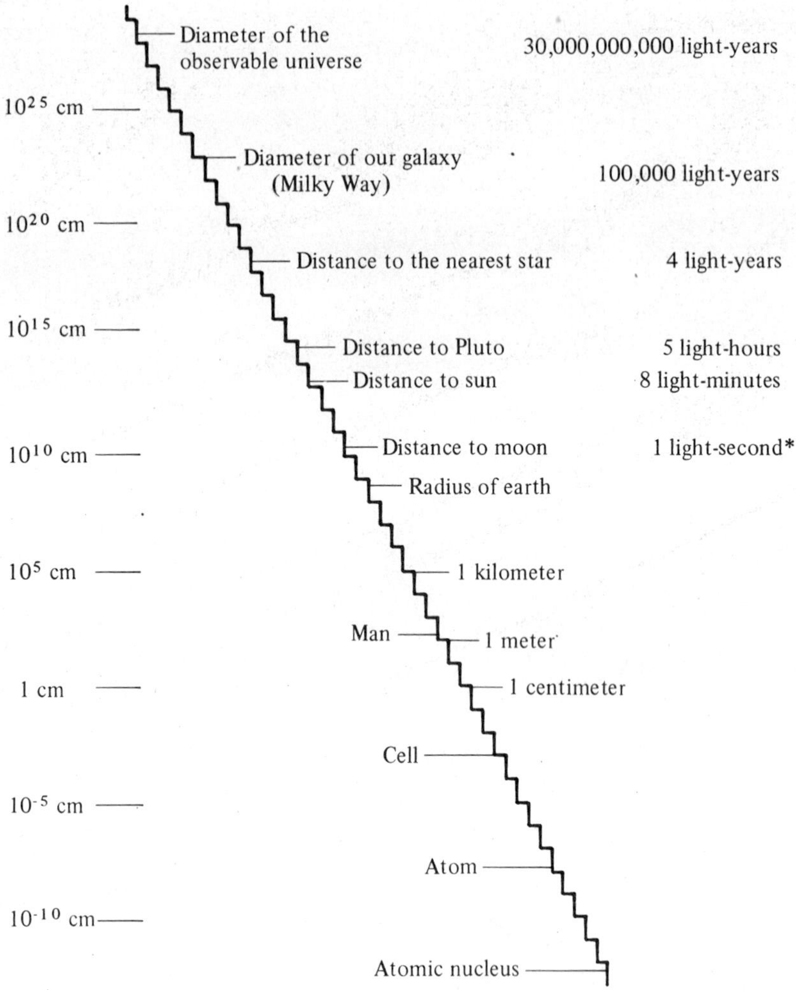

FIGURE 9 Range of sizes in physics (each step corresponds to a factor of 10 increase in size. *1 light-second = distance that light travels in 1 second = 3×10^{10} cm or 186,000 mi. (National Academy of Sciences, *Physics in Perspective,* Vol. I, p. 84. Washington, D.C., 1972)

originate; what is its age; of what is it made? It involves the study of such fascinating phenomena as black holes, gravitational collapse and neutron stars, the "Big Bang," and so on.

Then there is the more human scale of the laboratory, one which permits experiments and studies of matter under carefully controlled conditions. Laws governing macroscopic matter are well developed and understood. Among the most elegant laws of nature are the laws of thermodynamics, which describe the thermal behavior of matter in terms of a few parameters characterizing each substance. Another example includes the detailed laws of statistical mechanics governing particulate structure of matter. Our general understanding of the mechanical, electrical, magnetic, and optical properties of most normal forms of bulk matter is also well established in terms of a few parameters for each material. However, there are exotic materials such as superconductors and superfluids for which the classical theories have not been adequate. They have turned out to be examples of macroscopic quantum fluids, a realization of the quantum theory associated primarily with the microscopic behavior of matter.

Indeed, in recent years, dazzling variations of the more familiar forms of materials have appeared, presenting challenges to established theories and many opportunities for application. Included are new types of semiconductors, amorphous materials with a great variety of tailored properties, polymeric and organic complexes, and two-dimensional as well as one-dimensional materials providing special opportunities for fundamental studies.

Finally there is the microscopic domain in which one attempts to identify the building blocks out of which matter is composed. In this domain, matter is reduced to collections of molecules and these to atoms. Atoms are reducible to electrons and a nucleus, the latter composed of neutrons and protons. Neutrons and protons now appear to be made up of "ultimate" building blocks, the quarks, a concept that arose out of considerations of internal symmetry introduced to account for the unexpected behavior of nuclear particles.

Although the general laws describing the behavior of matter at the atomic and molecular scale are well established, much remains to be done to arrive at a detailed description of the properties of matter on the basis of fundamental physical principles, quantum and statistical mechanics, except for very simple systems. High speed computers and new mathematical techniques are now advancing our knowledge of matter at the atomic and molecular scale. However, it will be a long time before our ability to understand matter at the molecular level will have advanced to where we can understand the building blocks of living systems in terms of basic physical laws.

ASTROPHYSICS AND COSMOLOGY

In earlier times, our knowledge of the universe beyond the earth depended entirely upon the perception of visible light, which represents only one very small portion of the broad spectrum of electromagnetic radiation that is emitted by matter under various conditions. The earth's atmosphere is an effective absorber of much of the radiation that falls upon it, and for this reason there are broad portions of the spectrum that can be detected only by specially designed instruments located above much of the atmosphere. During the past 25 years or so, however, remarkable instruments have been devised which allow us to see the cosmos as it appears in radio waves, in infrared and ultraviolet radiation, and in X-rays and gamma rays (Figure 10).

The sources of electromagnetic radiation are molecules, atoms, electrons, and the nuclei of atoms. Each of these forms of matter can exist in a number of different energy levels, or states, which correspond to different vibrational patterns, or different rates of rotation, or different configurations of the electron cloud surrounding atoms and molecules. The absorption of radiation raises the object to a higher energy "excited" state, and radiation is emitted when the object undergoes a change or transition to a lower energy state. The particular wavelengths or frequencies of the absorbed or emitted radiation are characteristic not only of the kind of transition involved but also of the specific kind of molecule, atom, or nucleus undergoing the transition. For example, the pattern of radiation emitted by iron atoms is different from that emitted by carbon atoms, and the same is true for different kinds of molecules and different atomic nuclei. The radiation pattern also depends on the temperature, pressure, and other conditions of the matter of which the atoms or molecules are a part. Thus the observation of electromagnetic radiation over a broad region of the spectrum provides a number of different and complementary windows through which the matter in the stars and galaxies of the universe can be studied.

In addition to the electromagnetic radiations, other radiations provide an increasing amount of information concerning the cosmos. For example, the charged nuclear particles in the cosmic rays are a direct source of information about galactic and extragalactic events. Also the ability to detect neutrinos offers a promising possibility for an entirely new means of astronomical observation. Furthermore, gravitational waves could be used as signals of catastrophic events if their detection could be successfully demonstrated.

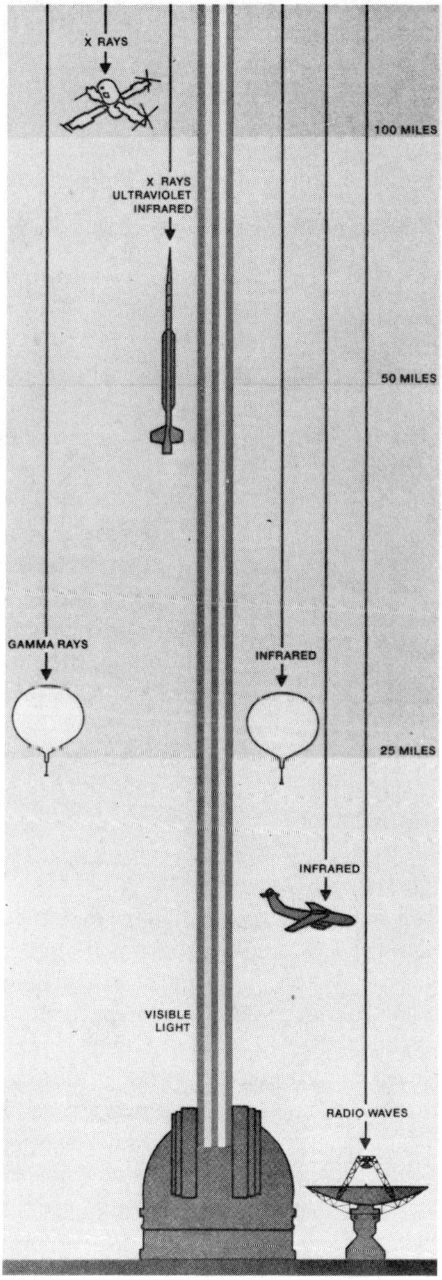

FIGURE 10 Electromagnetic radiation. Visible light and radio waves penetrate the earth's atmosphere and can be observed by ground-based instruments. For unobscured observation of most other parts of the electromagnetic spectrum, instruments must be carried high up in, or above, the atmosphere.

COSMOLOGY

During the early years of this century, most astronomers believed that all of the visible universe, both stars and nebulae, were contained within one great system, the Milky Way galaxy. By 1929, however, Edwin Hubble had established the fact that many of the observed nebulae are actually "island universes" or independent galaxies much like our own, which are vastly more distant than had previously been thought, and that they appear to be flying apart from each other like pieces of shrapnel from a great explosion.

Our present picture is that the universe contains many billions of galaxies that are distributed throughout space rather uniformly in all directions out to the limits of our ability to observe them; and that the cosmic expansion first noticed by Hubble had its origin in a primordial explosion, the Big Bang, which occurred some 10–20 billion years ago.

Will the universe keep on expanding forever, or will the present expansion gradually slow down and perhaps eventually reverse into a collapse? If it does collapse back into a giant fireball, will there then be another Big Bang? Is ours a "one-shot" universe, or does it oscillate endlessly between expansion and collapse?[1] These questions are central issues in cosmology, and although they may seem mind-boggling, they do lend themselves to scientific analysis. Much of the excitement that pervades contemporary astrophysics arises from the power of the new observational astronomy and of modern theory to search out the cosmological evidence.

The Age of the Universe

Hubble's key observation was that the observed galaxies all appeared to be flying away from each other at speeds that were proportional to the distances between them. That is, galaxies twice as distant were receding from each other at twice the speed. By running this mutual expansion of all the galaxies backward, it is possible to arrive at the "Hubble time" when all of the galaxies would have been packed together in the primordial fireball of the Big Bang. Hubble's original calculations indicated a time of about 2 billion years since the Big Bang, but subsequent studies have steadily lengthened this estimate of the Hubble time until it is now placed between 10 and 20 billion years. The large uncertainty in this estimate still remains to be resolved, but it is noteworthy that the same range of values has been found for the ages of the oldest observed stars, and also for the ages of the chemical elements, which are determined by measurements of radioactivity.

The Hubble time would be expected to give the correct age for the universe only if it had been expanding at a constant rate since its beginning in the Big Bang. If the expansion has gradually been slowing down, however, then the actual age of the universe must be less than that indicated by the Hubble time. It is reasonable to expect that the mutual gravitational attraction of all the matter in the universe must work against the universal expansion and thus tend to slow it down, but by how much?

Density of Matter in the Universe

A second approach to the age problem is to relate the expansion of the universe to the total amount of matter and energy that it contains. Gravity acts upon all forms of matter and radiation (including light), and mutual gravitational attraction will eventually stop the universal expansion if the average density of such mass-energy throughout the universe is great enough. Conversely, if the average density is less than a certain critical value, then the universe will continue to expand forever—an open universe. At present, the estimates of mass-energy in the *observed* forms of stars, gas, and dust fall far short of the critical density needed to halt the expansion and produce a closed universe.

If the universe is in fact closed, then there must be a great deal of matter present that has not yet been observed. Some preliminary evidence from the new X-ray astronomy points toward the presence of substantial amounts of matter in very hot, dilute intergalactic gas. There are also several other ways, not yet confirmed, in which large mass may be detected if present—in black holes, for example. However, arguments based on calculations of the abundances of elements produced by a Big Bang tend to favor an open universe of low density and these arguments include all matter whether in galaxies or intergalactic gases.

Ages of Stars

Many facets of astrophysics also bear on cosmological questions. The ages of stars are an important clue. For example, our present model of stellar evolution indicates that a massive star will consume only about 10 to 15 percent of its hydrogen before it expands into its "red giant" phase. This standard model also predicts that a certain number of neutrinos should be observed on the surface of the earth as a result of the thermonuclear reactions occurring in the core of the sun, but the most persistent efforts to detect these neutrinos have yielded only one-third or so of the predicted number. Some possible explanations for the solar neutrino puzzle would yield a model for the sun which might burn hydrogen for 20 billion years instead of only the 10 billion years predicted at present. Estimates of the age of the galaxy based on the ages of its oldest stars would then have to be revised correspondingly upward.

This uncertainty emphasizes the fundamental importance of understanding the sun, since we can hardly trust our modeling of other stars if the solar model is in doubt.

Three-Degree Background Radiation

During the first few minutes of the Big Bang, temperatures were billions of degrees, and energy appeared in the form of intense, indistinguishable fields of matter and radiation.[2] As the universe expanded, the temperature decreased and the wavelength of the radiation increased—both in direct proportion to the expanding scale of the universe. After some 100,000 years, the temperature had fallen to about 4,000°K (degrees Kelvin = degrees Celsius + 273.15), and the intensity of the radiation was no longer great enough to prevent atomic nuclei from capturing electrons and thereby becoming neutral atoms. At the time of this "decoupling" of matter and radiation, the radiation was no longer scattered by free electrons, and the universe became transparent. As though a fog had been dispelled, the universe was filled with blinding light.

Since the time of decoupling, the universe has expanded by about 3,000 times. As a consequence, the wavelength of the radiation present at that early time should now be increased by the same factor of 3,000 to a present wavelength of about one millimeter. The spectrum of such radiation—a cool, fading afterglow of the primordial fireball—would have a characteristic temperature of about 3° above absolute zero (3°K). This relic radiation was actually discovered in 1964–65, and its existence is perhaps the strongest evidence we have for the Big Bang picture of the universe. The 3°K background radiation has since been observed with remarkable precision in many different experiments: from the ground, from balloons, and even from a U-2 aircraft. Even though the intensity of this background radiation is only about a hundred billionth of the thermal radiation from this sheet of paper, it accounts for most of the radiant energy in the universe, with an energy density exceeding that of all starlight and cosmic rays.

The universal background radiation appears to be remarkably uniform in all directions, with the exception of a slight systematic variation that can be attributed to the motion of the solar system through the universe. Measurements indicate that the sun and its planets are moving toward a point in the constellation of Leo at a speed of about 390 kilometers per second relative to the background radiation.

A satellite mission, the Cosmic Background Explorer (COBE), is now being prepared to make refined measurements of all aspects of the background radiation. Results of this mission should provide important new knowledge of the nature of the early universe.

HIGH-ENERGY ASTRONOMY

The branches of astronomy that are based on space technology are still in their pioneering stages; therefore, great improvements in observational capabilities can be expected within the near future. This is particularly true of high-energy astronomy, now the most rapidly developing of the new astronomies. High-energy astronomy's principal window to the cosmos, the X-ray region of the electromagnetic spectrum, provides a view of many of the most bizarre and violent phenomena on the celestial scene (Figure 11). Spectacular variability is almost universally characteristic of X-ray sources, with time scales ranging from milliseconds to years. As an example, there are now more than 30 known sources called X-ray bursters that typically flash at a power level of a million suns for about 1 second, but are quiescent 10 seconds later. One such source, known as the rapid burster, repeats its explosions about a thousand times a day, going off like a string of cosmic firecrackers.

X-Ray Stars

The X-ray sources within our own galaxy include white-dwarf stars, neutron stars, and (theoretically) black holes. Each of these three kinds of curious objects is believed to be an alternative end-point to the process of stellar evolution. When a star has exhausted its nuclear fuel and can no longer generate the central gas pressure that had previously sustained the crushing burden of its overlying mass, the star begins to shrink inward under the force of its own gravitation. For stars of relatively small mass, such as our sun, the shrinkage ends with the formation of a white dwarf, which is no larger than the earth but is so dense that a spoonful of its matter would weigh about a ton.

Neutron Stars

The end point for stars of medium or large mass can be either a neutron star or a black hole. In the first case, the inner core of the star is crushed down in a fraction of a second to form a neutron star, and simultaneously the star's outer layers are blown off in a gigantic explosion called a supernova. A neutron star is essentially a giant nucleus of about 10^{57} tightly packed neutrons in a volume only 10–20 kilometers in radius and with a density of about a billion tons per cubic inch. In its collapse, the star's original magnetic field can be concentrated to perhaps a million

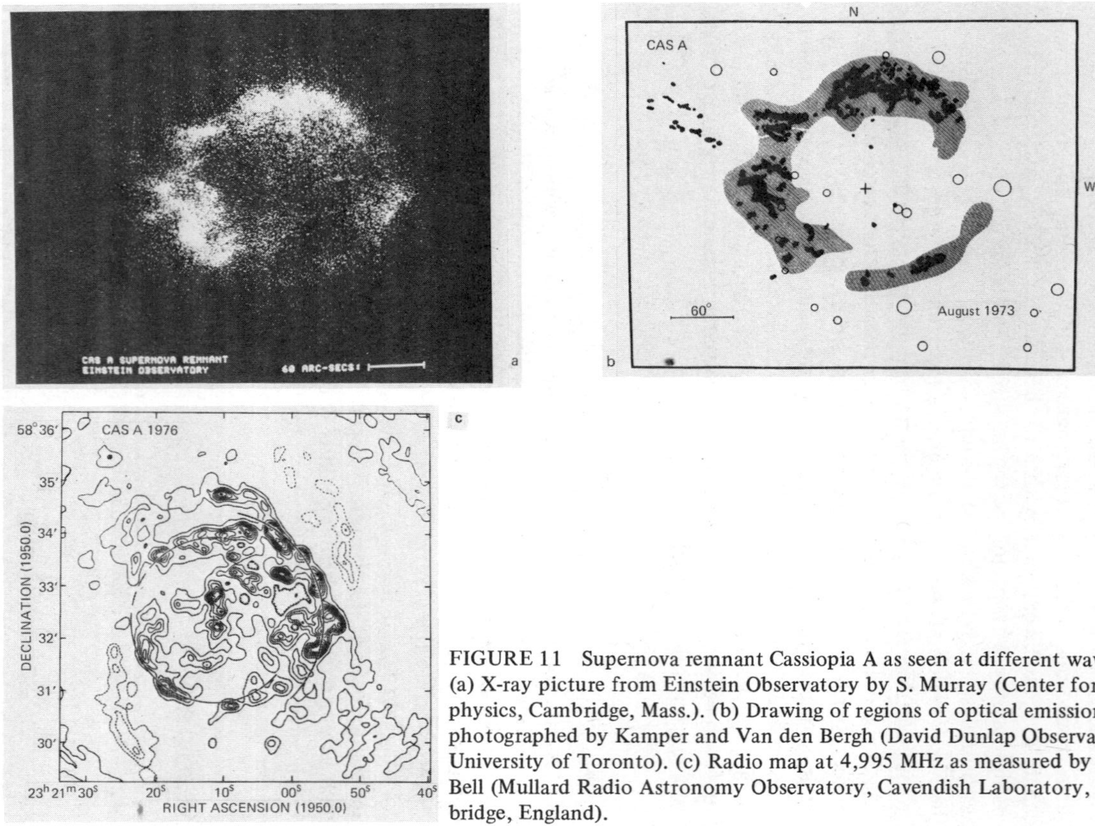

FIGURE 11 Supernova remnant Cassiopia A as seen at different wavelengths. (a) X-ray picture from Einstein Observatory by S. Murray (Center for Astrophysics, Cambridge, Mass.). (b) Drawing of regions of optical emission as photographed by Kamper and Van den Bergh (David Dunlap Observatory, University of Toronto). (c) Radio map at 4,995 MHz as measured by A. R. Bell (Mullard Radio Astronomy Observatory, Cavendish Laboratory, Cambridge, England).

million times its initial strength and its rotation speeded up to a few or even tens of revolutions per second. Electrons locked in the star's magnetic field can be whipped around the star at nearly the speed of light, with the result that they beam out radiation of every wavelength from radio waves to gamma rays, which is called "synchrotron radiation" for reasons given later.

Pulsars

It is believed that radiation is more intense from some regions on the surface of a neutron star than from others. Thus, as the star rotates, the radiation appears to vary in intensity, or pulse, depending upon whether these regions or "hot spots" face the observer on earth or not. Such pulsating sources of radiation, or pulsars, were first discovered in 1967 and now number several hundred. The first X-ray pulsar detected was the neutron star at the center of a supernova remnant, the Crab Nebula. The Crab pulsar emits strongly in every region of the electromagnetic spectrum. No other pulsar has been observed in both the radio and X-ray regions, although a pulsar in Vela emits radio waves and gamma rays quite strongly. Two recent discoveries of pure gamma-ray pulsars add to a diversity of pulsars that we cannot yet explain. The pulsar process must be more complex than the simple description given above.

Binary Systems

When a neutron star is coupled closely to a normal star to form a binary pair, it can accrete gas by intercepting the "stellar wind" of particles thrown off by its larger companion or by direct gravitational attraction from the companion's surface. The powerful gravity of the compact star attracts the gas toward its surface with great kinetic energy, and the hot infalling gas is funneled by the star's magnetic field onto an area of only about one square kilometer at each magnetic pole, which thus becomes a source of intense X-rays. If the magnetic axis of the star is not the same as its axis of rotation, the "hot spots" at the magnetic poles will sweep around the sky and thus beam X-rays in pulsar fashion as the star spins. At present, some 14 X-ray pulsars are known, with rotation periods ranging from less than 0.1 second to 16 minutes.

Black Holes

When the nuclear burning cycle ends, the gravitational forces for stars of sufficient mass become so strong that no equilibrium condition such as that of a white dwarf or a neutron star is possible. In such cases, the final collapse is thought to occur in less than a second and to result in the formation of a black hole—a region of space in which matter, figuratively speaking, has been crushed completely out of existence. The name was chosen to express the fact that theoretically nothing, not even light, can escape from the surface of this hole in space because of its immense gravitational field. Black holes resulting from the collapse of stars would typically have masses 3–50 times that of the sun, and diameters 18–300 kilometers.

The gravitational force is believed by most scientists to be correctly described by Einstein's general theory of relativity, and the idea of black holes is so fundamental a consequence of this theory that astronomers look for every possible clue to confirm their actual existence.[3] Since no radiation can escape from the black hole itself, we must search for effects produced outside the hole by its powerful gravitation. A black hole creates a kind of gravitational whirlpool in space that draws any nearby matter toward it. The combined centrifugal and gravitational forces cause the swirling particles to form a flat, gaseous accretion disk that can be millions of kilometers in diameter. Within the disk, frictional forces cause the individual bits of matter to spiral gradually inward until they are finally swallowed up by the black hole. The frictional forces at the inner edge of the disk heat the swirling gas to such high temperatures that as much as 80 percent of their thermal energy is radiated away as X-rays.

Astronomers have so far identified four X-ray sources which are likely candidates for black holes. The X-ray emission from these sources has the noisy, flaring character that would be expected from the intensely turbulent inner region of the accretion disk. In this inner region, any hot spots that develop will beam out exceptionally intense X-rays as they orbit around the hole; and since the orbital period, or time per revolution, is in the range of only a few thousandths of a second, X-ray astronomers are searching for trains of pulses that recur with such short periods. Observation of such X-ray pulses would be strong evidence for the existence of black holes and would reveal important properties of the gravitational "warping" of space close to the hole.

HIGH-ENERGY EXTRAGALACTIC SOURCES

The sources of X-rays, gamma rays, and neutrinos beyond our own galaxy include the hot gas that forms a diffuse medium in clusters of galaxies, and also several kinds of discrete sources such as the nuclei of unusually active galaxies.

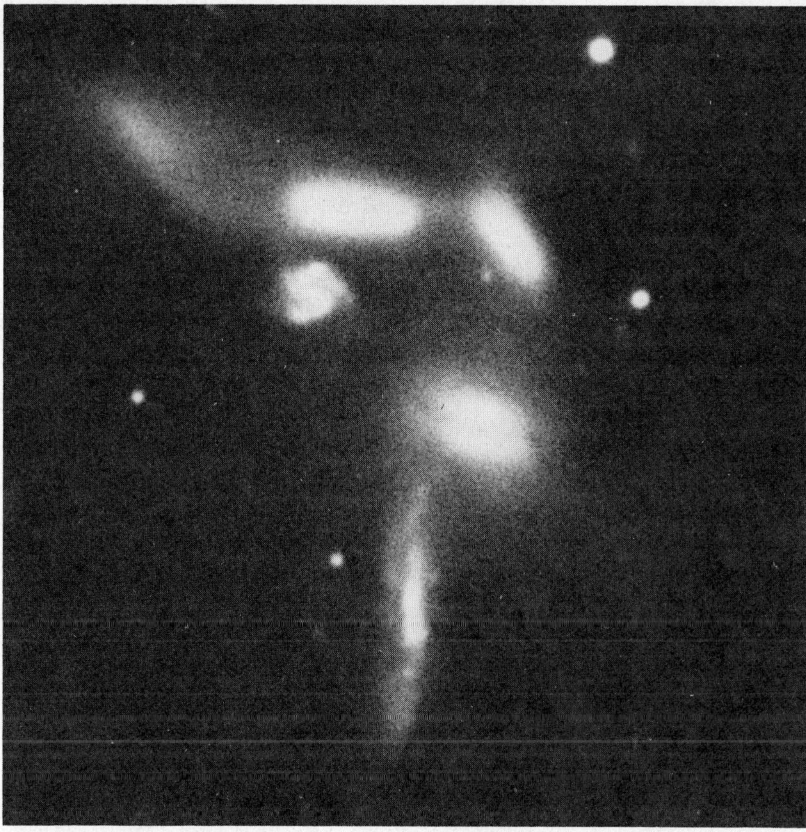

FIGURE 12 Group of five galaxies in Serpens with unusual connecting clouds. (Hale Observatories)

Cluster Gas

Most of the observed galaxies are found grouped with others in clusters of various sizes (Figure 12). There is also evidence for a clustering of clusters. Such superclusters or groupings represent the largest aggregations of matter in the universe. Our own Milky Way is a member of a local group consisting of 21 galaxies, while the largest galactic clusters, such as those in Virgo and in Coma, contain thousands of galaxies, all gravitationally bound together and traveling at high speeds within a roughly spherical volume that is not much larger than that of the local group.

The space between galaxies in such clusters appears to be filled with a diffuse gas that can be heated to very high temperatures in several different ways including gravitational infall. The X-rays emitted by this hot gas carry information about its composition, density, and temperature, and thus about the dynamics and evolution of the cluster as a whole. On the largest cosmic scale, there is some preliminary evidence that superclusters may also be enveloped in enormous clouds of hot gases emitting X-rays. As stated previously, early studies have tended to indicate that the average density of matter does not appear to be great enough to halt the universal expansion and thus to form a closed universe. However, the matter present in the diffuse gas clouds in clusters and superclusters may account for a great deal of mass that was previously invisible but that now can be seen in the light of its X-radiation.

Discrete Sources

A number of localized sources rival the richest galactic clusters in the intensity of their X-ray emission. These include the central regions, or nuclei, of certain galaxies in which cataclysmic events appear to be occurring—Seyfert galaxies and others which radiate strongly at radio frequencies—and the still more powerful starlike quasars and related BL-Lacertae objects. These sources were identified during the first major X-ray astronomy mission, HEAO-1, but their detailed study and the discovery of other such sources must await more sensitive instrumentation of the kind described in the last part of this section.

GAMMA-RAY AND NEUTRINO ASTRONOMY

Gamma-ray astronomy has developed more slowly than X-ray astronomy because the required instruments are intrinsically much larger and heavier. Even though the instruments that have been used so far in balloons and in small satellites have had only limited sensitivity, important results have already been achieved. Gamma-ray emission is generated primarily within the disk of our own galaxy, and it reveals the structure of the disk and the interactions of cosmic rays with the ambient interstellar medium. Some two dozen point-sources have been found so far, but few of these correspond to sources observed at other wavelengths. Exceptions are the pulsars in the Crab and Vela, plus several other pulsars. It comes as a great surprise that these sources generate so much of their energy as gamma-ray pulses.

Since many processes that generate gamma radiation will also generate neutrinos, these new observations suggest that neutrino astronomy holds great promise for further exciting developments. Neutrino astronomy divides itself into three energy regions: low energy neutrinos from stars (about 1 MeV or million electron volts); intermediate energy neutrinos (about 10 MeV) from gravitational collapse; and ultra high-energy neutrinos (greater than 1 GeV or billion electron volts). In the Homestake gold mine experiment to detect low-energy solar neutrinos not as many were counted as were expected, as mentioned before. Future experiments are being planned to resolve this discrepancy.[4]

It is expected that in the final stellar collapse to form a neutron star or a black hole more energy will be radiated in the form of neutrinos than anything else. Currently, experiments are operating in the Homestake gold mine and in the Soviet Caucasus that should see the intermediate-energy neutrinos from a collapse anywhere in our Galaxy. The ultra-high-energy neutrinos are produced whenever high-energy protons collide anywhere in the universe. There are both discrete sources such as quasars and pulsars and diffuse sources such as the collisions between cosmic rays and the interstellar gas. Limits on these sources are beginning to be obtained by the detectors used in the intermediate energy range; however, definitive measurements with angular resolution capable of picking up point neutrino sources will probably have to wait for *dumand*, the *D*eep *U*nderwater *M*uon *a*nd *N*eutrino *D*etector, for which detailed design studies are just beginning. *dumand* would consist of a cubic kilometer array of detectors located approximately six kilometers underwater.

One interesting spinoff of neutrino astronomy is that these large arrays of detectors may provide the best means to determine limits on proton lifetimes. This is a crucial question in the grand unification of theories to be mentioned later. Also, proton decay may even be related to the origin of matter in the Big Bang.

HYPERACTIVE NUCLEI OF GALAXIES

Quasars were recognized in 1963 as possibly the most energetic objects in the universe. So great are their luminosities, or total energy emission, that theorists speculated about the need for new physics to explain them. The hundreds of quasars and other quasarlike objects that have been discovered since 1963 appear to form a general class of violently active extragalactic objects in which the light of stars is overwhelmed a hundredfold or more by nonstellar light. As the catalog of quasars and other active galaxies (Seyfert galaxies, radio galaxies, etc.) has expanded, a sense of evolutionary continuity relating all of these objects has begun to emerge. The common link is believed to be a gravitational concentration of enormous masses of stars and stellar debris in the central regions, possibly into black holes having masses millions or even billions of times that of the sun.

Quasars are optically brilliant objects, and many are also powerful radio, infrared, and X-ray sources. Even though the energy source in such objects may be concentrated in a region only a light day across, the associated radio structure sometimes extends over several million light years. The power output of the energy source is as startling as the rapidity with which it can vary. As an example, in 1975 a certain quasar flared to unprecedented brilliance in both the optical and radio bands. Within a few weeks its brightness had risen to 10,000 times that of the entire Milky Way galaxy and just as quickly had subsided.

Since a black hole is intrinsically an absorber rather than an emitter of matter and energy, how can it produce the enormous power of a quasar? The mechanism is thought to be the capture by a massive black hole of the gaseous debris of stars that have collided in the densely packed central core of a galaxy, or of stars that have come so close to the black hole that gravitational tidal forces have ripped them apart. It would require the accretion of only one or two star masses per year for a black hole of a hundred million solar masses to power a quasar, and this process might continue for several million years before the reservoir of stellar material in the nucleus of the galaxy was exhausted. Most of the observed quasars and other powerful radio sources are very far away in distance and thus in time, and this fact can be interpreted as evidence that these immensely luminous objects were much more common when the universe was young, dying out within perhaps a billion years after they were originally formed.

Although the black-hole model of active galactic nuclei has the virtue of being a powerful energy-generating mechanism, this idea is not accepted by all theorists. Perhaps the best that can be said of it at present is that it is the least untenable theory yet offered. The next decade must bring far more sensitive and detailed observations in every part of the spectrum before we can hope to solve the quasar mystery. As an example of the problems involved, recent studies tentatively suggest that the quasar 3C-273, one of the most powerful radiation sources in the entire universe, produces a very large fraction of its energy in the form of energetic gamma rays. This is an astounding and baffling observation.

RADIO ASTRONOMY

The ability of the mirror of an optical telescope to separate two close images, or its resolving power, increases in proportion to its diameter and decreases in proportion to the wavelength of the radiation it is used to detect. Since the wavelength of celestial radio waves is thousands or more times longer than that of optical radiation, it would be totally impractical to build a radio antenna that could match the resolution of even a modest optical mirror. The radio astronomer's solution to this problem is an instrument called an interferometer, which consists of two or more radio telescopes that are separated from each other by a certain distance. When the signals from the two telescopes viewing the same source are combined, the resolution obtained matches that of a single telescope whose diameter equals the distance of separation. The separation does not, of course, affect the instrument's signal-gathering power, which depends only on the combined surface areas of the individual radio dishes.

Early interferometry was limited to telescopes located close enough together to be linked by cable or microwave communication. With the development of very accurate clocks, however, it has become possible to transfer time precisely and to combine, in a computer, signals recorded on tape at widely separated telescopes; a technique called "aperture synthesis." With only the size of the earth to limit telescope separation, very long baseline interferometry (VLBI) has pushed resolution in the radio spectrum to better than a thousandth of a second of arc, far beyond anything achieved optically (one second of arc subtends an angle that is 1/3,600 of a degree). Any problems in matching atomic clocks can be eliminated by transferring phase information between telescopes via real-time satellite communication, a method whose feasibility has been clearly demonstrated with a Canadian satellite.

Plans are now being developed for a steerable three-meter radio dish antenna to be carried aboard the space shuttle and to be used in combination with one or more ground-based telescopes such as the dish located in Arecibo, Puerto Rico. It will provide continuously variable baselines, a great advantage for aperture synthesis.

The taped-signal technique has allowed astronomers to combine VLBI arrays of several existing telescopes to gain more detail in the synthesized data. The gain is related to the possible number of pairings of telescopes: two provide a single pair, three make three pairs, four make six, and five make ten combinations. A more widespread geographical distribution permits a fuller map of the sky, and operations are planned for transcontinental and eventually transoceanic combinations stretching from Massachusetts to Hawaii and south into Texas and Mexico.

INFRARED ASTRONOMY

The infrared astronomer is interested in objects with temperatures less than about 3,000°K, the sun, by example, has a surface temperature of 5,700°K. In the near-infrared region of the spectrum we see very cool stars and the "tail" of the emission from ordinary stars. The stellar emission decreases toward longer wavelengths, and the infrared contribution comes mainly from dust particles heated by nearby stars.

Within the galaxy, the major sources of infrared radiation are regions of ionized hydrogen of great extent and optical brightness: dense molecular clouds, dark nebulae where starlight is obscured by dust that reradiates infrared, and young stars surrounded by dust shells. All of these objects are associated with stellar birth. At the other extreme of stellar evolution, dying stars may eject large amounts of dust and become very bright in the infrared. The mass loss may take the form of a slowly growing dust shell around a cool star or of an explosive ejection accompanying the outburst of a nova or supernova. Such reprocessing of material to the interstellar medium has an important role in the evolution of the galaxy.

The center of our galaxy is especially bright in the infrared, comparable, in fact, to the visible luminosity of the entire galaxy. Thermal emission by dust must be the source of this infrared radiation, but the heating process is not understood. Many distant galaxies radiate infrared from their nuclear regions at 10,000 times the luminosity of our galactic center. The mechanism involved must be closely connected with the total process of hyperactivity in galactic nuclei, but it is again not clearly understood. In fact most quasars, Seyferts, and other extragalactic objects radiate more energy in the infrared than in any other part of the electromagnetic spectrum. This means that infrared astronomy may eventually be one of the most effective ways to study these exciting objects.

NEW INSTRUMENTS

Space Telescope

Scheduled for launch on the Space Shuttle in 1983, the space telescope will offer such fundamental advantages over ground-based instruments that it will become the main tool for the deep-space studies that are essential to cosmology. The power of a ground-based telescope to detect and resolve very faint objects is ultimately limited, not by the quality of its optics, but rather by the shimmering of the earth's atmosphere and the background light of the atmosphere's airglow. A location in space makes possible the sharpest imagery and also extends the range of possible observations into the ultraviolet and infrared regions of the spectrum. The space telescope will provide a resolution of about 0.1 to 0.05 of an arc-second, and will detect objects as faint as the twenty-eighth or twenty-ninth magnitude—at least 100 times dimmer than can be observed from the earth's surface. The high cost of placing the 2.4-meter mirror and all of its accessory equipment in space requires that the space telescope be serviced and its instruments updated over a period of 15 years, and perhaps even a decade. This concept has become feasible only with the advent of the space shuttle.

Ground-Based Telescopes

The largest ground-based telescopes now operating are the five-meter reflector on Mount Palomar and the six-meter reflector on Mt. Semirodnika in the Soviet Union. It is doubtful that larger mirrors could be successfully cast or that such mirrors could be mounted to operate without distortion. Instruments consisting of multiple mirrors are not limited by these constraints. Now nearing completion atop Arizona's Mount Hopkins is the first large Multiple Mirror Telescope, the prototype of the next generation of very large optical telescopes. This instrument will use six 1.8-meter mirrors to match the light-gathering power of a 4.4-meter telescope at less than one-third the cost of a conventional mirror and its dome housing. Tests have demonstrated that a laser sensing system can be used to bring all six mirrors into a common focus. Designs are also being studied for multiple mirror telescopes with equivalent apertures up to 25 m. A direct scale-up of a monolithic mirror of such size, even if it were technically feasible, would cost perhaps $2 billion and require some 50 years to complete.

Radio Telescopes

A very large array (VLA) is now nearing completion on the plains of San Augustin near Soccoro, New Mexico (Figure 13). It will be the world's largest aperture-synthesis radio interferometer. With 19 of the 27 antennae now complete, along with about one-third of the trackage and electronics, the VLA already surpasses in sensitivity any radio astronomy facility in the world. The full array of 27 antennae will be ready for operation in 1981. Its agenda will include quasars, radio galaxies, black holes, and stellar and cosmic evolution.

It is also important to move toward shorter radio wavelengths to gain resolving power. A very successful millimeter-wave antenna serves the Five Colleges Radio Astronomy Observatory at Amherst, Massachusetts. Its efficiency will be further increased by the use of cryogenically cooled detectors. At the Owens Valley Radio Observatory of the California Institute of Technology, a mirror 10 meters in diameter has attained a resolution of 27 arc seconds at a wavelength of 1.3 millimeters. This excellent performance was achieved through the development of some remarkable mirror-fabrication techniques.

Infrared Telescopes

As noted earlier, most of the infrared spectrum must be observed from aircraft, balloons, rockets, or satellites. The technology of cryogenically

FIGURE 13 Very Large Array radio telescope being built near Socorro, N. Mex., by the National Radio Astronomy Observatory under contract with the National Science Foundation.

cooled detectors and mirrors is very difficult to perfect, and small cooled telescopes have been used in only a few rocket flights. In 1981, a small helium-cooled telescope will be launched aboard Spacelab 2. Also planned for the same year is the Infrared Astronomical Satellite (IRAS) a joint project of the Netherlands, the United Kingdom, and the United States. It will carry a larger (60 centimeters) cooled telescope and detectors to cover a large fraction of the infrared spectrum. It should be able to detect a photostellar region like the molecular cloud in Orion out to distances of about 10,000 light years and to provide a catalog of perhaps a million infrared sources.

X-Ray and Gamma-Ray Instruments

The first High Energy Astronomical Observatory (HEAO-1) has been in orbit since mid-1977 with an array of about 80 square feet of X-ray detectors. The catalog of X-ray sources compiled through HEAO-1 will exceed 1,000, and the new observatories now being planned may extend this catalog to 100,000 or more. The recently launched HEAO-2 (Einstein Observatory) satellite is now carrying the first true X-ray telescope aboard a satellite. Its 60-centimeter diameter nested mirror has special instrumentation for refined wavelength measurements. Since it will be possible to perform only a small fraction of the high-priority observations suited to this satellite during its expected life of two years, X-ray astronomers are already well into design studies of an Advanced X-ray Astronomy Facility (AXAF) for the mid-1980's. AXAF will have a 1.2-meter X-ray telescope launched as a free-flyer from the space shuttle, and its lifetime is expected to be 10–20 years.

A group of instruments has been selected to fly on a Gamma Ray Observatory (GRO) in 1984. The payload of about 15,000 pounds will permit gamma-ray astronomers to develop instruments that are 10 times as sensitive as those of the first generation. The prospect is that GRO will make gamma-ray astronomy a full-fledged partner of the other new astronomies.

CONDENSED MATTER

Condensed matter science is concerned with the electronic and atomic properties of the solid and liquid phases of matter and with the ways in which these substances respond to mechanical forces, to heat, to electric and magnetic fields, and to radiation. The subject area is extremely broad, ranging from questions of a chemical nature concerning natural or synthetic structures, to such topics as the distortions of electronic and atomic symmetry that result in the properties of magnetism and superconductivity. Previous fundamental advances in the field have led to the discovery of the transistor, solid-state masers and lasers, high-temperature superconductors, solar cells, superconducting junction technology for ultrafast computers, and many other important solid-state devices.

The accumulated knowledge of past years has resulted in our present very detailed understanding of simple crystalline solids and of liquids. Indeed, the electronic properties of many crystalline solids are now so well understood that the properties themselves can be used to elucidate the nature of the bonding between the atoms in the material.

All solids and liquids can exist in different phases, that is, different structural arrangements of their atoms and electrons. The most significant conceptual breakthrough in this field during the past 10 years has undoubtedly been the vastly improved understanding of phase transitions. Examples of such transitions are the change from a magnetic to a nonmagnetic phase in solids, and the evolution of a gas phase into a liquid. The key to this achievement has been precise description of the relationships between the scales of length, energy, and time that are characteristic of the particular change in phase. The basic ideas developed in this work are also applicable to other phenomena, for example, to the electrical conductivity of noncrystalline materials.

The past 10 years have also seen many important experimental discoveries in condensed matter science. Among the most significant of these have been the discovery of the superfluid (frictionless-flow) phase of the light isotope of helium, ^3He, and the discovery of charge-density waves in solids. A charge-density wave is a periodic spatial variation of the electron density in a crystal that is different from the spatial structure of the crystal lattice itself—a sort of electron-supercrystal. The possible uses of this new phase of matter are now being studied.

Present work in the field is proceeding on a broad front. There is renewed interest in the study of disordered or amorphous materials and glasses, an area where many fundamental questions concerning electronic and atomic structure remain to be answered. The study of surfaces has also drawn intense interest, in part because of the new experimental techniques that ultrahigh vacuum technology has recently made possible. A good deal of progress has been made in understanding the properties of defects and of grain boundaries in crystalline solids. In addition, new methods of growing crystals have been developed that allow a solid to be built up essentially one atomic or molecular layer at a time. Studies of novel materials have resulted in the observation of complex and fascinating behavior, and some of these new materials have already begun to find practical applications.

In all of this work, progress has been linked closely with the development of more powerful radiation sources and more sensitive detection instruments. These include sources of synchrotron radiation in the ultraviolet and X-ray regions of the electromagnetic spectrum, high-resolution electron microscopes, and intense sources of neutrons and of energetic ions, as described on page 189.

AMORPHOUS SOLIDS

The remarkable progress that has occurred during the past 50 years in the understanding and practical application of crystalline materials has been based on the special properties of their regularly repeated (periodic) structure. Although our understanding of the far more complex amorphous materials has been much slower to develop, the past decade has seen an accelerating pace of discovery and a growing awareness of the exceptional importance of amorphous materials for both scientific and technological purposes. The significance of contributions in this field was recognized by the 1977 Nobel Prize in physics.

The arrangement of the nearest neighbors of any particular atom in an amorphous solid is often quite similar to that in the corresponding crystalline material. This regularity does not extend out to more than a few atomic separations, and thus overall periodicity is lost. In contrast to crystals, which have both short- and long-range order, amorphous solids have only short-range order. One consequence of this difference is the fact that many of the limitations on composition and on atomic arrangement that hold for crystals do not apply to amorphous materials, with the result that an enormous range of such materials can be prepared. Thus new material properties become accessible that are uniquely characteristic of amorphous materials, new kinds of chemistry occur, and an entirely different theoretical approach is required for understanding these materials.

The next decade may well see a revolution in materials science and its applications, as work with amorphous solids begins to come to fruition. Three examples may help to illustrate some of these possibilities: Metallic glasses now commercially available have magnetic properties that make them very attractive as transformer-core materials; they also have unusually high resistance to corrosion and radiation damage, or unusually high ductility (see p. 305). Very low loss glass light-guides are now being manufactured that are suitable for communicating by light waves. Amorphous semiconducting films are now being prepared on an experimental basis, which may lead to economically feasible generation of electricity from sunlight.

SURFACES

The study of surfaces involves determination of their atomic and electronic structure and energy levels, the strengths of their chemical bonds, and the interrelationships among these. The goals of such studies include an understanding of surface chemical reactions, of crystal growth, and of the interface (surface-to-surface) phenomena that occur in electron devices. The progress made in this field during the past decade has largely depended upon advances in ultrahigh vacuum technology. This is because clean surfaces are very reactive chemically and thus require a high-vacuum environment if rapid contamination is to be avoided.

Surface experiments typically consist of directing a beam of electrons, ultraviolet radiation, X-rays, atoms, or ions at the surface; then measuring the energies and angular distributions of the particles that are reflected (scattered) back from the surface or knocked out of it. The sample under study can usually be moved to different locations within the vacuum chamber so that it can be prepared, tested, and then studied in an efficient sequence of operations that minimizes contamination.

Theoretical modeling of surfaces has advanced in concert with the development of more sensitive experimental techniques. Recent successes include the deduction of detailed atomic configurations from the scattering of low-energy electrons, and insight into chemical bonding and atomic geometry from the radiation-induced emission of surface electrons. These and several other new experimental techniques are still in their early stages but are now developing rapidly. It is possible to foresee major advances in the understanding of surface processes as the accumulating experimental results begin to reveal the systematic underlying trends. The knowledge gained in this work will clearly have direct application to such complex surface phenomena as catalysis, corrosion, lubrication, and crystal growth, as well as to the interface phenomena in electronic devices and in electrochemical processes.

NOVEL MATERIALS

The study of such relatively simple materials as metals, semiconductors, and magnets was the central focus of condensed matter research during its early years. With the present understanding of these materials, there has been increasing emphasis during the past decade on the search for novel materials that exhibit properties of unusual scientific or practical interest. Examples of such materials are those in which the structure is built up of successive layers or of linear chains of atoms. There are also new materials that consist of rigid clusters of atoms, or of two different groups of atoms in which one of the groups can move relatively freely through the stable lattice formed by the second group. A brief outline of some of the more promising of these new materials is given in the following paragraphs.

Layered Materials

The class of materials known as graphite intercalation compounds consists of layers of carbon atoms that have sandwiched (intercalated) between them alternating layers of other kinds of atoms or molecules. Interest in these compounds was greatly increased by the recent discovery of certain forms that rival copper in their ability to conduct electricity. With a better understanding of the manner in which the intercalated atoms are taken up and ordered between the graphite layers, it may be possible to develop economical new materials of light weight that can perform some of the same functions as metals.

Earlier work based on the use of alkali atoms as the alternate layers in an intercalated structure has resulted in the development of a new class of batteries that will soon be available commercially. In addition to this practical application, these materials were found to have remarkable properties even when the intercalated atoms were not present in the structure. For example, at low temperatures the mutual interaction of the electrons within the material resulted in the formation of spatially periodic arrays of electrons called charge-density waves. In a sense, the electrons formed a crystal structure of their own that was not related in any simple way to the underlying crystal structure of the material itself. In fact, there is some preliminary evidence that this electron crystal can move as a whole within the material under the influence of an electric field.

Another example of a layered material is a compound in which layers of alumina (aluminum oxide) are interleaved with layers of alkali ions that occupy only a fraction of the available sites within the layer. As a consequence of this fractional occupancy, the ions are relatively free to move within the alkali layers, and the ionic motion is rapid enough to allow these materials to be used as solid electrolytes (charge carriers). The study of such materials is a necessary step toward the eventual goal of lightweight and reliable batteries made up completely of solids.

Chains and Clusters

A different approach to the search for new metal-like materials consists of forming organic solids in which chains of molecules that accept electrons are alternated with chains that donate electrons. Work in this area has led to the synthesis of organic solids that are good conductors of electricity. The chain character of these compounds gives rise to a number of different forms of electronic instability, the study of which has resulted in important theoretical insights into the nature of such materials.

Work during the past several years has resulted in the synthesis of a very interesting series of cluster compounds, so called because the electrons within the material are confined to certain clusters of atoms. Such materials often exhibit superconductivity (the flow of electric current without resistance) at higher than customary temperatures. However, when magnetic ions are interspersed within the materials, the incompatibility of magnetism and superconductivity becomes graphically apparent, with the material alternating between superconducting and magnetic phases as each group of electrons in turn comes to dominate the behavior. Because of their potential importance for electrical power transmission and magnetic confinement in fusion reactors, study of these materials will be valuable in the continuing quest to raise the temperature at which superconductivity begins.

There are many other examples of important new materials that might be mentioned here, including the development of metals that can absorb and evolve hydrogen efficiently. In any case, the search for new materials with novel properties is certain to continue during the coming years, perhaps with special emphasis upon organic compounds and upon mixtures of organic and inorganic materials.

ATOMIC AND MOLECULAR ENGINEERING

Traditional methods of crystal growth have primarily involved melting and recrystallizing solids at high temperature or growth of crystals from solution. At present, several new techniques are being developed in which thin films are grown from a controllable beam of atoms, ions, or molecules so that new materials can essentially be engineered a layer at a time. Because of the major advances in analytical tools, primarily electron microscopes, it is now possible to study defects, diffusion, and other microscopic phenomena in much greater detail.

One of the more promising methods for growth of new materials and structures is molecular beam epitaxy. With this vapor-growth technique, single-crystal films can be grown, layer by layer, on an underlying crystalline substrate. Controlled deposition makes it possible to vary the chemical composition in predetermined ways and thus to fabricate unique structures with electrical and optical properties not normally found in nature. To date this technique has been applied mainly to the controlled modification of layered semiconductor structures that have important technological applications. However, one can envision superconducting, metallic, and magnetic structures grown in this fashion. For example, when superconductivity occurs at an elevated temperature in a material, the superconducting phase is usually less stable than other phases, and the material is more difficult to grow. However, it has been demonstrated that epitaxy facilitates the growth of the desired phase so that it appears possible that such unique metastable compounds may be fabricated using this new growth technique.

Along somewhat different lines, the study of diffusion phenomena at surface interfaces has led to the discovery of solid phase compound formation. The diffusion of metals into semiconductors, for example, very often results in compound growth proceeding into the semiconductor. Solid-state epitaxy is thus a new method for producing desired structures at interfaces. Interfacial interactions and diffusion are also important in the oxidation process. Greater understanding of these processes is of considerable technological interest for a variety of metal-oxide-semiconductor devices.

The properties of amorphous semiconductors have also received a good deal of attention during the past decade. Recently, vapor-growth techniques for amorphous silicon and germanium have been developed that allow the controlled modification of these materials in a manner somewhat analogous to their crystalline counterparts. The detailed physical mechanisms responsible for the properties of the amorphous materials are not yet known. As noted above, these materials are of much scientific and technological interest.

Another new technique of atomic engineering is ion implantation. This method involves implanting foreign ions into a host material to a predetermined depth, which depends on the kind of ion used and its energy. This process also leads to structural damage and hence to a disordered layer in the material. This technique has already found uses in the "doping" of semiconductors with foreign atoms and in the production of metastable and disordered structures and of new alloys.

Very thin films, of a thickness comparable to atomic dimensions, exhibit phenomena that are unique to these two-dimensional systems. As an example, ultra-thin metal films serve as an important testing ground for our understanding of metallic conductivity. In addition, thin (two molecules thick) liquid-crystal films are being used to study the phenomenon of melting in systems of two dimensions, while incomplete single layers of inert gases on crystalline substances are also being used to probe these phenomena. Ultra-thin magnetic films also have unique properties. Better control of the growth processes of these thin-film structures promises further advances in these areas.

The work described above represents only a few examples of this rapidly expanding field, which is important to such areas of technology as microfabrication of integrated circuits, development of high-strength materials, and exploration of new techniques for the generation of energy.

SYSTEMS FAR FROM EQUILIBRIUM

The different phases of matter—for example, liquid, crystalline, magnetic, superconducting—are usually studied in systems that are in or near states of thermal equilibrium with their surroundings. It is also interesting to study the behavior of systems in states far from thermal equilibrium, since such situations arise quite commonly under natural conditions. Examples are the response of a gas to a sudden increase in pressure that will eventually cause it to condense to the liquid state, and the evolution of convective-flow patterns in a liquid layer strongly heated from one side. The latter system will eventually reach a state of turbulent flow, the detailed characterization of which remains an important unsolved problem in hydrodynamics. In many cases, the behavior of a system can be drastically altered by making only a small change in some external condition. The subsequent evolution of the system is usually quite complex and has been difficult to predict.

Examples of such instabilities occur not only in phase transitions and hydrodynamic flow but also in the dynamics of chemical reactions and in the high-temperature plasmas used in laboratory thermonuclear fusion devices. The present studies of instabilities in systems far from equilibrium bring together the disciplines of condensed matter physics, chemistry, nonlinear mechanics, and hydrodynamics in quite novel ways. The next five years will see a major new effort at understanding these problems, aided by the theoretical methods developed in the study of phase transitions.

FUTURE DIRECTIONS

The studies of amorphous materials, surfaces, novel materials, atomic engineering techniques, and systems far from equilibrium described in the previous sections are among those that are easily identified as ripe for progress during the next five years. However, these selected topics provide only a sampling of the much broader range of past achievements and future research opportunities that exist in condensed matter science. The exact directions that the field will take during the next five years are of course impossible to predict, but rapid progress can reasonably be expected in both scientific understanding and technological application. The field is marked by increasing cross-fertilization of ideas among the conventional disciplines of chemistry, solid-state physics, materials science, and metallurgy, and this trend can be expected to intensify as the work evolves toward studies of more exotic materials and more complex phenomena.

Condensed matter research shares with the other fields of study on the structure of matter a continuing development toward instrumentation of greater power and sensitivity, and thus of greater cost. An inevitable consequence of this development is the trend toward centralization of major research facilities at fewer laboratories, and this in turn has led to important changes in the ways in which the work of the field is carried out.

MOLECULAR AND ATOMIC STRUCTURE

Since the familiar macroscopic forms of matter are made up of molecules, and molecules are made up of atoms, studies of these fundamental building blocks of matter are among the most important in all of physical science. The entire field of chemistry is concerned in one way or another with the structure of molecules. Chemical studies provided the earliest information about molecular form, and the continuing evidence provided by such studies remains very important. Subsequent to the classic experiments of Rutherford establishing the model of the nuclear atom, much of the information about the structure of atoms was derived from observations of the characteristic patterns, or spectra, of visible light that are emitted by different species of atoms. Such observations marked the beginnings of spectroscopy, which continues to this day as the principal experimental technique for studying the smallest particles of matter.

Since spectroscopic studies will form an important part of the content of this section, it will be useful to recapitulate briefly a few fundamental facts of spectroscopy:

- Atoms and molecules can exist in any of a number of discrete (quantized) states, each of which represents a different energy level of the particular system in question.
- These states correspond either to rotational motion of the system, or to vibrational motion, or to different configurations of the electron cloud around an atom or a molecule.
- Changes from one state to another are called transitions, and each transition is accompanied by either the absorption or emission of electromagnetic radiation of a precise wavelength (or frequency).
- The absorption and emission of radiation occurs in distinctive patterns (spectra) that can often serve to identify both the specific kinds of atoms or molecules that are involved and the particular kinds of transitions that have occurred.

MOLECULAR SPECTROSCOPY WITH SYNCHROTRON RADIATION

The most important recent advances in the spectroscopy of molecules in the ultraviolet and X-ray regions have resulted from the use of new synchrotron radiation sources. Electron synchrotrons and storage rings provide a variable-frequency X-ray source whose intensity is about a million times that of a conventional X-ray tube.

Fluorescence Spectroscopy

The variable-frequency feature of synchrotron radiation is particularly helpful in fluorescence spectroscopy, which makes use of the radiation that is re-emitted after absorption of X-rays by heavy elements in the molecule. This technique can distinguish the fluorescent radiation from different elements, thereby greatly increasing the sensitivity and reliability of detecting elements that are present in extremely small proportions. Also, since the synchrotron radiation is produced in precisely repeated, very sharp pulses, it can be used to measure accurately the lifetimes of excited states as short as a nanosecond (one billionth of a second).

EXAFS *Studies*

The technique known as EXAFS (extended X-ray absorption fine structure) has been developed rapidly into a useful technique for molecular-structure research because of the availability of synchrotron radiation. EXAFS studies involve the absorption of an X-ray photon that ejects an electron from the absorbing atom. The details of this process depend on the local environment of the absorbing atom in a molecule and thus provide information about short-range molecular structure, even when the structure is not highly ordered. An example is the identification of the elements surrounding a metal atom in a noncrystalline biological system. Very small amounts of metal play a central role in the function of important biomolecules (for example, iron in hemoglobin). The capabilities of EXAFS also hold out great promise for a deeper understanding of catalysts, those substances that greatly accelerate chemical and biological functions without appearing to react themselves. The role of catalysts in chemical reactions is of course vitally important for industrial chemistry as well as biochemistry.

MOLECULAR SPECTROSCOPY BY MAGNETIC RESONANCE METHODS

Nuclear Magnetic Resonance

Nuclear magnetic resonance (NMR) methods make use of the fact that the nuclei of many kinds of atoms are small magnets that can take on any of several different orientations in the presence of an applied magnetic field. Transitions from one orientation to another can be induced by radio waves, and the particular frequencies at which such resonance transitions occur serve to identify important elements of molecular structure.

As an example, hydrogen atoms are a significant structural element in organic substances, and the frequency at which these atoms reorient or flip is quite sensitive to the arrangment of their neighboring atoms in a molecule. After some years of experimental study, an enormous body of NMR information has been collected that correlates specific NMR frequencies with particular structural units of organic molecules. As a result, it is now possible to identify these units quickly in other molecules simply by seeing their characteristic frequency "signatures" in an NMR spectrum.

Carbon is also an essential element in organic substances, but its common isotope, carbon-12, has no magnetic moment and thus does not resonate. There is, however, a rare isotope, carbon-13, which does have a magnetic moment and can therefore be used to identify structural elements by the NMR technique. Similar methods can be used to examine the surroundings of other kinds of atoms through the use of suitable isotopes of fluorine, phosphorus, nitrogen, and about 100 other isotopes. Structural problems that once required years of study by chemists can now be disposed of in an afternoon with the aid of an NMR spectrometer.

Because of its particular value for organic chemistry, NMR plays an important role in illuminating biological function. It has been used, for example, to elicit the role played by magnesium atoms in chlorophyll and has also provided valuable insights into the behavior of enzymes, the structure of DNA, RNA and hemoglobin, and so on. NMR is also a powerful tool for monitoring the synthesis of organic compounds; this may in fact be its largest single use.

Electron Paramagnetic Resonance

Electron paramagnetic resonance (EPR) is a spectroscopic technique that closely parallels NMR because electrons are also tiny magnets that can take on either of two orientations in the presence of an applied magnetic field. Molecules that contain an extra electron (i.e., not paired with another electron) can therefore exist in either of two energy states in such a field. The energy difference between the two states is much greater than that of atomic nuclei because of the much stronger magnetic moment of the electron, which means that the resonance frequency is correspondingly higher (in the frequency range of microwave radar).

The EPR method has been widely applied to problems in physics, chemistry, and biology. One of the most important areas of application is called spin-labeling. Although molecules containing an unpaired electron are usually very reactive, those known as free radicals are stable. When a stable free radical is attached to another molecule, the result is a spin-labelled system that is detectable by EPR. A spin-label attached to a component of a cell membrane, for example, provides a direct probe of the membrane's structure. From the characteristics of the EPR signal, it is possible to deduce where in the membrane the spin-label is located, its rate of migration, and much additional information. The use of EPR with spin-labels has probably been the most informative nondestructive technique applied in recent times to the study of the structure and function of membranes in living organisms.

There are many important biological and biochemical processes during which free radicals are formed. An example is green plant photosynthesis, the process by which the energy of sunlight is used to convert carbon dioxide and water into the carbohydrates, proteins, and fats upon which all animal life, including that of man, depends. EPR studies have been used to demonstrate that the primary light-conversion step in photosynthesis involves a special pair of chlorophyll molecules. As a result, there has been considerable success in replicating natural photoreactive chlorophyll, and synthetic chlorophyll special pairs are now being examined for their possible application to solar energy conversion.

LASERS AND MOLECULAR STRUCTURE

A laser is a very intense source of light of a single wavelength.[5] This is made possible by the fact that atoms in the same excited state can be stimulated by light to emit light of the same wavelength. Laser light makes it possible to distinguish effects associated with very slight differences between molecules; for example, the effects of replacing one atom in a molecule with a different isotope of the same element. This is not only a powerful tool for high-precision structural studies but also a possible practical technique for separating the different isotopes of a single element.

Picosecond Spectroscopy

Many events of great scientific importance take place in an incredibly short period of time. The fastest ordinary chemical reactions occur in about 10^{-12} seconds, or a picosecond. The lifetimes of many intermediates in chemical reactions are also of this magnitude, and the rate at which energy is transferred from one part of a molecule to another is again on the same time scale or even faster. Thus the picosecond spectroscopy made possible by the laser has become an extremely important experimental technique. Experiments can be conducted with intense laser pulses only a few tenths of a picosecond long, so that one can record a flash spectrum of the molecule, and from that spectrum construct a picture of the structural changes taking place during these very brief periods in its life. This gives entirely new insights into chemical processes.

Perhaps the most interesting work of this kind concerns molecules of biological interest. For example, it has long been a mystery how the energy is transferred within molecules of chlorophyll to bring about the process of photosynthesis. When light is absorbed by chlorophyll, a small fraction of it is re-emitted (fluorescence) as a result of the energy-transfer mechanism, and this makes it possible to study the mechanism by means of picosecond spectroscopy. Although the interpretation of the results in hand is not yet unambiguous, the data have helped to define the problems of energy transfer more clearly, and the field is now being vigorously explored.

LASERS AND ATOMIC STRUCTURE

As previously stated, lasers can produce light of unprecedented intensity, directionality, spectral purity, and coherence; and certain kinds of lasers can be tuned over a broad range of wavelengths. These properties make lasers very powerful tools for probing the intricate structure of atoms. The extreme accuracy attainable in such studies also serves to test the validity of fundamental physical principles. In the following paragraphs we give only a few examples of the achievements of recent laser experiments.

Giant Atoms

Lasers are now being used to study atoms in which one of the electrons has been raised to an extremely high excited state and thus moves in a very large orbit. Because the volume occupied by these giant atoms can be a million or more times that of the unexcited atom, they exhibit vastly different properties. Such giant atoms occur naturally in regions of space where stars are believed to be forming, and also in such important laboratory environments as those associated with fusion experiments. Knowledge of the structure of giant atoms is thus expected to have both basic and applied significance.

Tests of Physical Principles

The precision of laser experimentation has already resulted in considerably more accurate determinations of two fundamental constants of nature: the speed of light, and the characteristic energy with which electrons are bound to atomic nuclei (the Rydberg constant). There are also in progress stringent tests of the time-dilatation effect (the slowing down of clocks) of the special theory of relativity, and of the basic electromagnetic theory called quantum electrodynamics (QED). In addition, a particularly interesting synthesis of atomic physics and elementary particle physics is now occurring in experiments at several laboratories, which are designed to test the landmark theory that posits the fundamental unity of the weak and electromagnetic interactions (see p. 187). The evidence from high-energy physics experiments is strongly supportive of this unified theory, but so far the results from the "table-top" atomic physics experiments now in progress have been indecisive.

TRAPPED IONS AND ELECTRONS

Since the early recognition of the atomic nature of matter, scientists have sought ways to isolate individual atomic and subatomic particles for detailed study. The problems involved in implementing such confinement schemes are formidable, but techniques that have been evolved during the past two decades have now led to practical realization of the goal. One of the new techniques makes use of a combination of electric and magnetic fields to confine either electrons or ions for periods as long as several days.

By studying the response of a single electron successfully confined for many hours to changes in the confining electromagnetic field, it has been possible to carry out the most precise measurement ever made of an intrinsic property of an elementary particle (its magnetic moment), and thereby to provide the most rigorous testing yet achieved of the theory that describes the electromagnetic interactions of charged particles—quantum electrodynamics.

The recent development of a novel method of radiative cooling holds promise of extending such measurements to even greater precision. One can also foresee the use of such trapped, low-temperature ions in practical applications such as frequency standards, where the frequency stability could be expected to exceed that of present devices by a factor of 10 or more to perhaps 1 part in 10^{17}.

EXOTIC ATOMS

Exotic atoms are atoms in which one, or more, of the usual constituents has been replaced by a particle not ordinarily found in matter because it is unstable. Since the lifetimes of such particles are at most only a few millionths of a second, they must be created by particle accelerators, then quickly trapped to form the desired exotic atoms. In this work the techniques of high-energy physics are combined with the precision spectroscopy of atomic physics to yield important information about the basic forces of nature.

Several kinds of exotic atoms have already been studied, and other kinds are possible. In what follows, however, we limit ourselves to the description of only one example.

Muonic Atoms

The particle called the muon appears to be exactly like an electron in every respect except its mass, which is about 200 times greater. When a muon replaces an electron to form a muonic atom, the greater mass forces it to take up an orbit in the atom that is about 200 times smaller. This has two important consequences: (1) Since the muon is moving in a region where it is affected almost solely by the positive electric field of the nucleus, its observed behavior can be compared with the very precise calculations that are possible as another important means of testing the validity of basic electromagnetic theory. (2) In heavier atoms, the muon's orbit becomes so small that it actually spends a large fraction of its time within the nucleus itself, and in so doing it probes both the size of the nucleus and the distribution of its electric charge.

FUTURE DIRECTIONS

It was not so long ago that many scientists considered atomic and molecular spectroscopy to be a closed subject, still valuable for its many practical applications, but no longer of direct scientific interest in itself. That this judgment was premature should be abundantly evident from the foregoing descriptions. In fact, the important developments in technique and instrumentation in recent times are only beginning to be exploited; and if the new facilities that will be needed to capitalize upon the exciting research opportunities can be developed, the prospect seems excellent for major advances in our understanding of atomic and molecular structure, and for important new practical applications. (The plans and need for new facilities are described in the Research Instruments section of this chapter.)

To conclude this section on molecular and atomic structure, we speculate briefly about two general lines of research that represent goals to be aimed for or limits on what may ultimately be achievable in this field of study.

Imitations of Life

It is instructive to compare some of the physical and chemical processes carried out by living organisms with those of modern technology. As an example, living cells are able to synthesize incredibly complex molecular structures at ordinary temperatures and pressures, and starting with very simple reagents. In contrast, chemical technology typically employs high temperatures and pressures, and violently reactive agents. As illustrations, compare the manufacture of rubber by a rubber tree with the corresponding brute-force industrial processes; or the heroic commercial methods required to fix atmospheric nitrogen with those of nitrogen-fixing bacteria. One can speculate that a more complete understanding of these biological processes at the molecular level may eventually enable us to mimic chemical processes in living organisms and thus to reap the benefits of their high efficiency and great synthetic power.

Molecular Electronics

In an even more speculative vein, there are fascinating long-range possibilities in what has been called molecular electronics. The ultimate goal here would be to synthesize molecules that would act as individual circuit elements, such as conductors, resistors, capacitors, etc., and then to combine these elements into amplifiers, memory devices, and so on. Such devices would serve the same purposes as present-day electronic devices but would be very much smaller in size and in power consumption, and very much faster in operation.

The example of electrical conductors can serve to illustrate these ideas. The chemical processes that constitute the metabolic activity of living cells involve the transport or movement of electrons as an essential feature. However, nature does not use small bits of metal wire to conduct electrons from one place to another. Instead, there is a carefully graded series of electron-transport proteins within the cell to serve this purpose. (Two such proteins are cytochrome and ferredoxin.) With a more complete understanding of how these proteins carry out their electron-transport tasks, it may become possible to develop practical electron conductors on the molecular level.

RESEARCH INSTRUMENTS FOR CONDENSED MATTER AND MOLECULAR AND ATOMIC STRUCTURE

An important theme that has run through this discussion of the sciences of condensed matter and molecular and atomic structure is the way in which research has been furthered by the development of new and more advanced instruments. The use of more powerful radiation sources and more sensitive detection instruments leads to increased understanding of the structure and atomic dynamics of molecules and condensed matter.

X-rays have provided the classical technique for studying the structure of systems that have atomic spacings characteristic of solids, liquids, and molecules, because these spacings are of a size comparable to X-ray wavelengths. X-ray measurements detect the positions of heavy atoms and also, along with measurements using ultraviolet radiation, yield valuable information on electron structure. In more recent years, neutrons have been used for similar purposes because they, too, have a comparable wavelength and provide a particularly sensitive means for determining the positions and momenta of the lightest atoms, such as those found in organic materials.

Circular electron accelerators (synchrotrons) and storage rings, such as those designed for high-energy physics research, are the most intense sources of radiation in the ultraviolet and X-ray regions because electrons traveling at high speeds in a curved path emit a broad spectrum of electromagnetic radiation.

Proton accelerators may be used to produce an intense source of neutrons since the collision of energetic protons with a target of heavy material such as uranium produces a burst of neutrons through a process called "spallation." A pulsed spallation source of this kind is now under construction in England, and a more modest version is being constructed in the United States.

On a smaller instrumental scale, studies of condensed matter and of molecular and atomic structure will continue to depend upon the techniques provided by electron microscopes, ion sources, lasers, and magnetic resonance devices. As before, each of these techniques will be used in structural or spectroscopic experiments to disclose different aspects of matter, but now with the greater sensitivity and resolving power that precision measurements of more complex materials will require.

As an example of this trend toward increasingly elaborate and costly instrumentation, the rather simple and inexpensive nuclear magnetic resonance (NMR) equipment that was commercially available some 10–15 years ago made the NMR technique readily accessible to the individual scientist. In contrast, a complete NMR facility suited to present-day studies may cost as much as half a million dollars and may include components that are not commercially available and that must be developed by the research group for its own specific purposes.

Similar remarks are applicable to what might be described as integrated facilities, that is, facilities in which no single large item of equipment is involved but in which a number of smaller instruments collectively constitute a major facility. For example, some of the most important experiments in surface physics require the use of an array of several different types of surface probes (electron and ion probes, electron microscopes) used under such high-vacuum conditions that not only the measurements themselves but also the preparation of the sample must be carried out in the final vacuum environment. A second example occurs in the field of very small-scale electronics, where a number of different techniques must be brought to bear on the fabrication and study of structures with dimensions on a scale of millionth of a meter or less. The trend toward more costly instrumentation and toward centralization of facilities has important consequences for the style of research in these fields of study.

NUCLEAR STRUCTURE

THE NUCLEUS AND ITS CONSTITUENTS

Most of the apparent structure and behavior of atoms is determined by the diffuse cloud of negatively charged electrons that surrounds the atom's other constituent, the nucleus. However, the mass of the atom is carried almost entirely by the nucleus, since even the lightest nucleus, a single proton, has a mass more than 1,800 times greater than that of an electron. Because the atom as a whole is electrically neutral under normal conditions, its positive electric charge, which is determined by the number of positively charged protons within its nucleus, also determines the matching number of electrons that the atom will have. Thus the specific molecular and chemical properties of matter are ultimately attributable to the large relative mass of each nucleus and to its total positive electric charge.

The chemical elements found in nature exist because the nuclei of their atoms are either completely stable or else so nearly stable that their lifetimes are longer than the time that has elapsed since they were first created. Most of the possible combinations of protons and neutrons that form nuclei are in fact unstable, and one of the important problems in nuclear research is that of understanding the connection between stability and nuclear structure. In a sense it appears to be purely accidental that of the two nuclear particles it is the proton that is nearly stable, while the neutron is not. Had the opposite been true, the universe as we know it could not exist because ordinary atoms and molecules of hydrogen would not exist as a stable substance.

Nuclear Forces

Since protons all have positive electric charge, they must exert upon each other powerful forces of electrical repulsion that would tend to blow them apart. But since this does not happen, what holds the neutrons and protons (nucleons) in the nucleus together? The answer is that the repulsive electrical force is overpowered by a much stronger force called the nuclear force. This force has associated with it an interaction energy, called "strong interaction," which acts to bind the neutrons and protons together in the nucleus. The strong interaction has the unusual property that it acts only over a distance that is comparable to the sizes of the smaller nuclei, about 10^{-13} cm, but within this range its strength is about one hundred times greater than that of electrical interactions.

The second unusual property of the nuclear force is that a nucleon is attracted only to a few of its neighbors even when there are more nucleons present within the range of the force. Because of this saturation property and the short range of the forces, nucleons are not bound into a large nucleus any more strongly than they are into a small nucleus. In contrast, the repulsive electric forces act over long distances and are additive: the more positive charges, the greater the force. Thus as the number of protons increases, the repulsive electrical force tending to push a proton out of the nucleus eventually becomes as large as, or larger than, the nuclear force tending to hold it in, with the result that very heavy nuclei containing many protons are unstable. This instability is counteracted to some extent by the addition of disproportionate numbers of neutrons to heavy nuclei to act as glue, but eventually the repulsive electrical forces become decisive.

THE VALLEY OF STABILITY

In principle, any combination of neutrons and protons would form a nucleus if they were brought together within the range of their mutual strong interactions. As we have noted, however, most such nuclei are unstable. The kinds of instability that occur can best be understood by starting from the naturally occurring stable nuclei.

If the positions of the various possible nuclei are plotted on a contour map showing contours of their masses in terms of the number of protons and the number of neutrons, as shown in Figure 14, the result is a valley, the Valley of Stability. The strongest mutual binding (lowest energy in accordance with $E = mc^2$) and thus the most stable nuclei are found at the bottom of the Valley. Those nuclei on the slopes of the Valley are unstable if they can drop to a lower position by radioactive decay. There are essentially two different kinds of instability of the lowest energy (ground) state of a nucleus: instability by beta decay, and instability by nuclear-particle decay.

Beta decay occurs when an unstable nucleus can change into a more stable form by converting one of its neutrons into a proton, or vice versa. In either case, the excess electric charge is carried off by the emission from the nucleus of a beta particle, with excess energy carried off by the beta particle and a neutrino. Beta particle is another name for either a negative or a positive electron (a positron), while the neutrino is a particle having no mass and no charge. Thus the total number of nucleons in the nucleus remains fixed, but the number of protons either increases or decreases by one to form a nucleus of a different element.

The second kind of decay process usually involves the emission from an unstable nucleus of a composite particle consisting of two protons and two neutrons, which is known as an alpha particle (and is also the nucleus of a helium atom).

Alpha-particle decay rates can be very slow, with lifetimes up to billions of years. This is why elements heavier than lead (82 protons = element 82) are still found in nature; all such elements are unstable because of the mutual electrical repulsion of the protons in their nuclei, but some are so long-lived that they have not yet had time to decay since they were originally formed.

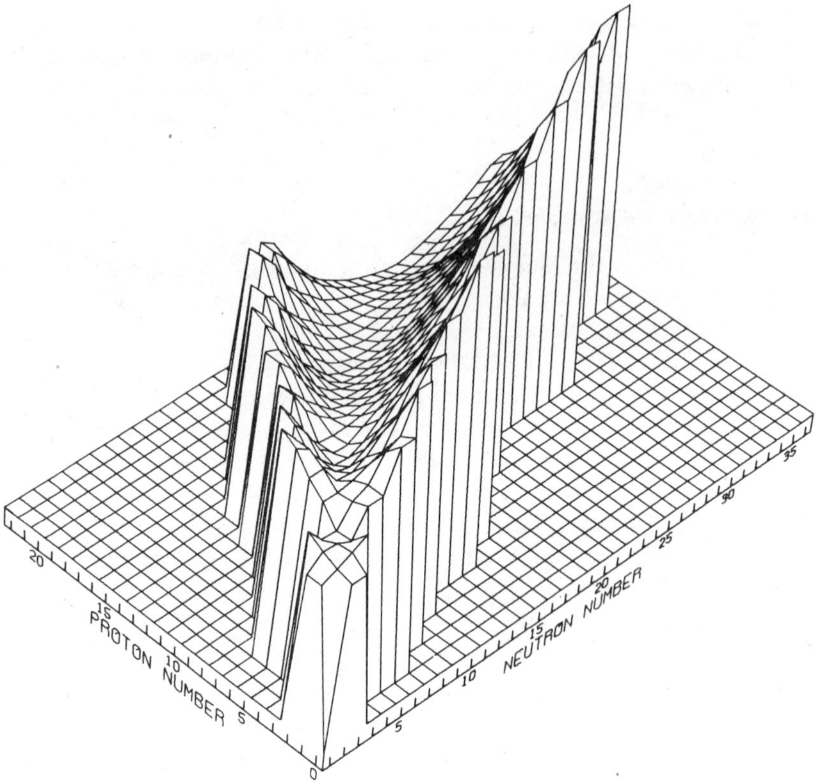

FIGURE 14 Valley of Stability. The bottom of the Valley of Stability is occupied by the stable nuclei. This figure shows the shape of the valley through element number 22. Nuclei on the slopes of the valley are radioactive, decaying to stable nuclei through any of the several decay processes. The highest slopes of the valley consist of the 6,000 or so nuclei predicted to exist but not yet discovered. These would be the least stable nuclei and would thus generally undergo the most rapid decays. (Computer construction courtesy of Jef Poskanzer. From "Exotic Light Nuclei," by Joseph Cerny and Arthur M. Poskanzer, *Scientific American*, June 1978.)

Artificially Created Nuclei

Although the naturally occurring nuclear species are confined to the lower regions of the Valley of Stability, it is possible by bombarding nuclei with charged particles such as protons, deuterons (the nuclei of heavy hydrogen), and alpha particles of high enough energy to overcome the electrical repulsion forces to create new nuclei that lie beyond the end and higher on the slopes of the Valley. The availability of intense sources of neutrons (nuclear reactors) has also been an important source of such exploration. Since the neutron carries no electric charge, it is not subject to the repulsive electrical forces and can thus be introduced into nuclei at low energy. In this way, many unstable isotopes have been produced, including those that decay into the transuranium elements neptunium (element 93) and plutonium (element 94). The recent development of powerful heavy-ion accelerators has made it possible to produce short-lived nuclei with considerably greater numbers of protons. The nucleus with the largest proton number observed to date is element 106.

Accelerators are now available that yield high enough energies and a wide enough variety of particle beams to explore the slopes of the Valley of Stability over wide ranges. For example, the normal isotope of sodium (11 protons) has a nucleus with 12 neutrons, but isotopes of sodium having as many as 24 neutrons have been produced.

The exploration of the rugged terrain beyond the Valley of Stability is also of great importance. There may be other "gorges" of stability far removed from the Valley we now know about. It may be possible to form super-heavy nuclei, having masses far beyond those observed. There may even be nuclei with peculiar elongated shapes, with densities twice that of normal nuclear matter, or with other unusual properties that would represent entirely new forms of nuclear matter. Exploration of this unfamiliar nuclear terrain has now begun in experiments at existing heavy-ion accelerators, and the future development of even more powerful machines can be expected to extend this promising field of study.

NUCLEAR SIZES, SHAPES, AND DENSITY DISTRIBUTIONS

Certain gross features of nuclei can be described in terms of the intuitive concepts of size and shape. These properties may be characterized by either the spatial distribution of electric charges and electric currents (charges in motion) in nuclei, or by the distribution of the nuclear matter itself. For a nucleus having rotational angular momentum, or spin, the electric currents produce a static magnetic dipole moment, like a simple bar magnet with a north and south pole. Magnetic moments can be precisely measured by several methods, using techniques that also have had important applications in many areas of material and biological science.

The scattering of high-energy electrons from nuclei is a powerful means for studying the distribution of electric charge in nuclei, since electrons interact only through the electromagnetic field. Further, since the electromagnetic interactions are relatively weak compared to the binding forces of nuclear particles, electron-scattering experiments can measure the structure of the undisturbed nucleus.

To measure the distribution of nuclear matter (rather than electric charge) requires the use of nuclear particles that interact strongly with both neutrons and protons. Since this strong interaction usually causes the probing particles to disrupt the nucleus, this method yields only a limited amount of information about the undisturbed nuclear state. The earliest information about the sizes of heavy nuclei was obtained from measurements on alpha-particle decay, and it has also been possible to obtain information on nuclear sizes from the scattering of neutrons and protons.

THEORIES OF NUCLEAR STRUCTURE

Nuclei present a wide range of empirical properties against which theoretical notions of their structure can be tested. Each nuclear state is characterized by its energy, by its rotational angular momentum (or spin), and by a number of other distinctive properties (quantum numbers). In addition, there is a wealth of information on decay and scattering processes, on electromagnetic transitions, and on nuclear reactions between different nuclear states.

The effort to extract order from this abundance of information proceeds in two related steps. First, qualitative patterns or systematic trends in the data are sought. Second, the particular empirical properties (such as spin, etc.) that underlie the systematic features of the data are identified and built into a simplified description or model of the nucleus. A theoretical model of the nucleus can be thought of as a kind of caricature wherein a few selected aspects of nuclear structure are given particular emphasis.

The formulation of models of nuclear structure in terms of individual nucleons is not expected to be as straightforward as is the description of atoms in terms of electrons, because there is no massive center of force to play the role of the nucleus in an atom. Nevertheless, a model of the nucleus that is a direct lineal descendant of the atomic shell model plays a major role in the theory of nuclei.

Nuclear Shell Model

The basic assumption of the many-particle nuclear shell model is that most of the nucleons combine to form a relatively inert inner core, a nucleus within the nucleus. The significant properties are then those of the remaining "valence" nucleons—those that occupy the lowest-energy orbits or shells outside the core and are free to interact and scatter among the orbits accessible to them. This simple picture has long been known to account for many qualitative features of nuclei, but a thorough test of the quantitative power of the model has taken much longer to achieve because of the enormously complex calculations that many-body problems require. As a result of advances in the technology of computers and in shell-model computations over the past decade, the interacting many-particle shell model has now been applied in detail to nuclei containing up to 40 nucleons and also to heavier nuclei with no more than four or five valence particles. In addition to fitting some of the energy levels and a variety of other properties and transitions with remarkable fidelity, the shell-model calculations have succeeded in reproducing some of the systematic trends of the collective models discussed below.

Collective Models

When many nucleons accumulate outside the central core, the nucleus may assume a permanently deformed shape roughly like that of a football. This deformed nucleus can rotate in space in different ways that correspond to different levels of excitation. In addition, the shape of the deformation can also change periodically in time, leading to different vibrational states of excitation. Rotation and vibration are the typical collective properties of nuclei in the sense that they involve the cooperative movement of many nucleons. Collective models can account for some of the characteristics of a large class of nuclei.

Further tasks of nuclear theory are to relate the collective models back to the shell model and also to relate the basic elements of the shell model (the binding energy and density of the core and the orbits and interactions of the valence nucleons) to the forces that act between free nucleons. Although progress has been made in both of these undertakings, much work remains. In particular, the many-body foundations of the shell model are poorly understood, and it has not yet been possible to use the qualitative insights gained from the success of collective models to reduce the complexity of shell-model calculations.

NUCLEAR REACTIONS

So far we have concentrated on the individual properties of nuclear states. A second major area of study concerns the various processes that occur when nuclei collide with each other or are struck by other kinds of particles that are capable of disintegrating them. When such nuclear reactions are induced by energetic beams of light ions (those consisting of up to four nucleons), two general kinds of processes can occur: direct reactions, in which projectile and target nucleus interact peripherally and the reaction products emerge almost instantaneously; and compound-nucleus reactions, in which projectile and target nucleus coalesce to form a larger nucleus that does not release the final reaction products until a very much longer time has passed. The study of light-ion induced reactions has provided a great deal of information about the relations between similar states in nuclei that lie close to each other in the periodic table. However, such studies reveal little of the response of nuclei to larger disruptions or excitations.

In order to explore these more complex and radical excitations, nuclei must be bombarded with heavier ions so that many nucleons can participate in the collisions. The following examples indicate some of the important new processes that have been observed in such work:

- Nuclear states are formed in which the nuclear rotation corresponds to very high values of angular momentum. Such states are significantly different from those previously studied in detail and should offer new insights into nuclear structure.
- In other collisions, the nucleons settle down into long-lived configurations that are simple but very unlike those found in the normal ground states. A striking example of such configurations called fission isomers has already been studied in heavy-ion experiments.
- New species of unstable nuclei are formed, particularly proton-rich isotopes, which extend the range of proton numbers over which nuclear properties can be explored.
- Heavy-ion studies have revealed a qualitatively new kind of reaction process that is intermediate between the direct and compound-nucleus reactions. Although these deep inelastic processes occur on the very brief time scale of direct reactions, they are not peripheral collisions because there is a large conversion of kinetic energy into internal heating or excitation of the colliding nuclei. These new processes can hardly fail to shed important new light on the nature of nuclei.

Under suitable conditions, nuclei can be excited by the beam particles into collective vibrational motion. The best-known example is the giant dipole excitation in which the neutrons in a nucleus oscillate as a whole against the protons. Recently, several more complex giant vibrational excitations have been discovered in studies of the scattering of electrons, protons, and alpha particles from target nuclei. It is clear that such vibrational excitations must play a significant role in the deep-inelastic processes described above, but the precise nature of that role remains to be elicited.

NUCLEAR FORCES AND NUCLEAR STRUCTURE

Viewing the nucleus simply as a collection of mutually interacting protons and neutrons confers fundamental importance on the strong or nuclear force that acts to bind these nucleons together. Scattering experiments over a wide range of energies have provided a detailed empirical description of the proton–proton interaction. The neutron–proton scattering data is much poorer and so, correspondingly, is our understanding of this interaction. However, indirect information obtained from the structure of light nuclei has led to the tentative conclusion that the nuclear forces obey the symmetry principle of charge independence, which means that they do not distinguish between protons and neutrons and thus affect both kinds of nucleons equally.

Pions

The electromagnetic interaction between two electrically charged particles is described in terms of a force-carrying particle (or field quantum), the photon, which is said to be exchanged between the two interacting particles. In a similar way, the strong interaction between two nucleons is believed to be carried or mediated by the exchange of a field quantum of a different kind. In contrast to the massless photon, the field quantum of the strong interactions must have an appreciable mass in order to account for the short-range nature of this nuclear force. This quantum has turned out to be the particle known as the pi-meson or pion (however, see p. 200).

The existence of pions as the intermediaries in nucleon interactions introduces an additional factor into the description of nuclear structure. One can no longer hope to obtain a complete description in terms of the motions of protons and neutrons alone but must also include the contribution of pions to the structure. For example, since some pions are electrically charged, their motions create electric currents within the nucleus which affect its electromagnetic structure. For many years there has been unambiguous evidence that some such mechanism was needed to account for the magnetic properties (moments) of certain light nuclei. Only recently, however, have precise experimental measurements been found in close agreement with the predicted effects of pion currents.

THE NUCLEUS AS A MICROSCOPIC LABORATORY

The wide variety of states in which nuclei can exist provides unique opportunities for exploring the fundamental interactions. Of the four basic forces in nature (strong, electromagnetic, weak, gravitational), all but gravity play a measurable role in nuclear structure.

Strong Interactions

The first evidence of strong interactions arose from studies of nuclei, and fundamental questions about these interactions continue to be investigated by the methods of nuclear physics. Precise quantitative measurements of nuclear forces provide an essential test of any theory of the strong interactions. Furthermore, the important internal symmetry principle of the charge independence of nuclear forces mentioned earlier has direct consequences for nuclear structure, and through study of these consequences the principle itself is being tested to increasing accuracy.

Weak Interactions

The earliest evidence of the weak interactions came from observations of beta decay of nuclei, and the study of nuclear physics has also provided much of the subsequent information about this basic force. It was an experiment with radioactive nuclei that first verified the remarkable fact that the weak interactions govern the only processes in nature that distinguish absolutely between left- and right-handedness (parity violation). Nuclear studies have been central to the efforts to unify the weak and electromagnetic interactions, and nuclear physics experiments are stringently testing the validity of this proposed unification.

FUTURE DIRECTIONS

In the years ahead, nuclear physics research appears likely to develop in both breadth and depth. Because the great variety of nuclei provide a very complex system requiring much further study, work will continue in determining nuclear properties with higher precision. In addition the availability of new techniques will open up many new research opportunities that are likely to provide the major impetus for progress in this field.

Nuclear Spectroscopy

Studies will continue in many areas of nuclear spectroscopy, which deals with the simple modes of excitation of nuclei, their properties and mutual relationships, and the nuclear models that embody their salient features. The questions here are primarily those that have motivated previous work, but that can now be studied at much greater levels of precision and sophistication. How are the various nuclear models related to each other and to the force between free nucleons? Will the underlying assumptions of these models allow their application to states of greater energies and angular momenta or of larger deformation? Are the currently recognized symmetries of nuclei all that are required? These are typical of the many questions that await sustained theoretical study and experimental work with accelerators of high precision and intensity but relatively low energy.

Extreme States of Nuclei

Our understanding of the structure of the lower energy states of stable nuclei is based on such extensive work that drastic changes in the overall picture are not likely. Instead, new phenomena are much more likely to occur in nuclei that are in some extreme condition, and the study of such extreme states is expected to be a central focus of future work. The advent of accelerators of higher energy and greater beam diversity will be an important stimulant in producing and examining such states. Some of the foreseeable systems to be studied are listed below.

- Nuclei on the edge of stability, with a large imbalance between neutrons and protons. These neutron-rich or neutron-deficient nuclei can be produced in several ways and studied on-line with isotope separators. The shapes, moments, and energy levels of some extraordinary isotopes of sodium and argon have already been studied in this way.
- Nuclei with extremely high masses, particularly superheavy nuclei that are stable or quasi-stable.
- Nuclei under extremely rapid rotation. Nuclei possessing angular momenta of 40 or 50 units are forced into unusual rearrangments of internal structure by the very large centrifugal and Coriolis forces.
- Nuclear matter in new states of very high density, as may occur in high-energy collisions between heavy nuclei.
- Nuclear structures formed from unconventional particles, for example, strange particles or antiprotons.

INSTRUMENTATION FOR NUCLEAR RESEARCH

For several decades the tools of nuclear science, the electrostatic generator, cyclotron, and electron accelerator facilities, have provided a wealth of information on nuclear structures and nuclear dynamics. However, some of the most challenging scientific problems require capabilities that are beyond those of current nuclear facilities but could be attacked by a modest upgrading of them. And, although high-energy accelerators (greater than a few billion electron volts) have been constructed for elementary-particle research, they are becoming increasingly valuable tools for the acceleration of heavy ions for nuclear science research. Machines for producing heavier ions of high energy are now under development for this purpose.

Within recent years, intense beams of pi-mesons (pions) have become available for the study of nuclear structure. The strong interaction of pions with nuclei provides an effective means for measuring the forces acting at short distances from nuclear particles. In addition, the transfer of a large amount of energy from pion to nucleus can result in the creation of unusual nuclei that lie far from the Valley of Stability. This work is still in its early stages, and in the next few years it should produce important insights into nuclear structure.

The very strength of the pion–nucleon interaction becomes a liability when the intention is to study the nucleus in its undisturbed state. For such studies, the much weaker electromagnetic interaction is the preferred probe. In particular, beams of electrons and photons (gamma rays) with energies up to a few billion electron volts serve as a very sensitive means for resolving the fine details of nuclear structure. Some studies of this kind have been carried out under special arrangements at existing large electron accelerators, but there is an important need for an electron accelerator more closely matched to the purpose and dedicated to experimental studies of this kind.

Studies of nuclear structure would also benefit greatly from experiments in which the nucleus is probed by beams of neutrons of greater intensity than now available, by neutrinos, and by high-resolution laser spectroscopy. Each of these areas is now being investigated for its scientific potential and for the kinds of instrumentation that would be required for successful experimentation. In the case of neutrino beams, the required intensity would be very large because of the weak neutrino–nucleus interaction. At present, there are no sources of neutrinos of the required intensity and correct energy range.

PARTICLE PHYSICS

The structural units of matter in its usual forms have been described thus far as electrons, protons, and neutrons. From these, nuclei, atoms, molecules, and ordinary macroscopic matter are constructed. As late as the middle 1940's these units were thought to be indivisible and were therefore called elementary particles.[6] In addition to those known particles, there was at that time evidence from cosmic rays and radioactive decay of the existence of other forms of matter—the pion, the neutrino (one of the products of beta decay), the muon (a heavy electron), the positron (a positively charged electron), and, of course, the photon.

Throughout the subsequent decades, many new kinds of particles were discovered, most of them related in some way to the two nuclear particles, the proton and neutron. These discoveries came slowly at first, then at an ever-increasing rate. By 1960 the number of known particles had grown from 4 to about 30, and most physicists had begun to have serious doubts that so many different kinds of particles could all be elementary in any meaningful sense.

Although the list of particles has now increased to more than 200, evidence has been accumulating during the past decade in support of the idea that *none* of these nuclear particles, not even the proton itself, is truly elementary. Instead, they all appear to be composite structures built up from simple combinations of only a few kinds of basic constituents called quarks. Quarks have not been directly observed in experiments and are thus still hypothetical particles, but the circumstantial evidence for their existence has now become so various and persuasive that there are not many doubters left.

FORCES AND INTERACTIONS

The forces holding the building blocks of matter together have been described as gravitational, electromagnetic, and strong interactions. Just as the photon is the quantum or packet of electromagnetic energy, the pion was originally thought to be the quantum of the interaction energy associated with the strong force of the nucleus. This is another concept that has been modified by recent developments but, nevertheless, the pions are a form of matter participating importantly in strong interaction physics.

The electron, positron, and neutrino are emitted by nuclei in the process of beta decay, or radioactive decay. This decay is ascribed to some kind of interaction with protons and neutrons, just as the emission of light by atoms is ascribed to the electromagnetic interaction. However, this interaction cannot be represented as one or a combination of the three forces mentioned above. It is a fourth form of interaction, called the weak interaction since the magnitude of the interaction energy required to account for the rate of beta decay is found to be very much smaller than that associated with the electromagnetic interaction. This means that beta decay processes occur very slowly.

In terms of the interactions that they are subject to, it is clear that, except for photons, particles fall into two general classes: those that are not subject to the strong interactions and those that are. Those not influenced by the strong interaction are called leptons, and include electrons, neutrinos, and muons, the latter behaving in every way like a heavy electron.

The particles undergoing strong interactions, which include nucleons, protons and neutrons, pions, and many others to be described later, are called hadrons, from the Greek form *hadr-* for "strong."

MATTER AND ANTIMATTER

As far as we now know, every particle has associated with it an antiparticle. For example, the positron is the antiparticle to the electron. The antiparticle has the property that it can annihilate the particle—this means that when the two combine they disappear and can be replaced by photons; and photons in turn can create another particle–antiparticle pair. One can imagine a universe of material made up entirely of antiprotons, antineutrons, and positrons composed to form antiatoms and antimolecules; in other words, antimatter.

The search for simplicity of the laws of physics led naturally to this concept of antimatter and to the simple principle that a universe of antimatter should be indistinguishable from a universe of matter. This symmetry concept is borne out by experience except for a very subtle unexplained property of weak interactions that does distinguish matter from antimatter. However, it should be kept in mind that an unexplained phenomenon appearing to be subtle and unimportant may signal a future revolution in science.

DISCOVERY OF THE HADRON FAMILIES

A series of unexpected discoveries concerning the hadrons began in the mid-1940's with the observation of some particles produced by cosmic rays that behaved in such an unusual way that they could not be identified with any particle known at the time. This behavior was reproduced a few years later in the laboratory by particles produced in collisions between hadrons (protons, pions, and neutrons) at very high energy. Because of their unaccountable behavior they were called strange particles.

Every type of strange particle is unstable and decays in a short time (approximately 10^{-10} seconds or less) into the ordinary, or nonstrange, particles. This lifetime, while short on the ordinary scale of time, is still very long on the nuclear scale and is therefore ascribed to the weak interactions.

There are two types of strange particles, one called the baryon, for which, roughly speaking, one of the decay products is a nucleon and the other called a meson, which, again roughly speaking, never decays into a nucleon. More precisely, a baryon can decay into an odd number of nucleonic particles (usually one) while a meson can only decay into an even number (usually zero).

Further investigations of the consequences of high-energy collisions between particles have led to an enormous proliferation of new kinds of hadrons, now numbering over 200 different species. This is a virtual zoo requiring a great effort by a large number of experimenters to untangle and identify the particles and the relationships between them. Although they probably occurred in large quantity in the early universe, most such particles are seldom observed now outside the laboratory because they have short lifetimes and the conditions necessary for their creation occur infrequently. All of the particles decay into combinations of lighter particles through a chain of events ending in the stable building blocks of ordinary matter and neutrinos.

QUARKS AS BUILDING BLOCKS

As the zoo of over 200 baryons and mesons of different masses, both strange and nonstrange, revealed itself through high energy experiments, certain regularities in the patterns of masses and other properties of these particles became apparent, indicating that there must be an underlying structure, just as the regularities in the behavior of the chemical elements or the regularities in the patterns of stable nuclei indicated the important underlying structures of those systems.

An early step in arriving at a description of families of hadrons was the recognition that symmetry could account for many features of the observed regularities. This symmetry was a generalization of the simple symmetry between proton and neutron already recognized as a fundamental aspect of nuclear structure. As we have noted, the proton and neutron differ only because one is electrically charged and the other is not. Their strong interaction properties, that is, their characteristics as hadrons, are the same, although there is a slight mass difference presumed to be caused by electromagnetic effects.

In an analogous way, similarities were found among other groupings of particles and these similarities were regarded as an expression of a basic internal symmetry among the particles. In its original form, this symmetry could be accounted for by treating all baryons as composites of just three particles, the quarks. The name quark arose from the quotation "Three Quarks for Muster Mark" from *Finnegan's Wake* because there were just three kinds, now called "up," "down," and "strange" quarks.

At the time this model was proposed, baryons corresponding to all combinations of quarks except that of three strange quarks had been observed. From the model, not only was it predicted that such a baryon should exist, but also it was possible to estimate the mass of this unknown particle. This prediction set off a frantic scramble among the high-energy laboratories to produce and identify the particle, already named the "omega-minus." It was soon found, with a mass within one-tenth of a percent of that predicted. This great triumph lent strong credence to the basic symmetry ideas underlying the quark model but it did not convince physicists of the existence of the quark, especially since no one had seen one.

While baryons were accounted for as combinations of three quarks, the very different characteristics of mesons could be accounted for if each meson was assumed to be made up of a quark–antiquark pair. The existence of an antiquark to go with each quark was a natural and necessary consequence of the assumption of symmetry between matter and antimatter.

Charmed Quarks

The notion that a fourth kind of quark must exist first arose in connection with attempts to relate the weak and electromagnetic interactions. The fourth quark was called a "charmed" quark by the theorists who posed the hypothesis, and charm emerged as a reality as the result of experiments establishing the existence of an entirely new kind of meson, called the "J" or "Ψ," or even the "J/Ψ," in 1975.[7] This meson has a mass greater than that of three nucleons, and a lifetime for decay into other mesons that is much too long to be accounted for as a pair of normal quarks, but that is in accord with the predictions of the charm hypothesis, if it is assumed to be a combination of a charmed quark with a charmed antiquark. Predictions were also made about the masses to be expected for excited states of this system. These predictions were quickly confirmed by experiments.

It is also to be expected that a charmed quark could combine with one of the original three antiquarks to form a charmed meson, and such a particle, the D meson, has also been found (Figure 15). Similarly, charmed baryons are on the menu, and there is recent evidence for their existence.

Six Quarks?

The existence of four quarks is generally accepted by physicists as being well established. However, as a result of another surprise, in 1977, there now appear to be more. At that time a meson having a mass more than nine times that of the nucleon was discovered, requiring the introduction of a fifth quark into the picture.

Finally, the unity of the theory of weak and electromagnetic interactions turns out to call for an even number of quarks. Therefore it is generally assumed that evidence for a sixth quark remains to be discovered.[8]

Where Are the Quarks?

Many experiments have been carried out in an effort to detect a free quark but have not been successful at the time of this writing.[9] The list of

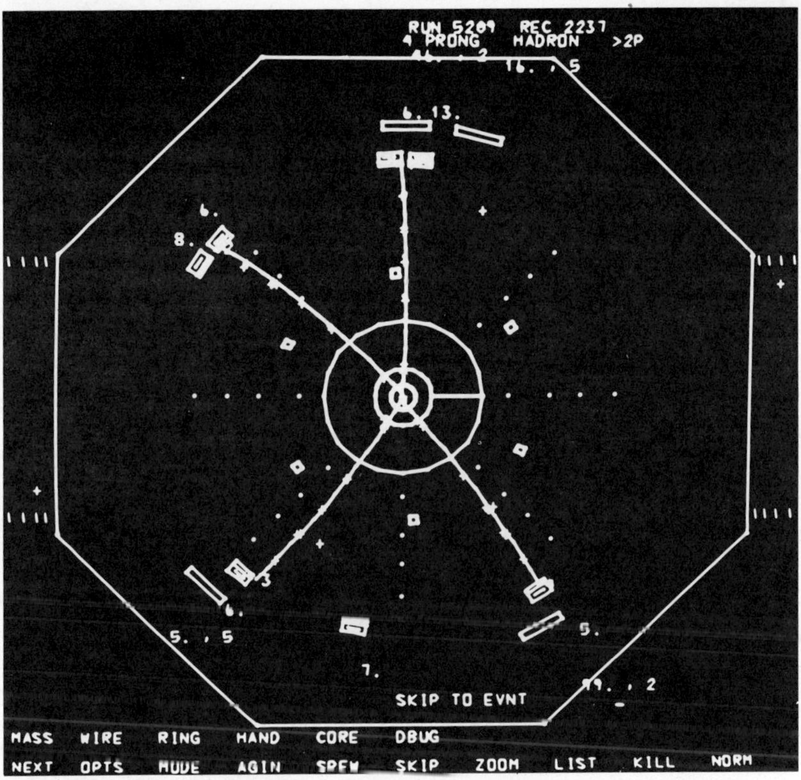

FIGURE 15 Decay of a charmed particle is reconstructed in a computer-generated display. The innermost circle represents the beam pipe in which electrons and their antiparticles, positrons, collide to produce other particles. The particle track at 12 o'clock has been identified as that of a negatively charged K meson; and the track at 7 o'clock as that of a positively charged pion. These particles are thought to be the decay products of a D^0 particle, bearing a property called charm. The D^0 particle decays too quickly for it to be detected directly. (Jon Brenneis. From "Fundamental Particles with Charm," *Scientific American,* October 1977.)

experiments includes not only those carried out at accelerators but also those involving analyses of seawater and of moon rocks, magnetic levitation of small metal balls, cosmic-ray studies, chemical reactions, and so on.

The key to identifying quarks is their remarkable assumed property of having electrical charges of one-third or two-thirds of the "indivisible" unit of electrical charge, that is, the charge of the electron. This eccentricity of charge means that the electrical charge of a quark cannot be neutralized in ordinary matter so that a quark should exist until it meets another quark with which to combine. This also provides a means for identifying the particle.

Since the chemical properties of atoms and molecules are controlled by the electrostatic forces holding them together, it is clear that quark matter would have quite different properties. One can speculate about the possibilities and they are quite interesting, indeed. The only difficulty is that no one has been able to collect even one quark.

Evidently the quarks are bound together to form baryons and mesons in an unshatterable way. Of course, the energy required to smash them into quarks may simply be much higher than is presently available. Or it may be that quarks are a mathematical fiction, yielding the correct properties in a mathematical way, but not real otherwise.

Neither of these explanations is satisfactory and there is a growing general belief among physicists that the explanation lies in an unusual direction. If the quarks are bound together by a force that increases in strength as the distance between them increases, much as would happen if they were bound together by rubber bands, then one could understand that they could never be separated. Theories along these lines are being pursued vigorously and, in the near future, it will probably be possible to propose definitive tests to validate them.

Such forces would represent the basic strong interaction, which binds the quarks together. Just as the electromagnetic fields binding electrically charged particles together can be described in terms of quanta, the photons, so this basic strong interaction field must have its quanta, which are called "gluons" because they glue the quarks together (logic of nomenclature at long last!). Gluons as particles would necessarily have very unusual properties in order to account for the rubber bandlike forces proposed to exist between quarks. Although a free gluon may not be observed until quarks become available to generate them, there are indirect ways of detecting their existence within particle structures and these effects are being investigated.

LEPTONS AND WEAK INTERACTIONS

The leptons provide the only examples of observed, truly elementary particles; they behave in all but some very subtle ways as "point" particles, that is, as structureless objects.

The existence of muons, or heavy electrons, came as a special surprise since there seemed to be nothing in nature requiring it. One famous physicist is quoted as saying, "Who ordered that?" when it was discovered. Another made a habit for years of keeping "μ" written in a corner of his blackboard to remind him of the mysterious role of this heavy electron in physics.

Another question that has often been raised is: If there exists one heavy electron, why not more? The question recently has been answered: A lepton, called the "tau," having a mass 3,500 times that of the electron, or twice that of the nucleon, has recently been discovered.

Some inkling of the role in nature of the muon results from the knowledge that the neutrino associated with the muon is a different particle from the one associated with the electron. Since leptons undergo interaction in pairs, this pairing off of an electron with one kind of neutrino and the muon with another suggests that these pairs of particles may have some fundamental significance in the structure of the physical universe.

The statement that leptons undergo interaction in pairs applies both to their electromagnetic and to their weak interactions. In the case of the former this may be seen by noting that an electron–positron pair is annihilated, or created, in interaction with a photon. For the weak interactions the appropriate neutrino is always associated with the emitted electron or muon in beta decay.

Since hadrons are subject to the weak interaction, so must be the quarks that compose them. The suggestion that there must be a fourth quark, the charmed one, arose from the desire to give the weak interactions of quarks a structure analogous to that of the weak interactions of leptons, thereby requiring that quarks also occur in pairs.

All of this leads to a rather simple picture of nature: Matter is made up of two parallel families of elementary particles: the leptons and the quarks. And there is a one-to-one relationship between them (Figure 16). The quarks, being subject to strong interactions, are buried in hadrons, but the leptons are free to roam. When the tau lepton was discovered, it seemed inevitable that another (fifth) quark would be found, and it was. The assumption that there must be a neutrino to go with the tau lepton then leads again to the need for a sixth quark, as already mentioned.

Unification of Weak and Electromagnetic Interactions

It has been recognized for some years that there are certain similarities between the weak and the electromagnetic processes, but attempts to formulate unified theories and experiments designed to demonstrate their validity ran into serious difficulties until recently. Now it can be said that a very promising unified theory, making use of the concept of pairing of particles as above, has been developed.[10]

One of the earliest questions was why, if these phenomena are so similar, is the weak interaction so very much weaker than the electromagnetic interaction? From the beginning the answer to this question was assumed to be related to the nature of the quantum of the weak field. This would be

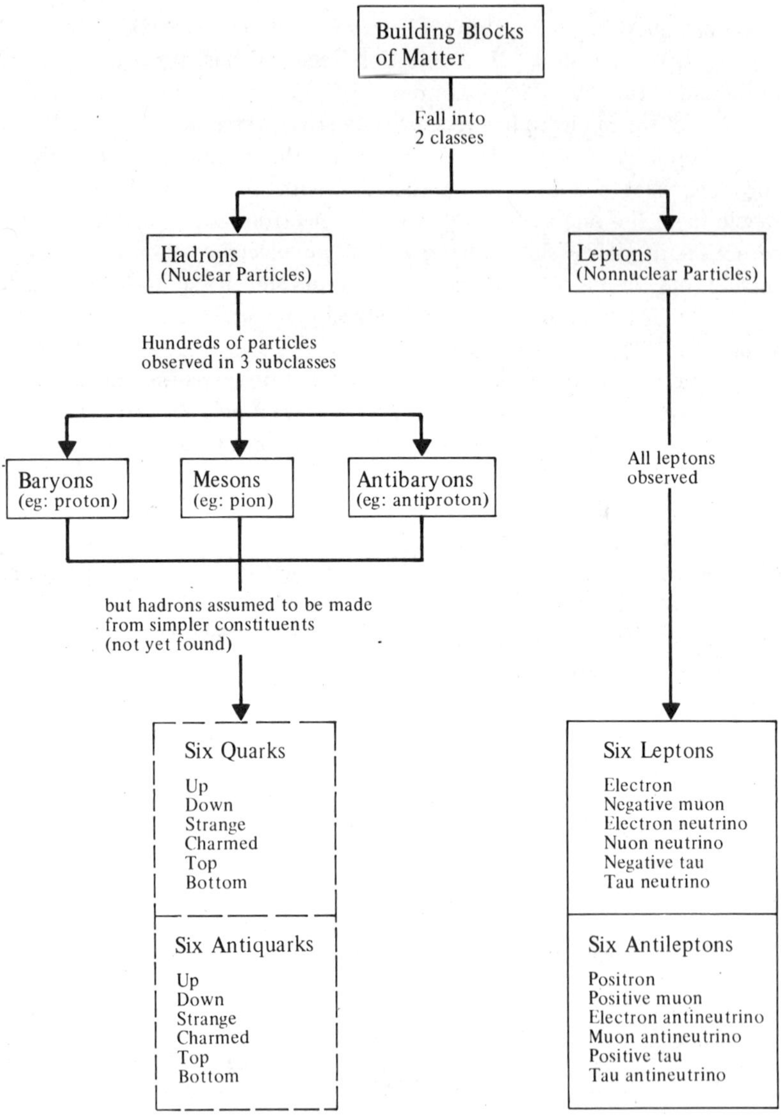

FIGURE 16 Basic building blocks of all matter. The evidence for quarks remains circumstantial.

the analog of the photon except that it would need to be electrically charged if it were to couple to an electron and neutrino the way a photon couples to an electron and positron.

If this charged quantum called the W^+ or W^- has large mass, the effective beta-decay coupling would be correspondingly weak or small. However, there were serious difficulties until a recent reformulation of the theory that included an electrically neutral *W*. This implied the existence of weak decay processes which did not involve charge, such as the emission of neutrino–antineutrino pairs. These are known as "neutral weak current effects." Such effects can be, and have recently been, observed in high-energy neutrino experiments.

Thus a promising theory has now emerged which, among other things, predicts the mass of the *W* to be about 75 times that of the proton. There is at present no accelerator capable of producing a particle of so large a mass, but efforts in this direction will be one of the principal objectives of proposed intersecting beam machines that will be able to attain the required level of energy.

CREATION AND ANNIHILATION OF PARTICLES

There are two ways in which the unstable forms of matter are produced, by decay of more massive unstable forms or by creation processes at high energy. Several decay processes leading to stable particles are familiar to us, for example the decay of an atom in an excited quantum energy state to a lower quantum energy state with the emission of a photon. This photon, having an energy equal to the difference in energy between the atomic states, may be said to be created in the process. By using the equivalence of mass and energy, one may also say that the total mass of matter was unchanged, the mass of the atom in the excited state being greater than the mass of the atom in the lower state by just the right amount to account for the mass of the photon.

In a similar way, electrons and antineutrinos are created in the beta decay of a nucleus, when it is energetically possible. The positron, which is a stable particle, is also produced, along with a neutrino, in beta decay when the unstable nucleus has a larger positive charge than the product nucleus.

Positrons don't persist in matter because they are annihilated by contact with ordinary matter containing electrons, their antiparticles. This process, and its reverse, electron–positron pair creation, is an example of the processes involved in the creation and annihilation of particles.

We know that a positron and electron can swallow one another, leaving behind nothing but the energy represented by the two masses. This energy appears as electromagnetic radiation or in the form of other particles. One can say that "matter" (the electron and positron) has been annihilated. Similarly, a photon of energy greater than twice the electron mass can, under the right conditions, be converted into an electron–positron pair; thus matter is "created."

Of course one can take the view that photons, having mass, are also matter and that there has been no change in the amount, or mass, of matter; there has simply been a change of form. Thus we arrive at the point of view that matter is conserved, but it may appear in a great variety of forms as long as they all have the same total energy.

Exotic particles may be created by such processes or by collisions between energetic particles. The mass of the pion is about 270 electron masses and it can therefore be produced in collision between protons if they have enough energy to produce this mass by a process like

$$\mathrm{p} + \mathrm{p} \longrightarrow \mathrm{p} + \mathrm{n} + \pi^{+}$$

which means that two protons collide to produce a proton, neutron, and positively charged pion. Protons of sufficient energy occur in the cosmic radiation falling on the earth from outer space, and it is for that reason that pions were first observed in cosmic ray experiments. The pion decays into a muon and neutrino in about 10^{-8} seconds. Muons, which have a lifetime 100 times longer, were discovered much earlier than pions, also in cosmic rays. The muon decays into an electron and two neutrinos.

PRODUCTION OF PARTICLES IN THE LABORATORY

In order to produce pions in the laboratory by the process just described, it is necessary to have machines capable of producing about 400 million volts to yield a proton beam of adequate energy. Since the pion is the lightest of the mesons, machines of much higher energy are required to produce the other mesons and baryons. With sufficient energy (a machine of over 5 billion volts) it is possible to produce antiprotons and antineutrons paired off with protons and neutrons, respectively. At even higher energies the antiparticles of the strange baryons and other members of the zoo are produced.

One may ask how it is possible to identify and work with particles whose existence is so transitory. For those of very short lifetime, observations and measurements are made on the decay products. Particles of lifetimes 10^{-10} seconds or greater are observable although that time is only long enough for a light signal to travel three centimeters. However, at high speed the particles actually travel much farther than this because of the relativistic effect of time dilatation predicted by Einstein: A moving clock appears slower to a stationary observer than does a stationary clock. This effect becomes large at speeds close to the speed of light. Because these particles are produced in very high-energy collisions, they move with such high speeds. The relativistic effect can be so large that, for example, it is possible to produce intense beams of pions over long distances despite their short lifetime (at rest) of 10^{-8} seconds. Such beams are used to bombard targets to produce other kinds of particles.

RESEARCH INSTRUMENTS FOR PARTICLE PHYSICS

The primary beams of high-energy particles required for particle production experiments and other high-energy studies are produced by large particle accelerators that are direct descendants of the atom smashers developed in the early days of nuclear physics. As we have seen, the more deeply one probes the substructure of nuclei, the greater the energy required—to take advantage of the shorter wavelengths of higher-energy particles so that structural details can be "seen" over short distances and to have enough energy to knock apart the strongly bound part of a system or to create particles of high mass (in accordance with $E = mc^2$).

The primary beams may consist of either electrons or protons. The accelerators may be either circular machines (synchrotrons) or linear machines. Since electrons copiously emit energy in the form of synchrotron radiation when moving at high speed on a curved path, there is a point of diminishing return in trying to accelerate electrons in circular machines.

In the United States the highest-energy (22 billion electron volts or GeV) electron accelerator is located at the Stanford Electron Accelerator Center (SLAC), and there is a lower-energy (about 15 GeV) electron synchrotron at Cornell University. The highest-energy proton synchrotron at the Fermi National Laboratory usually operates at about 400 GeV although it has been pushed to 500 GeV. Additional proton machines are the 30 GeV Alternating Gradient Synchrotron (AGS) at Brookhaven National Laboratory and the 12.5 GeV Zero Gradient Synchrotron (ZGS) at Argonne National Laboratory.

Developments for the future include the construction of an energy doubler at the Fermi Laboratory to increase the proton energy available to 1,000 GeV = 1 TeV (1 trillion electron volts). At the other accelerator laboratories, several colliding beam devices are planned that increase the effective energy even more for reasons to be given later. To increase the available resources needed to carry out these plans, it has been decided to shut down the ZGS at the end of fiscal year 1979.

Electron and proton accelerators using a fixed target are very effective because the primary beams have a very high rate of interaction with a target and therefore provide a large number of events for measurement.

They also produce intense secondary beams of particles moving at high speed. However, there is a disadvantage to these machines associated with the high speed of the protons, namely, much of the energy of the primary beam goes into the motion of the secondary particles and not into their mass.

To realize maximum use of a machine's energy, the beam can collide with another high-energy beam moving in the opposite direction instead of making it strike a fixed target. When two similar beams of the same energy collide there is no wasted energy; all of the collision energy can go into the mass of the produced particles. Devices making use of this principle offer the ideal way to explore for particles of extremely high mass.

As a means for following up the discovery of new particles with precision measurements, colliding beams do have limitations. The most serious is the rather small interaction rate resulting from the low "target" density, since the target of one beam is the other beam. This can only be overcome by substantially increasing both beam intensities, but under no circumstances can an event rate comparable to that of fixed target machines be anticipated. The other limitation is that the secondary particles produced are moving too slowly to be useful as beams. If they have short lifetimes, they decay quickly because there is no appreciable time dilatation.

Colliding beams of electrons with positrons, each with 4 GeV of energy, are provided by the facility called SPEAR at SLAC. PEP, also an electron–positron colliding beam machine, with each beam having an energy of 15 GeV, is also under construction at SLAC. A facility comparable to PEP is operating at Hamburg, Germany. The only proton–proton colliding beams presently in existence are provided by the intersecting storage rings (ISR) at the international facility, CERN, in Geneva, Switzerland. The energies of the two proton beams are 31 GeV each. To produce the mass equivalent to the 62 GeV available at the ISR would require a fixed target machine of about 2,000-GeV energy.

Construction has just begun at Brookhaven National Laboratory on an intersecting proton beam machine designed to produce colliding beams of 400 GeV protons. At the same time, an effort is under way at CERN to use the 400 GeV Super Proton Synchrotron (the equivalent of the Fermi Laboratory machine) to circulate protons and antiprotons in opposite directions and bring them into collision. Thus the region of very high masses, such as that of the hypothetical *W* to be described later, will be explored.

MEASUREMENTS AND INSTRUMENTS

Measurement of the properties of the particles produced in collisions of the beams with targets or with each other is the principal objective of the science of particle physics. Such measurements including numbers, directions, momenta, masses, and energies of particles must be made with high precision. It is also necessary to separate events at incredibly close time intervals. The instruments required for the purpose become increasingly complex as the energy increases. There are at least two reasons for this. One is that the number of particles produced in a collision increases, making it more difficult to sort out the particles of interest. The other is that the higher the energy, the more difficult it is to measure the momentum of a particle by bending its trajectory in a magnetic field. In order to capture the necessary information, the instruments must increase in size as well as in complexity.

Required instruments include particle detectors, huge magnets, very fast response electronic circuits and on-line computers to digest the incoming information. The detectors depend on the ionization that charged particles produce in going through matter, while the detection of nonionizing particles, such as photons, neutral pions, or neutrinos, depends on their production of charged particles, which are then detected. A charged particle passing through a gas or liquid leaves a trail of ionizing particles that can be made visible as a vapor trail in a cloud chamber, a trail of bubbles in a bubble chamber, or as sparks in a spark chamber. As the need to observe a multitude of these trajectories more and more accurately has grown, special electronic devices have been developed, the most recent being the drift chamber, which operates inside of strong magnetic fields to give a very precise measure of points along particle trajectory (to better than 0.1 millimeter), and handles many trajectories at the same time. This information is combined with that obtained by other methods, such as total energy calorimetry, which measures the energy absorbed in a mass large enough to absorb all of the energy of a jet of many particles. With these and other techniques it is possible to collect information at an enormous rate. It then takes powerful, fast computers to analyze these data to the extent that the experimenter knows what has happened in his apparatus. The final step, scientific interpretation of the data, requires the most careful attention of the best scientific minds.

A major experiment at a high-energy accelerator requires a large team of physicists backed up by engineers, technicians, and computer personnel to plan and design an experiment, develop and design the apparatus, design and prepare the data-acquisition system, and analyze and interpret the data. One experiment may involve 30 or 40 physicists, plus a comparable number of support people and require two years of planning and preparation, another two years for running the experiment, and at least a year for analysis and interpretation. This is a far different picture from that of the lone scholar in his laboratory, but the scholarly consequences in terms of the depths of insight are different, too, and such formidable efforts are required to reach to these depths.[11]

CONCLUSION

THE ROLE OF MEASUREMENT

Precise measurement is the hallmark of the remarkable advancement in understanding the physical universe in modern time. Quantitative description of matter leads to mathematical formulations of the laws of nature, which, after many tests of their validity, may tentatively be regarded as established truth. Reproducibility and predictability of data go hand in hand with precision of measurement. Understanding grows as the character of observation shifts increasingly from the qualitative to the more precise and quantitative. What is perceived as a color becomes a precisely measured wavelength of light. A shape is related to carefully measured angles of cleavage planes of a crystal. Such measurements attach numbers such as length, time, and angle to physical phenomena, and these numbers may be remeasured many times to test their reproducibility and validity, thus helping lay the foundation of a science.

SIMPLICITY OF THE LAWS OF NATURE: THE ROLE OF SYMMETRY

Precisely measured numbers alone are not enough, however. A complete description of matter requires that it be related, according to a set of rules formulated in terms of simple mathematical expressions. These rules transcend the boundaries of scale, and it is an axiom of science, consistent with a vast amount of experience, that they are based on principles that apply everywhere and for all time. The simplicity of the mathematical expressions may be manifest in either their structure or the statement of symmetry implied, or both, when the two are related.

The elementary concept of symmetry is that of geometrical form. For example, the geocentric and heliocentric models of the universe take the sphere as the natural geometrical form. The eventual realization of the role of spherical symmetry in nature is expressed in terms of the homogeneity of the space around us—no matter which way we turn, the character of empty space is the same. This leads directly to the pervasive fundamental law of conservation of angular momentum, which plays as important a role in the structure of atoms as it does in the structure of the planetary system.

There are many other important geometrical expressions of symmetry in physical science, including the description of the structure of crystals, the behavior of conduction electrons (those carrying electricity), and the structure of molecules. The whole understanding of the chemical behavior of the elements turned out to depend on a very general internal rather than geometric symmetry property of electrons, related to their interchangeability—the fact that two electrons are indistinguishable. This concept of a symmetry resulting from the interchangeability of particles has been generalized, as in the case of the neutron and proton, with remarkable consequences for our understanding of the fundamental building blocks of matter.

UNITY OF THE PHYSICAL SCIENCES

The fact that the complex behavior of the physical universe is governed by a set of universal principles and mathematical laws implies unity among the various fields of physical science, but unity at a deep theoretical level not readily apparent. Nevertheless, because of a commonality of the underlying principles, there is a unity of method both in theory and experiment. The result is that the subfields interact in a supportive and synergistic fashion. Physical science has an organic quality, with the whole being much more than the sum of the parts.

The unity of method and structure is manifest in the understanding of cooperative quantum behavior of many-particle systems, whether they be electrons in solids, atoms in liquids, nucleons in nuclei, or neutrons in neutron stars. Symmetries and "broken" symmetries play very similar roles in the theories of elementary particles and the theory of macroscopic phenomena such as magnetism. The properties of the elementary particles play an essential role in theories of the origin of the universe, and some recent ideas arising from the unified theories mentioned below suggest a previously unanticipated picture of the way in which matter may have been created in the beginning.

This unity of method is a reflection of an even deeper unity of fundamental principles. The history of physics has been marked by a struggle for ever deeper unifying principles. Examples of successes are manifold: Newton's connection between the earth's gravity and the motion of planets around the sun; the unification of heat and energy by Clausius; the unification of electricity, magnetism, and optics by Maxwell and Faraday; the unification of the corpuscular and wave theories of light by quantum theory; the unification of chemistry and physics by quantum mechanics; and the current, apparently successful unification of electromagnetic and weak interaction theories. Recent work has led to the hope for an early extension of this last concept to include the strong interactions of particles.

Einstein sought the unification of gravitational and electromagnetic theories as his ultimate aim. Through the study of quantum effects of gravitational fields, we may see his hope realized on a much grander scale than he ever imagined by the unification of gravitation, electromagnetism, weak and strong interactions—all of the known forces of nature.

OUTLOOK

The following outlook section on structure of matter is based on information extracted from the chapter and covers trends anticipated in the near future, approximately five years.

ASTROPHYSICS AND COSMOLOGY

Rapid developments in astronomy and space science during the past few decades have led to an entirely new perception of the cosmos. Observations of almost the entire electromagnetic spectrum from radio and infrared through visible and ultraviolet to X-ray and gamma rays have been made possible by new ground-based instruments and space probes. Similarly, important progress in the study of cosmic rays has also been made possible by new instrumentation and the mounting of instruments in space vehicles.

The detection of 3°K radiation has provided convincing support for the Big-Bang concept of the origin of the universe. The evidence for the existence of neutron stars confirms theories of stellar evolution and the strong suggestion of evidence for black holes tends to confirm one of the more remarkable predictions of Einstein's general theory of relativity.

The prospects are for continued rapid advances in this field. Observation and measurement are just beginning at both the infrared and gamma radiation ends of the spectrum. The planned gamma-ray space observatory will permit astronomers to make measurements with about 10 times the sensitivity now possible. The space telescope scheduled for launch in the space shuttle, the new ground-based large multiple-mirror telescope, and the very large array radio interferometer will permit much deeper penetration into the universe.

These and other new instruments may help resolve some of the important cosmological and astrophysical questions concerning the structure of the universe, for example, whether it is open or closed, and the actual existence of black holes.

Also significant questions concerning stellar processes such as nuclear synthesis of the elements will be addressed by cosmic ray experiments. Neutrino detection methods may help resolve questions about energy sources of the sun raised by recent experimental results.

CONDENSED MATTER

Research on matter in the large, on the macroscopic scale, has an impact on all the ways in which materials are used. There have been an enormous number of advances in our understanding and application of the science of condensed matter. Some of the most recent developments showing promise for substantial progress in the next five years are studies of amorphous materials, surfaces, and exotic new materials, as well as atomic and molecular engineering and study of systems far from equilibrium. The trend is toward dealing with systems of increasing complexity at higher levels of precision.

An important theoretical advance has been our improved understanding of phase transitions, such as transitions from gases to liquids or nonmagnetic to magnetic phases in solids. This fundamental work is applicable to a remarkably wide range of phenomena.

Since instrumentation has become more elaborate and expensive, ways of sharing apparatus among institutions are being explored. Already possibilities for qualitatively new experiments on condensed matter with major centralized facilities are being realized. For example, the new research frontiers provided by synchrotron radiation sources of high-intensity ultraviolet and X-radiation, which are powerful tools for studying the structure of condensed matter, are being approached by several new facilities presently under construction.

Intense neutron sources that can be used to study different aspects of the structure of condensed matter represent another example. They also are a very powerful tool for the study of atomic dynamics of liquids and solids. Great progress in this field has been made in Western Europe, which for some time has had the most intense steady source of neutrons, at the Institut Laue-Langevin in France. In addition, a new high-intensity pulsed source is under construction at the Rutherford Laboratory in England. Somewhat more modest steady sources have been available in this country for many years, and a considerably more modest pulsed source is under construction.

Since condensed matter science serves as a technology base both for other fields of science and for industrial development, it is essential to maintain our strong capability in the field.

MOLECULAR AND ATOMIC STRUCTURE

Progress has been rapid in the study of the structure of molecules, especially large organic molecules such as those occurring in biological systems. Developments in instruments and spectroscopic techniques paralleling those used in investigating condensed matter have been largely responsible for this. Exploitation of nuclear magnetic and electron paramagnetic resonance methods has been especially fruitful, but, again, increased sophistication of these techniques has raised the cost of instrumentation so that new institutional arrangements to replace the one-scientist–one-instrument tradition must be explored.

Synchrotron radiation sources and intense neutron sources also hold great promise for research in this subfield. The use of neutrons is now at its most primitive stage, beginning with small angle scattering to obtain gross features of large molecules. An order-of-magnitude increase in the intensity of neutron beams is needed to realize the potential of this technique.

Because of the coherence, intensity, and spectral purity of the light they produce, lasers have become a powerful tool for high-precision molecular spectroscopy, making it possible to observe molecular fragments in, for example, chemical reactions occurring in very short time intervals of 10^{-12} seconds—a picosecond.

The continued deepening of our understanding at the basic level of processes taking place in and between molecules, especially large organic molecules, may make it possible to mimic some biological processes.

A number of elegant developments in atomic physics make possible very precise tests of some of the fundamental laws of nature. Again, lasers have greatly increased the precision of atomic spectroscopy. They have also made it possible to isolate and study a single atom at a time, and to investigate the structure of giant atoms in highly excited energy states, which occur naturally only in outer space. Laser techniques also may enable the separation of isotopes of an atom for practical applications.

Very unusual atomic forms, exotic atoms, made up of unstable particles that can be produced by high-energy machines are being studied. These, along with other forms of advanced atomic spectroscopy, are making it possible to determine the fundamental constants of nature with ever increasing precision.

NUCLEAR STRUCTURE

Here, too, the physicist has moved into an era of great understanding. The availability of high-energy particles enables the study of already known nuclei and the production of heretofore unfamilar ones. Pion, electron, and photon beams are being used to probe the nucleus. Nuclei, and therefore atoms, of much greater mass than those occurring naturally may be produced. Production of such nuclei and even of grossly deformed nuclei with beams of fast heavy ions is just beginning. It may be expected that the discovery of new states of nuclear matter will shed further light on fundamental questions concerning combinations of large numbers of particles.

The problems of nuclear theory have much in common with those of condensed matter, especially those associated with quantum fluids, and there has been a successful sharing between these fields.

PARTICLE PHYSICS

Of the many particles known, one group continues to be regarded as elementary, i.e., indivisible. They are the leptons: electron, muon, the tau particle, and the neutrinos associated with each. The "zoo" of more than 200 other particles called hadrons differ explicitly from the leptons in that the forces between them are very strong, like the forces holding protons and neutrons together in nuclei. Hadrons are not believed to be elementary; indeed, very recent developments have convinced most physicists that they are made up of quarks, although they have not been seen as separate entities. The recent developments in particle physics have resulted from a beautiful interplay between theory and experiment—theory governed to a considerable extent by the idea that there is a simplicity and symmetry in nature waiting to be revealed. For example, there appears to be a match-up between the quarks and the leptons; thus, the discovery of the tau particle, a form of the electron 3,500 times heavier, is associated with a new quark suggested by the discovery of the upsilon.

The quest for simplicity and symmetry also stimulates the quest for unification of apparently different kinds of fields and forces, such as the gravitational and electromagnetic fields. Although success has not as yet been achieved, there has been a recent important advance in tying together weak interactions with electromagnetism; weak interactions being responsible for beta radioactivity of nuclei and electromagnetic interaction for the decay of excited states of atoms by emission of light.

It is expected that there will be rapid progress during the next few years in unraveling the relationships between particles and between forces. We seem to be on the threshold of tying together the weak, electromagnetic, and strong interactions. Theoretical progress in avoiding some of the poor approximations used in solving these problems is expected. Whether new particles yet to be discovered will modify current ideas about quarks cannot be predicted. There will be a major effort to search for the *W* particle, the quantum of weak interaction, at the international accelerator facility in Geneva. The *W* is believed to have about 75 times the mass of a proton, and no existing machine is capable of providing the energy to confirm its existence. However, a method capable of doing this, a beam of protons colliding with a beam of antiprotons, is being developed. The discovery of this particle would be a great step forward in the unification of the fundamental forces of matter.

Influences and Trends in Physical Science.

The present health of U.S. physical science depends on three major factors that have played a decisive role in establishing the present remarkable strengths of the field. The first is the quality of the personnel. We have benefited from a rich diversity of scientific manpower, selected on a worldwide basis, and from very special educational opportunities.

The second factor is the nature of available instruments and facilities. These depend on the industrial and technological environment and the technical ingenuity of the scientists, as well as on financial resources.

Finally, there is the difficult to define but important element of style. This includes balancing risk and certainty in the selection of a task, choosing between simple and complex methods and techniques of accomplishing it, and the judicious selection and use of needed resources. Style also relates to the way in which the interaction between experimental and theoretical aspects of the science takes place. It is influenced by the objectives of the work, whether it is oriented toward basic knowledge or applications. It reflects the degree of commitment, motivation, and faith in a successful outcome on the part of investigators.

The traditional setting for fundamental research is the university, and the burden for maintaining the thrust of the research as well as the quality of the personnel falls on universities. A current trend toward reducing this traditional role has significant implications for the future. For one thing, inflation, funding that has leveled off, and fewer academic jobs have reduced the opportunities for academic work and careers. This situation is affecting what has been for three or four decades the very high quality of our scientific manpower, and emphasizes the need for new institutional arrangements.

One such arrangement that is having an important influence on some academic institutions is the addition of a number of "permanent," nonacademic, postdoctoral research positions, thereby changing the balance between the faculty, research associates, and students. This trend has been given strong impetus, not only by the changing job market, but also by the growth in research activities centered at large and complex off-campus facilities, since the team effort associated with these activities usually must make use of some experimental full-time scientists.

The growing need for more complex, larger, and more expensive instruments has been manifest for many years in fields such as astronomy and high-energy physics. However, in many other fields as well, the advanced status of research requires more sophisticated instruments in the laboratory and also an increasing need for large centralized facilities.

Except for astronomy, where the need for centralized facilities has been generally recognized, scientists have usually resisted the transition from the small one-man laboratory to the large centralized facility. The idea of the independent scientist and a few of his students working together with apparatus that each can comprehend completely and control is rooted in tradition and continues to be attractive. However, as physical science has become more complex, it has become increasingly evident that the complexity of the apparatus and extent of the facilities must increase if certain research opportunities are to be fully exploited. This leads to a major shift in style—to team research, to the commuting professor, to the student resident at a facility hundreds or even thousands of miles from his university. Such adjustments were made in the field of particle physics and in space sciences some years ago; they are currently taking place in nuclear physics and are about to occur in some special areas of condensed matter science.

Of course, team effort is not new to engineering development. It has been the necessary style in applications of technology on a large scale. However, the need to speed up the translation of the results of recent scientific research into high technology, as in the development of nuclear fusion energy, is introducing new relationships, leading to increasing importance of interdisciplinary work between the basic sciences and engineering.

The ubiquitous computer has also profoundly affected the style of physical research as has almost every other information-related human activity. High-powered computers, which make possible the solution of scientific problems that cannot be treated by other methods, have become powerful scientific tools.

Recording and analysis of experimental data have taken on new dimensions through the development of small computers. Computers linked with scientific apparatus are now essential whenever large amounts of data are involved. They also make possible continuous monitoring of experiments so that adjustments can be made during the course of the experiments. Furthermore, much of the analysis of a complex experiment can be completed within a short time after the data are taken.

In summary, it can be said about the prospects for the science of structure of matter that while the field is in good health in the United States now, there are trends suggesting a weakening of the fabric in the immediate future. Their cause and effects must be given close attention if we are to overcome them.

REFERENCES

1. Gott, J.R. III. Will the Universe Expand Forever? *Scientific American* 234(3):62–79, 1976.
2. Weinberg, S. *The First Three Minutes: A Modern View of the Origin of the Universe.* New York: Basic Books, 1977.
3. Kaufman, W. *Cosmic Frontiers of General Relatively.* Boston: Little Brown, 1977.
4. Hartline, B.K. In Search of Solar Neutrinos. *Science* 204(4388).42–44, 1979.
5. Schawlow, A.L. (ed.). Lasers and Light. San Francisco: W.H. Freeman, 1969.
6. Drell, S.D. When is a Particle? *Physics Today* 31(6):23–32, 1978
7. Schwitters, R.F. Fundamental Particles with Charm. *Scientific American* 237(4):56–70, 1977.
8. Glashow, S.L. Quarks with Color and Flavor. *Scientific American* 233(4):38–50, 1975.
9. Nambu, Y. The Confinement of Quarks. *Scientific American* 235(5):48–60, 1976.
10. Weinberg, S. Unified Theories of Elementary-Particle Interaction. *Scientific American* 231(1):50–59, 1974.
11. Kirk, W. Parity Violation in Polarized Electron Scattering. *SLAC Beam Line* (Stanford Linear Accelerator Center). Report No. 8, October 1978. (An example of the nature and difficulty of experiments in high-energy physics.)

BIBLIOGRAPHY

Eisenberger, P., and B.M. Kincaid. EXAFS: New Horizons in Structure Determinations. *Science* 200(4349):1441–1447, 1978.

Estrup, P.J. The Geometry of Surface Layers. *Physics Today* 28(4):33–41, 1975.

Ferris, T. *The Red Limit: The Search for the Edge of the Universe.* New York: Wm. Morrow, 1977.

Field, G., D. Verschinier, and C. Ponnamperuma. *Cosmic Evolution.* New York: Houghton Mifflin, 1978.

Friedman, H. *The Amazing Universe*. Washington, D.C.: National Geographic Society, 1975.

Mermin, N.D., and D.M. Lee. Superfluid Helium 3. *Scientific American* 235(6):56–71, 1976.

Physics in Perspective, Volume I and Volume II, Part A. (NRC Physics Survey Committee). Washington, D.C.: National Academy of Sciences, 1972.

Rhodin, T.N., and D.S.Y. Tong. Structure Analysis of Solid Surfaces. *Physics Today* 29(10):23–32, 1975.

Tauc, J. Amorphous Semiconductors. *Physics Today* 29(10):23–31, 1976.

II TECHNOLOGY

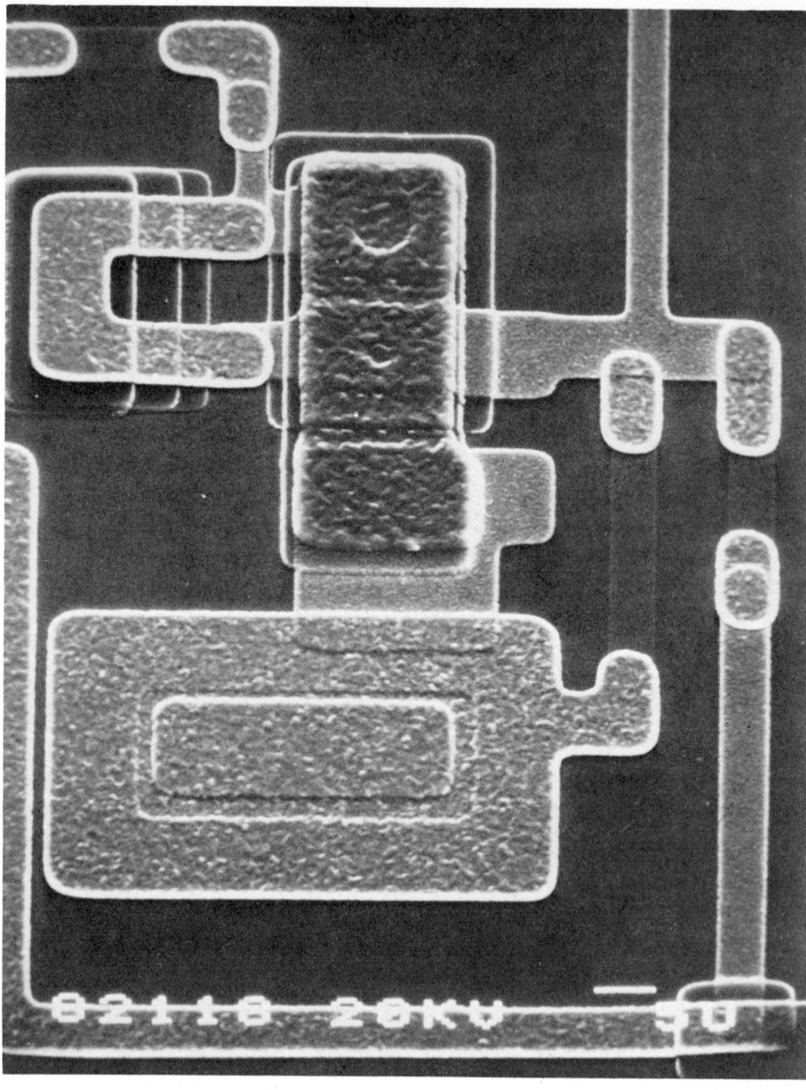
20KV
5U

4 Computers and Communications

INTRODUCTION

Over the next five years, we can expect that technical factors will continue to drive computer development. The rapidly falling cost of computational power already has led to amazingly inexpensive and versatile computer circuitry and to significant though less dramatic reductions in the cost of data storage, input, and display equipment. All this is making for a swift development in breadth, and allowing computers and computer-controlled equipment to penetrate even more broadly into the daily reality of factories, offices, and homes. In the next few years, the computer uses we already see about us—reservation systems for airlines and hotels, microcomputers that control cooking times and temperatures in microwave ovens, chips that adjust the flow of fuel to automobile engines—will expand rapidly.

These near-term uses, in which millions of microcomputers may come to be used during the next decade, will also have an impact on many large systems. Telephone automatic dialing systems that allow call-backs and waiting for busy signals are already in use. The technology that makes possible today's cash-dispensing terminals will grow, as will office text-editing and document preparation and distribution systems. Continually improved computer capability and lower cost are the keys to this expansion.

◀ Micrograph of experimental Josephson circuit, which operates at speeds approaching the limits imposed by the speed of light. (International Business Machines Corp.)

CURRENT TECHNOLOGY AND SYSTEMS

ORIGINS FROM SOLID-STATE PHYSICS

Mechanical aids to computation such as the abacus have been in use since ancient times. And a great many machines for adding and multiplying had already been proposed by the end of the eighteenth century. Many of the most fundamental ideas used in today's computers can be traced back to the nineteenth-century British inventor Babbage, who with Admiralty support attempted to build a computer organized in a manner very similar to today's.

After World War II, a whole generation of vacuum-tube based computers was developed that routinely accomplished large computational tasks. But practical computers of today's high speeds are very much the product of sophisticated electronic technology.

HISTORY OF INTEGRATED CIRCUITRY

Early in the history of electronics and radio, experimenters recognized that certain materials had special electrical properties. The crystal detectors—the "cat's whiskers"—of the first radios and the selenium rectifiers of the 1930's, prized because they allowed electrical current to pass in only one direction, were soon in wide use.

As the quantum theory of crystalline materials subsequently developed, physicists came to understand that semiconducting elements like selenium and germanium constituted a potentially useful borderline between the conducting materials, such as aluminum and copper, which pass electrons freely, and the insulators, which do not. Research on these semiconducting materials in the mid-1940's revealed that it might be possible to build electrical amplifier circuits by forming junctions between semiconducting zones of different properties and that solid-state amplifiers of this sort might replace the inefficient and bulky vacuum tube. Starting from their observations on the use of semiconductors, Shockley, Bardeen, and Brattain of Bell Laboratories mounted a systematic theoretical and experimental thrust toward this goal. By December 1947 their work, and in particular their ability to manage the science of materials, resulted in their creating the first transistor.

This Nobel Prize-winning breakthrough was at once recognized as a revolution in electronics. Not only were transistors much smaller and more reliable than the vacuum tubes they replaced; but, more important, they did not require a hot cathode as a source of electrons, and hence consumed very little power. This revolutionized the form and function of everything electronic, from the portable radio up. Moreover, the transistor's advantages in size, reliability, and low power-consumption made it practical to put together much more complex electronic assemblages than had previously been attempted. This fact is a technological cornerstone of the computer age.

STEPS TO LARGE-SCALE INTEGRATION

Between 1947 and 1957 germanium was the dominant semiconductor material in transistors. But it was soon realized that silicon, another attractive semiconducting material, had one great advantage: When it is heated in the presence of oxygen, a layer of quartz forms upon its surface. Quartz is one of the best-known insulating materials and also is relatively impervious to most of the environmental impurities that can degrade transistors. The possibility of encapsulating silicon in very thin quartz films simply by heating, and then of etching controlled patterns through the quartz, led in 1960 to silicon planar technology, the basis of all current microelectronic technology (see p. 310). In this technology, small regions with special electrical properties are formed in silicon by diffusing impurities, such as phosphorus, through holes etched in the quartz layer on top of a silicon wafer. Metallic conductors connecting these regions are then deposited in patterned strips atop this same quartz layer. This technique, which can be used to create very complex circuits with individual components of microscopic size (e.g., 0.004 millimeters), is the basis for today's sophisticated large-scale integrated (LSI) technology.

CHIP SIZE AND COSTS

Solid-state technology has surged forward during the past decade, revolutionizing the computer industry. A key factor has been an increasing ability to install complex logical functions on a single chip. (In the manufacture of semiconductor devices, large sausage-shaped single crystals of silicon are sliced into thin circular wafers with the circuitry then developed on their surfaces. After processing these wafers are cut into small chips, about one-fourth inch on a side.) Starting from a one-transistor-per-chip technology in 1960, the number of transistors on a single chip has approximately doubled each year. Chips manufactured today routinely contain tens of thousands of transistors and other components, complete with electrical connections. Today's LSI chips can, for example, store 64,000 binary bits of information (Figure 17) and make them available to a computer within a small fraction of a microsecond. A complete microcomputer, with processor, memory, and input–output circuitry, can now be built on a single chip. Moreover, LSI chips are manufactured by a kind of photolithography, in which cost is relatively

FIGURE 17 64K-bit chip stores information roughly equivalent to 1,000 eight-letter words. (International Business Machines Corp.)

independent of the pattern being developed. As the capability of chips has risen dramatically, chip cost has remained nearly constant.

More complex functions are attained merely by imprinting a more complex, finer-grained pattern on each chip, producing a larger number of individual electronic elements. Smaller elements tend to increase operating speeds and decrease power requirements, which makes the dominant trend to finer structures especially important. To maintain the momentum of this trend, it is necessary to maintain the accuracy with which parts are shaped and the accuracy with which successive layers of microscopic parts are aligned. It also means increasing the number of layers of metal and glass that can be accurately built up on the surface of a chip. Where processes involving five and six layers dominated in the late 1960's, new chip manufacturing processes typically involve 8–12 layers. This trend may be expected to slow since it is difficult to manufacture many layers without raising chip costs and lowering chip yield (i.e., the number of chips that pass electronic inspection at the end of the manufacturing process).

Components used in today's high-volume complex integrated circuits are as narrow as four micrometers, and new optical image projection systems are expected to reduce this to one–two micrometers over the next several years. Historically, the size of integrated circuit elements, i.e., single transistors, has been cut in half every six years. Progress at about this rate can be expected to continue through the next decade, putting circuits with one-micrometer dimensions into mass production by about 1990. Nonoptical techniques (electron-beam writing and X-ray lithography) may provide even smaller structures, those in the submicrometer range.

Chip costs can be controlled and even reduced by decreasing the number of defects on a processed wafer of silicon through improved process control and manufacturing cleanliness. A significant factor here is the broader application of new processing techniques, such as projection masking, plasma etching, and ion implantation.

An additional favorable cost factor is the use of larger silicon wafers, which allows more chips to be produced per processing step. Wafer sizes have increased from the two-inch standard diameter of 1968 to three–four inches today. Equipment for processing four-inch wafers is now generally available; further expansion to five- or six-inch wafers in the early 1980's is expected. However, upgrading production facilities will require much present equipment to be replaced by larger, more expensive equipment.

The rapid growth within the solid-state industry of smaller devices, greater complexity, and more functions per chip is well illustrated by Figure 18, showing the rising number of bits per solid-state memory chip. This development has taken us from 1,024 bits per chip in 1971 to 64,000 bits per chip in 1978, has kept chip cost approximately constant, and has thus decreased cost per bit by a factor of roughly 16. It seems reasonable to

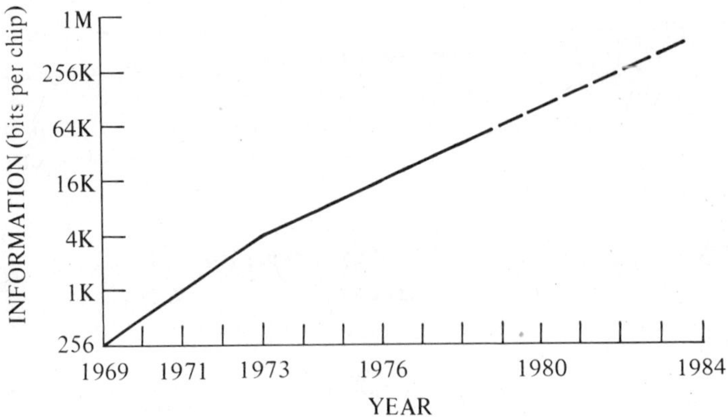

FIGURE 18 Growth of semiconductor technology.

extrapolate this trend, and to anticipate 256,000-bit memory chips in the early 1980's and million-bit memory chips by the mid-1980's.

The development of microcomputers shows a similar trend. Microcomputers have evolved from minimal units involving 2,000 transistors in 1971 to devices that today integrate 30,000 transistors on a chip. Here also progress will continue for the foreseeable future.

Speed has also increased substantially, from the microsecond speeds that typified early integrated circuits to today's fastest commercial chips, which are capable of adding a pair of decimal digits in less than five-billionths of a second. This development should also continue, but at a decelerating pace unless the industry progresses to a new technology, such as the Josephson devices discussed below.

ALTERNATIVE TECHNOLOGIES

Throughout its history the silicon integrated circuit has faced potential competition, but thus far none of these competitors has become important. Nevertheless, it is worth having a look at two of these alternative technologies, one oriented toward storage applications, the other toward extremely high-speed computation.

The first of these is the so-called magnetic bubble technology. This technology exploits the fact that discrete regions of deformed magnetic fields can be formed within thin films of crystalline garnet and then moved under the influence of external magnetic fields along precisely controlled paths. These special magnetic regions or bubbles can be used to store information in much the same way as the magnetized spots in a conventional magnetic tape; except that in a bubble memory it is the stored spots that are moved magnetically, thus the medium that contains them need not move. This allows much faster data motion and makes for much more reliable devices. Unlike silicon circuit memories, bubble memories retain information even when power is turned off; in this respect, they are similar to magnetic discs or tapes.

Another advantage of bubble memories is that their extremely simple physical structure may enable very high storage densities, with several million bits stored per garnet chip. Their disadvantage is that they do not offer the same type of ultrafast information access, within 0.1 microsecond, that silicon memory devices do. Rather, they must cycle through a potentially large number of stored words of information to reach a particular item. This requirement increases typical information retrieval times for magnetic bubble devices to hundreds of microseconds. They thus represent potential all-electronic substitutes for electromechanical bulk storage devices and for the cheapest forms of electronic storage, rather than for fast memories.

The first commercial bubble memories appeared in 1976, some eight years after research had first demonstrated their feasibility. Bubble devices storing 256,000 bits are expected to be available in 1979, and 1-million bit structures are expected a year or two later.

Josephson junction technology is a second alternative to today's silicon chip. It represents a superspeed circuit technology that may begin to be used, at least for special applications, within the next decade. Josephson devices have been shown in the laboratory to switch at speeds of fractions of one-billionth of a second. Moreover, the power dissipated by devices of this kind is very low. The most likely initial use for this technology is in special ultraspeed applications beyond the capacity of silicon devices; for example, in extremely high bandwidth multiplexers for satellite data reception. The rate at which Josephson devices are perfected will be determined by the extent to which resources are committed to this development during the next few years; but it does seem unlikely that any commercial computer will use this technology during the next five years.

It is worth noting that the finite speed of light limits computer performance in a manner that is steadily becoming more significant as circuit speeds increase. Present-day circuits switch in about one-billionth of a second, which is the time required for light to travel about one foot. Josephson circuits could switch in the time it takes light to travel an inch. Since light is the fastest of all physical signals, ultraspeed computers will also have to be physically small.

We have discussed some of the more radical alternatives, but many variations are also possible within the area of semiconductors themselves.

These include logic designs based on gallium arsenide and on silicon on sapphire and a great variety of process innovations and new logic designs.

MICROCOMPUTERS AND MINICOMPUTERS

By 1971 semiconductor technology reached a point at which it became practical to build a complete small computer central processing unit—or microcomputer—on just one or two silicon chips. When combined with appropriate memories, themselves consisting of only a few LSI chips, the resulting microprocessors were capable of performing a very wide variety of high-speed control or computational functions. The advent of inexpensive microcomputers has made it practical to employ computer control in hundreds of applications, from sewing machines to television games, which had previously been out of reach for size, cost, or other reasons. Total design costs are kept low because a microprocessor consisting of just a few standardized chips could be used in a great variety of applications, ensuring that a single design could be produced in hundreds of thousands of identical copies.

The first microcomputers had simple overall structures involving relatively few transistors. Evolution since these first designs has been rapid and is expected to continue for many years. Microcomputers have already developed to a level of function overlapping that of yesterday's minicomputers. The computer size and cost spectrum now runs from the largest machines costing several million dollars to the one-chip machines costing less than five dollars.

Directions of Microcomputer Evolution

The evolution of microcomputers is dominated by two major design considerations. The first is that the manufacturer must choose between offering a fixed processor versus a compatible family of chips that can be combined in many ways. The second is the size of the arithmetic quantities that can be manipulated directly on a single chip.

Fixed-design microcomputers are now being delivered by several manufacturers. Generally these are intended for specific control applications, where very large volume use is anticipated, for example, the control of automobile engine spark timing, fuel injection, or exhaust gases. As performance requirements for these high-volume applications expand, single-chip microcomputers will grow in capability and in memory size.

Microcomputers oriented toward complex control applications and toward data processing are evolving into families of compatible chips, allowing very flexible ranges of small systems to be built. One popular approach is to provide chips that can be connected end to end to form arithmetic processing units capable of manipulating quantities of any desired size. In addition to general-purpose microcomputer chips, such families will include memory chips, chips to control peripheral devices such as printers or data-storage discs attached to computers, communication controllers, and special processors, such as high-speed arithmetic units.

Another major trend is toward the development of more sophisticated basic microcomputer designs featuring increased word size. The first microcomputers operated on only four binary bits (essentially one decimal digit) of information at a time. With such short words, many cycles were required to handle multidigit arithmetic data. Consequently, these first microcomputers were relatively slow and also hard to program.

The second generation of microcomputers, still in wide use today, provided units capable of processing 8-bit words (essentially two decimal digits) in parallel. Recently microcomputers have been extended to 16-bit words. These new microcomputers, which are also more advanced architecturally, exploit the new semiconductor technology that allows the manufacture of semiconductor devices 10 to 50 times more complex than the first such devices. Some of these 16-bit microcomputers are more than 10 times as fast as the earlier 8-bit ones.

Micros and Minis

Today microcomputer performance overlaps a significant portion of the range that has been covered by minicomputers; distinctions between mini- and microcomputers based on hardware performance will tend to blur over the next five years. Increasingly, the distinction between micro-, mini-, and larger computers will depend on the peripheral devices to which they are attached and the software systems with which they are offered. Minicomputers will be those small computers that are supported by substantial peripheral devices, including high-speed discs and printers, and extensive software systems; while microcomputers will be those small computers, consisting of a very few chips, which are buried within some larger device and not separately visible. High-performance computers, some perhaps consisting of only a few dozen chips, may manage very large storage systems and numerous attached peripheral devices representing most of a system's total bulk and cost. Moreover, such large systems will typically be provided with extensive applications-oriented software, translation programs for languages, and complex multi-user operating systems. Microcomputers typically are supplied with only minimal software support packages; minicomputers are in between.

We can also expect microcomputer technology to reach out in a number of important specialized directions. Special signal-processing applications can exploit many of the design and logical control techniques developed in connection with microcomputers, and this will allow an increasing part of currently analog-signal processing to be accomplished digitally. This makes possible great improvements in the fidelity of signal recording and transmission, a technical development that may eventually lead to home sound- and video-recording systems of the highest quality. Another specialized direction growing in importance is represented by fast arithmetic units to supplement microcomputers.

Over the next five years, microcomputers of truly impressive performance, incorporating advanced designs, will appear. It seems only a matter of time until microcomputers possess the 32-bit word length now employed in many large data-processing systems. Indeed, in the next decade, micros will possess the computing capabilities of all but the largest existing machines. The microcomputer, by immensely extending the range of economical computer applications, will become a major computational workhorse for the future.

VERY LARGE COMMERCIAL AND SCIENTIFIC COMPUTERS

An important trend in commercial data processing is toward increased use of large, constantly active data bases. Therefore, effective use of memory hierarchies—i.e., data-storage systems that combine very fast electronic memories for storage of transient data with very large but much slower rotating disc memories—will be a central design concern. Inexpensive, magnetic-bubble memories or systems based upon arrays of charge-coupled devices may win a place for themselves in the price and performance range somewhere between fast electronic storage and bulk disc storage. If three-level memory systems involving fast electronic, magnetic bubble, and bulk disc memories come to be. widely used, hardware and software systems that help manage the resulting problem of data motion between memories of different speeds will be required.

The Reliability Problem

Organizations are becoming increasingly dependent upon the uninterrupted functioning of their computers. The trend to larger volumes of on-line data in normal commercial operations increases this dependency. Hardware and software reliability will therefore become a central issue in computer system design and can be expected to absorb a good deal of design attention. To achieve hardware reliability in the present environment of fixed design costs and falling circuit costs, designers will probably tend to use redundant circuitry more and more. The problem of software reliability is not amenable to any equally easy solution, but useful techniques of data duplication, which allow file recovery after hardware or software failure, are starting to develop. Special hardware and software features that ease the need for such data-backup operations will be considered intensively.

Distribution of computational tasks over computer networks is another approach to system reliability (see p. 236). Consider the case of a distributed system that interconnects several computers. If one machine fails, its tasks can be transferred to the others in the network. This achieves many of the benefits of circuit redundancy.

Competing Designs

Scientific computation exhibits an insatiable demand for raw computational power. In the search for ways to meet this demand, the computer designer may find it appropriate to use computing circuitry much more lavishly than would otherwise be reasonable—for example, duplicating elements so that many computations or computation fragments can proceed simultaneously.

Two main types of general-purpose scientific computers have been developed: single data-stream computers and vector processors. The fastest single data-stream computers, which execute a series of instructions to produce a single data result, have now attained instruction rates in the neighborhood of 20–40 million instructions per second. Vector processors, which execute similar sequences of instructions but apply them in parallel to produce multiple outputs, can generate results 10 times as fast. In a single-stream processor, computation speed can be increased, up to a certain point, by adding circuitry.

This added circuitry can serve a variety of purposes. Arithmetic operations such as multiplication can be accelerated by combining the action of large numbers of individual circuits; multiple arithmetic processing units can be provided and their use scheduled by special hardware. Overlapped or look-ahead computers containing multiple arithmetic units can be programmed in the same straightforward way as strictly serial computers, but gain additional speed by internally overlapping the various subphases of instruction execution. If pursued vigorously, the use of overlapped instruction scheduling should allow serial computers capable of executing 100–300 million instructions per second, as compared with the present 20–40 million instructions per second, to be developed by the mid-1980's.

However, this overlapped serial approach to computer design does have its limits. As the number of circuits used in a computer grows beyond 100,000, the marginal effectiveness with which computation can be accelerated by integrating additional computing elements diminishes. In this range the alternative vector–processor design begins to appear attractive, especially since the inherently repetitive design of vector processors allows relatively low-cost, mass-produced chips to be used. Moreover, since vector processors apply each instruction to many data items, their requirement for high-speed instruction handling is diminished.

The main problem of such vector processor systems is not their design or production, but the software problems connected with their use. Present problem-solving approaches cannot easily be adapted to run efficiently on vector processors. Straightforward program adaptation often leads to programs that use only a small percentage of the computing power actually available from a vector processor, but no very systematic program translation procedure for doing better than this is available. Moreover, our theoretical understanding of programs and computational procedures appropriate for vector processors is still slight. For this reason, high-speed overlapped computer architectures, which can be programmed effectively using well-established current techniques, are bound to remain attractive for scientific and commercial processors, in spite of the advantages of simplicity and ease of manufacture that favor vector processors.

Microcomputers, which are produced in very large volume and attain very high levels of integration, provide exceptionally cheap arithmetic computing capability. For this reason, it has been suggested that large-scale scientific computations be performed by large networks of microprocessors working in parallel. However, the problem of programming such systems effectively is even more severe than that of vector processors. Nevertheless, the strikingly low cost of large-volume microcomputers emphasizes the considerable significance of design cost in the total cost of large computers. Given that the quantities of computers produced and sold will range from about a thousand for intermediate sizes of machines down to a dozen or so for the very largest supercomputers, a concerted effort toward automation in chip and computer design may significantly lower costs and increase performance.

Current design procedures for integrated circuits still tend to incorporate many manual steps, driving design costs to a level that is sometimes estimated as high as $100 per transistor. This cost level is clearly unacceptable in a technology that is making it possible to put hundreds of thousands of individual transistors on a chip. We need much more thoroughly automated design practices, ones that will in time resemble the computerized translation techniques we use to generate machine-level computer codes from programs written in less detailed programming languages.

SOFTWARE: PROBLEMS AND TECHNIQUES

The rapid rise of computer science has now made the words algorithm and computer program familiar to most educated people. But there are several other words that almost, but not quite, capture these concepts: procedure, recipe, process, routine, method. Like them, an algorithm is a set of rules or directions for getting a desired output from a specified input. A program is the statement of that algorithm in some well-defined language that a computer is able to handle either directly or after mechanized translation. Although each computer program represents an algorithm, the algorithm itself is a mental concept that exists independently of any specific representation.

The distinguishing feature of both algorithms and programs is that they cannot tolerate the slightest degree of vagueness. In other words, they must be so well defined that a mere machine can follow them as written.

The difficulty of software, that is, program, development springs directly from this and from the fact that large programs are among the most complex objects that mankind has ever attempted to build. In this regard, programming contrasts sharply with other kinds of engineering development. No matter how complex, most other engineering projects will always be governed by a few dozen physical principles, whose mastery will open up a relatively smooth road to a valid design. By contrast, completion of a large programming project can require the mastery of thousands or tens of thousands of interwoven details.

Software engineering aims to control this avalanche of detail by discovering principles that allow programming to be approached systematically, organized in standardized ways, and reduced in mass by elimination of redundant details. Related goals are to find effective ways of using a computer to check the internal consistency of large programs, to pinpoint discordant details, and to define formal techniques whereby programs can be proved correct or the generation of incorrect programs rendered impossible. But progress toward these goals has been limited, so that programming costs have been a steadily growing fraction of the costs experienced by private and governmental computer users.

PROGRAMMING LANGUAGES

Early in the history of computing, the buyer of a computer expected to receive only the basic hardware. He counted himself fortunate if that hardware was fast, not too expensive, reasonably reliable, and service was available. In the mid-1950's, however, a dramatic change began. Customers began to realize that the effective use of computer hardware also required software packages, including both operating system programs to regulate the flow of work through a computer and specialized language translators that could convert formal programming language into detailed sequences of instructions for a particular machine. Except in the case of very inexpensive (mini- and micro-) computers and some very large one-of-a-kind systems, it became conventional for the computer manufacturer to supply this software. The quality and diversity of this software has become an important factor in deciding between competing computers. Recently computer purchasers have even come to expect vendors to supply major items of application-oriented software: data-base management systems, inventory control systems, or typesetting systems, to name but a few. This trend will continue.

To understand the techniques and difficulties of developing programs, one can think of the processes as an operation whose stages resemble the three main phases of architectural development: stage 1 corresponds to an architect's rough rendering; stage 2, to detailed architectural plans showing all dimensions, walls, shafts, and plumbing lines; stage 3, to full fabrication drawings showing every bolt-hole and electrical fitting.

• Stage 1. Program development usually begins with a statement of a problem in rough terms: Read the information file on company employees; then read the pay-period records on hours worked and overtime; correlate these records with the employee records, and print out a payroll check for each emploees,

• Stage 2. A programmer must then convert that rough specification into a running computer program. To do so many problems of representation and of method must be faced and solved. A typical representation question would be: In what format should the employee information file be laid out?

Method questions would resemble the following: By what sequence of actions will one convert hours worked and overtime into a paycheck for an individual employee, complete with the necessary tax, insurance, and other deductions? How will error situations, e.g., overtime records that do not correspond to any employee in the employee file, be handled? What backup files will be maintained to guard against record loss, and how will they be maintained?

• Stage 3. The final step is conversion of the sequence of instructions produced by the programmer into a minutely detailed sequence of machine instructions, which answers such questions as: In what precise order will the subparts of expressions like NET PAY=GROSS PAY (1.00−TAX RATE) −INSURANCE+BONUS be evaluated within the computer? Where, within the memory of the computer, are the various quantities that appear in such expressions to be stored? In what pattern, and to what memory locations, is the computer to jump between its elementary instruction subsequences?

Stage 2 in the above sequence is normally accomplished manually, and stage 3 automatically. A crucial question is where to draw the line between these two levels of detail. For example, how are data layouts and detailed instruction sequences to be chosen: automatically, by the computer, or manually, by a programmer?

Were efficiency no problem, the designer of a programming language could reduce the distance between stage 1 and stage 2 by systematic use of "very high level" programming languages. These languages make significant abstract structures—such as mappings, patterns of characters, curves and surfaces in space, all of which can be represented in many ways within a computer—directly available to the programmer. They also allow free use of new techniques for the combination of processes, such as parallel exploration in many directions at once and the direct programmer manipulation of programs as objects. However, free use of such powerful but abstract operations creates programs that do not always use computers in ways that attain maximum efficiency.

The opposite approach is to use a so-called assembly language, much more similar to the internal instruction language of the computer itself. Such a language makes it possible for a programmer to control all of the computer's features and to attain efficiency in a particular application. Use of a high-level language will eliminate much expensive detail, but by the same token will give the programmer less control over the machine operations and reduce the extent to which coding skill can be used to generate highly efficient sequences of computer instructions.

Efficiency-Related Considerations

Until recently, efficient use of computational power has been crucial. For this reason most of the programming languages in broad use today are sufficiently detailed to be translated easily into efficient internal instructions for a computer. However, the rapidly falling cost of computation can be expected to motivate more aggressive attempts to replace highly detailed programming with more abstract approaches. A crucial technical question is the extent to which the process of translation from condensed, somewhat abstract programs to internal machine language can be made more sophisticated. Today's program translation routines look at the texts that they translate rather myopically, concentrating on one detail at a time. To do better, one needs to develop translation routines able to uncover those significant overall facts about particular programs that good human programmers exploit. For example, to choose the most appropriate layout for a collection of data, one needs to determine all the operations a program will apply to these data, the order of these operations, and the places in the program at which all or part of the data is no longer required.

These questions belong to the discipline of program analysis and optimization. Much attention has been devoted to this area, and it is developing, but its problems are inherently difficult. Although progress to date has been slow, the current research in program optimization techniques should lead to increased understanding of this area in the next few years. Such research seeks to develop systems that will efficiently and reliably transform programs written in powerful, abstract programming languages into efficient machine codes.

COMPUTER CONTROL OF CONCURRENT PROCESSES

Despite the difficulties outlined above, programming language designers have several substantial accomplishments to their credit. There can be little doubt, for example, that by reducing the mass of detail required to put business and scientific application programs into a computer, the developers of the COBOL and FORTRAN programming languages greatly accelerated the growth of the computer industry.

The much more recent development of languages oriented toward the control of concurrent processes is another success story, and one that is still unfolding. These languages are important to those classes of programs that must function in real time and manage many peripheral devices in parallel. This requirement typifies computer operating systems. These regulate the flow of work steps through a computer and see to the simultaneous reading of input, printing or display of output, and movement of data between storage devices. The requirement also typifies military software, which may have to regulate computer systems consisting of dozens of small and large computers distributed over many command posts, aircraft, ships, and weapons. Software of this sort has been notoriously difficult to produce, delaying major computer applications or making them prohibitively expensive. Recently, however, programming languages that make concurrent-control software considerably easier have begun to appear.

These languages embody several significant ideas. The first is that of coordinated parallel processes. These arise naturally out of the need to regulate external devices having their own timing constraints. Consider an operating system that manages rotating electromechanical data-storage discs. When an input/output operation addressed to a disc is executed, a long wait for particular data to appear will be followed by a period in which these data must be quickly transferred. A natural way to organize this is by parallel processes. In this case, it is natural to think of an input/output process as an activity that monitors a turning disc until a critical position is reached, at which point the process is activated and transfers data. However, this process must alternate its activity with that of other computational, communication, and device management steps, which are in turn activated, suspended, or resumed. A second important notion is that of shared data structures, through which processes acting in parallel can communicate with one another.

These two ideas made conveniently and directly available bring a higher degree of organization to the writing of concurrent software packages; they markedly ease the writing of parallel control software, and allow flexible software systems to be created for a great variety of applications. This exemplifies the ability of programming language design to supply programmers with clean conceptual approaches to their practical problems, thereby accelerating the application of computers to new areas.

THEORETICAL COMPUTER SCIENCE AND THE CONTRIBUTION FROM MATHEMATICS

Computer science has been able to draw many of its most fundamental concepts ready made from mathematics. This fact has greatly accelerated the practical development of computing. The computer industry, for example, could not have developed as swiftly as it did without the principles of Boolean algebra from nineteenth-century mathematics. Similarly, programming found its universal possibilities and its limits laid out at its very beginning, in the work of the great mathematical logicians Godel and Turing. Generally speaking, those theoretical computer science areas that have been able to draw upon the mathematical tradition have advanced surely and rapidly.

ALGORITHM DESIGN AND EFFICIENCY

The contribution from mathematics has been particularly helpful in the design and analysis of algorithms, which, again, are patterns of steps or directions that govern the actions of a computer.

It turns out that for any desired result many possible patterns of steps or algorithms—some relatively obvious, others highly ingenious—can be used. However, the total number of elementary steps required will vary enormously with the algorithm used. For important problems, the right approach may be billions of times more efficient than a more obvious approach. Thus the choice of an algorithm to achieve a desired end is in many cases the single decision having the most impact on computer efficiency, one that can vastly exceed the impact of machine speed or code quality.

To see why this is so, consider an important but elementary problem: that of searching for a given name—for example, the name of a particular client of some large public agency—within a computer's memory. Computer memories are normally organized in sequentially numbered small subareas called cells or words. Suppose that we need to deal with 1 million such names, occupying 1 million cells. If a name is not found, the computer is to respond: "No such customer." This is, for example, a very important operation in computerized cash-dispensing systems.

If the names are held in random order within the cells, then, in searching for a given name, the computer must examine every cell, comparing the name in this cell with the name sought. No cell can be ignored, since any omitted cell may contain the sought-for name. A million comparisons will therefore be required to ensure that a particular name is not present, and half this number of operations, on the average, to locate a name that is present. Even on one of today's fast computers, this will take about a second.

Now suppose that the names are kept in alphabetical order. The search can begin by looking at the entry in the middle of the table. If the desired name is alphabetically before this middle entry, the entire second half of the table can be eliminated and need not be searched; similarly, if the name is alphabetically greater than the middle entry, one can eliminate the entire first half of the table. Thus a single comparison yields a search problem that is only half as large as the original one. The same technique can now be applied to the remaining half of the table, and so on until the desired name is either located or proved to be absent. Since one, doubled 10 times, is 1,024, another 10 doublings gives a number more than a million. Therefore, to search an alphabetized table of a million entries by this so-called binary search method, only 20 comparisons are required. By proceeding in this more efficient way, the same computer could handle at least 50,000 different name searches in a second, rather than just one search.

As with all algorithms, two questions arise in connection with this binary search procedure. In how wide a range of circumstances will it retain full efficiency? Is binary search actually the best possible search procedure, or can we devise some other still more ingenious search method? It is clear that if an unchanging collection—e.g., names or words in a dictionary—is to be searched repeatedly, the initial labor of setting up an alphabetical arrangement is easily worthwhile, since the time saved in later searches will more than recover the cost. Suppose, however, that the collection is changing rapidly because of insertions and deletions. Is it then worth maintaining the alphabetical order. If so, how can this order be maintained most efficiently? Maintenance of order is not a trivial matter, since a naive attempt to insert a single new element into the middle of a million alphabetized elements could make it necessary to move as many as 500,000 elements up or down one place.

It turns out, however, that there are special arrangements of alphabetical data that allow arbitrary insertions and deletions to proceed just as rapidly as binary searches, so that alphabetical order can be maintained. The idea is to lay out the data items as the twigs of a branching, treelike arrangement of data. The details of this arrangement, and of the pattern of actions needed to maintain it as insertions and deletions are made, are by no means obvious. This whole approach must be regarded as a significant algorithmic invention, intellectually comparable to the discovery of a mathematical theorem.

Algorithm Analysis

Algorithm design has established itself as a major branch of computer science. Something like a thousand ingenious algorithms and groups of related algorithms have been discovered and described. This includes algorithms for many kinds of numerical computation, for searching and sorting, manipulating algebraic and logical formulae, developing optimal plans and schedules, compressing and expanding data, checking for and correcting errors, translating computer languages, working out effective geometric layouts for computer and other integrated circuits, calculating properties of important abstract mathematical structures such as graphs, and dozens of other purposes. The overall aim of algorithm design is to analyze every computational task and to find the best possible way to accomplish it. We can expect this effort, which makes effective use of mathematical tools and ideas, to continue to grow as one of the most stable and productive branches of computer science.

Since many algorithms can be devised for any task, we need to be able to calculate the efficiency of competing algorithms. In doing this, computer science has used all the most highly developed enumerative tools of mathematics, including techniques originally used in combinatorics, number theory, probability theory, and mathematical analysis. These applications have in turn begun to have measurable impact on mathematics, one that is bound to grow; for example, computer science has been important in the remarkable revival of combinatorial mathematics during the past two decades.

To calculate the efficiency of an algorithm that performs a given task is one thing; to prove that the algorithm is the most efficient among all possible ways to do this task is considerably harder. To see this, again consider the problem of searching. The binary search technique is quite efficient; in particular, the work it expends in searching a table of given size grows only slowly with the size of the table. A table of a million entries can be searched with only 20 probes, and only 30 probes are needed to search a billion entries. But is this the best method? In fact, there is a technique, even more ingenious than binary searching, that is able to search a properly laid-out table of a billion or a trillion names just as fast as it can search among a thousand or a million names. It deliberately scrambles the names in a predetermined manner that allows a name to be looked up, on the average, in a fixed number of probes independent of table size. The point is that the designer of algorithms must never be content with an algorithm that is merely good. Until it can be proven mathematically, there is always the possibility of a much better algorithm.

LIMITS OF CALCULABILITY

A rigorous mathematical proof that a given algorithm cannot be bettered will always tend to be difficult, since it must examine all possible ways of accomplishing a given task. Nevertheless, by using the theoretical tools originally developed by logicians for proofs of undecidability and unsolvability, it has been possible to give such proofs in a number of cases. These same tools have been used to prove certain problems inherently difficult or impossible, and thus to define the theoretical limits of computability (Figure 19).

Proofs of this kind are particularly interesting when they show that significant problems arising in practice are inherently difficult. The bin-packing problem is an example. Here we are given a collection of sticks of various lengths and then required to arrange them end to end into groups of no more than a yard in length (although groups of less than a yard in length are allowed). The objective is to end up with as few groups as possible. (This problem arises, for example, when strips of paper of various widths have to be cut from yard-wide rolls, with as few rolls cut as possible. Another example is that of loading trucks without exceeding a stated weight limit.) Small problems of this kind can, of course, be solved by enumerating all possible groupings into two, three, or more subgroups and by testing each such group to see if it meets the criterion that no group should have a total length exceeding one yard.

However, for a case that is even moderately large (e.g., 100 sticks) this approach is totally unworkable, since the number of possible groupings is enormous. For example, the total number of ways in which 100 sticks can be separated into two groups is 2^{100}, or approximately

$$1{,}000{,}000{,}000{,}000{,}000{,}000{,}000{,}000{,}000{,}000.$$

This raises the question of whether there exists any algorithm that will find a best possible solution in all cases without having to examine some

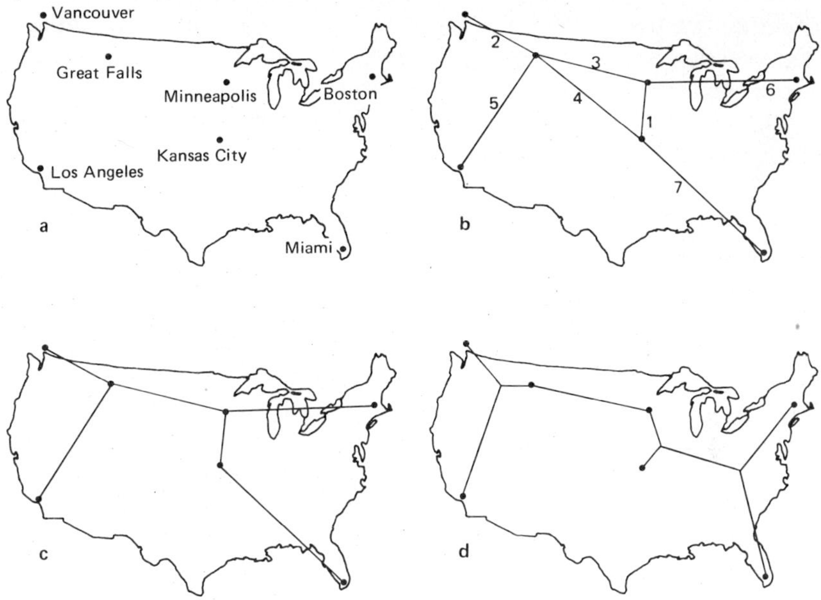

FIGURE 19 Algorithm efficiency illustrated by the spanning-tree problem. The problem is to find the shortest network of lines connecting a set of points. For example, the shortest railroad system linking a set of cities (a). If the lines meet only at cities, the problem is solvable with an efficient algorithm. First, the closest pair of cities are joined, then the next-closest, and so on (b). Lines joining cities already connected indirectly are then omitted, producing the optimum spanning tree (c). An even shorter network is possible if lines can meet outside of cities. However, there is no efficient algorithm for finding the points where the lines should meet. The optimum network (d) was found only by exhaustive search. (From "The Efficiency of Algorithms," by Harry R. Lewis and Christos H. Papadimitriou, © January 1978, Scientific American, Inc.)

substantial part of the collection of all possible groupings. Very substantial theoretical evidence indicates that the answer to this question is "no," i.e., that the problem of best possible stick packing is inherently hard. However, quite good approximate solutions can be found in most cases, as can exact solutions in many, but not all, cases.

The stick-packing problem has been related to a very wide variety of other packing, graph-analysis, and arrangement problems, all of which have been shown to be equally difficult in a precise theoretical sense. It is likely that none of these problems can be solved much more efficiently than by a comprehensive search over an exponentially enormous space of possibilities. Other slightly deeper problems—for example, various problems arising in computer manipulation of mathematical formulae—are known to be unimaginably difficult. Such results remind us that the outer boundaries of the calculable are more restricted than initial enthusiasms for the raw speed of computers might suggest. Other algorithmic approaches, such as heuristics and probabilistic algorithms, may provide approximate solutions to some of these problems. Nevertheless, since these results apply to calculating devices and robots of every kind, they also warn us that the path to development of artificial intelligence cannot be broad or trivial.

Faced with theoretical obstacles of this fundamental sort, computer scientists have begun to search for ways in which the very notion of calculability could be relaxed so as to allow a wider range of possibilities. An interesting recently discovered possibility is that of working with algorithms that need not yield correct results in all cases, provided that the probability of error can be estimated rigorously in advance and can be shown to be extremely small. A number of interesting examples, such as testing an integer for primality or a polynomial formula for validity, have been found in which by allowing even very minute probabilities of error (e.g., one error in every 10^{1000} cases processed) much faster algorithms can be devised than seem otherwise possible. Algorithms of this type illustrate some of the imaginative new possibilities that algorithm designers may be able to exploit.

COMPUTERS AND COMMUNICATIONS

The enormous advances in integrated chip technology have combined in recent years with major advances in communication technology to create a computer-communications environment that promises to have major commercial and social impact. This technological development has made large computers more usable by decentralizing access to them and moving the facilities to the everyday locations of their actual users. It has also helped the trend toward interconnecting computers from different locations into computer networks. This trend—fueled by rapidly falling computer equipment costs and also by a somewhat less steep decline in communications costs—will continue over the next five years.

These developments make it possible to connect banks, retail stores, travel reservation systems, credit verification services, and inventory control systems within a unified data-communication environment. This, in turn, has given rise to a number of experimental national and international computer networks, some of which have been in operation for some time: ARPANET, TELENET, and TYMNET in the United States and DATAPAC in Canada. Other such networks, only just beginning, are TRANSPAC in France, EPSS in England, EURONET elsewhere in Europe, and DDN in Japan.

Improvements in communications technology that have strengthened the trend toward the network use of computers include digital encoding and transmission of signals, grouping of signals into intense bursts that can share a common channel with other such bursts, new switching techniques, and satellite transmission. The basic technique of digital transmission, i.e., transmission of precisely controlled streams of bits rather than waveforms as with voice or TV, allows digital computers to be interconnected over compatible digital communications channels. It also allows digital error-correction to preserve the accuracy of information transmitted over long distances. Burst multiplexing and switching increase the efficiency of communication channel use by allowing single channels to be shared among multiple data sources and receivers. A variety of useful multiplexing techniques such as electronic fast-connect circuit switching and highly flexible time-division networks have been developed, but in recent years the packet-switching technique has become particularly important. In packet switching, data to be transmitted are partitioned into small bundles, called packets, each of which carries its own destination address and can therefore make its independent way through a communication system. Pulse-code modulation converts analog signals, such as voice or TV signals, into streams of bits, thus allowing analog traffic to flow along the same transmission channels used by computers.

SATELLITE COMMUNICATIONS

Communications satellites have developed enormously since their initial appearance in the late sixties (see p. 41). In 1970, 14 satellites had been placed in geostationary orbit, and 96 additional satellites will probably be in orbit by 1980. The decade since the launching of the first stationary communication satellite has also seen a 10-fold increase in the communication capacity of satellites, an almost 100-fold increase in their effective radiated power, and a roughly 100-fold decrease in costs. Increasing satellite power has been particularly significant since this has decreased the size and cost of satellite earth stations, from the 30-meter diameter, $3,000,000 station required for the original INTELSAT system, to the 5-meter, $30,000 station available today. Earth–station costs will continue to drop as we go to higher frequency (12–14 and 20–30 billion cycles per second) systems. Among other advantages, low earth–station costs will allow direct satellite communication between business establishments.

In addition to these important advances, the next five years will see significant advances in the following areas: larger satellites with greater on-board power generation and with on-board transmission-processing capabilities, multiple-spot satellite communications beams, and more sophisticated access and channel sharing schemes.

LOCAL COMMUNICATIONS

Satellite communication and long-haul cable and microwave systems address the long-distance communication problem successfully. But a great need remains for technical advances in short-distance communications. The local communication problem is no longer that of connecting the telephone central office to another central office, but rather economically connecting these central nodes to large numbers of individual end-users. Currently, the most prevalent local communication technology is the same one in use since the invention of the telephone and telegraph over a hundred years ago: signal transmission over pairs of copper wires. Still, wire-based communication techniques will continue to improve and will supplement the medium-range wireless and long-range satellite digital communication techniques.

One local communication problem receiving a great deal of attention today is in-plant communication, i.e., the flexible transmission and switching of large volumes of voice, image, and data traffic among a single contiguous group of user-owned premises, such as adjacent factories or offices. Current in-plant schemes commonly involve a time-shared wire or group of wires, ranging in length from a few tens of feet to a mile or so, connected to a group of terminals or other signalling devices that share access to a single long-distance, possibly broadcast, channel. A common idea in such systems is to provide a means for burst communication at full channel capacity, but to share this capacity among many data sources and receivers under the control of some mechanism that allocates the channel on a demand basis. A properly designed system of this kind can replace the large numbers of. wires that are ordinarily used for local data and telephonic transmission. This technology has great potential for in-house communication in office buildings, factories, or ships.

We can also expect a continuing development of ground radio systems for computer communication; some will divide a geographic area into local cells, while others will make use of a packet-switching technology like that suggested for long-distance communications. In the near future these will be attractive to data-communication users who must remain mobile, such as doctors, taxis, and ships.

OPTICAL FIBERS

Fiber optic transmission systems are a relatively new technology that may enhance the information-carrying capacity both of local wired communication networks and of somewhat longer-range networks (see p. 310). These systems use modulated light sources to transmit digital information along fine fibers of optical material. Compared with copper-wire systems, they offer large bandwidth (large information-carrying capacity), low weight, small diameter, freedom from crosstalk, and low cost. Fiber optics provide an information-carrying capacity as high as that of expensive shielded video cable at a cost comparable to that of the conventional copper-wire pairs used for telephones. Their ability to carry enormous amounts of data in a physically small cable suits them ideally for use in metropolitan areas, where underground conduit space is limited and difficult to expand. However, optical fibers still require special handling, protection, and splicing equipment, and are not yet fully ready to compete with copper wires.

Experimental communication links using optical fibers have already been implemented at all of the standard wideband data communication speeds, ranging from 1.5 to 274 million bits of data transmitted per second. In principle, fibers can accommodate still more enormous rates of data transmission, although at present the optical dispersiveness of fibers, as well as the limitations of the transmitting and receiving technology, do not permit more than a hundred million pulses per second on a single fiber. Current research is aimed at development of more powerful and longer lasting light-emitting elements, more complex fibers that will reduce the attenuation of optical signals below their present level of approximately 10 percent signal loss per kilometer, better fiber-splicing techniques, and more adequate optical amplifiers for use as receivers.

ELECTRONIC MAIL

The sending of messages directly between computers or terminals has a number of advantages over today's letter mail. A message can be delivered reliably and almost immediately, can be read by its recipient at leisure on a terminal, and can be acknowledged. Messages can be filed for later retrieval, answered and automatically addressed back to the original sender, and copied for retention by a sender or for transmission to third parties. Multiple addresses can be easily specified, and address lists for special purposes easily generated and maintained by each sender. Messages can be sent, not to a specific terminal, but to a person wherever he happens to be located; the receiver can retrieve his mail using a terminal wherever he happens to be. All of these facilities exist today on a number of systems.

The use of such systems is expanding rapidly, particularly within corporations. Indeed, most of the currently important computer-based electronic mail systems are used to connect dispersed locations of single firms. Rapid growth in the interconnection of these systems will soon permit electronic mail between firms. These electronic mail systems already handle computer-generated information and information directly keyed into the system. The development of facsimile scanning equipment, inexpensive high-resolution display terminals and printers, and new methods for efficiently encoding graphic information is also strengthening the capacity to transmit printed and handwritten documents. As such devices develop, they may lead to integrated communication services in which not only data and facsimile, but also voice and video, are carried on a unified network. Unified transmission of different types of traffic is already here, but the development of well-defined unified applications is taking longer than predicted.

SECURITY AND ENCRYPTION

The sharing of an electronic communication system by many people, and also the transmission of great volumes of data over easily intercepted satellite channels, create serious issues of privacy and require special techniques both to guarantee security and to authenticate message authorship. This is largely a software, rather than a hardware, problem and has been addressed through improved encryption techniques. Such techniques protect computer communications both from unauthorized reading and from attempts to insert, delete, or alter their contents. Such protection will become increasingly important as electronic transmission of mail, stock market transactions, and banking transfers begin to be more common.

Promising research is now under way in the development of electronic signature systems that can be used to sign or notarize transactions—for example, in messages between stockbrokers. Certain types of arithmetic computations, which are inherently difficult to reverse, probably can be made the basis for unforgeable electronic signatures. Such signatures can be added to any messages and used by the sender to authenticate his presence while a message's receiver can use a corresponding decryption technique to verify the source of a message.

OFFICE SYSTEMS

The past few years have demonstrated the usefulness of office typewriter systems based on computer-driven machines with televisionlike display screens. Savings in labor and speed of operation have justified the installation of such systems at many large- to medium-sized offices. These systems, which allow a great deal of flexibility in text editing, make it possible to change a particular word automatically wherever it appears in a manuscript or completely rearrange paragraphs and sentences. Other new systems can verify and correct the spelling of every word in a manuscript. Office directories are being made available on computer displays; here the ease of making new entries, deletions, and fast look-ups, to say nothing of the savings in printing costs for directory reprints, are all very attractive. These new systems are already beginning to change the offices in which they have been installed and are starting to have noticeable impact on day-to-day office procedures, on the way in which information circulates within offices, and on the daily work of secretaries.

ELECTRONIC FUNDS TRANSFER SYSTEMS

Numerous forecasts proclaim that electronic funds transfer systems will shortly usher in an era of checkless banking, if not a completely cashless society. Almost as numerous are articles dismissing this perspective as illusory, arguing that the public does not trust computers, wants to have pieces of paper in hand as evidence of funds available, and prefers to delay payment of a bill until some time after it has been received. Neither of these polar viewpoints captures the fact that electronic funds transfer is not a single entity that will either take over or fail completely. Rudimentary electronic funds transfer systems, e.g., funds transfer by telegraph, have been with us for over a century. The question now is in what form and how rapidly such transfers will grow, in what areas they will grow most rapidly, and what their impact on commerce will be.

Certain environments in which the growth of electronic funds transfer has been particularly rapid are easy to note. One such is the telephonically—and soon to be electronically—verified credit card. Whenever a customer submits such a credit card for verification and it passes, the credit card company is in effect instantaneously creating money, giving credit to the customer and guaranteeing payment to the vendor, whether or not the customer has cash on hand at the moment. At the end of the month the customer receives a consolidated statement (which testifies to his continuing desire to see and validate a written statement before authorizing payment) and discharges his obligation, generally by writing a check on a bank. Further developments are possible here, including the presentation of monthly statements at a terminal within an electronic mail system and authorization of payment from a terminal without any paper check being necessary.

Large corporations are increasingly making credit arrangements directly with business customers rather than paying a bank to arrange them. Computers, which make it easy for an accounting office to do complex, bankerlike calculations, are one factor working towards this do-it-yourself financing. Of course businesses, like individuals, are eager to maintain their access to cash float; but this can be handled in a computerized business environment by appropriate systems of discounts or fees for early or late transfer of funds.

All this makes it plain that the growing use of electronic funds transfer will affect ordinary consumers, businesses, and banks in ways that should be better understood. For example, consumer privacy may be compromised by systems that allow all the details of a consumer's pattern of purchases to be marshalled centrally. The pattern of cash float—the interval between billing and payment—upon which firms now depend may change radically. These issues, which have already received considerable attention, need to be explored further, since they may have significant impact on the whole structure of major financial institutions.[1]

HOME COMPUTING

We can expect that within a decade most homes will contain at least one computer and that within two decades virtually every home will. The homeowner may not know that it is there if, for example, a computer chip controls the tuning of a color TV or the defrosting of a refrigerator. In this section, however, we concern ourselves with more explicit home use of computers. Many such applications are conceivable although speculative:

- The combination of computer and telephone may be useful for remote control of household appliances and heating systems, delivery of personal reminders, medical monitoring, security alarms, and baby sitting.
- A wide variety of useful information services may be delivered via home computer terminals. These can provide anything from encyclopedias or libraries of relatively fixed information to highly dynamic information on current news, market conditions, supermarket specials, and want ads. Hundreds of thousands of pages of useful, highly organized information may be accessible to the subscribers to such systems through their telephones, home computers, and attached displays.
- Computer games and clubs linked by computer may provide an attractive social milieu for many enthusiasts.

Home computers, which are steadily becoming more powerful, can be terminals communicating with larger and more powerful central computers. This raises the question of the degree to which home computers will offer extensive computational capability versus the extent to which they will remain limited devices depending strongly on a central facility. Related to this is the question of how extensively home computers will be tied to the nation's telecommunications network. Currently advertised home computers—over 100,000 have already been sold in the United States—are sold as stand-alone devices, with little emphasis on their communications capabilities or even on the fact that they are often linked on hobbyist networks. Nevertheless, it is possible that the use of home computers for communication will grow very large. They may, for example, come to be used as electronic mail stations.

It is also true that the home computer of the future is likely to provide a substantial computing capability in its own right. Available for only a few hundred dollars, future home computers may have speeds now associated only with large computers. Nevertheless the home computer will need to become easier to use if its use is to spread beyond hobbyists.

In time, the need for a link to a central computer will become greater since a central computer can provide major information services, such as daily news bulletins and market quotations. Moreover, natural language processing, graphic capabilities, and sophisticated error handling are all important for most people, and these require substantial quantities of memory likely to be available during the next five years only through large central computers.

Sophisticated software will be required for home computing, and this raises significant questions. Vendors of home computers now cater to a market of hobbyists who enjoy programming. Typically, such vendors put BASIC or some even more primitive programming language on their machines and let the buyer write useful programs. Very little by way of prefabricated programs is offered. However, the broader public will not want to program home computers but to use them directly and flexibly. This points to a very large market for software packages, including programs capable of dealing with various types of natural language queries. Development of such packages, and of the advanced programming tools needed to build them, should be spurred by the increasing use of home computers, but not much has yet been done. Likewise the legal protections required for the adequate functioning of a broad software market, e.g., protection for intellectual property maintained in electronic form, have not yet been adequately defined. Because of these unresolved problems and its inherent technical complexity, software is likely to be a thorny obstacle to the spread of home computing.

Traffic patterns on other communications networks may be substantially altered as home information systems become practical, particularly to the degree that they are linked to central or other remote home computers. Over time, this can have a major effect on communication facilities that have been designed to accommodate traffic patterns characteristic of voice communications. This could substantially change the economics of proposed systems. Although there is little understanding of what these effects will be, they cannot be ignored.

A related question is that of the balance between specialized, one-program devices like computer games, and multifunctional, general-purpose terminals. Depending on the direction in which software for home computers evolves, the trend may be either to single computers, or groups of identical computers, providing many services, or to a variety of separate devices that compute, but which individually provide very different services and are called by different names. Regulations established by government, as well as pricing decisions taken by the administrators of centralized computing services, will affect this outcome.

REGULATORY ISSUES

For computer communication systems to attain their full potential they will have to be ubiquitous, cheap, and very easy for most persons to use. A number of key regulatory uncertainties will have to be resolved for this to be possible. Are systems of computers and terminals data-processing systems (and so unlicensed) or communication systems (and therefore licensed)? Should electronic mail be viewed as a service, having the character of a national monopoly and therefore to be offered on a universal basis by a monopoly carrier, be it the Post Office or the telephone industry, or simply as a set of products to be developed and offered by a wide range of competitive suppliers? What will be the impact of electronic mail on the postal system? What charge structure is appropriate? How can the needs of small users be served best?

Distinguishing Computers and Communications

In the all-digital systems of the future, it will be hard to draw a clear technical distinction between communication services and computer services. Both will use computers to manipulate bits and to transmit them between points. Identical processing and transmission equipment will be used to pull segments together into a text, edit it, address it, and distribute it. Yet the Communications Act of 1934, by authorizing the FCC to regulate electronic communication, requires the commission to seek some legal formula distinguishing between digital communication and digital data processing if it is not also to be drawn into regulation of the computer industry. That the FCC should seek some such formula is understandable and proper, but it would be an illusion to believe that such a formula, whatever it is, can easily be applied to technical reality or that this formula could be effective for very long, considering the rapid and unpredictable course of technology.

While some portions of the communication/computer industry naturally require regulation, others are best served by fostering vigorous technical and economic cooperation. Regulation of limited and eventually congested satellite "parking space" and communication bandwidth is of course required. As with ground radio, a limited spectrum must be organized and allocated. Continued regulation is also likely to be required in the use of public right-of-way for information distribution systems, whether copper-wire pairs, coaxial cable, or optical fibers. In its role as guardian of the quality of public communication services, the FCC must also bear in mind the potential impact of sudden and heavy computer-generated demands for communication bandwidth on the services received by more conventional systems users. Moreover, alongside the electronic compatibility and standardization requirements of the past, a new and far more complex set of compatibility issues—this time related to software—is growing. However, in responding to these problems by new regulations, the FCC must also avoid hampering the use of new digital communication equipment or the provision of new computerized communication services.

Still another range of problems relate to the First Amendment. The Constitution guarantees the freedom of the press, but not particularly the freedom of business machines. Freedom of the press has been construed to include radio and television. Are computer communications also protected? In this connection the rather different legal history of the telegraph and of radio is interesting. If one looks back at the history of telegraph law, few references to First Amendment issues are seen. Yet such references are frequent, and have often been central, in the legal history of radio. The reason is that the telegraph has been thought of as a business machine, not a device to which a citizen would rush to express his concern. By contrast, radio was a means of expression from the beginning. Although computer communications were first regarded as adjuncts of business operations, there is good reason to believe that such communication will eventually encompass a great part of our society's means of interpersonal contact and expression. If home computer terminals become a major channel for mail, for consumer information, for education, and perhaps for political campaigning, they may end up being as important to the twenty-first century as the printing press has been for the past 500 years. If so, the First Amendment will become important to practices and arrangements for the regulation of computer communications.

Besides regulation and competition, another important force affecting the rate of development of new computer/communications services is the U.S. government's own large equipment and service purchases. The government has procured the most sophisticated technologies available for its national security and space programs, but not for its ordinary civil activities. Here government has a significant chance to promote technological innovation; to develop a technical base upon which more sophisticated applications must rest.

ARTIFICIAL INTELLIGENCE

Since this report is concerned with the next five years rather than the next few decades, we have focused most of our attention on practical technological developments that are most vigorous today. Even so, a short survey of the revolutionary long-term efforts to create humanlike capabilities in computers is in order.

The computer is not just an adding machine; it is a universal information-processing engine, one which may be capable of duplicating many of the most characteristic capabilities of the human brain. This possibility, intuitively grasped by such pioneers of the computer age as Turing and von Neumann, has continued to motivate and fascinate computer scientists.

But even if we admit construction of artificial intelligences and superintelligences to be the ultimate perspective of the computer age, it still does not follow that one or another branch of computer science should be emphasized, nor can we tell what discoveries, still hidden from us, will emerge, nor whether the necessary development will take decades or centuries.

Work on programming computers to display humanlike problem-solving characteristics is widespread and ranges from work aimed at modeling human performance in specific problem-solving areas, such as medical diagnosis, to programs whose search techniques are based on heuristics and "rules of thumb" and even to the use of computerlike models of thinking as a tool in cognitive psychology. The use of human intelligence in real world situations requires an intimate coordination of thinking with seeing or hearing and motor action. The branch of artificial intelligence research that seeks to design systems capable of sensory-motor coordination is generally called robotry.

While only a small fraction of artificial intelligence research has been directed toward sensory-motor tasks and robotry, this domain does provide a rather broad view of the nature of the research problems across the whole field. For this reason, and because of the growing practical importance of industrial robotry, we will look a little more closely at the current status of artificial intelligence research on robots.

ROBOTS

The general goal of research on robotry is the development of versatile robots that can perform varied mechanical tasks in homes, factories, offices, and out-of-doors. To be successful such robots require a broad range of abilities. They must be able to plan a sequence of actions and to simulate and test the execution of these actions in a model of the environment. Continuing correspondence of the model to a dynamically changing environment will depend on a robot's ability to acquire and analyze sensory information. When surprises occur, the robot must be able to plan and carry out corrective actions.

Even though these abilities do not seem to require high "intelligence," their realization in machines is a complex matter. Techniques powerful enough to permit development of truly flexible robots still elude us. However, specialized robots that can automatically perform many assembly, materials handling, and inspection tasks are now coming within reach.[2] These systems are much more versatile than the numerically controlled machine tools now in use; they are controlled by small, general purpose computers that respond to touch, force, and television inputs and are guided by general descriptions of the task. The evolution of robot assembly systems should henceforth be continuous. Ultimately these can be expected to develop into "automatic factories"—versatile computer controlled systems for processing raw materials into finished goods (see p. 320).

The more elusive goal of a general service robot will require programs capable of analyzing major sensory inputs and responding to them with something like human sophistication. Researchers are already attempting to duplicate all important sensory functions, including speech recognition, language analysis, and analysis of visual patterns. Machines are now capable of recognizing small numbers of words and short phrases spoken in isolation, but correct response to continuous speech is much more difficult. A recent five-year research effort demonstrated the feasibility of experimental systems capable of responding to spoken vocabularies of several hundred words, but much more basic work is needed before practical speech-recognition systems can be built. Challenges will continue to inspire work over many decades.

LANGUAGE AND ROBOTS

It is of course much easier for computers to deal with information in printed than in spoken form; but even for written material the subtlety of language puts truly flexible natural language processing somewhat beyond a computer's grasp. Translation between natural languages and internal computer languages cannot be accomplished by rudimentary dictionary look-up techniques. To understand natural discourse, one must have a reasonably good model of the content of the discourse. (To see this, think of what is needed to understand the following sentence properly: "The rancher, returning his pen to his pocket, stepped into the pen.")

The development of such models and their integration into language-analysis programs is a formidable task, on which only experimental starts have been made. However, there are information retrieval systems that allow one to communicate successfully with a computer using subsets of natural English dealing with restricted subjects. Meanwhile, researchers are currently experimenting with more sophisticated natural language systems possessing larger knowledge bases. Here it is reasonable to hope that within the next ten years computer teaching systems, where natural language capability is highly desirable, will be able to carry on adequate English dialogues with students in certain specialized subjects.

Large bases of knowledge concerning particular subject domains are also of central importance to evolving computer systems of the consultant type, in which fragments of knowledge collected from expert practitioners in a specialized field, such as medical diagnosis or ore prospecting, are collected and organized to attempt duplication of human judgments.

SIGHT AND ROBOTS

General purpose robots would certainly have to process visual information. The problem of visual perception by computer is a minefield of subproblems. At the simple end of the scale, systems already exist for comparing what the robot's eye would see with stored pictures; these are used today to guide a missile toward a target. The problem of interpreting objects in photographs collected routinely by weather and surveillance satellites is somewhat more difficult, but still feasible. Reliable techniques have been developed to automate parts of these photo-interpretation tasks. Similar systems can also be used to spot roads, bridges, or railroads in aerial photographs. Related techniques allow machine reading of printed text.

Somewhat more sophisticated vision systems are now being introduced into factories to control industrial robot arms during materials handling, inspection, and assembly. Research in machine vision is concerned with such issues as reconstruction of the three-dimensional shape of a solid body from the pattern of lines and corners that constitute its image on an electronic retina. Work has also begun on the detection of the presence and form of bodies, parts of which are obscured by other bodies. Visual abilities of this level of sophistication will be required by general purpose robots intended for household or outdoor use.

The widespread research effort to endow machines with abilities of this sort is contributing to the development of a computational theory of intelligence that is beginning to have an impact on psychological theory. Significantly, one of the most important scientific insights resulting from the work is the realization of how complex the processes underlying intelligent behavior must be and what a vast, but somehow patterned, storm of perception and selection must underlie every conscious act or thought. However, beyond this sheer complexity, there seems to be no conceptual barrier to the eventual synthesis of intelligence in machines, which may very well be the final outcome of the march of technology that we have described.

For now we can epitomize the attainments, hopes, and disappointments surrounding the field of artificial intelligence by glancing at one of its amusing, yet typical and challenging efforts—the design of chess-playing computer programs.

The first serious attempts to develop game-playing programs go back to the 1950's. Progress with simpler games (such as checkers) was swift and dramatic. Chess proved more difficult. In fact, computers still cannot match the ability of the human chess master to discern critical sequences of positions without explicitly examining a very large number of cases. Nevertheless, chess-playing programs have improved steadily. In 1977, a small but significant landmark was passed: CHESS 4.6, a program developed at Northwestern University, drew one game and won another (although losing its match) against British chess master David Levy. And these programs continue to improve.

OUTLOOK

The following outlook section on computers and communications is based on information extracted from the chapter and covers trends anticipated in the near future, approximately five years.

Over the past two decades, electronic computers, initially the esoteric tools of a small scientific community, have become familiar presences in daily life. As a result, the shopper who keys in for a quick infusion from a computer-controlled cash dispenser and the child whose exposure to computers begins with a video game are both beneficiaries of this new technology. More and more of the information upon which business and administration depend is available in machine-readable form and managed by computers. The technology that has made this development possible has drawn ideas from many areas ranging from mathematics to linguistics. But its driving force during the last decade has been the amazing decline in the cost of semiconductor chips. As chip costs continue to fall during the next five years, this trend will put more and more computing capability in the hands of the public. In turn, this will lead to further improvements in all the areas now touched by computers and an expansion into new fields.

OUTLOOD FOR TECHNOLOGY

The development of semiconductor technology has taken us from 1,024 bits of information per chip in 1971 to 64,000 bits per chip in 1978 and has kept chip cost approximately constant, achieving a 16-fold cost reduction. Looking forward, it seems reasonable to anticipate 256,000-bit memory chips in the early 1980's and a million-bit memory chips by the mid-1980's.

During the same period, microcomputers evolved from minimal units involving 2,000 transistors in 1971 to devices that today integrate 30,000 transistors on a chip. This progress has made it possible to enjoy the size savings and low cost of microcomputer control in hundreds of applications, from sewing machines to television games. Similar progress in size, costs, and applications will continue over the next five years. The microcomputer of the mid-1980's, often consisting of only a few chips, will be capable of truly impressive performance and will employ advanced architectures and likely the 32-bit word length used by the larger systems of today. This microcomputer will not only extend the range of economic computer applications, but will also become a major computational workhorse of the future.

Device speed has improved from the microsecond speeds of early integrated circuits to today's fastest commercial chips, capable of adding a pair of decimal digits in less than five-billionths of a second. This development should also continue, but at a decelerating pace unless the industry introduces new technology.

Two possible future technologies are bubble memories and Josephson junctions. Magnetic bubbles, which allow data to be stored very densely and moved magnetically, can provide fast, reliable bulk-data storage and retrieval. Josephson junction technology is a superspeed logic technology that may begin to be used, at least for special applications, within the next decade.

OUTLOOK FOR COMPUTER SCIENCE

Programming, the writing of instructions to tell computers what to do, has become a major element in the cost of computer operations. Software engineering is intended to control the avalanche of detail that makes programming so expensive by discovering principles that allow programming to be approached systematically, organized in standardized ways, and reduced in mass by elimination of all redundant details. Related goals are to find effective ways of using a computer to check the internal consistency of large programs, to pinpoint discordant details as fully as possible, and to define formal techniques whereby programs can be proved correct or the generation of incorrect programs rendered impossible. Progress toward these important goals has, however, been limited.

An important field of research is the development of new, more powerful programming languages that can make program development easier without losing efficiency. Improved techniques for writing programs to control many simultaneously active devices are also being developed. But since all this is basic research, some time will elapse before its impact is felt by either the industry or the public.

Computer science in the past has made use of a great many silent borrowings from mathematics. Theoretically work is continuing in many areas related to the mathematics of computers and is gradually transforming the *ad hoc* design approaches of a decade ago into a more mature, theoretically based undertaking. The possibility of using computers to create artificial intelligences displaying humanlike capabilities has continued to fascinate and inspire computer researchers. However, except in a very few areas, we are still far from being able to do this.

SOCIAL IMPACTS

The use of computers in communications has already begun to have an important effect on the way we live. The airline reservation systems, long-distance bibliographic searches, and credit-card verification schemes of today will expand during the next five years, as will office text-editing systems. The coming period will also see the spread of computing power to more people than ever before.

Centralized computers—connecting users by various telephone and direct-wire links—characterize both the early days of computer use and many systems still in use today. There is, however, a trend away from centralized computers toward decentralized systems in which significant computing power is located as close as possible to the actual users.

Computer technology has made many new ways of communicating possible. Examples include data networks, satellite and digitally encoded telephone communications, optical fibers, electronic mail, text editing and distribution systems, and electronic funds-transfer systems.

The role of these computer communications systems raises basic regulatory questions; for example, whether the First Amendment applies to computer communications, and whether security and privacy are adequately guarded by current encryption techniques. These questions, however, are overridden by the fact that progress has come so rapidly that computer technology and communications technology are converging. This raises important questions for an agency like the Federal Communications Commission as it seeks to protect the public interest.

REFERENCES

1. *EFT in the United States: Policy Recommendations and the Public Interest.* Washington, D.C.: National Commission on Electronic Fund Transfer, 1977.

2. Will, P.M., and D.D. Grossman. Experimental System for Computer Controlled Mechanical Assembly. *I.E.E.E. Transactions on Computers* C-24(9):879–888, 1975.

(Ted Jones)

5 Energy*

INTRODUCTION

The U.S. energy sector is changing rapidly. Oil and natural gas, which provide three-fourths of our energy and upon which depend most of our technology for transportation, space heating, and industrial heating, are being depleted and must be conserved. At the moment, and for the next few decades, the principal alternatives are coal and nuclear fission, mainly in the form of electricity.[1]

Both of these alternatives have limited futures given the current technologies. By the turn of the century or somewhat later, depending on energy consumption rates in the meantime, they in turn must begin to be replaced. We must then be well on the way to deploying sustainable, environmentally acceptable, long-term energy sources, compatible—in the interest of international stability—with the changing roles and aspirations of the developing countries.

Meanwhile we must use what is available—coal, nuclear fission, vigorous energy conservation, some solar energy, and small amounts of geothermal heat—to reduce our dependence on oil and natural gas. However, the rising concern with environmental quality, and the political and economic friction of making large industrial shifts, will make this difficult and will tend to focus attention on short-term expedients. It will

*This chapter benefitted greatly from the work of the National Research Council Committee on Nuclear and Alternative Energy Systems (CONAES). However, the contents of this chapter have not been reviewed by the Committee and do not necessarily reflect its views.

be necessary to take broader, longer views of this problem than those to which we are accustomed.

Many of our energy consumption systems cannot be substantially changed in one or two decades. Large urban areas, where most energy is used, have made their greatest growth in the past 40 years. During this period the real price of energy fell as large oil and gas discoveries were made worldwide, electric utilities took advantage of technical refinements and economies of scale, and the environmental costs of energy use were largely unknown or ignored. People thus paid no penalty for energy use that under today's conditions—and tomorrow's even more—would be considered very inefficient. Long-distance commuting by energy-inefficient automobiles became commonplace. Too much insulation was a waste of money. The size of the average house rose, and more and more were detached, with all four walls exposed to the elements. Most industries optimized their capital equipment with energy consumption as a minor consideration.

That is now changing. The price of oil has quintupled since 1973, with coal and gas prices not far behind, and U.S. dependence on oil imports has increasingly apparent political and strategic implications. Nuclear fission, once promoted as a source of clean and nearly limitless power, faces major and increasing difficulties, although it could have a productive future. While coal is abundant, it is ironic in this time of general concern about the environment to be forced back to a fuel earlier abandoned for most uses in part because of its potential for pollution.

Still, only coal and nuclear fission can be widely available over the next two or three decades, and we must find some acceptable combination of the two, with much less reliance on oil and natural gas. We must also do the work that will be necessary to put sustainable long-term energy sources in place.

SOME IMPORTANT TIME PERSPECTIVES

Difficult choices arise. Short-term expedients may foreclose serious long-term options, and the nation must take great care that its accommodation to today's realities does not limit its scope in dealing with the more distant future. For example, strong emphasis on expanding domestic oil production, without comparable attention to how it is to be used, may tend to deepen our dependence on oil and to intensify the harshness of the eventual and inevitable withdrawal.

Financial and political institutions are more responsive to the short-term view dictated by market rates of return and by electoral intervals than to the longer and broader view determined by the depletion of domestic resources and the international implications of growing reliance on imported energy. It will require an educated public and enlightened officials to temper this disparity through responsible efforts to avoid the future economic and political difficulties to which short-range expediency can commit us.

The shortest meaningful time perspective for serious social decisions in the private sector is determined by normal market expectations of rate of return on investments. This leads typically to time horizons usually of five years, and rarely as great as 10 years, in present industrial practice. Investments maturing later are relatively unattractive compared to more immediate payoffs. Exploratory research aimed at very distant returns can be justified, but the cost must be low by normal market rules. The political sector has a similar time perspective, determined by the intervals between elections.

For energy technology the shortest time perspective is given by the time it takes to develop and commercialize new major technologies: solar power, new and cleaner ways of using coal, more resource-efficient nuclear fission reactors and fuel cycles, nuclear fusion, and mature energy conservation technologies. This time varies somewhat depending on the technology, but many take 20 years or more, except under the most exceptional conditions and strong government incentives. Civilian nuclear power, for example, which was accorded remarkable federal priorities, was first seriously explored in the United States in the early 1950's and only by the early 1970's was well enough developed to contribute 2 percent of the nation's electric power. (But the industrial momentum built up during all that time led to a 12 percent contribution in 1978.) The magnitude of a development program similar to that of nuclear, and the length of time before it could repay the initial investment, place it clearly beyond the view imposed by the market.

This disparity between the economic and technological time perspectives leads to market exploitation of existing options and neglect—possibly foreclosure—of potential new ones. Governments can correct this imbalance by either underwriting the long-term development itself, as a nonmarket social good, or constructing appropriate market signals—tax and other incentives, regulations, and so on—to stimulate option development by the private sector. Governments in fact do both.

Several yet-longer time perspectives are also important. The time it takes to deplete a particular energy resource, so that more supplies become too expensive for their accustomed uses, should exceed the time it takes to develop and deploy alternatives or to adjust to a social and technological position of doing without. Thus, the perception that the resource depletion time for oil and natural gas in the United States may be comparable to or shorter than the time it will take to produce a replacement is a principal cause of concern. It leads, for example, to demands for new technologies on short notice.

Another important time perspective is even longer. The urban and industrial infrastructures of civilizations, which heavily influence our energy use, can persist for centuries. Cities, for example, last for centuries and change fundamentally only over generations.

MID-TERM ENERGY OPTIONS

In the next two or three decades, the range of energy options available to us is rather strictly limited to existing technologies. Oil, natural gas, coal in various forms, and nuclear fission must among them supply most of the nation's energy to 2000 and somewhat beyond. The declining supplies of oil and gas further limit the scope of our actions, forcing us to rely increasingly heavily on coal and nuclear fission for any real increase in energy supplies. Given the many environmental, political, and resource uncertainties plaguing coal and nuclear fission power, it is obvious that energy conservation must play a large and increasing role in the national energy programs. But no one source of energy or conservation alone will suffice.

DISCOVERING PRIORITIES: CONSERVATION FIRST

The broadest questions that can be asked about energy policy deal with the balance of emphasis between supply and conservation. In general, supply is relatively overemphasized, because the supply sector is simpler and better organized to provide its products and to receive and recognize rewards. Energy conservation tends to offer rewards that are received later in time by a diffuse and ill-organized population. Thus effective energy conservation requires either generating a strong public awareness or providing government financial incentives, or both.

Energy conservation consists of two different categories of activity. The first is relatively simple cessation of present waste: Turn down the thermostat instead of opening windows, turn out the lights when leaving the room, don't leave the car's engine running when you go back into the house on a last-minute errand. These steps reflect simple social concern and thoughtfulness, and should become more general as people are made aware of the costs of waste. The second category concerns deeper changes in the economic structure, the designs of what we use, and public incentives. It is here that we focus our principal attention.

Energy supply costs money, and the marginal cost of energy increases as more is demanded; however, in the relatively pristine field of energy conservation large savings can be captured at low cost. An analysis by Gibbons and his colleagues[2] suggests that even if real energy prices remain roughly constant from now to the year 2010, all sectors of the economy will find it profitable to raise their energy efficiencies significantly—by as much as 25 percent in some industries.

There are many good technological opportunities for using energy more efficiently and thus conserving; for example, leveling electricity demand over the day, storing energy, using electric automobiles in cities, developing new thermal cycles, and using new less-energy-intensive processes for industry.

Related to energy conservation is the question of how energy use and GNP (a convenient but imperfect measure of public well-being) are related. In the short run, reducing the number of Btu consumed reduces GNP, but the long-term possibilities are much better. As capital equipment is retired it can be replaced by equipment designed to use less energy, generally at higher cost.

The time scales for such transitions are different for different things. Well-built houses last a hundred years, but recent experience in residential energy conservation shows that old houses can be much improved; new ones require only half as much energy or less, for heating. That comes out to halving the energy requirement per unit of service in 100 years, or about 0.7 percent per year. Cars last on the average 10 years, and we are requiring that the fleet fuel-efficiency of new cars almost double between 1976 and 1986. The life of industrial plants is intermediate, averaging about 20 years; new equipment uses typically perhaps three-quarters as much energy as the old. Applying those simple numbers, we find that industrial energy per unit output could be approximately halved in about 45 years.

Further detailed analyses,[3] taking into account these and many other replacement possibilities, suggest that, overall, energy used per unit of GNP could be as much as halved in 40–50 years, with negligible effects on GNP. This means that conservation incentives, including rising energy costs, could be structured so that by the year 2020 half as much energy as is now used would suffice for today's national total of goods and services; that is, the same amount of energy could accomplish twice as much. The detailed strategy for doing this would involve frequent reviews and appropriate corrections from time to time.

The key to effective energy conservation is planning for the long term in the sense just described. Attempts to induce efficiency by replacing houses, cars, and factories before they pass their useful lifetimes requires that useful things go unused, which in turn generally implies drops in both energy consumption and GNP growth. As stated earlier and repeated here for emphasis, short-term energy consumption and GNP tend to be closely coupled (as experience in the 1973–74 oil crisis demonstrated); but over periods of decades large beneficial changes can be made.

An example of how one group perceives the range of possibilities will help to focus this discussion. Gibbons and his colleagues,[2] in a study for the National Research Council, projected U.S. energy consumption to the year 2010, using four scenarios representing a range of plausible movements in energy prices to that date. Table 1 gives assumed prices in 2010, with the 1975 prices listed for comparison. The scenarios were developed on the assumption of 2 percent average GNP growth between 1975 and 2010; a variant of scenario III, which allows for 3 percent average GNP growth, is included. Table 2 shows the general results of this scenario analysis. The highest and lowest scenario totals differ by more than a factor of 2. The panel found the whole range consistent with the assumed rate of economic growth; the variations in energy consumption

TABLE 1 Assumed Energy Prices for 2010 Demand Scenarios

	Energy Conservation Policy	Consumer Price ($/Million Btu)			
		Distillate Oil	Natural Gas	Utility Coal	Electricity
Scenario					
I	Very aggressive; deliberately arrived at reduced demand requiring some life-style changes	13.49	14.84	3.24	26.37
II	Aggressive; aimed at maximum efficiency plus minor life-style changes	13.49	14.84	3.24	26.37
III	Slowly incorporates more measures to increase efficiency	6.74	7.42	1.62	15.82
III_3	Same as III, but 3% average GNP growth	6.74	7.42	1.62	15.82
IV	Unchanged from present policies	2.81	3.09	0.81	7.91
Actual					
1975		2.81	1.29	0.81	7.91

TABLE 2 Energy Demand for 2010 (quads)

	Buildings	Industry	Transport	Total	Losses[a]	Primary Consumption[b]
Scenario						
I	6	26	10	42	16	58
II	10	28	14	52	22	74
III	13	33	20	66	28	94
III_3	17	46	27	90	44	134
IV	20	39	26	85	51	136
Actual						
1975	16	21	17	54	17	71

[a] Losses include those due to extraction, refining, conversion, transmission, and distribution. Electricity is converted at 10.500 Btu/kWh, coal is converted to synthetic liquids and gases at 68 percent efficiency.

[b] These totals include only marketed energy. Active solar systems provide additional energy to the buildings and industrial sectors in each scenario. Under the most aggressive assumptions (Scenario I) this amounts to 5 quads.

reflect the variation, from scenario to scenario, in energy prices and other incentives to substitute inputs such as labor and capital for energy in the economy.

THE EXAMPLE OF CITIES

It is in and near cities that most energy is used, and urban life shapes and is shaped by how energy is used, both positively and negatively. Air pollution peaks in cities, but so does the ease and efficiency with which heat, light, and transportation can, at least in principle, be provided. The ease with which cities can adapt to changing energy realities is therefore an important question.

There is a good deal of inertia in the nation's urban and industrial infrastructures. Buildings and industrial plants have lifetimes measured in decades. A realistic conservation strategy must therefore deal not only with the opportunities afforded by the new but also with the resistance to change of the old.

The kinds and amounts of energy demanded depend in complicated ways on ingrained features of urban structure and urban life. Most of the large urban areas in the United States had their greatest growth in the past 30 or 40 years, when energy prices were falling and energy efficiency thus mattered little. Settlement and transportation patterns, building standards, and industrial equipment all developed in this period in essentially their present forms, and to varying degrees they resist fundamental change. Basic transportation and settlement patterns, for example, may persist for centuries.

In the long run, one may speculate that with escalating energy costs, cities will slowly evolve into more compact forms, with less suburbanization or more focused suburban centers. Given the probable energy supply options, cities may come to depend more on electricity. Basic services may be supplied from larger energy units to take advantage of district heating using "waste" heat from nearby power plants and industrial facilities, as is now done in some parts of northern Europe. But there are many improvements that can now be made. Table 3 lists both short- and long-term problems and potential solutions, all of which might benefit from prompt attention.

MID-TERM SUPPLY STRATEGIES

The next important issue is how to develop energy sources. The difficulties of supply appear to lie primarily in oil and gas, but many of the alternatives being prepared are designed to provide electric power. Nuclear fission, nuclear fusion, geothermal energy, much coal combustion, most large-scale solar concepts, and others—all are either exclusively or mainly for electricity. Synthetic liquid fuels from coal or biomass could amount to only a small fraction of present oil and gas use (about 55 quads per year)*, in any event for several decades and possibly also in the indefinite future. The same can probably be said of synthetic fuels derived from domestic oil shale, even though the emerging *in situ* extraction processes may reduce the water consumption and general environmental damage that have

*The term "quad" is short for "quadrillion Btu's." It is used throughout this chapter as a means of putting different energy sources on an equivalent basis. One quad is equivalent to about 186 million barrels of oil, a trillion cubic feet of gas, or 40–50 million tons of coal (depending on type).

hitherto been serious deterrents to exploiting this huge resource.[4] Biomass from municipal, agricultural, and forest wastes could provide at most about 10 percent of present use of oil and gas, and even then only after the year 2000.[5] So-called energy farms could supply a good deal more in principle, but at the cost of undue competition with food production and ecological risks typical of any monocultured crop.

The prospects for producing coal-based synthetic fuels are limited also. For example, water supply may perhaps be a major limitation in the production of synthetic fuels from coal, and possibly oil shale. A very pessimistic view is presented by Harte and El-Gasseir, who assert that freshwater supplies will limit production of synthetic fuels from coal in the United States to about eight quads per year.[6] Critics of this work feel that insufficient account was taken of the potential for more efficient water use, including recycling, for interbasin transport, and for tapping new sources such as brackish underground water. All would agree, however, that water availability and competition from other water uses will impose severe restrictions on the siting of plants, and will require major technological developments in efficient water use and large water-related investments.

There are further problems in using fossil fuels, especially coal, some complicated by considerations truly global in scope. (We will see that nuclear power too has its problems.) The most urgent of concerns in burning fossil fuels is the continuing buildup of carbon dioxide in the atmosphere, apparently due in large part to fossil-fuel combustion, as well as widespread deforestation, especially in the tropics. A likely result of such a buildup, if it continues, is an overall warming of the earth's atmosphere, which holds the potential for shifts in global wind currents, rainfall patterns, and temperature distributions. The outcome of such climatic shifts are poorly understood; world agriculture, whose capacity is even now strained in many ways, could be severely disrupted. This possibility is being seriously studied, and a continued, coordinated effort at understanding this effect is a necessity.

The increasing demands of developing nations for their share of dwindling oil and natural gas supplies further emphasize the need for this nation to move away from reliance on fossil fuels. The current world oil supply and price situation already forces some nations to choose between slower economic progress and rapidly growing indebtedness.

The United States uses 55 quads of petroleum and natural gas annually. This seems considerably in excess of what is producible from politically secure reserves or from coal-derived synthetics over the long term. Faced with this apparent imbalance between the need for and the supply of liquid and gaseous fuels, we can either try to shift technological development more toward synthetic liquids and gas (mainly from coal with much better

TABLE 3 Energy Conservation

Problem	Relation to Energy	Consequences If Problem Is Unattended	What to Do?	Some Impediments to Implementation
Housing	Poorly insulated buildings cause heat loss, expense (this is but one example)	Housing less desirable because of increased living cost. Poor people suffer most	Insulate, retrofit, new housing codes, public education	Building codes, initial cost, lack of information, lack of low interest loans or tax rebates, landlord-tenant dilemma (who pays?)
Pollution	Mainly caused by high energy use	Bad health, loss of economic values, lowered quality of life	Stricter standards and better energy technology for cities, especially with respect to automobiles. Also cogeneration, district heating where appropriate.	Reliance on cars for commuting, ability of business to escape antipollution laws. Lack of information on pollution effects.
Transportation	Reliance on energy-intensive modes	Pollution, congestion, run out of energy resources	Higher taxes on parking, vehicles. Bikeways. Zone for high population density near public transport.	Special interest groups, commuters, desire for personal security and private transportation. Substandard mass transit.

			Buses (probably electric) and rapid transit extensions where feasible.	
Waste heat from power plants	Obvious	Poorer urban microclimate, higher energy use. pollution	District heating, cogeneration	Utility regulations, piecemeal planning and zoning, short time perspectives, lack of overall policy
Urban waste	Energy content equals several percent of city's needs	Environmental degradation, waste of resource	Recycle, pyrolize, methanate, etc.	Disorganization of recycle industry, organization of electric utilities and of urban refuse sector. Regional political fragmentation.
Cost of municipal government	High energy cost	High tax rate, services cut	Initiate program of energy awareness, audits, capital improvements	Lack of expertise, short time perspectives, lack of funds
Welfare	Higher energy costs hurt those least able to afford them	Worse health problems because of inadequate heating	Winterization programs for the poor, emergency assistance	Insufficient funds, poorly used

technology), or to restructure the technological basis of the civilization to reduce permanently the current heavy dependence on such chemical fuels. In practice, both these approaches will be necessary, but the balance of emphasis between them is complicated, important, and ill-understood.

Summing up, the central issues in U.S. energy policy are conserving, preparing for long-term shifts in energy sources, anticipating a possibly more electric future, and planning to phase out fossil fuels. These are all difficult, and opinions differ on the extent to which each is needed or feasible. However, they remain the key issues, and policy as yet displays no very coherent or comprehensive view of their relative importance.

Oil and Gas

Recent estimates of total remaining recoverable resources* of petroleum available in the United States at a cost not higher than about twice the present world price—that is, about $30 per barrel[7]—are estimated at 100–125 billion barrels. Present total domestic use is about 7 billion barrels per year, of which nearly half is imported. Although oil will not abruptly stop flowing, one may think in a general way of the recoverable resources as representing a little more than 15 years' supply, or 32 years' at the present import rate. But these figures are only approximate, because of uncertainty about both resources and projected consumption.

Oil obtained by secondary recovery—such as water or gas injection—have been already figured into these estimates; but in general tertiary recovery schemes—such as fire-flooding or use of detergents to release the oil from its rock matrix—have not, because both the cost of these methods and the outlook for substantial additions to reserves from them remain very uncertain.

Domestic oil-shale resources are estimated to contain 3,660 quads of recoverable oil (the equivalent of about 680 billion barrels); this is several times the size of the domestic oil and gas reserves combined. A technology for exploiting the resource does exist, and in principle production could begin almost immediately. But there are serious difficulties. Virtually the entire resource is concentrated in a small, arid, and ecologically fragile area on the western slope of the Rocky Mountains, and the problem of water supply, spent shale disposal, and air and water pollution are formidable. Also, the raw shale oil would need substantial refining even for use in large boilers and utility stations; among other things, it contains large concentrations of arsenic, for which the removal and disposal

*The quantities of resources are defined in terms of reserves and resources. In general, reserves are the amounts that can be extracted economically (that is, with current technology and at prevailing prices). Resources represent the total amounts known or estimated to be in place, without regard to the feasibility of extracting them.

technology is not yet available. One should not count on large contributions from this source for many years, though a small pilot program might not be out of place.

Tar sands represent a possible major resource to Canada and Venezuela, but domestic deposits are of much poorer quality and are unlikely to figure strongly in the U.S. energy economy.

A principal difficulty with all these unconventional sources of oil (and also those of gas) is that they are very capital-intensive and expensive to produce; it is therefore unlikely that they could more than partially offset the decline in production from conventional sources.

The total world resource of oil is figured by Moody and Geiger and others as about 2,000 billion barrels, about 50–60 years' supply, depending on both patterns of use by industrial countries and growth rates of developing countries.[8] The present high and prospectively higher oil prices have damaged many plans of developing nations, and have led to undesirable increases in the use of local noncommercial energy sources—more deforestation of India and Nepal, for example. The 2,000-billion-barrel estimate presumes that a number of large oil fields will be discovered, such as in Mexico.

The United States' ultimately recoverable resources of natural gas have been variously estimated at 500 trillion cubic feet, and maybe much more, depending on whether large amounts of methane can be extracted from the geopressured brines of the Gulf Coast at reasonable prices, perhaps $5 per thousand cubic feet or less. Prudence advises against counting on much from those brines. Present annual domestic use of natural gas being about 20 trillion cubic feet, we see a resource time of some 25 years. The prospects for importing large amounts of natural gas are problematical, because liquefying and shipping the gas is expensive and hazardous. A possible exception is the prospect of importing large quantities from Mexico by pipeline. Thus finding more gas, making substitute gas, and reducing consumption must all receive urgent attention.

These findings, agreed to by most analyses, coupled with the material in earlier sections of this chapter, pose a series of policy difficulties, most of which will be compounded when we consider coal a little later. As an example of such difficulties, note that with a prospective U.S. resource base for oil and gas of 15–30 years but shorter economic time perspectives, short-term optimization urges expansion of domestic production capacity as rapidly as prudently possible. But that decision perversely tends both to expand, or at least maintain, petroleum- and gas-dependent technologies and to exacerbate the global environmental problems of fossil-fuel use. Also, petroleum- and gas-dependent technologies will be all the harder to readjust later on even shorter time scales.

What seems missing, or at least insufficiently considered, is a strategy or series of possible strategies not only to reduce dependence on imported petroleum via short- and medium-term domestic expansion, but also to greatly reduce total dependence on gas and oil over approximately the next 25 years. Recent delays in gaining public acceptance of even much more modest adjustments to national energy policy suggests such strategies will be hard to implement and will require a better public understanding.

Regarding domestic policies to stimulate petroleum conservation in the short term and domestic exploration for the medium-term transition, fuel taxes have been suggested as effective instruments and require serious study. The long-term objectives of U.S. policy for oil and gas should be to have prices more nearly reflect the replacement cost of supplies. The actual U.S. technology for petroleum and gas exploration and development is in private hands and unexcelled.

Coal

Although coal has many unattractive features, it is the only large, assured fossil energy source available to the United States that can in principle be exploited fairly quickly.

The coal reserve base in the United States, perhaps 500 billion tons (4,000–5,000 quads), exceeds the petroleum and gas reserves. The total coal resource is several times larger than this. Nevertheless, coal does not appear as attractive for the long term as these numbers might indicate. In the first place, an energy growth rate of 2 percent per year (compared to the actual 2.5 percent growth between 1920 and 1972) would exhaust the reserves in one century, assuming that coal must account for almost all the increase. In the second place, the potential climatic impacts of global carbon dioxide emissions from fossil fuels and other sources could dictate a much earlier transition from so much use of fossil fuels; this will be discussed later. In the third place, coal has problems of its own.

The supply situation is not straightforward. The price of coal may tend to rise, for example, to keep just under the oil price by whatever are the incremental costs of transportation, pollution control, and so on. Labor costs in a rapidly expanding coal industry would be an especially important source of upward pressure on prices. Indeed, the unavailability of labor at almost any price could place a ceiling on the rate at which coal can be mined. The Tennessee Valley Authority recently contracted for eastern Appalachian coal that conforms to sulfur emission and other standards at $50 per ton, far above the typical steam-coal price of $20–$30.

Besides large problems of environmental damage on local, regional, and global scales, coal presents a host of lesser ones. Here are some:

- Legislative decisions requiring that sulfur oxides be scrubbed from the stack gases from all new large coal-burning plants (irrespective of the coal's initial sulfur content), which effectively keeps low-sulfur western coal out of eastern markets.

- The enormous need for upgrading railway tracks and roadbeds, if most coal is to continue being carried by rail.
- The problem of polluted-water disposal at the output ends of proposed coal slurry pipelines, if they become a dominant mode of coal transport.

Research, development, and demonstration related to coal suffered until the early 1970's from near-starvation; the Office of Coal Research, for example, was funded at only about $10 million annually through the 1960's. Nuclear options, at the same time, were funded generously. The results now appear in urgent demands for new coal technology on a too-short time scale.

Fluidized Beds Taking the pragmatic view that coal use must be increased for several decades despite a poor long-term outlook, we should attempt to use the best technology likely to be available. One promising approach is fluidized bed combustion, in which a bed of ash plus continuously fed pulverized coal is kept levitated by streams of air blown from below. The coal burns at a relatively low temperature in the bed, thus minimizing the formation of nitrogen oxides. Furthermore, sulfur dioxide absorbers such as crushed limestone can be added to the bed, and most of the sulfur is removed with the dry ash. Some remaining problems are:

- The carry-over of particulates and hydrocarbon into the hot gas stream is still not well enough known.
- The erosion and corrosion of steam-raising pipes that pass through the bed, giving very good heat transfer from the bed to the steam, may require the use of special steels.

Compared with typical problems faced by the nuclear power industry, these seem simple. The main impediments to resolving them are a legacy of very conservative attitudes in the coal-equipment industry and coal-burning utilities.

The other main desulfurization scheme, and the one most used now, is scrubbing the exhaust gases; this, unlike fluidized bed combustion, does not remove nitrogen oxides or some other pollutants. The use of regenerable scrubbing chemicals, as in the so-called double-alkali system, may be the best approach; it tends to minimize the large amount of polluted limestone slurry that comes from the once-through systems now used.[9]

Coal Conversion

The art of making clean synthetic fuels from coal progresses, but slowly. Fuels of this kind are envisioned both as substitutes for both petroleum and natural gas in applications that require liquids or gases and as cleaner burning substitutes for coal. They are not expected to contribute very substantially before the 1990's, and apparently will never be able to replace much more than half of even our current imports of oil and gas, largely because of the perceived limitations of water supply. Furthermore, there is some question about the net benefits to pollution abatement in using such fuels to generate electricity; this might only shift certain environmental impacts from the power plant to the synthetic fuel plant. As supplements to the supply of oil and gas, however, they may be important.

Synthetic Gas The main step in converting coal to gas efficiently consists of adding hydrogen to the coal. This is usually done by partial combustion in the presence of a hydrogen-rich material (usually water). The result is dissociation of the complex molecules in coal into simpler and lighter substances consisting mainly or solely of carbon and hydrogen, plus a variety of residues. In simple gasification processes coal reacts in air with steam to produce carbon monoxide and hydrogen, with heat for the process provided by burning of some of the coal. This produces so-called synthesis gas, which has a heat content about 10–15 percent of that of natural gas. This low-Btu gas is suitable for onsite use, but because of its low heat content is not economical to transport very far. If oxygen and hydrogen are used in the gasification process, a somewhat higher-quality gas can be produced, having about 30–35 percent of natural gas' heat content; such intermediate-Btu gas must also be used onsite, but may have advantages over low-Btu gas in some applications. Finally, synthetic gas, fully equivalent to natural gas and therefore useful in existing pipelines, can be manufactured by adjusting the hydrogen-to-carbon monoxide ratio of the raw gas to about 3 : 1 and passing the mixture over a catalyst at high temperature to yield almost pure methane.

Promising early technological developments in using gas from coal, ranking almost equally with fluidized bed combustors in timing and importance, are some combined-cycle systems using low-Btu gas from coal prepared and burned onsite to generate electricity. Technologies for producing low-Btu gas have been in hand for many years; present environmental and economic conditions require some technological improvements deemed quite feasible. Cost estimates are high, running to \$3–\$4 per million Btu, corresponding to \$17–\$23 per barrel of oil; but other gaseous fuels may be in such critical supply as to be disallowed for electricity generation in the late 1980's. The overall thermal efficiency of a combined-cycle system can exceed 50 percent from input gas to electricity, and 40 percent or more on the basis of raw coal. These efficiencies are satisfactory for a clean system.

Synthetic Liquid Fuels Many coal liquefaction approaches have been evaluated. The three that are favored at present—Gulf's Solvent Refined Coal II (SRC-II), Hydrocarbon Research's H-Coal, and the Exxon Donor Solvent (EDS) process—are hydrogenation processes that yield about 2.5-3.0 barrels of liquid per ton of coal. They all use approximately the same process conditions and enjoy roughly equivalent technical status. The SRC-II process is most advanced in terms of scale, with a Department of Energy-sponsored facility in Tacoma, Washington, operating at about 30 tons of coal per day and a 6,000-ton-per-day plant planned for operation in 1985. Plants of similar or larger size have been proposed for the other two processes.

If it were deemed necessary to begin deploying a liquefaction industry today, the best available technology would be the Fischer-Tropsch process, used in Germany during the Second World War and still in use in the Union of South Africa. This is a two-stage process; low-Btu synthesis gas is produced from coal, and then converted to a variety of liquid fuels. Compared to the more advanced systems under development, the process is low in efficiency and high in cost.

Present projections are for synthetic oil prices to be $20–$25 per barrel of fluid that contains many more biologically active (and therefore possibly toxic or environmentally hazardous) species, such as anthracenes and phenols, than does natural crude oil. This material can be chemically refined to open up the molecules into long chains similar to those in natural petroleum. But the cost will be higher, the yield lower because hydrogen must be added, and the plant necessarily more leak-tight than is the custom in present oil refineries. In addition, the chemical and physical stability of these fuels against forming tars and gums during storage is a continuing problem, since they can polymerize more readily.

In summary, while coal is almost sure to be used, it remains in a difficult technological condition, with very substantial environmental, technological, and social questions outstanding.

Nuclear Fission

Nuclear fission power can be viewed as both a mid-term and a long-term energy alternative. It is mid-term because it is available now in appreciable amounts, but this energy option will last only until sometime in the next century, given present technology using naturally fissionable uranium. It can also be long-term, because with breeder reactors the uranium (or thorium) resource would suffice for millions of years. Here we discuss fission as a mid-term option, chiefly to compare it with coal. The long-term aspects are discussed later in this essay.

Nuclear power in the United States faces an uncertain future. Fostered and promoted by the government for three decades, the industry and the technology have lately suffered a number of setbacks. These are reflected in the last five years' precipitous decline in utility orders for reactors, from 34 in 1973 to only 2 in 1978, and there are no signs of improvement in this situation. Thus, although most projections suggest that life in the next few decades would be very difficult without nuclear fission, even if the correspondingly increased fossil-fuel use proves acceptable, nuclear power may be barred from contributing significantly.

The reasons for the industry's pessimistic outlook are complicated. In general, utilities are inhibited from ordering reactors by several considerations. First by a wide margin is uncertainty about the future regulatory and political environment of nuclear power. The technology has lately become subject to public criticism on a number of grounds; organized antinuclear activist groups have demanded and received easier access to participation in the licensing and siting processes, and the consequent delays for litigation have raised the risks of financial losses and capacity shortfalls. In addition, some states and localities have placed legislative limits—some amounting to virtual bans—on various nuclear power activities. At the moment, this kind of opposition centers around the question of reactor safety—especially, as this report is written, in the aftermath of the Three-Mile Island reactor accident near Harrisburg, Pennsylvania, March 28, 1979—and waste disposal; opposition often also centers around nuclear fission's high technology per se and its supposed social attributes. The overall result is a lack of a reasonably sure and uniform siting and licensing policy for nuclear plants and related facilities, and periodic requirements for retrofitted safety devices on existing reactors. The utility industry is as averse to this institutional uncertainty as it is to technical unreliability; thus the present *de facto* moratorium on new nuclear plant orders in the United States.

Also, the breeder reactor development program, on which utilities have counted as a follow-on technology and as a source of plutonium for fuel when the nation's limited uranium resources are depleted, has been delayed. Recycling of spent reactor fuel has also been deferred. Both actions were taken in part because of the Administration's concern about weapons proliferation.

Another factor sometimes mentioned as a vital consideration is the fact that projections of electricity demand are significantly lower now than they were five or ten years ago. However, the rate of new orders is at present inadequate to satisfy even these lower projections.

Underlying the nuclear industry's troubles are serious social concerns.

As a technology for producing electricity, nuclear fission is at least as reliable and economical as coal, its closest counterpart in the production of baseload power.[16] But concern with its safety is persistent, and the issues of waste disposal, reactor safety, and weapons proliferation must be convincingly and publicly addressed if the technology is to have a productive future.[17]

Waste Disposal No feature of nuclear power contributes more uncertainty to its future than the problem of permanent disposal of nuclear wastes. While many promising technical options exist for geological storage of solidified wastes or spent fuel, the federal government until very recently has neglected their demonstration in a systems engineering sense. As in the case of catastrophic reactor accidents with many casualties, the possible scenarios of what could go wrong are hypothetical and speculative. The hazards cannot be tested by experience, and we are forced to rely on rather theoretical calculations, with only some of the individual steps in a long sequence testable by experiment. There is always room for the criticism that the theory may have overlooked something. In the case of reactor safety the accumulation of operating experience free of accidents with severe public consequences could eventually justify theoretical optimism. In the case of waste disposal, however, there is no wholly experiential way of validating the calculations, because the safe storage period extends over many generations. The main mitigating feature on the other side is that the consequences of a storage malfunction are much less severe, even under extreme assumptions, than the consequences of reactor accidents, hydroelectric dam failures, or breakdowns in many other present and proposed industrial waste systems.

Reactor Safety The hazards of accidents with nuclear power reactors are also important sources of public concern. The most severe accidents conceivable—for example, a loss of coolant resulting in a core meltdown followed by a breach of the reactor's containment—could be catastrophic indeed, with possibly thousands of deaths and property damage measured in billions of dollars, assuming the most unfavorable possible conditions of weather and population density. The calculated chances of such a catastrophic accident are exceedingly small—of the order of one in a billion per reactor-year. This assessment of the risk, however, is based not on actual experience, but on calculated probabilities of hypothetical events involving human fallibility. The uncertainties in such calculations are very large, and although the overall actuarial risk is almost certainly small compared to that of almost any other energy technology, the uncertainties surrounding the probabilities of the largest accidents is a serious deterrent to public confidence. The reality of these uncertainties and of the depth of public concern was illustrated in the Three-Mile incident, mentioned earlier. To be sure, a sequence of events rather similar to what actually happened has been studied, but not for that specific make of reactor, and not accounting for some of the human errors that compounded it—nor for the later human ingenuity that limited it. This uncertainty should be reduced rapidly in the next five years, as analysis of operating experience continues and the theoretical study of accidents is refined.

The question of civilian nuclear power's connection with the spread of nuclear weapons is equally difficult. This is not really an issue affecting light water reactors since they now use a once-through fuel cycle. Reprocessing of spent fuel for its fissile content, however, would, with the processes now available, separate plutonium for a short time in a form suitable for weapons manufacture. Some fear that a nation might be tempted to use such material in a clandestine effort to acquire nuclear weapons, or that a determined and well-organized terrorist group might be able to steal such material for its own purposes. Since the breeder, and thus the long-term prospects of nuclear fission, depend on fuel reprocessing, the resolution of this question will strongly determine the technology's future. Reprocessing facilities naturally must be extremely secure to minimize this risk. But, overstating the risk is unproductive; nuclear power systems are clumsy and not very private ways to acquire nuclear weapons, and it is not clear how this particular problem contributes to the large and complex issue of international security. It is important to recognize, for example, that civilian nuclear power, as an economical source of energy, may in contributing to economic development actually lessen international tensions.

The public debate about nuclear power, and the concomitant irresolution of goals and policies, may continue until we settle the question of whether enough coal can or should be mined to produce most baseload electric power. Much more definite answers should be available by 1985, but by that time the nuclear supply industry is likely to have attenuated drastically. Recovering it could then be very costly.

Nuclear power must be considered in the context of the other national and international energy concerns. It is a source of relatively cheap and environmentally clean energy, qualities hard to find in other available energy sources.

SOME OUTSTANDING ENVIRONMENTAL AND HEALTH EFFECTS

As suggested earlier, the industrial and resource problems of energy supply, while a cause for concern, do not fully express the difficulties this nation will face in providing energy over the next few decades. Energy production and use in the past have had unforeseen impacts on the human environment. Many of these remain poorly understood in detail, but some grosser cases of energy-related ecological damage, public health problems, and of a range of impacts on agriculture and industry are recognized; and regulatory remedies have been imposed in the forms of emission standards, mine reclamation requirements, and the like.

It is not yet possible to rank the various available energy sources precisely according to the relative severity of their impacts, and it may not be especially desirable to do so. Different energy supplies are used in different ways to meet different needs, so that their relative benefits—which must be set against their impacts—are not strictly comparable. However, there are grave reasons for special concern about the impacts of certain energy sources, given current energy projections. Especially important are the hazards of fossil-fuel combustion. Local and regional hazards, for example, arise from emissions of nitrogen and sulfur oxides and particulates into the air. By far the most worrisome global hazard is the prospective buildup of CO_2 in the atmosphere, in part as a result of fossil-fuel combustion.

Coal is by far the most destructive fuel in ecological and public health terms, and the likelihood of large increases in its use presents some unpleasant prospects; we know that even now we pay a rather substantial environmental price for its use, including some air-pollution related deaths and crop damage. We pay a large price also in deaths and injuries of miners, in water pollution from mine drainage, and in disruption of mined land and underground hydrological structures. All of these problems have been dealt with, with various degrees of success, by regulation; the cost and the difficulty of this regulation will grow along with the role of coal in the national economy.

The CO_2 problem is so important as to require special mention (see p. 34). This is an unavoidable consequence of fossil fuel use, and there is no realistic means of control other than limiting the use of these fuels.[10] Again, coal is the worst offender per unit of energy, though oil and natural gas, with about 80 and 50 percent, respectively, of coal's CO_2 emissions, also make important contributions.

Atmospheric CO_2 helps to regulate the temperature profile of the earth's atmosphere and surface. This happens because, although the atmosphere is transparent to the wavelengths of incoming solar radiation, the CO_2 and water vapor in the atmosphere absorb the longwave infrared radiation (heat) reradiated from the earth's surface. They thus trap heat and raise surface temperatures; this is usually referred to as the greenhouse effect. Woodwell calculates that today burning fossil fuels and oxidizing biomass, mainly by destroying tropical forests and the associated humus, each contributes approximately comparable amounts to the global atmospheric

CO_2 increase. The amount will double within two generations, if present trends continue; the result could be a significant climatic readjustment, perhaps a general warming of a few degrees Celsius, with several times as much near the poles.[11]

Government energy planning throughout the world virtually ignores this problem. But it presses now, because the fundamental changes that would constitute a remedy would take a long time. For example, global agriculture, by complicated geographic, social, and institutional arrangements, matches crops to particular areas. Experience coupled with simple analyses shows that total production decreases in times of changing climate, because neither the pattern of land use nor the fertility can change rapidly enough to accommodate. The system has inertia. If rainfall shifted from Iowa to Arizona, for example, the corn would not; it would stop growing in Iowa, but it would take many decades at least for suitable soil to develop in Arizona.

If these projections are valid, hard choices lie ahead, such as substantially reducing the combustion of fossil fuels worldwide, in the face of growing demand for them, especially by developing countries. The options of solar power (probably without using biomass, so as to maintain photosynthetic carbon uptake at the highest practicable level), energy conservation (especially in industrialized countries), and nuclear power thus take on special significance. But the first and most difficult task may be to make people all over the world aware of the problem. No very effective institutional mechanisms appear available to deal with such global matters.

Most air-quality degradation arises from conversion or combustion of fossil fuels, but we know little about the consequences. The effects are most apparent in and near cities, where much energy is used in relatively little space. Thus, patterns of energy use profoundly affect urban living; one uncertain consequence is that of urban air quality degradation on health.[12] Evidence abounds that persons with pulmonary and cardiac dysfunctions suffer notably, and die more readily, when the air is bad. But it is difficult to quantify the role of air pollution in deaths among this population.

One should also consider why people got sick in the first place. That is a very difficult question, involving economic status, social and dietary habits, and many other things. The combination of epidemiological, clinical, and basic etiological studies has so far been insufficient to provide very good answers. Compared with studies of the health hazards connected with nuclear power, the health hazards related to fossil-fuel use are very poorly known.

The emissions from fossil-fuel burning include sulfur and nitrogen oxides, among other things, and these interact with moisture and other atmospheric constituents and contaminants to yield smog and acid rain. Thus, these emissions are increasingly strictly controlled as the years pass. But the bases for setting air-quality standards are themselves inadequate, and even the list of controlled effluents has notable omissions. It is worth noting that, because of fossil-fuel emissions of sulfur and nitrogen oxides, in about one-sixth of the United States—defined approximately by a western boundary just west of the Mississippi River and a southern boundary at the latitude of northern Alabama—the rainfall is distinctly acid, with a pH 5.0 or less, and that, in a region about one-tenth that size in the northeastern United States, the acidity is severely high, a pH about 4.0 or less. These regions are considerably larger than they were 20 years ago. The consequent damage to forests and sports fishing in northeastern lakes has been estimated, probably conservatively, at $100 million annually, and corrosion of the surfaces of buildings and other structures at between $100 million and $500 million per year.[13] These are high costs to pay and expensive phenomena not to understand.

Yet another general environmental difficulty, related especially to energy, is the question of water supply. The topic has been often mentioned, but its gravity is not well appreciated. This matter was discussed earlier in this essay.

LONG-TERM ENERGY OPTIONS

Long-term prospects for using large amounts of oil and gas are not bright. We have seen that the supplies of petroleum and natural gas are limited, both nationally and globally, especially at costs anything like present ones. We have also seen that the prospects for making large amounts of such fuels from coal, oil shale, or biomass are not good. Finally, the CO_2-driven climatic problem appears to depend on the long-term rate of using all fossil fuels (or destroying biomass without replenishment), irrespective of other considerations.

Thus we turn to inquire what energy options *are* available in the long term—options upon which it is possible at least in principle to construct civilizations sustainable for the long term. Only two major classes of options exist: solar power in its various forms, and nuclear power, both fission and fusion. These can be, and are now, augmented in some places by relatively modest amounts of geothermal power, but the normally small heat flow from inside the earth makes solar power much more promising, except for some special locations. Tidal power is very limited; a dam built one kilometer offshore around the entire United States would, with 100-

percent-efficient low-head turbines, hardly generate enough electric power to meet present demand in Massachusetts, and would do so at environmental and capital costs likely to cause much complaining in other states. Much more power is available from waves, though, and the possibility of extracting it is being seriously studied in the United Kingdom, where suitable conditions exist—a small land area with concentrated population and industry, surrounded by stormy, shallow seas.

The long time it will take to develop and install solar or advanced nuclear power technologies, comparable to the time in which the classic fossil fuels must probably be phased out, provide a sense of urgency. A lively debate exists about whether the relative research and development priorities for the prospective long-term energy sources reflect proper social purposes. In this area, as in many other parts of the energy scene, we find people advocating options on social as much as on technical grounds. As we have seen, however, the larger danger may lie in having too few rather than too many good energy options.

In what follows, it is important to note that many of the most promising long-term options produce electricity as their most natural product. This fact bears strongly on the small-versus-large, diffuse-versus-concentrated debates now fashionable. All the nuclear systems are large; and while many of the solar technologies can be modular and small, the questions of interconnections and of backup energy sources bring us back to the necessity of accepting substantial centralization. In all but the most primitivistic views of future society we must accept substantial centralization of energy supply and delivery. Electric utility systems, for example, will remain with us.

Nuclear Fusion

To understand the development of controlled fusion, the hugeness of the task and the attitudes of those associated with the effort have to be appreciated. Perhaps it is merciful that the scientific difficulties now actually being solved were not recognized from the start. But now it seems quite clear that an acceptable fusion plasma—a gas of deuterium and tritium ions 10 times hotter than the core of the sun—can be produced and confined in a vacuum by strong magnetic fields, can be kept from touching any physical surface, and can be confined long enough for an interestingly large fraction of it to react—producing mainly helium, neutrons, and energy. Thus, in a scientific sense, controlled fusion appears possible. It has taken about 25 years and several billion dollars of effort in several countries, virtually without international hindrance of any kind.

Fusion Reactors But the most expensive and difficult parts—the technology and the practical engineering of a working reactor producing power—are still to come: huge superconducting magnets, vacuum wall materials exceptionally resistant to radiation damage, and so forth.

The basic technological feasibility of controlled fusion will probably not become really clear for at least another decade, and the engineering and economic feasibility even later. Fusion will come, if at all, later than breeder reactors and advanced solar power systems, both of which may be called for by about the end of this century. Still, fusion's most pressing uncertainties may be resolved by then, so it remains a contender for the long term.

Fusion reactors will probably come in large sizes, perhaps 1,000 megawatts electric, or about the size of large modern coal and nuclear power plants. Their high technology suggests that cost will also be high by today's standards. However, the scarcity of acceptable, well-developed long-term energy options gives the word "high" considerable flexibility.

Fusion Hazards Controlled fusion produces radioactive materials and other hazards. But the wastes consist not of highly radioactive fission products, but of tritium and of the activated reactor itself. All these hazards, taken together, seem much less than those of fission. Thus in an environmental and social sense, fusion would be intermediate between solar and nuclear energy.

The connection between controlled fusion power and nuclear weapons is much less relevant than in the case of fission. To be sure, a flood of neutrons emerges from any fusion reactor, and these could be used to produce plutonium for weapons, by surrounding the reactor with a uranium blanket. However, any nation that has mastered the sophisticated and expensive practical technology of fusion will have available many easier routes to fission (and fusion) weapons.

Alternatives At present, the most favored configuration to confine the plasma is the tokamak, a doughnutlike structure with very complicated external electric windings, conceived and first developed in the Soviet Union. More is known about this confinement system than about any other type, but it is still unclear whether a practical fusion reactor will be a tokamak or some other device yet to be developed. It would be best to develop other and possibly simpler toroidal confinement schemes, at least to the stage of determining their real feasibility vis-à-vis tokamaks.

Several nontoroidal experimental plasma confinement schemes have been proposed as prime candidates. The so-called magnetic mirror schemes suffer from particle losses at the ends, giving poor energy efficiency. The inertial confinement schemes, using either high-powered pulsed lasers or ion beams, must be severely pulsed through many millions of cycles, which introduces the difficulty of early fatigue of structural materials, mainly because of thermal stress.

Fusion has been developed to the point at which it requires large and expensive experiments; while unnecessary duplication of effort should be avoided, such experiments will mean expenditures at current and higher levels. As with the other major options, there is no cheap way out.

The effectiveness of R&D in this field has been fostered by almost completely open international research, and at present all parties are eager to cooperate more closely. An internationally planned and coordinated program would avoid costly duplication, speed the dissemination of results, and provide a basis for true collaboration on very large and very expensive engineering trials. Such an international activity is now under way (as of March 1979), chiefly under the aegis of the International Atomic Energy Agency.

Solar Energy

Solar radiation is a diffuse source of energy and large amounts of materials are needed for collectors, storage devices, and so on. To build equipment that can capture and convert solar energy to useful forms requires capital investments embodying nonrenewable resources that are far from free, even though sunlight itself is free. In a similar sense, the uranium to fuel breeder reactors is practically free, because the fuel costs are insignificant compared to the capital costs and the resources that the reactors represent. The real attractiveness of solar power, besides its ubiquity, is the relative ease with which it can be transformed for a number of uses. However, this attraction has often been oversold by various high-technology schemes. A more realistic view is imperative.

Solar power, by convention, includes not only direct conversion of sunlight into useful forms, but also hydroelectric power, winds, and biomass (organic matter). Discussion of this alternative thus tends to be extensive, since applications are diverse, and options therefore are hard to compare.

Low Technology The most immediately promising solar application is the production of low- and intermediate-temperature heat, from about 70°C for domestic hot water to about 200°C for a variety of commercial, agricultural, and industrial purposes. The simplest systems use flat-plate collectors like those on the increasingly familiar rooftop water heaters. More advanced ones use mirrors or simple lenses to concentrate the solar heat and provide higher temperatures. Most require little or no further science or advanced engineering; they will succeed if their design and construction are ingenious and simple enough to make them economically attractive. Their principal competition now is fossil fuels, mainly oil and gas. Another decade of rising oil and gas prices plus improvements in commercial solar systems should be sufficient for this technology to develop a strong commercial position, assuming that the systems' reliability turns out to be adequate. The federal government has acted to provide some economic stimuli to these low- and intermediate-technology items.

High Technology In the past the solar energy program, partly through a history of influence by NASA and other high-technology agencies, concentrated on tasks with formidable (but interesting) science and engineering problems. At this extreme, we find advanced photovoltaic conversion schemes, the power tower concept (in which a vast array of steerable mirrors focuses sunlight onto a boiler atop a several-hundred-meter tower), and ocean thermal electric conversion systems, which exploit the temperature difference between the surface and the deeps of tropical ocean water.

One of the most promising technologies is photovoltaic electricity generation, which is technically feasible in a variety of installation sizes ranging from the individual houschold to large central station generators (Figure 20). Photovoltaic generators are technically feasible today and have in fact been marketed for specialized applications, such as power sources for satellites or for electrical equipment in remote locations, where the high cost is justified. Wider applications must await cost reductions of roughly an order of magnitude, depending on what happens to the cost of other alternatives in the meanwhile. In one scheme, single crystals of silicon or gallium arsenide would be produced at much lower cost than today. It may pay to concentrate the sunlight with simple optical systems, since this would permit a higher cost per unit area for the cells. In many cases the cost of the optical system and supporting equipment may dominate the cost of the cells themselves, so that both need equal engineering attention.

The other photovoltaic approach aims at the scientific understanding, development, and eventual use of amorphous, or noncrystalline, photovoltaic materials, especially as thin films that can be evaporated cheaply onto inexpensive backing material such as mylar film. This is still in the research stage, because the science of noncrystalline materials is generally more difficult than that of pure crystals. Some success has been achieved, but experimental conversion efficiencies remain smaller than those of

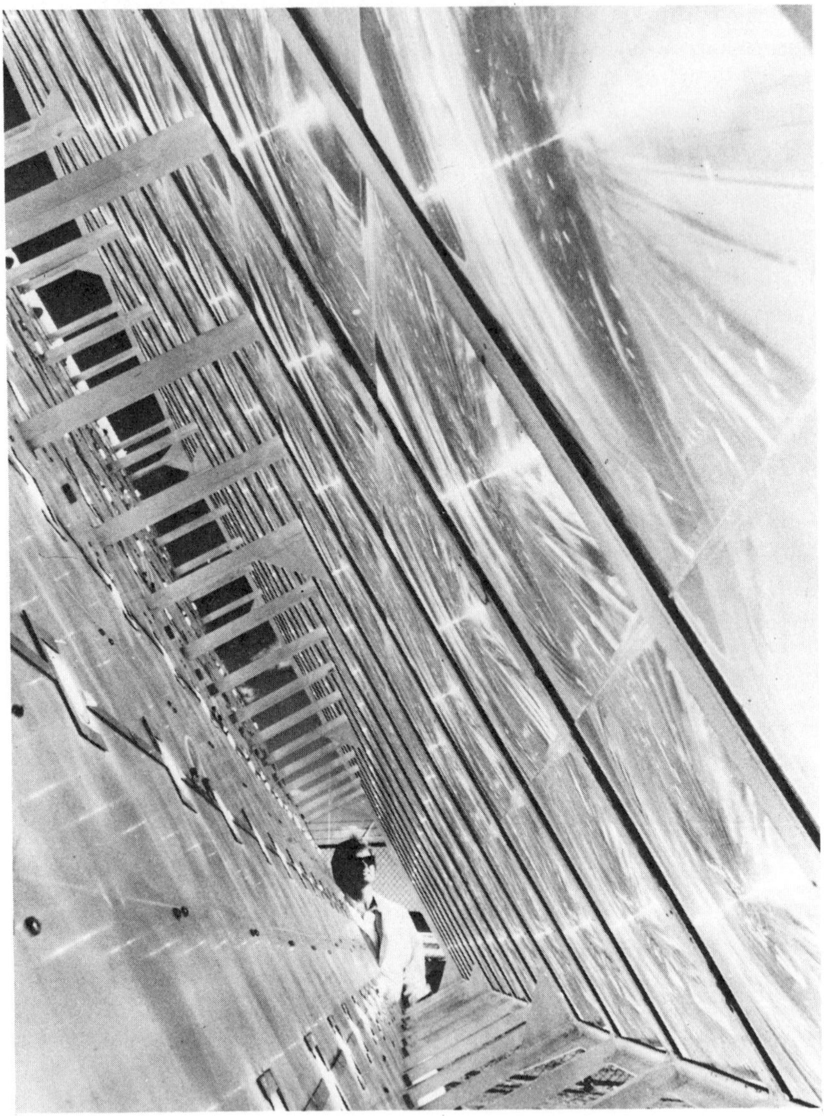

FIGURE 20 Plastic fresnel lenses (right) focus the sun's rays on silicon solar cells (bright spots on left) to convert sunlight directly into electricity. (Sandia Laboratory photo, Department of Energy)

crystalline photovoltaic material. Technical progress in both single-crystal and polycrystalline or amorphous cell materials has been very rapid in the last 5 years, based on the large background of knowledge derived from materials and solid-state research over the preceding 15 years (see pp. 164 and 313). There is now great optimism about the long-term future of photovoltaics, but according to a recent study[14] their market penetration is unlikely to exceed 1 percent by 2000. Even if the future economics were much more firmly established, the problems of creating a large industry and supporting institutional structure and integrating it into the existing electric grid would probably preclude much more rapid market penetration in the present century. Early deployment will be primarily for peaking capacity, and will require conventional backup. Larger market penetration in the next century will increasingly depend on the availability of economical energy storage systems.

The power tower does not look promising. A planned 10-megawatt pilot plant in Barstow, California, will cost $120 million. It will be hard to replicate much more cheaply, because its costs depend so much on those of concrete, steelwork, mirrors, and so on—all of which represent well-developed technologies for which significant cost reductions cannot be expected. A particularly difficult engineering problem is the boiler atop the tower; it must withstand large and rapid thermal fluctuations as clouds pass over the field of mirrors.

Both the ocean thermal energy conversion scheme and, even more, the solar-power satellite scheme are unlikely energy options; they are very capital-intensive and full of serious and poorly understood scientific and technical problems.

Windpower devices occupy an intermediate status, neither high technology nor low, and like most other solar technologies should be regarded as augmenting conventional power supply. The official line of development, which tends toward devices with vanes as large as the largest airplane wings, may not be the best approach. Much cheaper and more reliable devices can serve local areas; solid-state electric circuitry can match the electric output of any reasonable-size windmill to the frequency and voltage of power lines.

Hydropower is usually imagined as the environmentally and ecologically ideal way to produce electric power. But quite apart from the oft-stated risks of dam failures, the dams themselves chop up river systems into ecological bits and unnatural parts, with generally unfavorable consequences too large for the amount of power actually produced. This may be true also of low-head or small-scale hydroelectric installations. Exacerbating the problem of making good judgments on this matter, a recent Corps of Engineers study of power available from low-head hydro dams overestimates by a factor of several (perhaps 10) the average amount available, at least for the New England area.[15]

Regarding biomass, probably 5–10 quads of energy might be producible from farm, forest, and domestic wastes, much of it at a considerable gain in environmental quality; for example, removing the energy from urban wastes and animal feedlot wastes. The concept of obtaining much more energy from intensive silviculture, such as growing sycamore trees for five- to eight-year harvest ("energy farms") looks unattractive. First, we find here the typical ecological problems of monoculture crops. Second, there is the potential for direct competition with production of food, lumber, and pulpwood. Third, restructuring the nation's forestlands for this purpose would take decades. Cutting down scrub forests has been suggested; but the energy content of all the trees in New England would provide that region's energy needs for only about three years.

Concluding with the notable exception of heating and cooling of buildings, note once more how many of the options naturally produce electricity; this reinforces the opinions stated earlier that we seem to be heading for a much more electric economy.

Many solar energy schemes would benefit from having associated energy-storage systems for when the sun does not shine, and virtually all require backup sources of energy for when the storage capacity is exhausted. There are many ways of providing storage, and the need for it varies from application to application and technology to technology. The need for storage has been remarked upon, especially in connection with solar power generation. Without wishing to de-emphasize the importance of energy storage, we point out that until the solar-derived fraction of electric power exceeds about 10 percent of the peak demand, an extended electric grid could absorb the fluctuations in solar output by adjusting the output of its conventional generators. A solar-electric system can be begun with its eventual storage system not yet in place. In the long term the best form of storage in connection with solar energy is the production of fluid or gaseous fuels that can serve as substitutes for hydrocarbons. A wide range of possibilities exists, of which the most important relates to producing hydrogen, either by electrolysis of water or by various photochemical processes. No such process has advanced to the point at which it can be seriously considered for engineering development. This is an important and hitherto neglected area for basic research and exploratory development.

Long-Term Nuclear Fission Technology

The light water reactors (LWR) that now provide almost all of the nation's nuclear generating capacity, operated as they are without reprocessing and recycling of fuel, are not very fuel-efficient. In fact, one recent estimate of the nation's uranium resources suggests that in terms of today's typical

1,000-megawatt power plants fuel supplies will limit LWR capacity to only about 300 plants with relatively assured lifetime supplies of fuel.[18] Other estimates are higher but have less supporting evidence. Fuel supplies for LWR technology can be considered comparable to oil supplies. If nuclear fission is to contribute substantial amounts of electricity at reasonable prices much beyond the end of this century, more efficient ways of using nuclear fuel will be needed.

One way of using uranium ore more efficiently would be to extract the directly fissionable isotope uranium-235 from the nonfissionable part more completely. New technologies—in particular laser isotope separation—have the potential for doing this, but the outcome is not yet certain. In addition, redesigning the LWR reactor cores could improve fuel utilization by perhaps 25 percent.

In fact, development of the LWR depended almost from the beginning on the assumption that spent fuel would eventually be reprocessed and recycled in reactors. This alone would raise the usable amount of the energy potentially available from a given quantity of fuel from about 0.6 percent to nearly 1 percent, thus extending somewhat the life of the available uranium. However, the wisdom of reprocessing has become a matter of public controversy.

Breeders Breeder reactors, which can convert uranium-238 (the common, nonfissile isotope of uranium) to fissile plutonium-239 in quantities larger than are consumed in the reactors, can eventually take advantage of more than 70 percent of the energy potential of uranium ore. There are also thermal breeders, which can convert nonfissile thorium-232 (an element more plentiful than uranium) to fissile uranium-233, and in this way use nearly 70 percent of the energy potentially available from the thorium. This ability to free the energy potential in the so-called fertile isotopes uranium-238 and thorium-232 has a tremendous multiplying effect on available nuclear resources. This is much greater than the approximate factor of 100 implied by the relative fuel efficiencies just cited, because the use of breeder reactors greatly reduces the impact of resource prices on the cost of electricity. This, by making available ores too low in grade to be usable in LWR's, multiplies economically useful reserves of fuel resources.

Until recently, the U.S. breeder development program has concentrated on the liquid metal fast breeder reactor (LMFBR) and its uranium–plutonium fuel cycle. Other nations with breeder programs—notably France and the Soviet Union—have generally done likewise. Thus, at present, this breeder reactor concept is much more fully developed worldwide than any other. LMFBR's could operate for millions of years on the known reserves of uranium in the United States. Even the concentrations of uranium in the stored tailings from today's military and civilian uranium enrichment plants could support a substantial growth in LMFBR capacity for upwards of 200 years, virtually eliminating the environmental and occupational-health impacts of uranium mining and milling.

Unfortunately, breeder reactors require reprocessing and recycling of fuel to achieve their resource-conserving potential and this raises again the hazard of proliferation or theft of nuclear material usable in weapons. Largely for this reason the current Administration has decided to defer reprocessing and delay the planned demonstration of breeder reactor technology while it evaluates, on nonproliferation and other grounds, a range of alternative breeder fuel cycles and reactor concepts.

Advanced converters, which produce fewer fissile atoms than they consume but are more efficient in this respect than LWR's, could significantly extend nuclear fuel resources without requiring reprocessing. The Canadian CANDU heavy water reactor, for example, with alterations in its fuel, could exploit about 1 percent of the energy potentially available in its uranium fuel, compared to 0.6 percent in LWR's, operating on a once-through fuel cycle. This could be further extended by improvements in isotope separation. This improvement, however, would not be very significant in keeping the nuclear option alive unless the growth in electricity demand levels off rapidly after the end of this century or uranium resources turn out to be much larger than now anticipated. Under any other conditions, it will be necessary to have advanced reactors and fuel cycles that exploit the advantages of fuel reprocessing and breeding; time is already growing short for the development of such reactors, even if we consider them as insurance against continued growth in demand for electricity or limited fuel supplies. Advanced converters could also operate on the thorium fuel cycle; although this would require reprocessing in some form, yielding uranium-233 in separated form, which like plutonium-239 is a potential nuclear weapons material.

Despite the likelihood that the LMFBR would be more useful in the widest range of electricity growth and uranium availability and should therefore be developed in any event, we can envisage two possible roles for advanced converters. First, they might serve as a hedge against unforeseen difficulties in breeder development, even if they do offer only partial protection against electricity demand continuing to grow at an appreciable rate or uranium resources of suitable grade turning out to be as limited as they now appear. Second, and more probable, would be use of advanced converters as complements to LMFBR's; for example, if for proliferation reasons LMFBR's, along with fuel recycling and refabrication, were restricted to a small number of secure international sites. LMFBR's could then be used to breed uranium-233 fuel for converter reactors by irradiating thorium blankets. A given amount of LMFBR capacity could support many more advanced converters than light water reactors in this kind of arrangement.

If nuclear power's contribution is to continue much beyond the year 2000, one or more of the following possibilities must be realized by the century's end: major new uranium finds, deployment of advanced converters and later of associated thorium fuel cycles, or development and public acceptance of breeder reactors and associated reprocessing and fabrication facilities with adequate controls on proliferation. The breeder option, in the form of the LMFBR, must be maintained so that if needed it can be deployed early in the twenty-first century.

Even though the present nuclear sector is in trouble, any preparations for the future must recognize that a substantial industry based on light water reactors is in place, and it is unlikely to switch to different reactor systems without credible government assurances and assistance. Several activities sponsored by the federal government make good sense:

- Prompt demonstration of an acceptable waste-treatment method to establish public credibility. Technological options appear to be available; an example is the Swedish proposal to turn the wastes into ceramic, to encapsulate them in titanate–copper jackets (to minimize leaching from water intrusion), and then to entomb them in stable granite rock, with bentonite packing filler providing an ion-exchange medium to trap any leached ionic waste. Other technological options such as some salt deposits are possible.
- Cooperation with Canada in developing an advanced version of their CANDU reactors, which would have the ability to deliver about twice as much electric energy per ton of uranium ore, even without recycling spent fuel—a decided advantage at this time of uncertain uranium resources.
- Further development of the high-temperature gas-cooled reactor preferably in cooperation with the Federal Republic of Germany; a modest prototype exists in Colorado. The principal advantage again is better use of uranium resources. Also, this reactor may ultimately be used with a closed-cycle gas turbine and dry cooling, so as not to require large quantities of cooling water.
- Support promising and more proliferation-resistant technologies for reprocessing spent fuel, such as the Civex process.
- Maintain a breeder R&D program. Success with the second, third, and fourth activities will put off the time when decisions need be made about deploying breeder reactors and give more opportunity for the technology of large-scale solar power to prove or not prove itself. But as seen at present, there exists and will continue to exist a need for breeder development. While the liquid metal fast breeder reactor is closest to final development, enough good ideas for alternative systems exist, some using

mainly thorium instead of uranium, that several lines should be followed, and no exclusive choice should now be made.

SOME RESEARCH NEEDS

Research and exploratory development (R&D) are vital to future energy technology. This account has already included some specific areas—for example, the disposal of radioactive wastes—where attention is needed. Now we further illustrate some of the R&D that might be done, with examples in large part derived from material on earlier pages. It is impossible in the restricted compass of this essay to say anything about relative priorities, and no conclusions on such matters should be drawn from what follows.

Our first major topic was time perspectives. We found them generally to be too short for the timely emplacement of new energy technologies. This may also hold for energy R&D;[19] demonstrations and engineering of mid-term technology may be overemphasized to the neglect of more fundamental research that might optimize technology to fit reasonably both the near and long terms. Both the private and public sectors have large stakes in this process, and the balance of activity requires continual study and readjustment.

THE RELATION BETWEEN ENERGY AND ECONOMIC GROWTH

Both short- and long-term research issues arise in conservation, energy demand, and economic growth. In the short term, energy and the economy seem to be closely coupled, but over the long term we found reason to expect more flexibility. However, the relations between economic consumption and energy growth are not well understood. The key assumptions of current models should be examined so that the economic feedback of reducing energy demand by various plausible methods can be more accurately predicted. Also vital is knowledge of the distributional effects of energy pricing policies and regulations intended to reduce demands; changes in such policies will affect different regions and different economic classes in different ways, and compensatory actions must be considered. These and many other social and economic aspects of energy conservation offer a broad scope for basic work in economics and other social sciences.

ENERGY CONSERVATION

Conservation is a fertile field for innovative research. Of most immediate interest are measures that could be taken now and in the near future, at costs quickly paid back in energy savings. There are undoubtedly many applications for new and sophisticated control technologies for optimizing energy use in all sectors. Industries in the United States, for instance, have made important gains in energy efficiency in response to the recent substantial increases in oil prices. In the building sector, more knowledge and better technology would allow the setting of realistic and effective thermal performance standards; no solid scientific basis exists at present for predicting the thermal performance of building subsystems, let alone entire buildings. Because of the expected continuing upward trend in air travel, it would be important to understand the financial and institutional barriers to introducing the next generation of more energy-efficient jet aircraft engines. Comprehensive energy conservation experiments in representative industrial establishments, perhaps with the costs shared by the government and the companies involved, should be seriously considered.

In principle, there is considerable energy saving potential in the introduction of industrial cogeneration—the simultaneous production of electricity and process heat. This results in the more efficient use of the primary energy resources and reduces the need for centralized generating capacity. The barriers to more widespread cogeneration are mainly institutional and economic. Experiments with various institutional arrangments for combined production and use of electricity and heat might well be encouraged and supported by government.

COMBUSTION

Combustion is a basic process of our civilization. If more were known about precisely what happened when fuel is burned, we could build more efficient engines, power plants, and space heating systems; reduce pollutant emissions; and perhaps learn to use lower-grade, less refined fuels in some applications and thus save refining costs. For example, recent advances in combustion control promise to ameliorate emissions of nitrogen oxide pollutants from power plants.

Among the important problems in combustion are the interactions of chemical and flow phenomena in combustion; the high-temperature oxidation kinetics of hydrocarbons, alcohols, ketones, and other compounds; surface catalytic reactions involving oxidation-reduction, pyrolysis, and decomposition; nucleation and condensation of combustion products to form soot; the mechanisms in the formation of pollutants, such as sulfur and nitrogen oxides, hydrocarbons, and other substances; and materials properties for refractories and coatings resistant to corrosion and other kinds of damage in hot gas environments.[20]

CHEMISTRY OF COAL

Understanding of coal has lagged far behind that of petroleum, not only because coal is chemically more complex, but also because the growing use of petroleum and natural gas in the past reduced the economic and scientific incentives to study coal. The available data on the chemical properties of different coals are inadequate, as are procedures for analyzing coals. The chemical reaction paths in important coal conversion processes are incompletely known, as are the thermal and kinetic interactions in combustion equipment and synthetic fuel plants.[20]

Research on these topics can yield a number of important practical benefits. Coal conversion involves reacting some of the hydrogen-deficient coal with water (or some other material whose original source of hydrogen was water) to turn the solids in coal into liquids and gases. This process is expensive in both materials and energy. Improved, more selective catalysts would lower both the requirements for and expense of hydrogen and also facilitate removal of oxygen, nitrogen, and sulfur from the products. Improved surface characterization techniques would be useful in understanding observed kinetic behaviors in catalytic gasification and would thus allow improvement of catalysts for those processes. The inorganic pollutants emitted by coal combustion could be more precisely controlled if the high-temperature interactions of the the mineral matter, and the nitrogen and sulfur, were better understood. The problems of slagging and fouling could also yield to this improved understanding. In virtually all the ways coal is and may be used—direct combustion, gasification, and liquefaction—a better understanding of coal's chemical structure and reactions would contribute to lower costs and better environmental qualities.

R&D Related to Some Other Energy Sources

New clean fuels and clean energy conversion schemes deserve particular attention. These include energy storage, generation and use of hydrogen, and fuel cells. R&D on high-quality work on amorphous semiconductors will aid the development of photovoltaic materials for solar-electric systems.

ENVIRONMENTAL EFFECTS

We have already mentioned the global CO_2 problem, about which we are somewhat uncertain but very concerned.

More generally, we still only poorly understand the environmental impacts of energy use. For example, a vital question, with extremely important economic implications, is the relationship between certain fossil-fuel emissions and public health problems. Combinations of various pollutants—sulfur and nitrogen oxides, particulates, hydrocarbons, and metals—are probably more hazardous than each alone, but the effects of either individual or combined pollutants are very poorly characterized. Better knowledge here could allow more cost-effective control of air quality and probably improved public health. This is especially important given that increased energy use may result in lower-grade fuels.

Another important topic is water consumption by energy facilities, especially synthetic fuel and power plants. Obviously, withdrawing too much water from a river can have extreme environmental consequences; some rivers in the West already demonstrate the effects. Two main areas of research may help us to produce needed energy, while avoiding undue ecological damage and shortages of water for nonenergy use. First, basic work in the reactions involved in synthetic fuel production could decrease greatly the need for water in these processes. Second, water actually available over the long term in river basins and underground aquifers across the country must be known with greater certainty, and in much greater detail, to allow optimal siting of water-consuming facilities.

THE IMPLICATIONS OF A MORE ELECTRIC FUTURE

The likelihood of increasing electrification has been proposed a number of times in this chapter. From the standpoint of energy use, what does that imply? There is not room here to develop the concept very far, but the following changes appear likely and should be examined:

- In the long run, the manufacture of hydrogen with off-peak power might serve the dual purpose of supplying portable fuels and increasing the energy-efficiency of the electric power system. In fact, "hydrogen–electric future" is probably the best descriptive phrase. The hydrogen would be used in aircraft, in many industrial processes, and possibly in electric reconversion fuel cells.
- Substantially electrified mass ground transport, like that in present-day Europe, may play an increasing role.
- More efficient electric heat pumps might become the dominant means of heating and cooling buildings. These could run on an annual cycle, making ice in the winter for use in summer cooling.
- Cogeneration could become much more common in industry.
- Electricity might find wide use in automobiles, in either all-electric or partly electric cars. The former would require much improved batteries; in the latter, a small engine might run continuously at optimum efficiency to

operate a small electric generator and electric motors at the wheels, with a battery charged by the generator to handle peak demands.

CONCLUSION

This assessment of our current energy situation might seem to imply that nothing but disaster lies ahead. Not so; the danger is there, but not the certainty. Neither the United States nor the world is running out of energy. The flow of oil and gas will not cease overnight, and a dedication to the conservation of energy as a social norm will relieve the strains on a changing supply system. Finally, the potential of solar energy, fusion, and new forms of fission are almost infinite if they can be developed and prove socially acceptable.

The difficulty ahead is in reconciling immediate problems with these opportunities. Of the problems the most apparent is our great reliance on gaseous and liquid fuels, both of which are in finite supply. The use of oil and gas is woven into the technical infrastructure of our society—transportation, industrial processes, heating, chemical manufacture, and so forth. The task ahead is not simply recognizing that these fuels will in time become scarce or in estimating when that will happen, but rather of preparing our society for a transition to new forms of fuels and a more careful use of them.

One complication is in the quite different perspectives of those who must participate in that transition; the difference, for example, between the time it takes scientists and engineers to develop a new energy technology to a commercial level and the usually shorter times when business expects to make a profit on its investments. Also, it is difficult for all to agree that a quite different energy future is possible; that, for example, liquid hydrocarbon fuels are not necessarily the essential base for advanced technological civilizations; that we have simply designed our society around those fuels most available, most convenient to use, and, until recently, quite cheap. The fuels are still convenient, but no longer as available nor as cheap; and, therefore, we must begin to think of other ways to fuel our technology. As an example, if we are technically clever about it, hydrogen can be used to fuel airplanes; indeed, much of our transportation could in time operate on energy from stationary sources, such as electric power plants. However, there will continue to be some need for liquid and gaseous fuels in some small amounts.

In the end, the discussion returns to questions of social purpose, economics, environmental costs, and public understanding. Our energy future will depend in part on whether energy is available, on its forms, and on our technological ingenuity, but even more on decisions made by society on what technological options to use, to what purposes, and with what safeguards.

OUTLOOK

The following outlook section on energy is based on information extracted from the chapter and covers trends anticipated in the near future, approximately five years.

Over the next five years, the United States will confront basic questions regarding its future energy pattern. How rapidly will oil and gas resources, domestic and global, decline? Can growth in energy demand be tempered sufficiently to allow the smooth substitution of new energy sources for oil and gas? What form will these replacements take and what problems may their use entail?

The answers to these questions are shaped in part by time. For the mid-term transition from oil and gas to other energy forms, the United States has in reality very little maneuverability, in both energy supply and use. Its major alternative sources will be coal and fissionable uranium; growth in energy demand, while it can with sufficient time be smoothly and slowly reduced with little harm, cannot be cut sharply without causing serious economic disruptions.

For the long term, here assuming beyond 2000, the United States has a number of potential options: various applications of solar energy; fission reactors using more plentiful uranium isotopes; possibly fusion (though somewhat later than the other options if at all); and much improved efficiency in energy use.

Most of the mid-term and long-term energy supply technologies are electrical, and this suggests a change in the character of energy supplies, from increasingly scarce liquid and gaseous fuels to electricity. Technological ingenuity can transform such resources as coal and oil shale into gases and liquids, but at considerable costs, both economic and in several cases environmental.

That this country's energy pattern will change in the coming decades is certain. Oil production and discovery are unlikely to keep up with rising demands. World reserves of oil in 1978 were only about 2 percent above 1972 levels, though meanwhile prices had quintupled and world demand for oil had risen more than 15 percent.

The United States is particularly vulnerable. Its oil reserves and production have declined over the past 10 years as its demand for oil has grown, and it will likely be forced in coming years to draw more heavily on an increasingly limited world market for oil. The domestic gas outlook is even worse. The world has a good deal of gas, which is potentially available, but, except for that in Mexico and Canada, it must be liquefied and transported to the United States at very low temperatures in special tankers. Its expense and hazards may make imported liquefied gas unattractive compared to alternatives, the strategic consequences of reliance on foreign sources aside.

These facts suggest that we are overdue in considering ways of limiting U.S. dependence on oil and natural gas, by substituting other forms of energy where possible and by instituting vigorous measures to increase the efficiency with which energy—oil and gas in particular—is used. At the same time, possible means of securing stable access to energy for the indefinite future must be examined. This will require new technologies and the political will to develop and use them responsibly. In principle both can be done; in practice each will be difficult.

Financial and political institutions are more responsive to the short-term view dictated by market rates of return and electoral intervals than to the longer and broader view determined by domestic resource depletion and the risks of depending heavily on imported energy. Only an educated public and enlightened public officials can temper this disparity by responsible efforts to avoid the difficulties to which short-range expediency can commit us.

The biggest difficulties will be not purely, or often largely, technological, but rather matters of social decision—what levels of pollution to accept, how best to protect national security, and even what, in broad terms, constitutes a desirable society.

CONSERVATION: THE HIGHEST PRIORITY

Conservation appears to deserve the highest priority in energy planning over the next several years. First it can provide some quick and fairly cheap gains through lowered thermostats, more careful thermal control in manufacturing processes, various kinds of insulation, and some simple waste heat recovery. Second, very substantial technological improvements indeed will be economical and more important in the longer term—as buildings, vehicles, and industrial equipment are replaced or rebuilt. Recent econometric and engineering analyses suggest, for example, that rising energy prices and conscious conservation incentives over the coming 35 years or so could as much as halve the ratio of energy consumed to economic output in the United States, with negligible impacts on economic growth. Again, this must be a gradual process, or it could result in grave economic losses. Furthermore, it would require continuing determined effort and cooperation between public and private sectors—a more-than-difficult task.

MID-TERM SUPPLY STRATEGIES

Next to planning conservation policies, the second important issue that must be settled in the next five years is how to begin the transition away from oil and gas and toward energy supplies that are essentially infinite. The first step is to determine what is available now. The choices are rather strictly limited: coal and nuclear fission, with some help from solar energy and perhaps a little geothermal energy in suitable locations.

The two major energy sources, coal and fission, both have serious problems. Coal, for example, is a prolific source of pollutants and other hazards, including the long-term climatic risks of CO_2 emissions; also, while plentiful compared to oil and natural gas, coal is not infinite. Nuclear fission in its current form, the light water reactor, depends on a quite limited resource: the naturally fissionable isotope of uranium; it is also under political and regulatory pressure that perils the reactor industry. Neither coal nor nuclear fission using light water reactors is appropriate for the long term.

However, the promising long-term technologies—solar energy in its more advanced forms, nuclear fission in breeder reactors, and controlled nuclear fusion—are far from ready for commercial deployment in the United States; none is likely to make a major contribution before the end of the century. Given the speed with which the world's oil and gas resources are being depleted, there is serious question about whether any of these will become timely replacements for coal combustion and nuclear fission, let alone serving as vehicles for a smooth transition directly from oil and gas. Thus, in the next several years the nation has no choice but to use what is available, improving where possible our ways of using coal and nuclear fuel, while doing the work necessary to ensure the availability of more appropriate long-term technologies. The difficulty of the consequent industrial and economic shifts reinforces the vital importance of energy conservation.

While there is intense interest in synthetic liquid and gaseous fuels from coal and oil shale and several large demonstration projects do exist, synthetic fuel production is still on a rather primitive basis. Developing it to the commercial stage will take another decade at least. Even then it is likely to be limited to a small fraction of present oil and gas use, notably by the availability of fresh water to meet the large requirements of the processes.

This apparent imbalance between likely demand and the supply of liquid and gaseous fuels presents a choice that will have to be seriously addressed in the next several years. We can either try to improve coal-based synthetic fuel technology to the point at which it can serve our energy needs in much the manner of oil and gas today or restructure the technological basis of civilization to reduce permanently the heavy dependence on such chemical fuels by turning more toward electricity. In practice, both approaches will be necessary, but the balance between them is complicated and ill-understood.

SOME OUTSTANDING ENVIRONMENTAL AND HEALTH EFFECTS OF ENERGY USE

Energy production in the past has had unforeseen impacts on the human environment. The costs of air and water pollution, waste-disposal problems, and many occupational hazards in energy production and use are only now beginning to be appreciated. It is not possible to rank the various energy supply options precisely according to the relative severity of their impacts. However, there are grave reasons for special concern about the impacts of certain energy sources given current energy projections.

Especially important are those of fossil-fuel combustion, and particularly those of coal with its enormous chemical variety and complexity. As it increases its reliance on coal, the nation in the next several years will have to address the possible climatic consequences of continued emissions of carbon dioxide, and air pollution in general. Also, water availability and competition for other water uses will impose severe restrictions on the siting of energy facilities, and will require major technological developments in water use as well as large water-related investments.

LONG-TERM ENERGY OPTIONS

Only two major classes of long-term energy supply options exist: solar power in various forms, and nuclear power, both fission and fusion. These can be, and are now, augmented in especially favorable locations by modest amounts of geothermal power, but the normally microscopic heat flow from inside the earth makes solar power a better bargain in most places. The potentials of tidal power, and probably also wave power, are quite overemphasized.

Nuclear Fusion

The prospects for fusion support what can be described as guarded optimism. It seems clear now that an acceptable fusion plasma can be produced and confined in a vacuum by strong magnetic fields—a plasma not touching any physical surface and kept together long enough for an interestingly large fraction to react. Thus, in a scientific sense, fusion power appears feasible. It has taken about 25 years and several billion dollars to reach this stage.

The most expensive and difficult parts—the technology and the practical engineering of a working reactor producing power—are still to come. The basic technological feasibility of controlled fusion will not become clear for at least a decade, and the engineering and economic feasibility much later. Fusion will come (if at all) later than breeder reactors and advanced solar power systems, both of which may be called for by the end of this century. Still, it remains a contender for the long term.

Solar Energy

Solar power requires investments embodying nonrenewable resources that cost money, though sunlight itself is free. The real attractiveness of solar power, besides its ubiquity, is the relative ease with which it can be put to various uses, such as water and space heating. This attraction, however, has been ignored by various high-technology schemes.

Solar power, by current convention, includes not only direct conversion of sunlight into useful forms, but also hydroelectric power, winds, and biomass (organic matter) considered as an energy source.

The most immediately promising solar application is in producing heat in the 70–200°C range, for residential hot water and space heating and a variety of commercial, agricultural, and industrial purposes. The simplest systems use flat-plate collectors. More advanced ones use mirrors or lenses to concentrate the solar heat and provide higher temperatures. All have benefited in recent years from advanced science and engineering. Another decade of rising oil prices plus improvements in commercial solar systems should put this technology in a strong economic position. The federal government has acted to provide some economic stimuli.

At the high-technology end of the possibilities are advanced photovoltaic conversion schemes, the so-called power tower concept, in which a vast array of steerable mirrors focuses sunlight onto a boiler atop a several-hundred-meter tower, and ocean thermal electric conversion systems, which exploit the temperature differences between the surface and the deeps of tropical ocean water. Photoelectric generation has two main branches: crystalline and amorphous photovoltaic materials. Both show promise and both warrant continued effort. The power tower does not look promising; it is expensive, and significant cost reductions cannot be expected. Ocean thermal energy conversion and proposed solar satellites beaming power to earth represent similar attempts to apply high technology inappropriately.

Windpower devices occupy an intermediate status—neither high technology nor low—and like most other solar technologies should be regarded as augmenting conventional power supply. Hydroelectric dams present small risks of dam failures. They also chop rivers into ecological bits and unnatural parts, with consequences disproportionate to the power produced. This may be true also of small-scale hydroelectric installations. In the field of biomass, the equivalent of 900-1800 million barrels of oil probably could be produced annually in the unlikely event that all the nation's farm, forest, and domestic wastes were exploited. "Energy farms" would not only present the ecological problems of any monocultured crop, but also might compete directly with food and wood production.

Nuclear Fission for the Long Term

The light water reactors that now provide almost all the nation's nuclear generating capacity exploit only about 0.6 percent of the energy potentially available in their fuel. According to one estimate, reasonably assured domestic uranium resources would suffice to fuel only about 300 thousand-megawatt power plants.[18] The fuel could be reprocessed to recover plutonium and some still usable uranium, but these additions would increase the total available energy by a factor of only about 1.6 or 1.7. Conventional methods of reprocessing plutonium for re-use in light water reactors lead to its brief appearance in a form suitable for nuclear explosives. If this course were taken, the proliferation resistance of the nuclear fuel cycle would be endangered for at best a moderate benefit.

Advanced converter reactors, the most frugal of which would use only about half as much uranium (and might also use thorium) to generate the same amount of energy, even without reprocessing, could help. Their contributions would be significant for the long-term, however, only under some national circumstances, especially of low growth in electric power demand.

If nuclear power is to have a long-term future, it must be with breeder reactors. These reactors as now envisaged convert the more common nonfissile uranium isotope, uranium-238, to plutonium, then fission the plutonium to provide heat. Breeder reactors can also be designed to convert thorium-232 to fissile uranium-233. Using breeders, more than 70 percent of the uranium that is mined can be effectively used, yielding about 70 times the energy from a given amount of uranium as can be obtained from light water reactors, even with fuel reprocessing. This means that we could then afford to pay very much more for the uranium than before; at that price, the reserve base becomes large enough to last millions of years. This much reduced uranium need is reflected in a similarly reduced need for mining and ore processing. Thus, the hazards of radiation exposure in these activities, now subjects of lively debate in connection with light water reactors, will be almost eliminated.

Breeder reactors require fuel reprocessing, again raising the question of nuclear weapons proliferation. New techniques of reprocessing, however, can be designed so that the availability of plutonium-239 or uranium-233 in a form usable for weapons is much reduced.[21] Nonetheless, breeder fuel in its present form represents a proliferation risk.

All fission reactors produce nuclear waste, and the total amount (measured by its radioactivity) depends hardly at all on the reactor type. An adequate science and technology base for entombing these wastes in geological formations exists. With proper site-specific engineering and geological knowledge, the long-term public hazards arising from waste depositories would consist of low-level radioactivity less than that from natural background radiation or other nuclear fuel cycle activities. What is lacking today is the detailed engineering and assessment for specific suitable repositories.

RESEARCH AND DEVELOPMENT NEEDS

Several important fields for energy research are discussed in the main body of this chapter, particularly under the heading "Some Research Needs." However, even the synopsis just presented reveals some particularly pressing problems:

- The relationship between energy use and economic growth is obviously vital. The economic impacts of constraining energy demand cannot be very precisely projected, but some good beginnings have been made in recent years.
- Energy conservation—in industry, in transportation, in buildings, and in appliances and services of many kinds—still receives inadequate attention.
- Because of the necessity of reducing global dependence on fossil fuels in response to world environmental hazards, and the need to increase use of some fossil fuels—especially coal—during the period of transition to more benign long-term energy sources, we must learn to understand the chemistry

and combustion kinetics of these fuels better, in order to derive maximum benefit from limited use.

NOTES AND REFERENCES

1. Hayes, E.T. Energy Resources Available to the United States, 1985 to 2000. *Science* 203(4377):233–239, 1979. An excellent overview of the prospects for energy supply (although we do not agree with him on some of the economic and environmental assessments implicit therein).

2. CONAES Demand and Conservation Panel. Energy Demand: Some Low Energy Futures. *Science* 200(4338):142–152, 1978.

3. *Energy Modeling for an Uncertain Future, Supporting Paper 2* (NRC Committee on Nuclear and Alternative Energy Systems). Washington, D.C.: National Academy of Sciences, 1978.

4. Probstein, R.F., and H. Gold. *Water in Synthetic Fuel Production: The Technology and Alternatives.* Cambridge, Mass.: MIT Press, 1978, p. 151.

5. Burwell, C.C. Solar Biomass Energy: An Overview of U.S. Potential. *Science* 199(4333):1041–1048, 1978.

6. Harte, J., and M. El-Gasseir. Energy and Water. *Science* 199(4329:623–634, 1978.

7. At much higher prices, the alternatives (synthetic fuels, nuclear or solar power, etc.) would become relatively attractive and thus tend to drive oil out of many markets.

8. Moody, J.D., and R.E. Geiger. Petroleum Resources: How Much Oil and Where? *Technology Review,* March/April:39–45, 1975.

9. Pretreatment of high-sulfur coal, mainly by washing, has been underemphasized as a sulfur-control method.

10. There appears to be no practical way of scrubbing the CO_2 from any appreciable fraction of fossil-fuel emissions.

11. See G.M. Woodwell. The Biota and the World Carbon. *Science* 199(4325):141–146, 1978; Minze Stuiver. Atmospheric Carbon Dioxide and Carbon Reservoir Changes. *Science* 199(4326):253–258, 1978; U. Siegenthaler and H. Oeschger. Predicting Future Atmospheric Carbon Dioxide Levels. *Science* 199(4327):388–395, 1978; C.S. Wong, Atmospheric Input of Carbon Dioxide from Burning Wood. *Science* 200(4338):197–200, 1978.

12. *Air Quality and Automobile Emission Control,* Vol. 1: Summary Report (NRC Coordinating Committee on Air Quality Standards). Washington, D.C.: National Academy of Sciences, 1974, p. 11.

13. *Nitrates: An Environmental Assessment* (NRC Coordinating Committee for Scientific and Technical Assessments of Environmental Pollutants). Washington, D.C.: National Academy of Sciences, 1978.

14. *Principal Conclusions of the American Physical Society Study Group on Solar Photovoltaic Energy Conversion.* New York: American Physical Society, 1979.

15. Estimate of National Hydroelectric Power Potential at Existing Dams. U.S. Corps of Engineers, 1977. A better calculation goes roughly as follows: if every drop of rain falling onto New England passed through a 100 percent efficient hydropower turbine and no water flowed downhill otherwise, the total power generated would be about 18,000 megawatts (yearly average). From this, subtract 70 percent for evaporation and vegetative transpiration, to obtain 5,100 megawatts. The Corps of Engineers estimates that half this power could be generated, but that would require stopping the flow on half of every stream, no matter how large, small, accessible, or remote.

16. Rossin, A.D., and T.A. Rieck. Economics of Nuclear Power. *Science* 201(4356):582–589, 1978.

17. *Risks Associated with Nuclear Power: A Critical Review of the Literature,* Summary and Synthesis Chapter (NRC Committee on Science and Public Policy). Washington, D.C.: National Academy of Sciences, 1979.

18. *Problems of U.S. Uranium Resources and Supply to the Year 2010, Supporting Paper 1* (NRC Committee on Nuclear and Alternative Energy Systems). Washington, D.C.: National Academy of Sciences, 1978.

19. *Report of the Office of Science and Technology Policy Working Group on Basic Research in the Department of Energy.* Washington, D.C.: Office of Science and Technology Policy, 1978.

20. *The Department of Energy: Some Aspects of Basic Research in the Chemical Sciences* (NRC Committee on Chemical Sciences). Washington, D.C., National Academy of Sciences, 1979.

21. Levenson, M., and E. Zebroski. "Fast Breeder System Concept—A Diversion-Resistant Fuel Cycle," paper presented at the Fifth Energy Technology Conference, Washington, D.C. February 27, 1978.

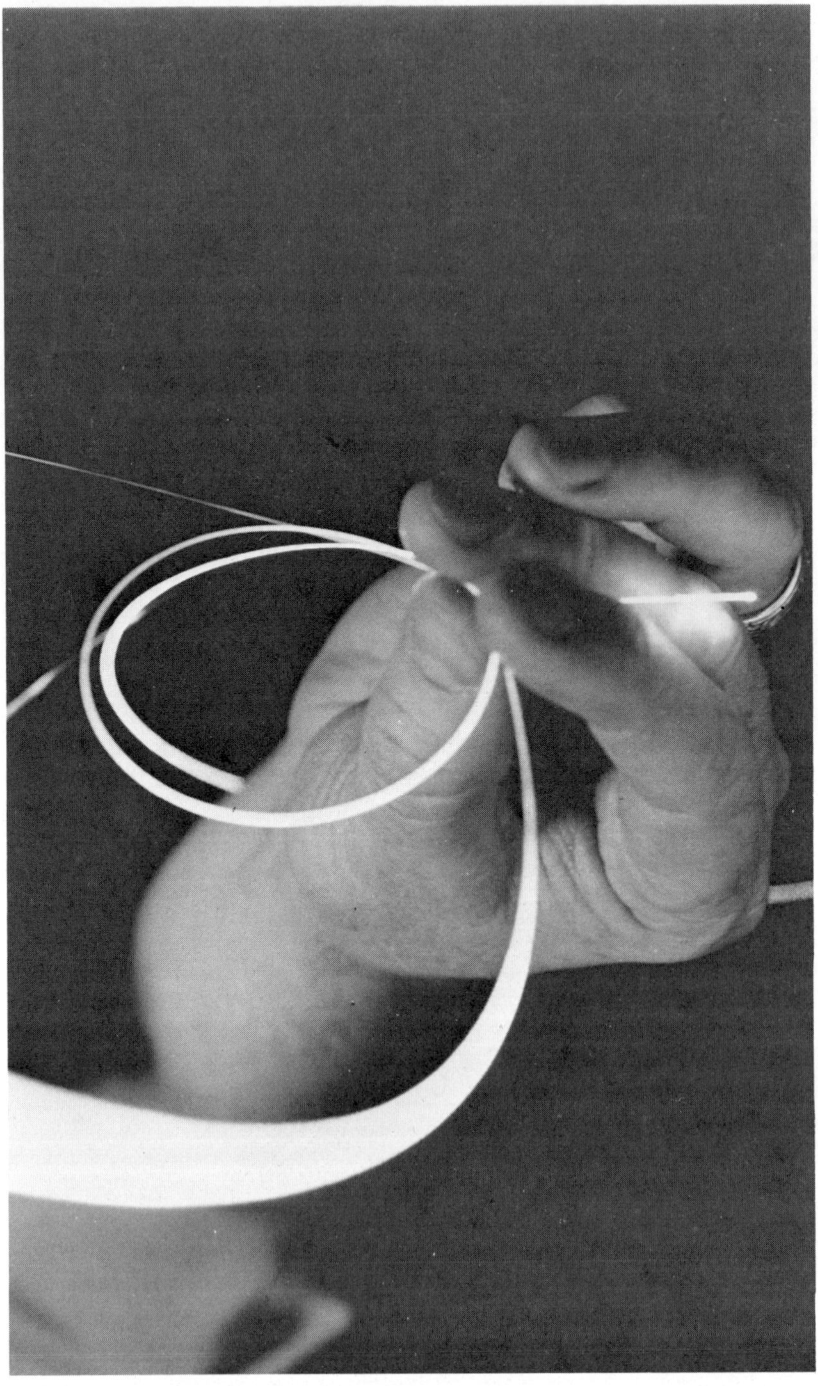

6 Materials

INTRODUCTION

The near-term future for materials will be shaped largely by our responses to forces already well recognized. Some of these forces are long-standing and amount collectively to supply and demand. Others have emerged only recently and are related to national problems that include energy, environment, governmental regulation, and productivity and innovation. Still others relate to worldwide political and institutional changes and growing global interdependence. We shall touch on some of these materials-related pressures later. Together these forces are creating clear-cut challenges and opportunities in materials science and technology.

Americans consume annually about 10 tons per capita of materials and 15 tons of coal-equivalent energy. It is difficult to project the particulars of future materials needs either nationally or internationally. For this country, however, it can be said that the needs will increase incrementally in quantity, but that the kinds of materials used will not change much during the next five or ten years. Beyond that period, the national origins and the kinds and amounts of the materials used could change markedly. New technologies change both what is usable and how it is used.

Expenditures on materials research and development in this country grew rapidly beginning in World War II and peaked in the 1960's. In

◀ Ultrapure glass fibers carry information in the form of light signals. (Bell Telephone Laboratories)

current dollars annual expenditures today are hovering at about five times the level of the 1950's, but are declining in constant dollars.

The results of the past 35 years of materials research and development have been spectacular. They have led to major new industries, including computers, plastics and synthetic fibers and rubbers, and nuclear power.

DEVELOPMENTS IN MATERIALS

The rapid evolution of new materials (see p. 166) that began largely during World War II is still under way. Research and development has become more selective, however, and a variety of business factors are inhibiting the movement of new methods and products from the laboratory to everyday use. In short, we are seeing a decline in the innovative risk-taking that characterized the rapid materials progress of 1950–70. Still, advances should continue to be stimulated by the interaction of industrial technology with basic research in the universities, government, and industry.

Selected developments expected during the next five years are treated here in terms of eight classes of materials: metals and alloys, energy-related materials, information-related materials, polymers, ceramics and other inorganic materials, composite materials, renewable materials, and biomedical materials. These developments will illustrate the achievements and near-term potential of materials science and technology. They are limited largely to engineering materials and structures; we have not, for example, attempted to treat the extraction of raw materials. Except in unusual circumstances, technological developments that will mature during the next five years will rely on research already in hand.

METALS AND ALLOYS

In basic metals like iron and steel, aluminum, and copper, striking advances on a broad front cannot be anticipated. We are more likely to see incremental improvements in specific properties, reduction in cost through processing innovations, and better tailoring of properties to meet specific needs. Technological innovation and gains in productivity in the basic materials industry in the five years to come will depend less on technical progress than on the generation and availability of capital to modernize plants and improve technological capability.

Ability to control the properties of metals depends in part on knowledge of their internal structure and composition. Metals normally are composed of grains made up of microscopic crystals in which atoms are spaced in a pattern characteristic of the particular metal. Although the grains have a regular, crystalline structure internally, they have irregular shapes. Irregular surfaces are created when adjacent crystals interfere with each other as they form from the molten metal. Impurities in metals tend to segregate at the grain boundaries, the interfaces between grains.

The early 1950's saw the development of tools like the transmission electron microscope that permitted direct observation of the internal characteristics that control deformation and fracture in metals. With these tools, metallurgists were able to clarify in detail the relationships among the composition, microstructure, and properties of metals and how they determine performance in service.

Superalloys

More recently, technological progress has made it possible to raise significantly the permissible operating temperatures for nickel- and cobalt-base superalloys in gas turbines. Superalloys are alloys intended for service above 750°C and are used routinely in jet engines (Figure 21). The higher operating temperatures of around 950°C which were achieved by designing microstructures that give the alloys unusual resistance to deformation and fracture—increase the efficiency of jet engines with significant improvements in performance and savings in fuel.

A further step has been the use of directional solidification to achieve unique high-temperature strength and performance in the nickel-base superalloys. Directional solidification of the molten alloy yields a structure that is strongly anisotropic—the properties of the metal are strikingly different in different directions. The part can be designed, therefore, so that it is strongest where it must bear the heaviest load. Further improvements have been made by the use of controlled solidification to produce single-crystal turbine blades and vanes of such alloys (Figure 25c). The absence of grain boundaries in these materials frees them of certain detrimental effects of grain boundaries on high-temperature behavior. Directionally solidified materials and single-crystal blades and vanes will be demonstrated in gas-turbine engines within the next five years. Both advances will increase the durability of turbines and may increase peak operating temperatures and thus efficiency.

Coatings with unusual resistance to oxidation have been developed to protect high-performance alloys in turbines and other hot environments. These coatings, together with the development of nickel-base alloys as mature engineering materials, have been key factors in the outstanding operating experience and long periods between overhaul of commercial jet aircraft engines.

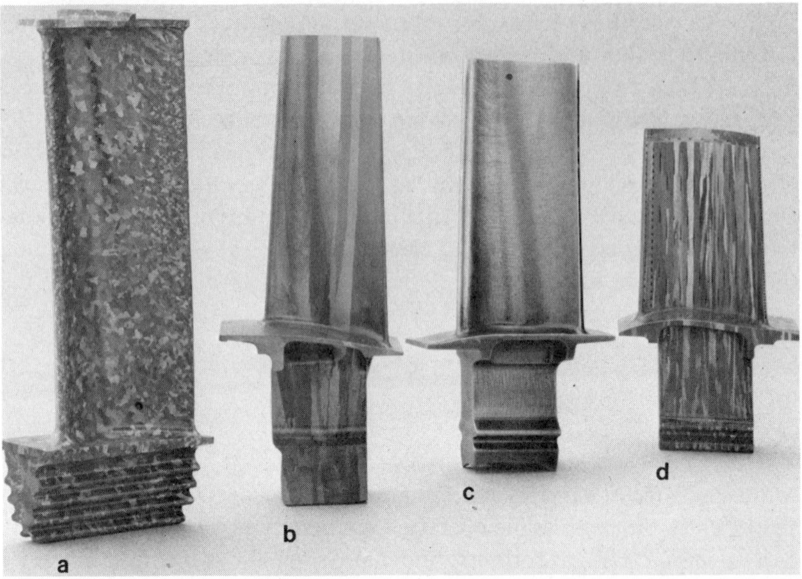

FIGURE 21 Gas-turbine blades etched to reveal grain structure. The solid cast blade of the 1950's (a) had a thrust/weight ratio of 1 and was used at 880°C. By the 1970's, the ratio was 1.9 and the hollow cast blade (b) was used at 950°C. For the 1980's, a single-crystal blade (c) will have a 2.2 thrust/weight ratio and operate at 970°C. The blade designed for the 1990's (d) will be made with directionally recrystallized columnar grains and will have a thrust/weight ratio of 2.9 and be able to operate at about 1,000°C. (A. R. Cox, Pratt & Whitney Aircraft Group, West Palm Beach, Fla.)

High-Strength Steels

Within the past five years, steels have been introduced that combine relatively high strength with costs only minimally higher than those of conventional low-carbon steels, which have been the workhorse of the structural steel industry. These high-strength steels are being used increasingly where higher strength-to-weight ratios are advantageous. In automobiles, for example, they may prove to be the most cost-effective means of reducing weight and thus improving fuel economy in the near term. It has been estimated that as much as 500 pounds per car of the new steels will be used in U.S. vehicles by 1985. The weight of these steels required to perform a given function in a car is expected to be 10–30 percent less than the weight of the materials replaced.

These high-strength steels are made by coupling the effects of minor concentrations of alloying agents with control of the rolling and quenching operations in the steel mill. The resulting metal has a fine-grained microstructure and correspondingly higher strength. One class of high-strength steels, the dual-phase steels, combines the ductility and fabricability of the softer, low-carbon steels with the strength usually associated with higher-carbon, hard steel. This unusual combination of properties is achieved at only modest increases in cost, and within the next decade the dual-phase steels should become a widely accepted class of structural steels. However, their availability from the U.S. steel industry currently is inhibited by economic factors, notably the cost of installing continuous annealing facilities.

Glassy Metals

A number of industrial organizations in this country and abroad are exploring commercial uses for a new class of materials, the glassy metals. Although metals are normally crystalline, certain alloys can be solidified in noncrystalline, or amorphous, form, like glass, by very fast cooling—100,000°C–1,000,000°C per second. These materials consist of metals like iron, cobalt, and nickel, alloyed with elements such as phosphorus, silicon, and boron. They can be solidified in glassy form in ribbons seven–eight centimeters wide and a fraction of a millimeter thick.

Glassy-metal magnets are very strong mechanically, and some show very low losses of energy of the kind that occur in other magnetic materials during magnetic cycling. This combination of properties makes the glassy metals good candidates for replacing iron–silicon alloy sheets in the cores of transformers used in the transmission of electrical power. It has been estimated that conversion to glassy-metal transformer cores could save the energy equivalent of 6 million barrels of oil per year.[1] The saving could be realized only gradually, however, because no more than about a tenth of the huge installed transformer capacity could be replaced annually.

Magnetic glassy metals in strips several centimeters wide can be purchased in development quantities today. The prospects for commercial use of the materials in the next five years, in applications such as magnetic shielding, are quite good. Glassy metals also resist corrosion exceptionally well. The hope of exploiting this characteristic undoubtedly will stimulate much research and development during the next five to ten years.

Corrosion

Corrosion of metals costs the nation billions of dollars annually. The problems range from rusting and deterioration of consumer products to failure of high-performance parts in hostile (e.g., hot, corrosive, erosive) environments. Corrosion is a problem particularly in the development of new energy-conversion devices in which hostile environments cause metal parts to deteriorate rapidly.

Many means of preventing corrosion have been provided by research on the mechanisms of corrosion; developments in protective coatings, paint systems, and surface treatments; and the availability of intrinsically corrosion-resistant materials, such as titanium. (It has been argued that the economic burden of corrosion could be reduced considerably by improved dissemination—and consequent wider application—of information already available on corrosion-prevention technology.) The progress now being made in surface science and technology offers the hope of acquiring new knowledge of the mechanisms of corrosion and the reactivity of metals at the atomic level—knowledge that could lead to new approaches to corrosion control.

Magnet Alloys

The cobalt–samarium alloys introduced recently for use in permanent magnets are finding a growing market. The materials exhibit unusually high magnetic force per unit of weight and so are especially useful in small, permanent-magnet electric motors where they replace a great deal of copper and also save weight. These alloys have stimulated new concepts in the design of electric motors that would be significantly smaller and lighter than is common today. The full-scale commercial use of the materials depends on the cost and supply of cobalt and samarium and the development of cheaper ways to form the alloys into magnet shapes.

Zaire is a primary source of cobalt, and the possibility of interruptions in supply is spurring a search for lower-cobalt or cobalt-free alloys for permanent magnets of equal efficiency. Alloys of rare earths are being sought as less costly replacements for samarium.

Newer Engineering Metals

Newer metals whose uses have grown during the past three decades include the reactive metals—titanium, zirconium, and hafnium—and the refractory metals—niobium, tantalum, molybdenum, and tungsten.

Of the reactive metals, titanium has enjoyed the greatest growth. The metal was introduced commercially in the late 1940's, and its light weight, strength, and corrosion resistance soon assured its future. Titanium is now entrenched as an aircraft structural material and has a growing market as a corrosion-resistant material for use in chemical plants. Recently the metal has been used in condenser tubes in steam-power generation, and it is being introduced as a strong, corrosion-resistant material for the blades in low-pressure steam turbines.

Production of titanium mill products in this country in 1979 will total some 20,000 tons. Although titanium is the fourth most abundant metal in the earth's crust (after aluminum, iron, and magnesium), current demand for the metal could strain world production capacity.

Zirconium and hafnium have significant uses in nuclear reactors for generating electric power. Zirconium is used to clad the uranium oxide fuel elements in light water reactors. The metal's essential characteristic in this application is its very low absorption of the neutrons that drive the fission process; the metal also has good corrosion resistance in high-temperature water. Hafnium, on the other hand, is an excellent absorber of the neutrons that drive the fission process and also has excellent corrosion resistance in high-temperature water. These properties make hafnium ideal for its present use in the control rods of light water reactors.

The refractory metals have been used for years in small parts in electrical and chemical equipment. With the advent of missiles and space vehicles, it appeared that these metals would be needed in larger-sized sheets, bars, and forgings, and the necessary production facilities were developed in the 1960's. However, many of the missile and space-vehicle applications either never materialized, because alternative materials were used, or required less of the metals than anticipated at first. Consequently, the nation's production facilities for refractory metals are underutilized at present. The facilities have proved valuable, however, in producing materials for hot-working dies and for chemical equipment.

Metal Science

Metallurgical science now is benefiting from new microanalytical tools such as auger spectroscopy and scanning and transmission electron microscopy. These powerful tools—by-products of basic research in physics—permit detailed study of the segregation of impurities that is known to control many of the critical properties of alloys. Perhaps more importantly, the new tools allow the character of solids to be probed down to the atomic scale. The resulting knowledge is accumulating rapidly and will lead to major developments in the understanding and control of surfaces—so important in corrosion and catalysis. The new knowledge also may lead to the ability to control and manipulate the composition and properties of internal surfaces (grain boundaries) in alloys. This ability would permit the properties of alloys to be tailored to particular uses to a degree not possible today.

ENERGY-RELATED MATERIALS

Power Devices

Better performance at lower cost will continue to be achieved in semiconductor power devices for converting alternating to direct current. These rectifiers, therefore, will gradually replace most of the older electromechanical equipment. Power devices generally will be made from silicon except for operation at the highest frequencies, where gallium arsenide may be superior. An important materials problem in the power area is the dissipation of heat generated in various types of equipment. The use of electrical insulators of high thermal conductivity, such as beryllium oxide, will remain important for this purpose. Alternative materials that may find use in some applications include copper and aluminum nitride and perhaps even diamond.

Dielectrics

Dielectric materials, which include polymers and ceramics, are insulators or nonconductors of electricity. These materials are important in equipment for generating, transforming, and transmitting electric power. They are also critical in telecommunications and various electronic applications. In the electric power industry, new insulating materials are needed for high-voltage transformers and capacitors, where they would replace the polychlorinated biphenyls (PCB's), which are environmentally unacceptable (see p. 452). Work is also under way on new dielectric structures for storing high-density energy to meet the requirements of modern electrical circuitry. Here, the role of surfaces at the interface of the dielectric and the conductor must be better understood. Also needed are practical ways to measure the degradation of dielectrics so that their service life can be forecast accurately.

Solar Cells

A considerable effort is under way to produce efficient, low-cost solar cells for converting sunlight directly to electricity. Silicon is the standard cell material; encouraging progress is being made in growing ribbons of single-crystal silicon by a continuous process, which would significantly reduce the cost of the cells. Some progress has been reported in exploring amorphous silicon (see pp. 164, 279). Highly efficient gallium arsenide cells have been made for use in sunlight concentrated by reflectors. More novel compounds of two and three elements, such as aluminum gallium arsenide, are also being investigated for this application. The cost of producing solar cells should drop significantly in the next five years, but they are unlikely to become a major source of electrical energy during this century.

Lighting

The development of more efficient lighting is being spurred by the cost of energy and by improvements in fluorescent-lamp phosphors and lamp-envelope (bulb) materials. Fluorescent lights with improved efficiency and color will find greater use in home lighting in the next five years. High-pressure sodium- and metal-vapor discharge lamps with superior color rendition and efficiency will be replacing mercury discharge lamps used currently in many outdoor and commercial applications.

Superconductors

A continuing search is in progress for materials that become superconducting at higher temperatures than those now available. Wire made of a superconducting material does not resist the flow of electricity and so conducts current without the loss of energy caused by the electrical resistance of conventional conductors like copper and aluminum. A superconducting material loses its resistance to electricity at a characteristic temperature called the critical temperature, and the critical temperatures of all known superconductors are not far above absolute zero. To reach such temperatures, the materials must be refrigerated with liquefied gases, and the higher the critical temperature, the lower the refrigeration cost.

A niobium–germanium compound has the highest critical temperature—about 23°K (-250°C)—of any material found so far, but it is too brittle to be fabricated. Superconducting wire made of a niobium–tin compound, cooled by liquid helium, with a critical temperature of about 18°K carries the circulating current in superconducting magnets now in use.

Superconducting magnets are available in sufficiently large size—15 feet internal diameter—to operate subatomic-particle detectors in research in nuclear physics. Superconducting magnets also are used in advanced energy-conversion test facilities to contain hot, corrosive plasmas by means of a magnetic effect. This application may be important for thermonuclear fusion and magnetohydrodynamic power generation.

Superconducting wire shows promise for use in the electromagnets in electrical generators and motors. In this application it would permit major reductions in the size and operating costs of the equipment.

Work has also been done on superconducting power transmission lines.

The problem here is the expense and difficulty of refrigerating long conducting lines.

INFORMATION-RELATED MATERIALS

Sensors

A variety of new sensor materials is on the horizon. Infrared detector arrays made from semiconductor crystals like indium antimonide, mercury telluride, and cadmium telluride will allow us routinely to "see" the world by the heat it emits, even in total darkness. Potential applications range from detecting tumors to locating sources of heat leakage from buildings and industrial operations. More recent development of the technology of thin-film infrared devices, as replacements for bulk-crystal devices, significantly extends the range of uses of infrared sensors. Also in prospect are new pressure-sensing materials, such as polyvinylidene fluoride and aluminum phosphate (berlinite). These piezoelectric materials, which generate small electric currents when stressed, will improve the performance of sound-wave sensors. This in turn will lead to developments such as microscopes that "see" by sound waves. Zirconium dioxide sensors have been developed and are being used to measure the oxygen content of molten steels, while zirconium dioxide and titanium dioxide sensors are now being used to measure and regulate the air/fuel ratio in advanced automotive engines. Control of the air/fuel ratio is a key requirement in the operation of vehicle exhaust-emission control systems.

Fiber Optics

Glass and quartz fibers have been developed through which information in the form of light signals can be transmitted several miles. Rigid control of composition and internal structure is essential in making the fibers. The sources of light will often be light-emitting diodes or lasers of high light-emission efficiency, and the light detectors will be made of silicon. The main advantage of these fiber-optic systems over conventional telephone lines is their much higher message capacity. Communication by fiber-optic transmission is well advanced and will see major growth in the next five years (see p. 238).

Electronic Displays

Electronic displays will continue to replace most of the mechanical devices used now to depict letters and numbers in cash registers, home appliances, instrument panels, and other equipment. Such displays will use light-emitting diodes (LED's), liquid crystals, and gas-discharge devices and will be driven by digital circuitry. LED's and gas-discharge devices emit light when stimulated electronically; liquid crystals change their reflectance of ambient light. Problems of performance of materials currently limit the utility of these devices. As the problems are solved, the devices will find many applications in addition to the already ubiquitous digital watches and calculators using liquid crystals and LED's.

Transistor and Computer Materials

Electronics is probably the most rapidly accelerating area in technology today, and silicon in the near term will remain the most emphasized electronic material. Our ability to control the composition, structure, and processing of silicon-base components has improved steadily over the years. As a result, almost annually for about two decades the number of electronic components on a single integrated-circuit silicon chip has doubled and the cost per component has fallen by nearly half (see p. 217). This momentum will continue for at least the next five years and probably longer. A very strong effort is under way in the development of new lithographic techniques employing electron beams and X-rays and of dry processing techniques such as plasma etching and laser-annealed ion implantation. These techniques will be used to provide increasingly complex circuits at lower costs. As circuits become more complex, continued improvement will be required in the quality of single-crystal silicon and in the control of process-induced defects.

Computer Memory

A technology is rapidly being developed for producing computer memory using magnetic substances such as gadolinium–iron–garnet (see p. 220). These bubble memories store information at very high density as microscopic magnetized domains. The production of these rather complex materials is demanding, but the opportunities they offer for low-cost, highly stable, mass memories will provide the driving force for their further development.

POLYMERS

Synthetic polymers—plastics and rubbers—are the fastest-growing class of materials. Since about 1950, U.S. production of polymers has grown at a rate exceeding that for any other material and now tops the production of steel in volume but not in weight. U.S. production of plastics in 1977 was 29 billion pounds (Figure 22), and production of synthetic rubbers

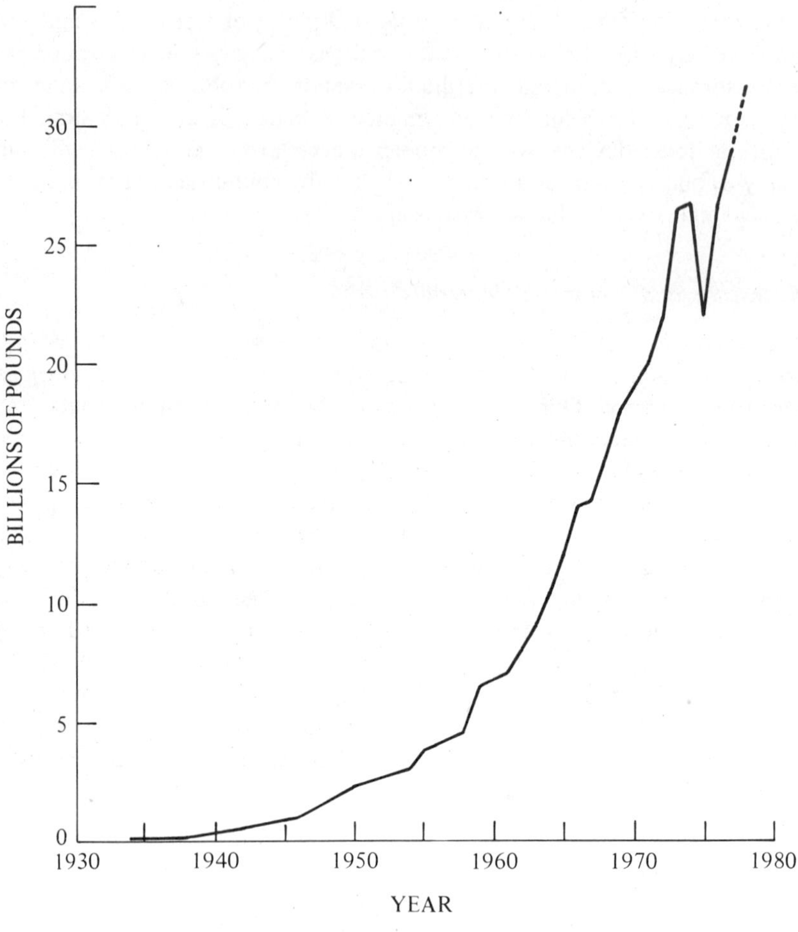

FIGURE 22 Growth of U.S. plastics production (1978 estimated 9 percent above 1977). (U.S. Plastics Production, *Modern Plastics,* January issue of years cited)

exceeded 5 billion pounds. This growth has been due not only to the development of new materials, but also to our steadily improving ability to couple materials, design, and processing so as to maximize properties and performance. In addition, plastics may offer manufacturing economies. One molded plastic part, for example, may replace a number of metal stampings at lower manufacturing cost and labor content.

Packaging, containers, and construction (siding and thermal insulation, for example) will remain major markets for plastics in the five years ahead. The auto industry is a growing market for plastics, which in part will be weight-saving replacements for steel. The plastics content of the typical

U.S. car, currently about 200 pounds, may reach 300–350 pounds in 1985 models. High-performance polymers—engineering plastics, silicones, and specialty plastics—will remain the fastest-growing segment of the polymer industry. Already plastics are replacing die-cast aluminum and zinc because in some cases they offer better performance at lower cost.

New and modified polymeric materials will be developed to achieve specific combinations of properties. Processing methods will be improved as well. Demand is growing for plastics with greater resistance to heat, oxidation, and light and for plastics with better electric properties. A range of coatings and fire retardants is being used to ease problems with the flammability of plastics and the substances emitted by them upon combustion.

Petroleum will remain the principal source of raw materials for polymers for many years, well beyond the time when its use as a fuel has begun to decline. Eventually, however, a shift is probable to other sources of carbon compounds, such as coal, shale oil, or renewable resources such as aquatic and terrestrial plants. For fossil sources of carbon compounds, the basic science and technology is largely in hand. For renewable resources, it is not.

CERAMICS AND OTHER INORGANIC MATERIALS

A major fraction of the materials and products used by an industrial society is inorganic and nonmetallic; ceramics are the best example of this class. Ceramics may be broadly defined as inorganic, nonmetallic materials processed or consolidated at high temperature. They are usually made from combinations of natural silicates, or compounds of silicon and oxygen with various metals, and oxides that are fused or sintered together. Ceramics include cement, bricks, tile, sanitary ware, china and dinnerware, glasses, porcelain enamel on metal, abrasives, and refractories. The technology of these materials evolves slowly, with progress relating principally to improvements in properties and production efficiency. There are, however, a number of scientific and technological challenges for the future, one example being the further development of the silicon carbide and nitride ceramics.

Silicon Ceramics

The silicon ceramics include silicon carbide, silicon nitride, and the SIALON's (compounds of silicon, aluminum, oxygen, and nitrogen). While these ceramics comprise one of the few classes of inorganic materials that do not occur in nature, recent research indicates that the chemistry of the silicon nitrides parallels that of the natural silicates. It should thus be

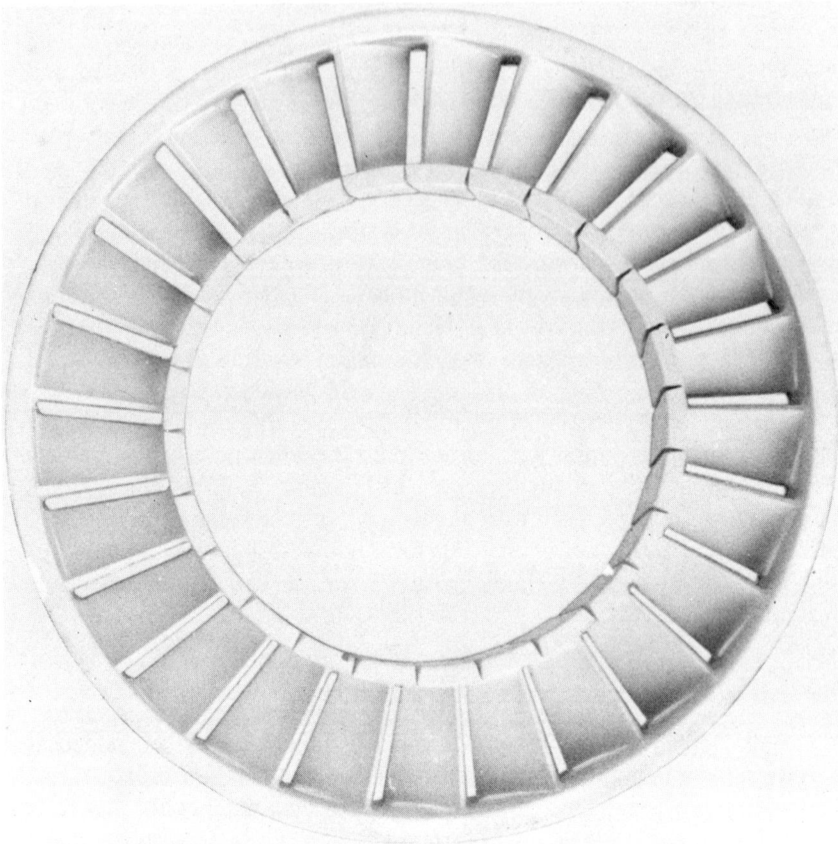

FIGURE 23 Turbine stator made of silicon nitride. (Ford Motor Company)

possible to synthesize a large range of silicon nitride-type compounds with unique properties. For example, some of the silicon and nitrogen atoms in silicon nitride can be replaced by aluminum and oxygen, as in the SIALON's. The next five years will see intensive effort to synthesize such compounds and to delineate their thermochemistry and properties as well as their potential in uses such as high-temperature structural materials, optical and electronic devices, and refractories.

The silicon carbide and nitride ceramics are generally stronger and more stable than the normal oxide-type ceramics at higher temperatures (Figure 23). They resist corrosion, erosion, and thermal shock exceptionally well.

Intensive development indicates that the silicon carbide and nitride ceramics offer substantial promise for replacing nickel- and cobalt-based superalloys in some high-temperature machinery, such as gas turbines, and would permit even higher operating temperatures than metals; the materials also are promising for use in ceramic heat exchangers. The higher operating temperatures that these materials would permit in energy-conversion processes would make possible higher efficiencies and savings in fuel. Success in this one area could yield far-reaching gains in power generation and in development of new generations of gas turbines. An all-ceramic turbine is under development as a power plant for automobiles, for example. Ceramics suffer from a tendency to fail unpredictably because of brittle fracture. Solutions to this problem depend in part on the development of improved, nondestructive means of detecting the minute flaws that lead to fracture. Reliability and predictability of service life are the key questions to be resolved if silicon carbide and nitride components are to be developed successfully. These materials and their development are an excellent example of the need for intimate interaction of materials, design, and processing in advancing the state of technology.

Optical Materials

New materials are being developed for military uses involving the acquisition and electronic processing of optical guidance and surveillance information. Two parallel needs—better sensors and better optical materials to protect them—comprise an extremely broad problem with nonmilitary implications in areas that include improved lighting and process-control equipment. Military equipment for acquiring optical information often must operate at high temperature and under severe mechanical pressures. Such equipment is limited at present by the need for optically transparent windows to protect the sensitive data-collection devices, or sensors, under these conditions.

COMPOSITE MATERIALS

A composite material generally consists of a matrix material through which a different, reinforcing material is distributed. The fiber composites originated with the glass-fiber reinforced plastics, in which the fibers have high strength, but relatively low stiffness. Glass-reinforced plastics have long been used in structural applications such as boat hulls, missile casings, and sporting equipment where a high strength-to-weight ratio is the predominant requirement.

In the 1960's, fibers of boron, graphite, silicon carbide, and other materials of high stiffness were developed. These newer fibers, used in both resin and metal matrices, promise fiber-composite structural materials of great strength and stiffness in a form capable of being fabricated into the most complex shapes. High-performance fiber composites already are used in aircraft. They have great potential for replacing metals in automobiles, where they may perform equivalent functions at weight savings of 50–70 percent. The key problem with high-performance composites in automobiles is cost. Much effort will be expended during the next five years to find new resin formulations and cheaper ways to prepare fibers and to develop methods of manufacturing composite parts in high volume at low cost.

RENEWABLE MATERIALS

Renewable resources are a potentially attractive source of materials, and a degree of attention is currently being devoted to the fabrication of new types of engineering or functional materials from resources like wood and paper. Certain wild plants also are being investigated as sources of materials. They include guayule, a source of rubber, and jojoba, a source of lubricating oil said to be the equivalent of sperm oil.[2,3] Both plants could be produced on desert scrublands. However, the economic feasibility of cultivating such plants as sources of materials is far from being established.

The use of renewable resources has certain broad limitations. Nonrenewable materials now account for some 90 percent of the tonnage of new supplies of nonfuel materials in this country. Attempts to reduce that percentage by shifting to renewable resources would begin at some point to create land-use conflicts with agriculture and other activities. Furthermore, intensive cultivation of renewable resources may require significant inputs of nonrenewable resources, such as fertilizers, pesticides, and mechanical energy.

BIOMEDICAL MATERIALS

Metals, polymers, ceramics, and composites should make further inroads in the next five years in replacing missing or defective limbs, joints, and other parts of the human body. The surgeon now has available materials of sufficient biocompatibility and durability for reasonably satisfactory use as soft-tissue implants in plastic surgery, hip and knuckle replacements, large artery replacements, hydrocephalus (cranial) drainage tubes, and implantable cardiac pacemakers. Materials for implantable teeth and bone sections and for replacing knee joints would appear to be nearing realization. This is not true, however, of materials for small-vessel (vein) prostheses and percutaneous (through the skin) electrical or fluid-conducting leads, as for constructing implantable lungs or kidneys.

Although substantial progress is being made in biomedical materials, they remain difficult to evaluate for use in humans. The problem would be eased by the development of laboratory tests that would predict the service life of implant materials more reliably than do present methods.

MATERIALS PROCESSING AND MANUFACTURING

Materials processing, broadly, is the conversion of raw materials into intermediate or finished products with useful shapes and properties. A distinction should be made, however, between primary processing and secondary processing or manufacturing. Examples of primary processing include the conversion of iron ore into steel (a bulk material) and mill processing of the steel into forms such as sheet and plate (engineering materials). Secondary processing includes the conversion of sheet, plate, or other semifinished steel into parts and finished products—machine tools, roller bearings, engines, aircraft.

PRIMARY MATERIALS PROCESSING

Productivity and product quality in the metals industries have been improved by important engineering advances during the past 50 years. Examples are continuous rolling of strip, automatic gauge (thickness) control, electronic inspection of bars and tubes, and computer control of rolling mills. The advent of the basic oxygen steelmaking process, the high-pressure blast furnace, and continuous casting initiated a revolution in steel productivity.

Full-scale adoption of technological advances in the primary processing of metals is extremely capital-intensive because of the scale of operations and the conditions and environment, such as high temperature and large mechanical forces, in which they must be conducted. The industry, therefore, approaches new technology cautiously and expects the corresponding new plant and equipment, once in place, to perform for a long time.

A number of forces lately have converged on primary metals processing to slow the rate of investment in new technology. Governmental regulations related to the environment and to occupational safety and health have required large capital investments that otherwise might have gone into production technology. As high-temperature processors, the industry has been affected significantly by the rising cost of energy and so has put major emphasis on energy conservation. Also, the growth of product-liability litigation has led to greater stress on nondestructive methods for detecting flaws in products. Shortages of raw materials like cobalt have led to price spirals in special metals. These forces have combined to depress the profitability and rate of innovation of most of the nation's basic metals-processing industries.

Improvement of Properties by Processing

The scientific basis of primary metals processing is sound in some areas, but less so in others. In melting and refining, a strong scientific base in thermodynamics, kinetics, and transport phenomena provides useful insights and guidance in processing operations. In deformation processing, such as rolling, the situation is less satisfactory. Macroscopic understanding of deformation processing is adequate as a guide to routine process planning and control, but a better scientific foundation is needed for future advances.

Metallurgists generally do not expect the discovery of broad new classes of alloys in the near term. Improvements in properties are more likely to come from learning to control and refine the structure of metals through more precise understanding and control of the steps in deformation processing. The high-strength steels mentioned earlier illustrate what can be achieved.

Improvement of the properties of metals through processing requires detailed knowledge of the degree of deformation and the temperature in all parts of the material in relation to elapsed time as the material deforms. This information must then be integrated with the kinetics of the development or change of internal structure in the material. Computer-based analytical methods for treating this problem are being developed.

In polymer processing, there are good possibilities for aromatic polyamides. Here, greater understanding of the structure and flow characteristics of the anisotropic fluid state and of the solidification of anisotropic fluids could lead to the production, through processing, of polymeric materials with stiffness comparable to that of metals.

SECONDARY MATERIALS PROCESSING

As we draw closer to the final manufactured product, we find a higher level of technological innovation in metals processing. Examples include casting and solidification, powder metallurgy, and microelectronics.

Casting and Solidification

In casting and solidification, great progress has been made, particularly in metals, by control of grain or crystal size and of the direction in which grains grow during cooling. This control is achieved by directional solidification: heat is removed selectively from the molten metal so that crystals grow in a given direction and parallel to each other to form a columnar structure. As a result, the boundaries of the grains in the metal are also approximately parallel, and the properties of the finished casting are significantly different in different directions. An even more recent advance is the casting and controlled solidification of single-crystal blades and vanes for aircraft gas turbines. As noted earlier, improved high-temperature durability is achieved in these materials by eliminating the grain boundaries entirely.

In a new process called rheocasting, the melt is cooled and agitated vigorously before pouring until it becomes mushy, with small crystals supported in the molten metal. The high solids content and lower casting temperature result in smaller, more uniform grains and denser, stronger parts. The lower temperatures involved also conserve energy, increase the life of dies, and make the entire process easier to control.

Other innovations in casting include the use of polystyrene patterns that vaporize when the molten metal engulfs them. These consumable patterns make it possible to produce complex shapes at low cost. Precision investment casting, in which the mold is first built up around a wax model of the part to be cast, has been refined to a high art, increasing both the size and quality of the castings that can be produced. Corresponding advances have been made in other foundry products through the use of molds of resin-bonded sand. This country leads the world in cast-iron technology.

Powder Metallurgy

Powder metallurgy—pressing and compacting of metal powders—has seen a revival with the advent of new consolidation methods and very fine metal powders, made by newly developed, rapid-solidification techniques. These powders are especially uniform in composition and microstructure and relatively free of otherwise embrittling microconstituents. The resulting products, after consolidation and heat treatment, have superior ductility and high-temperature properties. The new consolidation methods minimize porosity and thus yield solids of high density. With hot isostatic pressing, in which pressure is exerted uniformly from all directions, the most intractable metal or ceramic powders can be pressed into shapes with very nearly the desired final dimensions, minimizing finishing operations. A cheaper route to high-performance parts is provided by powder forging, in which preformed powder-based billets of about 20 percent porosity are forged to the final product. The new fine metal powders permit the production of metals sufficiently fine-grained to be superplastic—to undergo very large plastic deformation. Superplasticity, when built into metals like titanium that normally are difficult to form, allows them to be formed easily if relatively slowly. This property could be exploited to reduce the cost of making parts from such metals.

Microelectronics

A striking example of structure-oriented materials processing is microelectronics, particularly large-scale integrated circuits for computers (see p. 216). Probably in no other technology has processing been coupled so intimately with structure and has progress been so rapid and far-reaching. The early military and space applications of the technology moved quickly into the civilian economy. The transfer was possible because of the strong scientific base that already existed in solid-state chemistry and physics and because of the relatively modest investment required.

Computer-Aided Design and Manufacturing

Although this country is the birthplace of mass production, the vast majority of parts are made in numbers too small to justify major investment in assembly-line machinery. However, the manufacture of parts in relatively small runs is undergoing sweeping change. It began with the development of the numerically controlled machine tool in this country in the 1950's and accelerated as computers became more compact and inexpensive. Today, computer-aided design is exerting a major impact on manufacturing, and computer-aided manufacturing should begin to do so in the next five years.

Computer-aided design (CAD) employs powerful programs (software) for obtaining approximate solutions to problems inaccessible to precise mathematical analysis. CAD greatly extends the ability to analyze the design of a part for factors like stress and generation of heat. It permits parts to be designed with greater precision and reduces the likelihood of unforeseen failure. CAD equipment also can reduce designs to paper, bypassing the laborious efforts of draftsmen and thus increasing productivity in graphics and drafting.

An important aspect of CAD is that it is becoming possible to use its output directly, in combination with manufacturing data, to generate automatically the programs needed to optimize and automate the manufacture of the designed parts.

In full-scale use, computer-aided manufacturing (CAM) will be capable of providing: computer generation of the optimized production plans—selection of processes, equipment, tooling, operating conditions, etc.; optimized production control—dynamic scheduling of the work, maximizing and balancing the use of the manufacturing equipment, minimizing the time that parts in process lie waiting to be worked on, etc.; and automated machining of the parts by numerical control of machine tools. In numerical control, design data are combined with manufacturing data to produce a control program. The program is then used, via punched tape or small computer, to control one or more machine tools that produce the finished parts from raw stock.

The great capability of the computer is making it possible to design processing operations to a degree heretofore undreamed of. For example, precision forging to almost final shape, so as to minimize subsequent machining, is becoming possible with computer design of forging steps and dies. To achieve the full potential of the CAD/CAM approach in such operations, we must acquire new information on how materials flow, better understanding of friction and how to model it, and improved means of applying computer analysis to problems related to the plasticity of metals. Progress good enough to lead to widespread adoption of CAD/CAM methods in metalworking is possible in the next 5–10 years.

Full-scale use of CAD/CAM will stimulate advances not only in process modeling and design, but also in other areas: materials management and control, cost estimation, and inventory control. CAM also involves robotlike equipment, whose use to perform unpleasant or routine assembly or inspection operations is already a reality; current research is increasing this capability.

Laser Processing

The laser is an economical tool for manufacturing processes such as cutting, welding, and drilling. It can also be used to produce very fine-grained or amorphous surfaces by rapid melting and solidification—laser glazing. This technique is potentially effective because structural failures usually originate at a surface flaw. Laser modification of surfaces is being developed for valve seats and other components in the automobile industry. The use of lasers in metal processing can be expected to expand rapidly as the cost of the devices declines and as more engineers become aware of the unique capabilities of high-powered lasers.

Metal Removal

The development of high-speed metal removal processes—5,000–30,000 surface feet per minute—could result in major increases in productivity. With machine tools properly designed for high-speed cutting, it becomes practical to machine wrought-aluminum alloys at speeds considerably higher than are used today, since the rate of tool wear in machining these materials is not excessive up to speeds on the order of 12,000 surface feet per minute. However, to make such operations really economical, materials handling and tool changing must be automated so as to reduce noncutting time. With the constant improvement in cutting-tool materials that is taking place, practical speeds for machining cast iron and steel are evolving toward 2,000 to 3,000 surface feet per minute. Again, however, to make operation at such speeds economic, noncutting time must be reduced.

In addition, better understanding is needed of chip-segmentation mechanisms, tool-workpiece interface reactions, cutting-tool wear behavior, and lubrication effects. Such knowledge, coupled with a development program for high-speed machining equipment, could produce significant progress in materials processing in 5–10 years.

BARRIERS TO PROGRESS IN MATERIALS PROCESSING

While the United States continues to lead the world in computer technology, it is not at all clear that we will capitalize on this position to take the lead in computer-aided manufacturing. A stronger effort is needed throughout industry. The situation here contrasts sharply with that in West Germany and Japan, which have well-established national research programs in CAD/CAM with joint industry–government funding.

The requirement for large capital investment is inhibiting the full-scale adoption of new technology in materials processing in this country. For example, 60 percent of our machine-tool base is more than 10 years old, as compared with only 30 percent in Japan and West Germany. Moreover, the economic risk and long development time tend to inhibit sustained research and development in materials processing. An essential condition for real progress in materials processing and manufacturing is an economic climate that will prompt industry on its own to modernize, develop and adopt new technology, and improve its productivity.

Materials-processing technology also suffers from insufficient attention in our engineering colleges. Fewer than 10 percent of the materials faculty (who themselves comprise only a small fraction of the engineering faculty) are expert in materials processing and manufacturing. These fields do not enjoy the status accorded some other academic disciplines, and little current research in the schools is relevant to major developments in materials processing. The near absence in our universities of research in materials-processing and manufacturing technology denies the country a potential source of new ideas and innovation. Furthermore, it means that the universities are not exposing young people to current advances in the field.

RECENT CONCEPTS IN MATERIALS

The research and development and other materials activities of the past few decades have helped to crystallize two important concepts: the total materials cycle and materials science and engineering.[4] The materials cycle is a physical concept—materials flow from the earth through various useful forms and back to the earth in a closed cycle that is global in extent. Materials science and engineering is an intellectual concept—a coherent system of scientific and engineering disciplines that combines the search for insights into matter with the use of the resulting knowledge to satisfy society's needs for materials.

TOTAL MATERIALS CYCLE

The total materials cycle is driven by societal demand, and materials move within it in five stages:

- Extraction of raw materials: ores and minerals, rock, sand, timber, crude rubber.
- Processing of raw materials into bulk materials: metals, chemicals, cement, lumber, fibers, pulp, rubber, electronic crystals.
- Processing of bulk materials into engineering materials: alloys, ceramics and glass, dielectrics and semiconductors, plastics and rubbers, concrete, building board, paper, composites.
- Fabrication of engineering materials into structures, machines, devices, and other products.
- Recycling discarded materials or products to the system or returning them permanently to the earth.

The materials cycle provides a framework for dealing with a system of interacting parts. The flow of materials at a given point can be sensitive to economic, political, and social decisions made at other points. Materials shortages usually are found to be due not to worldwide scarcity, but to dislocations in the cycle that interfere with the arrival of materials at a given point in the usual amounts and at reasonable prices. A shortage may arise at one point, for example, because of inadequate processing capacity at another point; but countermeasures can be taken within the cycle, including stockpiling, recycling, and substitution of one material for another.

Materials, energy, and the environment interact strongly at virtually every point in the materials cycle. About one-half of the energy consumed by all manufacturing industries in the United States goes into the value added to materials in producing and fabricating them to the point of

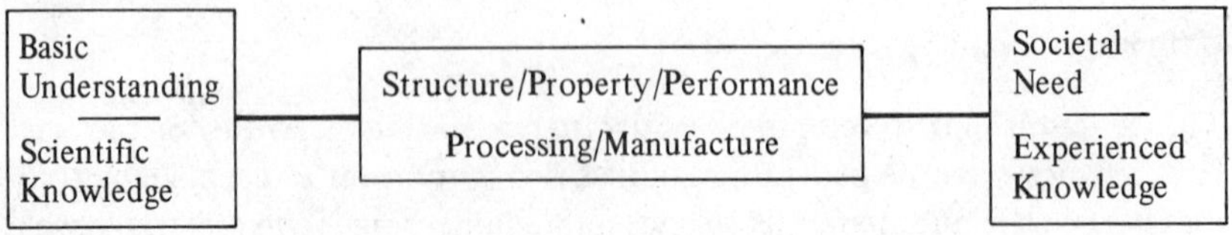

FIGURE 24 Materials science and engineering. Scientific knowledge flows to the right, while experienced knowledge flows to the left.

becoming engineering materials.[5] But materials also are crucial to making energy available in the first place. In fact, inadequacies in the performance of materials currently are the primary constraint on the efficiency, reliability and safety, cost-effectiveness, or in some cases actual realization of our advanced energy-conversion technologies—gas turbines, nuclear reactors, high energy-density batteries, fuel cells, magnetohydrodynamic generators, coal conversion, and solar-energy conversion.

The materials cycle offers exciting opportunities in materials research and development. Objectives for each stage of the cycle might be:

• Extraction and processing	Reduce energy consumption and pollution
• Fabrication	Designs that improve efficiency of construction
• Assembly	Designs that improve recyclability
• Operating life	Greater durability
• Recycle	Designs that ease disassembly and sorting

In this manner, functional design could be coordinated with materials selection and the other operations of the materials cycle so as to generally facilitate recycling, substitution, and conservation.

MATERIALS SCIENCE AND ENGINEERING

The central purposes of materials science and engineering (MSE) (Figure 24) are to probe the relationships of the internal structure and composition of materials to their properties and performance; and to use the resulting knowledge in producing, shaping, and otherwise processing materials so as to control their properties and achieve the desired performance in the finished product. MSE links fundamental understanding of the behavior of electrons, atoms, and molecules to the performance of products.

It is remarkable that almost all of the technical advances in metals and alloys described earlier were based on science developed in the nineteenth century (classical thermodynamics—especially as elaborated by J. W. Gibbs—and chemistry) in conjunction with some more recently developed experimental techniques (also based largely on old science). It may be that these applications awaited the formation of a large cadre of technologists with enough training to understand thoroughly, and so to apply effectively, the old science.

MSE promotes the application of basic science to the development, processing, and use of materials by establishing a two-way flow of information between the basic scientists at one extreme and the user/consumer of materials at the other. We have been accumulating empirical knowledge of materials for a long time. But only in this century—and at an accelerating pace during the past few decades—have scientists made substantial progress in acquiring the corresponding basic insights. The microstructure of materials has been revealed by optical microscopy, substructure by electron microscopy, crystal and molecular structure by X-ray diffraction, atomic structure by various spectroscopies, and nuclear structure by high-energy atom smashers.

With this new knowledge it has become possible to exploit the linkage of structure, properties, and performance. The strength and dimensional stability of polymers, for example, can be upgraded through methods of synthesis that yield highly ordered molecules that cluster into crystalline arrays. Transistors are made by manipulating the electronic structure of silicon and other semiconductors; they are produced on a large scale by methods that achieve exceptionally precise control of composition and internal structure.

Studies of the MSE process at work indicate that the two-way flow of information is most productive when basic understanding of a materials problem and the empirical need to solve it are mixed so intimately that it becomes difficult to tell which provided the initial impetus toward a solution. In the main, however, the initial impetus in the MSE process seems to arise more often from "technological pull" than from "scientific push."

The successes of the MSE approach should not be construed to mean that properties can necessarily be predicted from structure alone, nor performance from properties alone, except in a general sense. As a rule, the structure–properties–performance linkage must be worked out through the reciprocal flow of information—scientific and empirical—that characterizes the field of materials science and engineering.

NEAR-TERM ISSUES IN MATERIALS

The basic concerns of the United States about materials supply and demand have changed very little from decade to decade. They commonly involve adequacy of supply, prices, and national security. Long-term world supplies of the basic resources rarely have stirred alarm. Possible shortages of more exotic or less abundant materials have been recognized, but not as serious threats to world health and welfare. Concerns about supply and demand will persist, but any consideration of the immediate future for materials must take account of certain newer issues.

THE MOVE OVERSEAS

In basic materials, certain locations overseas offer advantages over this country, including richer ores, cheaper energy, cheaper labor, and in some instances tax advantages and readier access to capital. Regulation of pollution and workplace health and safety is generally less strict than in the United States, although it will not necessarily remain so. Except for copper, many of the overseas resources are owned and used by U.S. corporations, which are importing into the American market as well as into Europe and Japan. Overseas processing is expanding, and trends will develop toward primary processing of shapes at the source. The coming decade could see significant reductions in the domestic capacity of U.S. metals-producing industries relative to domestic demand. American companies also are manufacturing components and finished products abroad and importing them into the United States.

Companies in the United States and other nations will tend to shift basic materials-processing operations abroad in the years to come. The incentive in part is the hope of assuring supplies by strengthening local economies and thus cementing relations with governments that control basic resources. The beginning of this trend can be seen already. Our imports of aluminum increased 310,000 tons, or more than 50 percent, during 1976–77; imports of refined copper rose 218,000 tons, or more than 140 percent; and imports of iron and steel rose 5.5 million tons, or 44 percent.[6] During the same period, our imports of raw and processed materials increased in value to $20 billion annually, and the excess of imports over exports increased from close to zero to $5 billion.[7]

There is a world surplus of aluminum, copper, and iron. Companies that use those materials naturally seek them at the lowest price. The consequent rise in imports has several negative impacts. Employment in the primary aluminum industry in this country declined by 1,000 during 1975–77, in the primary copper industry by 5,000, and in the primary iron and steel industry by 80,000. These declines in employment have become a serious problem and led to the President's Nonfuel Minerals Policy Study.

The study is being conducted jointly by the White House Office of Science and Technology Policy and the Department of the Interior, and the findings are due in the fall of 1979. Other negative impacts of rising materials imports include pressure on the balance of payments and uncertainties about reliability of supplies.

Loss of Self-Sufficiency

A new element in the situation is the gradual decline in the nation's ability to supply itself from domestic resources (Table 4). The United States historically has been accustomed to a large measure of self-sufficiency. However, most nations rely on resources abroad, and national economies clearly can thrive on imported materials. Notable examples are West Germany and Japan.

IMPACT OF REGULATION

Federal regulation has worthy social objectives, but it should be systematically reexamined to ensure that a reasonable balance is struck between costs and benefits. The problem is analytically difficult. Nonetheless, there is preliminary evidence suggesting that the cumulative impact of regulation is quite significant for some industries.[8]

ENERGY

The recent rise in the cost of energy is another new element in the materials situation. It imposes a new constraint on our ability to make economic gains merely by substituting low-cost external energy for human energy. This constraint is particularly troublesome in materials. The materials industries tend to be energy-intensive, and the declining quality of resources, the need to control pollution, and the adoption of new, high technology all call for large inputs of energy. For example, to refine U.S. copper sulfide ores of 0.8 percent copper content requires about 20 percent more energy per ton of copper produced than to refine a Chilean copper sulfide ore of 1.5 percent copper content.[9]

Because of the cost of energy, technology that reduces its consumption is emerging in the materials industries. Examples include the chloride cell for refining aluminum, hydrometallurgical refining of copper, and electric furnace/scrap recycle for making steel. Materials fabricators and product manufacturers are selecting materials and processes so as to reduce energy consumption and increase productivity. In some instances, the materials and applications are new. Automobile makers, for example, are using more plastics, aluminum, and high-strength steels to reduce the weight of their cars.

TABLE 4 U.S. Net Imports of Selected Metals and Minerals as a Percent of Apparent Consumption[a]

Minerals and Metals	1950	1955	1960	1965	1970	1973	1974	1975	1976	1977[b]	Major Foreign Sources[c] (1973 to 1976)
Columbium	100	100	100	100	100	100	100	100	100	100	Brazil, Thailand, Nigeria, Malaysia
Mica (sheet)	98	95	94	94	100	100	100	100	100	100	India, Brazil, Madagascar
Strontium	100	98	100	100	100	100	100	100	100	100	Mexico, Spain
Manganese	77	79	89	94	95	98	98	98	98	98	Brazil, Gabon, South Africa
Cobalt	90	68	66	92	98	98	99	98	98	97	Zaire, Belgium and Luxembourg, Norway, Finland
Tantalum	99	100	94	95	96	87	87	81	96	97	Thailand, Canada, Australia, Brazil
Platinum group metals	74	91	82	87	78	87	87	83	90	92	South Africa, U.S.S.R., United Kingdom
Bauxite and alumina	55	73	74	85	88	92	92	91	91	91	Jamaica, Australia, Surinam, Guinea
Chromium	95	83	85	92	89	91	90	91	89	89	South Africa, U.S.S.R., Turkey, Rhodesia
Tin	82	80	82	80	81	84	84	84	85	86	Malaysia, Thailand, Bolivia, Indonesia
Asbestos	94	94	94	85	83	82	87	82	85	85	Canada, South Africa
Fluorine	33	55	48	77	80	79	81	85	79	80	Mexico, Spain, Italy, South Africa
Nickel	90	84	72	73	71	69	72	72	70	70	Canada, Norway, New Caledonia, Dominican Republic
Potassium	9	0	E	7	42	53	58	51	61	66	Canada, Israel
Gold	25	34	56	72	59	48	63	52	76	60	Canada, Switzerland, U.S.S.R.
Zinc	41	51	46	53	54	64	59	61	59	58	Canada, Mexico, Australia, Peru
Antimony	33	32	43	36	40	50	44	49	54	52	South Africa, People's Republic of China, Bolivia
Cadmium	17	20	13	20	7	41	46	41	64	51	Canada, Australia, Belgium and Luxembourg

Selenium	53	18	25	44	11	57	59	66	59	47	Canada, Japan, Mexico, Yugoslavia
Mercury	87	20	25	49	41	78	86	69	62	46	Spain, Algeria, Mexico, Yugoslavia
Silver	66	58	43	16	26	66	55	30	50	42	Canada, Mexico, Peru, United Kingdom
Barium	8	25	45	46	45	37	38	32	42	40	Peru, Ireland, Mexico
Tungsten	80	NA	32	57	50	66	68	55	54	38	Canada, Bolivia, Peru, Thailand
Titanium (ilmenite)	33	40	22	9	24	23	33	25	29	38	Canada, Australia
Vanadium	4	E	E	15	21	43	36	38	37	37	South Africa, Chile, U.S.S.R.
Gypsum	28	27	35	37	39	35	37	34	35	35	Canada, Mexico, Jamaica, Dominican Republic
Iron ore	11	18	18	32	30	35	37	30	29	33	Canada, Venezuela, Brazil, Liberia
Copper	31	17	E	15	E	8	20	E	12	17	Canada, Chile, Peru, Zambia
Lead	40	39	33	31	22	29	19	11	15	14	Canada, Peru, Mexico, Australia
Iron and steel products	E	E	0	7	4	10	7	9	7	13	Japan, Europe, Canada
Salt	E	E	2	5	6	6	7	4	7	8	Canada, Bahamas, Mexico, Netherlands Antilles
Aluminum	17	E	E	4	E	18	4	E	9	8	Canada
Pumice and volcanic cinder	3	2	3	5	11	8	7	4	2	5	Greece, Italy
Cement	E	1	0	3	3	7	4	5	4	4	Canada, Spain, Norway, Bahamas
Iron and steel scrap	2	−14	−24	−17	−25	−21	−19	−27	−22	−11	—

E = net exports; NA = not available.

[a]Apparent consumption equals the U.S. primary plus the secondary production plus net imports. Based on net imports (imports minus exports plus or minus Government stockpile and industry stock changes) of metals, minerals, ores, and concentrates.

[b]Estimate.

[c]Major foreign sources listed in descending order of amount supplied.

SOURCES: *Minerals Yearbook,* U.S. Bureau of Mines, various years; import and export data from U.S. Bureau of the Census. (From *Mining and Minerals Policy, 1977 Annual Report of the Secretary of the Interior Under the Mining and Minerals Policy Act of 1970,* p. 50. Washington. Government Printing Office, 1977.)

CONSERVATION

Conservation of materials may be defined as the reduction of losses from the materials cycle or of materials going through the cycle. Losses for common metals have been estimated to range from half to three-quarters of the amounts that enter the cycle.[10] Materials are lost at all stages of the materials cycle, from mining and production (tailings, slags), through product use (wear, corrosion), to ultimate disposal. Of the means available for reducing such losses, two of the potentially most effective are recycling and substitution.

Recycling

Recycling of materials, where it reduces costs, will grow in the next five years. Cost advantages will be evaluated on the basis of total-cycle costs. Manufacturers already recycle significant amounts of in-plant scrap. Industry also recycles large amounts of materials from discarded products, including automobiles and telephones. In addition, worn but intact products are overhauled and reused extensively where the practice is profitable, a good example being automobile parts. Recycling of products, as opposed to the materials they contain, offers marked potential for conserving materials, recovering the energy invested originally in manufacturing the products, and reducing the pollution resulting from manufacturing.

The Resource Conservation and Recovery Act of 1976 is designed particularly to stimulate the recovery of materials and energy from municipal waste. Recycling of municipal wastes will be stimulated by technology that leads to design for recycle. It is essential, for example, that recyclable material be easily identified. The point is demonstrated by the relatively high recycle rate of the all-aluminum beverage can. Stimulated by high resale value, collections of cans in 1977 totaled a record 6 billion, about one in every four sold.

The concept of a waste dump as a man-made ore body is intriguing.[11] Most municipal waste is a poor source of iron and aluminum—it is leaner in those materials than are useful ores. However, such waste may be a better source of other metals, once the burnable constituents are consumed and the iron, aluminum, and glass removed. Work has been done on recovery of materials from incinerator residue, and the approach warrants greater consideration.

The level of recycling to be expected in the next five years will depend on a number of factors, of which the most important, as implied earlier, is cost. Old scrap metal and wastepaper, for example, must compete with in-plant wastes of known quality. The reworking of discarded products is labor-intensive and may offer a relatively low rate of return. We have much of the technology required to recover materials from municipal waste, but problems remain in separating these materials—sorting glass by color, for example—so they can be recycled in the most useful forms. Difficulties with institutional arrangements and waste collection must also be resolved.

A most important problem requiring research and development during the next 5–10 years is the recycling of plastic scrap (Figure 25). Landfill is rapidly becoming expensive and unavailable, so the scrap must be disposed of in some other way. Processes are needed to separate waste plastics satisfactorily for reuse or conversion to chemical materials or, alternatively, to convert them pyrolytically—at high temperature without burning—to a petroleum equivalent for use as fuel.

Substitution

Substitution of one material for another historically has been motivated by cost reduction, specific functional advantages, or supply considerations. Economic incentives certainly have been a main driving force. The rapidly increasing substitution of plastics for metals and glass in many consumer and industrial applications provides a model for characterizing the substitution process.

In general, substitution tends to be an evolutionary process. It results not only from the advent of new materials, but also from improvements and tailoring of older materials for specific uses. Indeed, substitutions stimulate competitive development and innovation in the materials they threaten.

The chief characteristic of a sound climate for materials substitution is a stockpile of materials technology to draw on. In all phases of the materials cycle, research and development generate the knowledge required to expand the range of alternatives and options for materials and process selection procedures. In that vein, it is important to restress the earlier observation that changes in any one phase of the materials cycle have impacts on the entire cycle. Major shifts in minerals or other materials must be meshed with capacity aspects of other phases of the cycle and can have, for example, major consequences in the recycling phase.

POLITICAL AND INSTITUTIONAL FACTORS

Shifts in the flow of mineral supplies can stem from political and institutional changes as much as from exhaustion of reserves. World consumption of industrial raw materials has been growing at record rates since World War II, with the United States joining Western Europe and

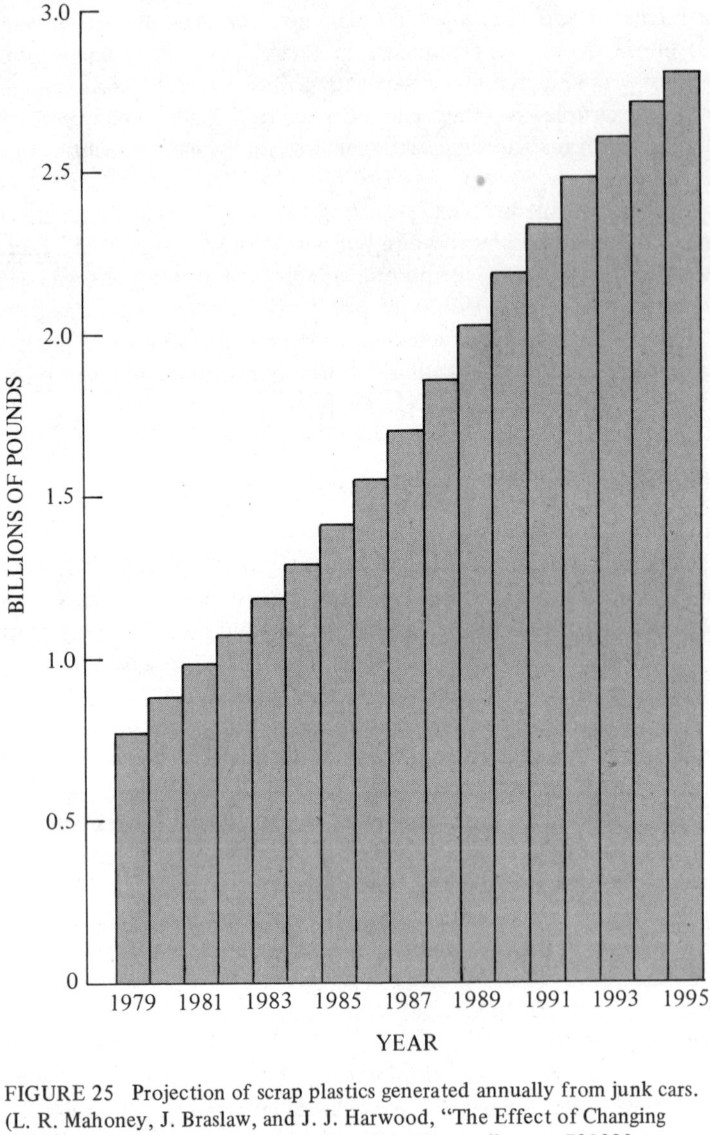

FIGURE 25 Projection of scrap plastics generated annually from junk cars. (L. R. Mahoney, J. Braslaw, and J. J. Harwood, "The Effect of Changing Automobile Materials on the Junk Car of the Future," paper 790299. Warrendale, Penn., Society of Automotive Engineers, 1979)

Japan in their increasing dependence on imports. Canada, Australia, South Africa, and the Soviet Union remain large suppliers of world markets, but a large proportion of the remaining higher-grade mineral resources is located in the developing countries of Africa and Latin America.

Coupled with the actual or impending shifts of the centers of production has been the disappearance of the traditional forms of long-term mineral concessions granted to companies based in the United States or Europe. The arrangements that have replaced those concessions are not necessarily stable. In addition, nations have claimed unrestricted sovereignty over the development of their resources, including the right to revise conditions unilaterally without penalty or recourse. Developing countries aspire to a new economic order; they wish to improve the terms of trade for primary products through joint actions among suppliers or through commodity agreements executed by both producers and consumers.

Technology Transfer

Technology is viewed widely by the developing nations as a key to industrial development. Their goal has been to gain greater access to and control over technology so they can play a larger role in the development of their materials and minerals resources and manufacture of products. Technology transfer has become a point of serious contention between the industrialized and developing nations. Our industry and our government are being pressed continually to establish new policies and programs involving the transfer of technology. Industry is encountering increasing difficulties in maintaining established private and public technology-transfer processes in the light of the new governmental systems, codes, and institutions proposed by the developing nations.

Where these changes in political and institutional arrangements will lead the world in the longer term remains uncertain. But it is apparent that they create a potential for serious dislocation of the orderly process of exploration and the development of new reserves.

DIFFICULTIES OF PROJECTION

We noted at the outset that it is difficult to project future materials needs, and the light metals provide classic examples of the perils of doing so. Aluminum, titanium, magnesium, and beryllium were each at one time or another touted as a "metal of the future." But only aluminum has become a large-volume, low-cost metal. Titanium and magnesium have become more important as metals, but the largest tonnages of them are sold in nonmetallic form: titanium dioxide as a pigment and magnesium oxide as a refractory material. Beryllium has remained an expensive specialty metal used in alloys and nuclear reactors.

The patterns of supply for these metals have also been diverse. Aluminum's emergence as a major metal depended totally on one ore—bauxite—and the metal could be produced only so long as natural cryolite for the refining process was available from Greenland. Today we have every reason to expect that clays and anorthosite eventually could supplant bauxite, and we have synthetic cryolite as well. Titanium is an abundant element, available commercially in two mineral forms. Magnesium is produced primarily from seawater rather than from terrestrial sources. The imported beryllium ore, beryl, has been augmented by a domestic mineral that was not even known as a commercial source of beryllium 30 years ago.

The light metals also illustrate the many potential paths that the materials cycle provides for meeting societal needs. The economic goal is not to supply society with aluminum, for example, but with a lightweight, low-cost material with excellent corrosion resistance and other useful properties. Rarely is a material uniquely able to meet a given functional need. The question, rather, is which material best meets the need at a given time with due regard to the entire cycle from extraction to recycle or disposal.

MATERIALS AND THE NATION

We must conclude that developed resources always appear to be so limited that concern is ever present for their depletion or exhaustion. But our total unexploited physical endowment remains large contrasted with current needs. Ingenuity remains the driving force required to transform these resources from crustal promises into economic realities. Ingenuity must be aided, however, by better information, careful analysis, innovative technology, and functioning political and economic institutions attuned to support technology, innovation, and industrial productivity. Universities, industry, and the government must all have the perception to support materials research and development to the extent necessary to turn basic ingenuity to best advantage.

Exploration and development of mineral and materials resources are no longer left solely to private industry. Government has intervened because of concern about accessibility of foreign supplies and the assurance of continued domestic sources and because minerals development has large potential environmental impacts to which the public is increasingly sensitive. The importance of our basic minerals and materials industries to the economic welfare and security of the United States makes the creation of a healthy climate—social, political, economic—for these industries a matter of public policy that warrants serious attention. The issue of continued investment in minerals and materials development both domestically and abroad will be more critical during the next five years than the question of physical adequacy of resources.

Continuing concerns about materials availability, supply, and costs have focused new attention on the role of materials in industrial operations and in national affairs. The recognized interaction and interdependence of materials, energy, and the environment are a needed catalyst for the understanding of the pervasive force of materials technology throughout our world. Certainly, the trends in materials supply, availability, and costs will bring additional pressures and intensify others. Research and development programs on materials, with emphasis on conservation, recycling, substitution, and the management of materials, can provide opportunities to offset some of these pressures. The systems approach to materials, in the context of a materials cycle, and research on materials processing and manufacturing may be key elements in the response of research and development to the materials–resources challenges of the future.

OUTLOOK

The following outlook section on materials is based on information extracted from the chapter and covers trends anticipated in the near future, approximately five years.

New pressures and issues have begun to affect the outlook for materials. Energy, environment, and transportation have become matters of national priority. New governmental agencies are involved in research and development and in regulatory activities in these fields. Uncertainties about industrial productivity and innovation and the business climate for risk-taking permeate materials technology and other fields. Decisions based on relative and near-term payoffs are shaping the research environment.

The health of materials processing and manufacturing technology in the United States is arousing concern. This is particularly so with respect to foreign competition and trends toward moving basic materials operations overseas to take advantage of richer ores, cheaper energy and labor, and less stringent environmental requirements.

In short, the forces that mold the materials enterprise are shifting to an uncommon degree from the purely technological to social and economic factors.

DEVELOPMENTS IN MATERIALS

It is difficult to forecast materials needs either nationally or internationally. For this country, however, it can be said that the needs will increase incrementally in quantity but that the kinds of materials used will not change much during the next 5–10 years. Subsequently, both the national origins and the kinds and amounts of the materials used could change markedly. New technologies change both what is usable and how it is used.

The rapid evolution of new materials that began during World War II is still under way. Research and development has become more selective in recent years, however, and business factors are inhibiting the movement of new methods and products from the laboratory into everyday use.

Metals and Alloys

Striking advances cannot be expected on a broad front in basic metals such as iron and steel, aluminum, and copper. More likely are incremental improvements in specific properties, reduction in cost through innovations in processing, and better tailoring of properties to meet specific needs.

Unique high-temperature strength and performance can be achieved in cobalt- and nickel-base superalloys by directional solidification of the molten alloy. Further improvements in high-temperature behavior have been demonstrated with single-crystal turbine blades and vanes of such alloys. Both types of materials will be demonstrated in gas-turbine engines within the next five years. They will permit longer performance life and perhaps higher peak operating temperatures and consequent higher efficiencies in the engines. Rapidly solidified superalloys, now under active development, hold even greater potential for the longer term.

Certain alloys, when cooled at very high rates, solidify in noncrystalline, or amorphous, form, as opposed to the crystalline structure found normally in metals. The properties of these amorphous metals make them good candidates for use in power-transmission transformer cores, where they could yield significant savings in energy. Amorphous metals have good prospects for commercial use in the next five years in other applications, such as magnetic shielding.

Energy- and Information-Related Materials

The cost of producing solar cells for converting sunlight directly to electricity should drop significantly in the next five years. However, solar cells are unlikely to become a major source of electrical power during this century.

Infrared detectors made of semiconducting materials will allow us routinely to "see" objects by the heat they emit, even in total darkness. Potential uses range from detecting tumors to locating sources of heat leakages from buildings and industrial operations.

Communication by fiber-optic transmission will see major growth in the next five years. Light signals from light-emitting diodes or lasers already can be transmitted several miles through glass or quartz fibers and detected by silicon-based devices. The main advantage of fiber-optic equipment over conventional telephone lines is its very high message capacity.

The number of electronic components on a single integrated-circuit silicon chip will continue to increase in the next five years, and the cost per component will continue to fall. New processing methods will provide increasingly complex circuits at lower cost.

The so-called bubble memories for computers, based on magnetic materials like gadolinium–iron–garnet, are being developed rapidly. The materials are difficult to produce but they offer opportunities for low-cost, highly stable, mass memories.

Polymers

Production of synthetic polymers—plastics and rubbers—will continue to grow rapidly, with emphasis on the development of new materials to achieve specific combinations of properties. The already large market for polymers in automobiles is likely to grow 50 percent in the next five years as manufacturers substitute plastics for metals to reduce the weight and thus the fuel consumption of cars. Petroleum will remain the principal source of raw materials for polymers well beyond the time when its use as a fuel has begun to decline. In the longer run, however, a shift is likely to other sources of carbon compounds, such as coal, shale oil, and, ultimately, renewable resources.

Ceramics

Silicon nitride and carbide ceramics offer promise for replacing superalloys in gas turbines and also may be used in ceramic heat exchangers. In such uses these materials permit higher operating temperatures than metals and thus even higher efficiencies than now are common in energy-conversion devices. The major problem to be solved with these ceramics is their tendency to fail unpredictably by brittle fracture, but much progress should be made in the next five years in improving their properties and reliability.

Recent research indicates that it should be possible to synthesize a large range of silicon nitride-type compounds with unique properties. The next five years will see intensive studies of such materials. Potential uses include high-temperature structural components, optical and electronic devices, and refractories.

Composite Materials

Considerable progress should be made with high-performance composites—resin and metal matrices reinforced with fibers of boron, graphite, and other materials of high stiffness. Such composites have great potential for aircraft, where they are already used to some extent. They are beginning to be used in automobiles to save weight.

Renewable Materials

Plants that could be raised on scrublands and in the coastal oceans are being studied as sources of materials such as rubber and lubricating oil. The economic feasibility of this approach is far from established, however. Also, a strong shift to renewable resources would begin at some point to create land-use conflicts.

MATERIALS PROCESSING AND MANUFACTURING

In the processing of metals, great progress has been made in casting and directional solidification. Precision investment casting has been raised to a high art, and corresponding advances have been made in other foundry techniques. This country leads the world in cast-iron technology.

The properties of metals can be improved by modifying their internal structures through more precise understanding and control of the steps in deformation processing. An example of what can be achieved is the high-strength microalloy steels produced by controlling the conditions of rolling and other processing operations. Automobiles of the 1985 model year may contain as much as 500 pounds of these weight-saving steels per car.

Powder metallurgy is seeing a revival because of new consolidation techniques that produce parts of high density and nearly the desired shape, thus minimizing machining and scrap generation. The cost of fabricating high-performance structural parts will be reduced by compressing very fine metal powders, made by new techniques, in large, hot isostatic presses. A more recent development is atomized powders that are especially uniform in composition and microstructure. These newly available powders permit the production of metals that are superplastic and thus more readily formable than they would be normally.

Laser treatment of surfaces to improve the wear-resistance of metal parts can be expected to expand rapidly as the cost of lasers decreases. The technique is being developed for automotive valve seats and other components.

The development of high-speed metal removal processes promises major increases in manufacturing productivity. Basic studies, coupled with a development program on high-speed machining equipment, could produce significant progress in materials processing in 5–10 years.

Computer-Aided Design and Manufacturing

The manufacture of parts in relatively small numbers is undergoing sweeping changes based on computer-aided design (CAD) and computer-aided manufacturing (CAM). Full development is yet to come, especially with CAM. An important aspect of CAD is that it is becoming possible to use the output of that process directly, in combination with manufacturing data, to generate automatically the programs needed to optimize and automate the manufacture of the designed parts by CAM.

NEAR-TERM ISSUES IN MATERIALS

The political and institutional factors that affect the worldwide flow of raw materials have been changing markedly. The long-term results are uncertain. It is apparent, however, that these changes potentially could seriously disrupt the exploration and development of new reserves.

American and other companies will tend to shift basic materials processing operations overseas in the next decade. The aim in part is to assure supply by improving relations with governments that control basic resources. Also, certain locations overseas offer advantages that include richer ores and cheaper energy and labor. This trend can have negative impacts: loss of jobs in the domestic materials industries, greater pressure on the balance of payments, and uncertainties in deliveries of materials.

The rising cost of energy could be especially troublesome to the materials industries, which tend to be energy-intensive. Also, large inputs of energy are required to offset the declining quality of domestic ores, to control pollution, and to use new, high technology. In some cases the materials industries are easing the problem by adopting technologies that reduce energy consumption. Examples include the chloride cell for refining aluminum and hydrometallurgical refining of copper.

Conservation, Recycling, Substitution

Conservation of materials may be defined as the reduction of the amounts of materials that flow through the materials cycle or the reduction of losses from the cycle. Such losses occur in all stages of the cycle, from mining and production (tailings, slags), through product use (wear, corrosion), to ultimate disposal. Of the available means of conservation, two of the potentially most effective are recycling and substitution.

Recycling of materials, where it reduces costs, will grow in the next five years. Industry already recycles large amounts of in-plant scrap and of materials from discarded products, including automobiles and telephones. Various products are now recycled by overhaul and reuse. This practice offers marked potential for conserving materials, recovering the energy invested originally in manufacturing the products, and reducing the pollution resulting from manufacturing.

Substitution of one material for another historically has been motivated by cost reduction, specific functional advantages, or supply considerations. The chief characteristic of a sound climate for substitution is a stockpile of materials technology to draw on. The rapidly growing substitution of plastics for metals and glass in many consumer and industrial applications provides a model for characterizing the substitution process.

Materials and the Nation

We must continue to examine the physical adequacy of resources and potential shifts among materials through time, but crises do not seem imminent in the near future. A more important issue may be the need to create an atmosphere—economic, social, and political—that will assure a strong future for the country in minerals and materials resources and a competitive industrial base in a rapidly changing materials world.

REFERENCES

1. Norton, E., Electric Power Research Institute, private communication.

2. *Guayule: An Alternative Source of Natural Rubber* (NRC Board on Science and Technology for International Development). Washington, D.C.: National Academy of Sciences, 1977.

3. *Products from Jojoba: A Promising New Crop for Arid Lands* (NRC Committee on Jojoba Utilization). Washington, D.C.: National Academy of Sciences, 1975.

4. *Materials and Man's Needs* (NRC Committee on the Survey of Materials Science and Engineering). Washington, D.C.: National Academy of Sciences, 1974.

5. Pick, H.J., and P.E. Becker. Direct and Indirect Uses of Energy and Materials in Engineering and Construction *Applied Energy* 1:31–51, 1975.

6. *Mineral Commodity Summaries 1978.* Washington, D.C.: U.S. Bureau of Mines, pp. 4, 46, 84, 1978.

7. Andrus, C.D. *Annual Report of the Secretary of the Interior under the Mining and Minerals Policy Act of 1970.* Washington, D.C., 1978, p. 32.

8. *Proceedings,* Council of Economics—AIME, 1978. Denver, Colo.

9. Page, N.J., and S.C. Creasey. Ore Grade, Metal Production, and Energy. *Journal of Research—U.S. Geological Survey* 3:9, 1975.

10. *Metal Losses and Conservation Options in Processing, Design, Manufacture, Use, and Disposal of Products.* Office of Technology Assessment, Congress of the United States, draft report, February 1979.

11. Blum, S.L. Tapping Resources in Municipal Solid Waste. *Science* 191(4227):669–675, 1976.

III SCIENCE AND THE UNITED STATES

(James P. Blair. © National Geographic Society)

7 Demography

INTRODUCTION

DEFINITION, METHODS, AND TOOLS

Demography is the quantitative study of human populations. Its basic materials are derived from censuses, vital statistics, and sample surveys. Its methods are observational, empirical, and statistical. Its techniques include the use of advanced mathematics to refine raw, often incomplete data into interpretable trends and comparisons.

The field can be divided roughly into formal demography and the broader area of social demography, or population studies. Formal demography is concerned principally with analysis of the three major determinants of population change: births, deaths, and migration. Social demography studies population change in social, economic, and physical settings, and, therefore, considers other variables such as occupational structure according to sex and age, income levels for different population groups, and housing patterns.[1]

DEMOGRAPHIC DETERMINANTS

Demographic patterns such as those of fertility, marriage, and migration originate in millions of individual actions. These in turn are triggered by such factors as individual aspirations, education, kinship, personal finances, and health.

The individual actions reflected in demographic statistics affect the economic and social environments in which people spend their lives. A sharp increase in the birthrate, for example, concentrates large numbers at the critical thresholds of the life cycle—entering and leaving school, entering the job market, establishing families, renting and buying houses, and eventually retiring and living on income from sources other than current earnings.

Employment and business conditions may affect the age of marriage and of childbearing. Changing life-styles affect family structures. The desire for independence by the elderly, along with the provision of some basic financial security, has in part stimulated the growth of one-person households.

Changing industrial and transportation technology, climate, tax, and labor laws, and business practices are among the major factors affecting where people live. New settlement patterns, in turn, have created new demands for public and private services and have led to competition among cities, states, and regions—between the "sunbelt" and the "frostbelt," for example, and between central cities and suburbs. A recent and unprecedented reversal of migration patterns away from large central cities is forcing reassessment of community planning and affecting other economic, social, and governmental activities.

Demographic trends are composites of individual actions, and national trends are composites of regional and local events. The national trend toward empty classrooms and unemployed elementary and high school teachers also subsumes rapidly growing communities and regions with increased loads on schools and other public services for the young.

Similarly, improvement in some of the highly desired aspects of life have not been equally distributed. The differences between men and women in expected longevity have widened, although life expectancy for both men and women has increased (see p. 383).

Finally, the form in which information is presented can lead to different interpretations. It may matter, for example, whether actual numbers of people or percentages are discussed. For example, between March 1975 and March 1976 more people moved into nonmetropolitan areas than out. However, a larger percentage of the nonmetropolitan residents moved to metropolitan areas than the reverse—3.0 and 1.8, respectively.[2]

USES OF DEMOGRAPHY

Population size and composition largely determine how funds are apportioned by federal and state governments to counties, municipalities, and other government units. They determine representation in national, state, and local legislative bodies. Furthermore, the rights, duties, and powers of municipalities depend in part on the sizes of their total populations and the sizes of defined segments, such as minorities, unemployed, children, the elderly, rural residents, and others.

SOURCES OF INFORMATION

The major sources of data about the American population are the decennial census; the registration of births, deaths, marriages, and divorces; the registration of immigrants; and sample surveys sponsored by government or by private organizations.

A census of population has been taken by the federal government at 10-year intervals since 1790. After 1980 a census will be taken once every 5 years.

The census is the only single source of information concerning population changes and characteristics for every state, county, municipality, township, neighborhood, and city block. But it primarily reports characteristics of the population at the time of the enumeration. Information concerning intercensal changes is derived by using previous census data as benchmarks and updating them with current data from other public records.

Registration of births, deaths, marriages, and divorces is done by state or local agencies. Although the primary purpose of registration is to meet legal requirements and to provide official records for individuals, these data are compiled on a monthly and annual basis.

Sample surveys by both government and private agencies are important sources of information. For instance, the Current Population Survey, conducted monthly by the Bureau of the Census and based on interviews of 60,000 households, provides monthly estimates of employment and unemployment rates and other information about important population developments. The increased use of computers for tabulation of data has greatly extended the usefulness of raw data.

PROBLEMS IN DATA COLLECTION AND ANALYSIS

Although individual responses to the census questions are compulsory, errors are possible in the information supplied or recorded. Methods have been developed to measure and to reduce the effects of such errors, but concerns about privacy and costs limit the degree to which such errors can be completely eliminated.

Error estimates themselves are subject to error, because they depend to some degree upon assumptions. For example, registration of births and deaths is presumed to be virtually complete, but data on marriages and divorces less complete. Finally, registration of immigrants is beset with a number of difficulties aside from the problems of counting undocumented aliens.

A major problem is that of missing data. Despite efforts to reach every person in the United States, it has been estimated that about 2.5 percent of the population, 5.3 million persons, were missed in the 1970 census.

Data from sample surveys also are subject to sampling, response, and recording errors, as well as errors caused by incomplete coverage of sampling units.

To say that data are subject to error is not to deny their utility for many purposes. Good practice calls for allowances for error along with results of a survey. While techniques for minimizing errors and for estimating the size and nature of those that remain have been developed, they continue to deserve the fullest attention.

Because of the political importance of demographic information, it must be reliable, current, and comprehensive. Many significant demographic shifts occur at the regional, state, or local levels. The capacity of state and municipal agencies to collect, analyze, interpret, and disseminate demographic data often falls short of their requirements. Furthermore, future censuses will require more assistance from local agencies than formerly, and it is important that these agencies be effective and efficient.

DEMOGRAPHIC PROGRAMS OF THE FEDERAL GOVERNMENT

Analytical work of high quality is being done by the federal government, through the Bureau of the Census, the National Institute for Child Health and Human Development, the National Center for Health Statistics, and other agencies. These agencies address questions related to their special missions; but no centralized federal agency has responsibility for monitoring and analyzing overall demographic trends in the United States. Congress, however, is examining the effects of population change; recently, for example, through the Select Committee on Population of the House of Representatives.

DEMOGRAPHIC TRENDS IN THE UNITED STATES

Birthrates have been generally declining for the past 100 years. Americans have been leaving their central cities since the 1920's, but the present rate of migration out of some metropolitan areas is unprecedented. The westward drift of the U.S. population center goes back to the founding of the nation, and in the past 200 years this center has moved from Baltimore to St. Louis. For the past 40 years there has been a slight movement of that center toward the South.

These changes can be interpreted in various ways. For example, if low birthrates and increasing life expectancies continue, the age structure of the American population will be transformed, raising the proportion of the elderly and lowering that of the young. In 1977 the largest 5-year age group was the 15–19-year-olds; in 2000 it will probably be the 40–44-year-olds. This change will affect social and economic structure in ways that are still not certain.

Migration patterns of the 1970's will affect cities and suburbs, the places left behind, and destinations. Many large cities, as well as some suburban areas, are losing young professionals and small towns are gaining them. Many middle-class black and white families are moving out of large cities.

There are exceptions to every aggregate demographic trend. While fertility rates are declining overall, they are higher for some groups within the population than for others. There is outward migration from metropolitan regions, but a large number of people are also moving in. Migration patterns leave some large cities in the north and north-central regions with a larger proportion of poorer, older, and minority populations; but some of these same cities are also experiencing urban renovation efforts by middle-class young families.

RATE OF GROWTH: FERTILITY, MORTALITY, AND IMMIGRATION

The annual population growth rate of the United States is calculated on the basis of gains and losses from births and deaths and international immigration and emigration. This growth rate has steadily declined from more than 3 percent before the Civil War to less than 2 percent between 1920 and 1960 and less than 1 percent in 1978. In numbers, the U.S. population grew by 1.7 million in 1978, or about 500,000 less than the growth in 1970. In 1970 there were 203 million Americans; in 1979 about 220 million. The decline in population growth has been steady but not smooth, marked by sudden drops and surges, such as the sharp increase in the number of births between 1947 and 1964, the years of the baby boom.

The gain or loss from annual births and deaths has and continues to be the largest factor in national population change. People in the United States, and in most major industrialized countries, are having fewer children and living longer. With the recent decline in fertility, immigration is becoming more important; it now accounts for nearly 20 percent of the total annual increase in the number of people in the United States, compared with about 11 percent in the early 1950's.

BIRTHRATE AND FERTILITY

The Baby Boom

After decades of decline the U.S. birthrate rose sharply after World War II, peaking in 1957 and declining below 1947 levels in 1964. Since then, the rate of decline has been rapid and has out-distanced recent declines experienced by other industrial countries. Between 1957 and 1978, the rate of childbearing among all women of childbearing age was halved and the annual number of births declined from 4.3 to 3.3 million.

The very rapid rise in birthrates after World War II and the equally rapid decline in the 1960's were unpredicted. The effects of these fluctuations remain, however, in the form of the very large number of babies born during the baby boom and the much smaller number following them. In all, the 42 million babies born between 1955 and 1964 set a 10-year record which has not been equaled. These babies comprise the nearly 41.5 million 14–23-year-olds in 1979 and will be the 41.3 million persons 35–44 years old in 2000. By 2030, when they will be 65–74 years old, their numbers will have been reduced to about 32 million—more than twice the number of 65–74-year-olds in 1978.

Current Fertility

Women now in the midst of their childbearing period are likely to have two children as compared with three for their mothers. In 1964 slightly more than half of all newborns were first or second births. In 1977, almost three-fourths of all births were first or second children.[3] Women are also likely to complete their families in 7 years as compared with 10 for their mothers. As marriages have been delayed, the mother's age at birth of the first child has been increased, while the age at which she has her last child has declined. Finally, the birthrate for women 35–44 years old declined by two-thirds between 1960 and 1976.

There are other indications that the birthrate will remain low or decline. One such indication is the increased number of fertile couples who seek sterilization as a form of birth control. Among one-fourth of the couples of reproductive age, either the husband or the wife has been sterilized.[4] Moreover, some 11 percent of women in their early twenties expect to remain childless.

Teenage Fertility

Birthrates differ considerably among various groups and fluctuate within those groups. For example, the fertility rates among women aged 15 to 17 are different from those aged 18 and 19, and both these rates are different from the fertility rates for women over 20 and under 15. Furthermore, the rates differ among ethnic groups.

In 1966 there were 629,554 births to mothers under 20, and in 1977 fewer than 600,000. Of the 1977 births, approximately 214,000 were to mothers 15–17 years old.[5]

Currently the birthrate for all teenagers appears to be declining. The birthrates among 18- and 19-year-olds declined from 121.2 per thousand in 1966 to 85.7 per thousand in 1975. This decline more nearly parallels the declining birthrate among women in their early twenties, the prime childbearing ages. Among 15–17-year-old mothers the birthrates per 1,000 women were 35.8 in 1966, 39.2 in 1972, 36.6 in 1975, and 34.6 in 1976. The rate peaked in 1972 and has declined since then. Birthrates for girls under 15 account for less than 0.5 percent of the total number of births with little fluctuation from one period to the next.[6]

Despite this decline in the number of births to teenage mothers between 1966 and 1975, the teenagers' share of births rose from 17 to 19 percent of the total number of births between 1960 and 1976. The 15–19 age-group increased considerably in numbers in the latter part of the period as the children born between 1957 and 1961 matured. The increased proportion of births to teenagers was further accentuated by the decline in the birthrate among older women.

The birthrate among teenage black women tended in the past to be higher than that for teenage whites. Between 1966 and 1975 the gap in birthrates between black and white 15–17-year-olds narrowed as a result of a decline among blacks and a slight increase among whites. There is some recent indication that birthrates among 15–17-year-old white women are now declining. Among the 18- and 19-year-olds, both blacks and whites, the birthrate had declined considerably. In sum, birthrates among blacks and whites of both age-groups have declined, and the rate of decline has been greater among black teenagers (Table 5).

There has been an increase in sexual activity among teenagers of all groups, but at the same time there has been an increase in the use of contraception and abortion. The latter is believed to account for the declining birthrates among teenagers.

The lower on the economic scale the teenager, the more likely she is to become pregnant, the lower her chances of completing her education, and the less likely she is to have proper medical attention. Also, the younger the mother, the less likely she is to be married. Furthermore, a higher

TABLE 5 U.S. Birthrates for Women under 20, by Age of Mother and Race (live births per 1,000 women)

Age of Mother and Race	1977	1976	1974	1972	1970	1968	1966
10–14 Years							
White	0.6	0.6	0.6	0.5	0.5	0.4	0.3
Black	4.7	4.7	5.0	5.1	5.2	4.7	4.3
TOTAL	1.2	1.2	1.2	1.2	1.2	1.0	0.9
15–17 Years							
White	26.5	26.7	29.0	29.4	29.2	25.7	26.6
Black	82.2	81.5	91.0	99.9	101.4	98.9	97.9
TOTAL	34.5	34.6	37.7	39.2	38.8	35.2	35.8
18–19 Years							
White	71.1	70.7	77.7	84.5	101.5	102.0	109.6
Black	147.6	146.8	162.0	181.7	204.9	201.3	209.9
TOTAL	81.8	83.1	89.3	97.3	114.7	114.9	121.2

SOURCE: *Monthly Vital Statistics Report,* Vol. 26 (No. 5 supplement), p. 9, 1977; and "Final Natality Statistics, 1976 and 1977." Washington, National Center for Health Statistics.

degree of risk is associated with childbirth among the very young, and the younger the mother, the more likely her child will suffer birth defects.

MORTALITY

Mortality rates .continue a steady decline among infants, women in childbirth, and the elderly. There has been a decline of 13 percent from 1950–75 in death rates for those 65 and over. This decline at older ages, signifying a new trend, is due mainly to lower death rates from cardiovascular diseases (see p. 387). The result of lower death rates from these diseases is that between 1950 and 1975 life expectancies after age 65 rose from 12.8 to 13.7 years for males and from 15.0 to 18.0 years for females. These changes in mortality rates at older ages have, of course, increased the projected number of the aged in future populations. The most recent estimate by the Bureau of the Census is 31.8 million people 65 and over by 2000, or 4–11 percent above projections made between 1964 and 1975.

There are differences in life expectancies between men and women, among residents of different states and different sections of a state, and among races. The difference in life expectancies at birth of males and females has steadily widened, from 4.5 more years for females in 1940 to 7.8 years in 1975. Similarly, there is a difference of nearly 8 years in the life expectancy at birth between those living in the most favored state (Hawaii) and in the least favored (the District of Columbia). The highest state average for women of "all other" races is equal to the lowest state average for white women. A study in Chicago showed that there are differences of as much as 10 years in the mortality rates of small areas within the city. Despite substantial improvements in recent years for blacks and whites, a black female baby born today with a life expectancy of 72 years has the same average life expectancy as her white counterpart born in 1950. Even for those 65 years old and over, the average number of remaining years differs by as much as 2 years between the highest and the lowest state averages.

IMMIGRATION

Net legal immigration into the United States has averaged about 400,000 per year since 1965. To that must be added an unknown number of undocumented aliens. The law provides for a maximum quota of 170,000 immigrants from the Eastern Hemisphere and 120,000 from the Western Hemisphere, and for specified groups of nonquota immigrants, such as the recent admission of refugees from southeast Asia.

While net immigration into the United States now accounts for an increasing fraction of total annual growth of the population, data on immigration are the least developed of our population statistics. There are several problems. The legal definitions of an immigrant are complex: Not all immigrants can be identified as such when they enter the country, and little is known about the persons who leave the country. The movements of undocumented aliens across national boundaries add other complications. There is not now, nor has there been, a set of immigration statistics commensurate with their importance to the United States.

AGE STRUCTURE

Declining mortality rates for the elderly and the concomitant increase in their numbers combined with current low fertility rates and fewer children have increased the proportion of the elderly in the United States. This has resulted in an aging population. By 2000, the population 65 years and older will have increased some 36 percent since 1977. The population of the very old, those 80 and over, will increase especially rapidly, with their numbers almost doubling, from 4.8 million in 1978 to 8 million in 2000.

Countries with high fertility rates, where each age-group is followed by successively larger ones, have a pyramidal age structure with a large number of young at the bottom and a small number of the elderly at the top. As the sizes of the different age-groups in the United States gradually

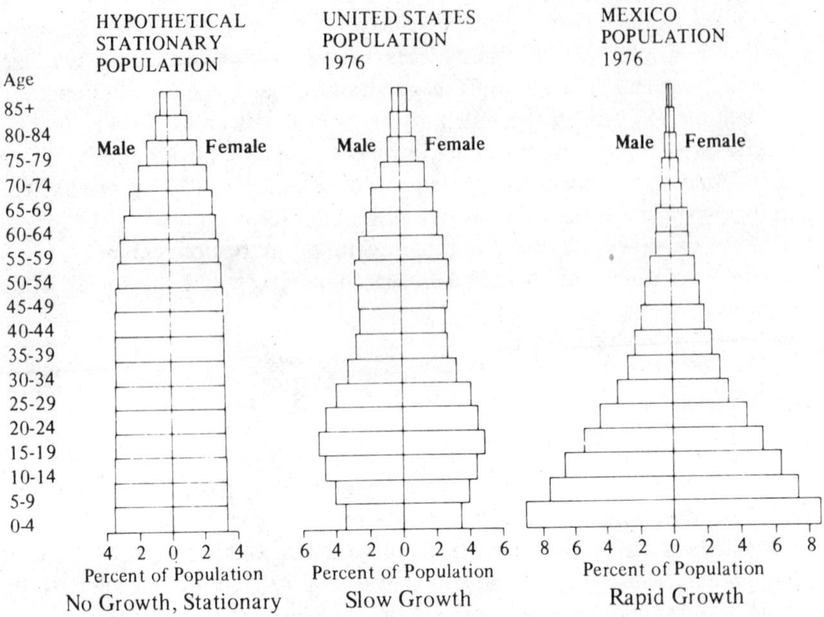

FIGURE 26 Stationary population age-sex structure and contrasts. (Population age-sex structures for Mexico and the United States reproduced from Arthur Haupt and Thomas T. Kane, *Population Handbook,* Population Reference Bureau, 1978, p. 14. Hypothetical stationary population age-sex structure calculated from New Zealand Life Tables, 1970-72. Wellington, N.Z. Department of Statistics, June 1976, Tables 1 and 2 for the non-Maori population. [From *Population Bulletin,* Vol. 33, No. 3, 1978. Courtesy of Population Reference Bureau, Washington, D.C.]).

are becoming more equal, the age structure is taking a more rectangular shape with a small number of very young at the bottom and proportionately more elderly at the top (Figure 26).

EFFECTS OF CHANGING AGE STRUCTURE

Just as "boom" babies crowded schoolrooms and universities throughout the 1950's and 1960's and are now putting pressure on the labor market, they could, when they reach retirement ages, strain pension systems and other services for the elderly.

The changing age structure affects the dependency ratio, the index for measuring the relative size of the working-age population (20–64 years old) in relation to the nonworking-age population (under 20 and over 64 years old). The 1978 dependency ratio was 78.7 nonworking age per 100 persons of working age. In 1960 it reached a high of 91.5/100, reflecting the effects of the baby boom.

The U.S. Social Security system is currently being affected by increasingly large numbers of the elderly in relation to the working population. In 1975 the number of persons 65 and over was 18.9 per 100 persons of working age. By 2010 it is predicted that the ratio will be 20.9/100 and by 2030, as the baby boom group reaches retirement age, the ratio will increase sharply, by 60 percent, and change the elderly/working-age ratio to 33.6/100.[7]

Because of the sharp increase in the size of the working-age population, dependency ratios should improve by 1984. In the long term, however, current low fertility rates will reduce the number of people paying into social security and the number receiving benefits will greatly increase. Therefore costs to the workers will go up.

INTERNAL MIGRATION

The average 20-year-old American moves nine times in his lifetime.[8] Many of these moves are and will continue to be outward from the cities and westward, continuing an American tradition. What is new is the current movement toward the South, for historically net migration was away from the South to the North and West. In general, 15 states, 8 of them in the South, which had a loss of population to other states in the 1960's, are gaining population in the 1970's. The states growing most by migration are Florida, Texas, Colorado, Arizona, and California. The largest losses were in Ohio, New York, and Illinois.

The historic movement of blacks from the South to other parts of the country also has been reversed in recent years, and there is now a movement of blacks to Maryland, Florida, Texas, and Virginia. However, at the same time, blacks continue to move out of Alabama, Mississippi, Arkansas, Louisiana, and the District of Columbia.

CHARACTERISTICS

Those who move usually differ from those who stay behind in socioeconomic status, age, and education. The movers tend to be young, usually in their twenties, finished with formal education, beginning their careers, and newly married but with no children. However, there is also a large number of relatively well-off retired people who move to the warmer climates of Florida and the Southwest.

METROPOLITAN EXODUS

The net flow of people from metropolitan areas to small towns and rural areas has received a great deal of recent attention. It eased the "urban

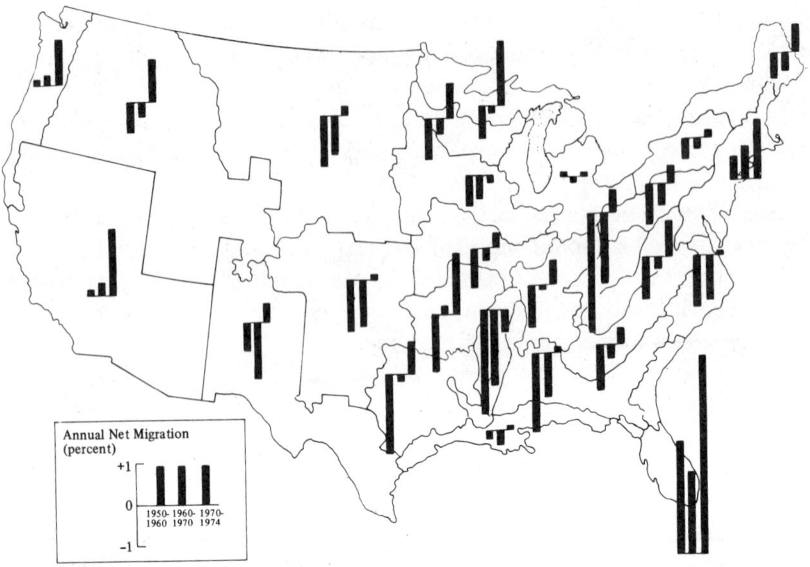

FIGURE 27 Annual rates of net nonmetropolitan migration for 26 U.S. regions. (Calvin L. Beale and Glenn V. Fuguitt, *The New Pattern of Nonmetropolitan Population Change,* CDE Working Paper 75-22. Madison, Wisc.: Center for Demography and Ecology, University of Wisconsin, 1975. [From *Population Bulletin,* Vol. 30, No. 3, 1976. Courtesy of the Population Reference Bureau, Washington, D.C.])

explosion" concern of the 1960's best illustrated by the fear of a "Boswash" conurbation. In the 1970's the comparatively higher population growth rate for nonmetropolitan areas reversed the situation of the 1960's, when the growth rates for metropolitan areas were four times as high as for nonmetropolitan areas (Figure 27).

Such aggregate comparisons may be a bit deceptive. Three-fourths of metropolitan population growth in the 1960's came from an excess of births over deaths and only one-ninth from the arrival of individuals from nonmetropolitan areas.[9] Other factors involved were immigration into the United States and expansion of metropolitan land areas as more people owned cars and as the interstate highway system spread. Regional migrations were a minor part of metropolitan growth during the 1960's.[10]

Moreover, loss of metropolitan populations through migration is not new to American society. In the 1960's, for example, 40 percent of metropolitan areas had more people leaving than arriving. But the effects of these migrations were cloaked by the much larger population gains from excess of births over deaths.

Seen in that light, the recent trend toward migration beyond metropolitan areas was foreshadowed in the 1960's and made evident in the 1970's by the continuing decline in fertility rates. The trend is ubiquitous. Most nonmetropolitan areas are gaining migrants, with the major exceptions being the tobacco and cotton belts from North Carolina to the Mississippi Delta and some of the Great Plains areas. Nearly two-thirds of all nonmetropolitan counties gained migrants in the 1970's, compared with one-quarter in the 1960's and one-tenth in the 1950's.[11]

About three-fifths of the nonmetropolitan gain is occurring in counties that are contiguous to the metropolitan areas.[12] At the same time there are over 500 nonmetropolitan counties losing population. Most of these are heavily agricultural and/or predominantly black.

Population losses in the large cities are mirrored to an extent by population losses for entire metropolitan areas, that is, a central city and its suburbs. During the 1950's, only one metropolitan area, St. Joseph, Missouri, had lost population. In this decade, eight metropolitan areas have already lost population. Nationally, metropolitan areas continue to increase their population, but at a slower rate than nonmetropolitan areas. Moreover the 15 fastest growing metropolitan areas from 1970 to 1975 were all in the South and West, including 9 in Florida.

Migration into nonmetropolitan areas and the general movement toward the South and West, where population densities have been generally lower, are resulting in a geographic shift in population densities.

CITY AND SUBURB

The proportion of Americans living in central cities relative to the suburbs has been declining for several decades. In 1920, two-thirds of metropolitan area residents lived in central cities and in 1975 less than half did so. A decline in the total number of people living in the nation's central cities is new, stemming in large part from the continued migration of whites out of the cities. The loss is now national and no longer limited to the cities of the north-central and northeastern regions of the country. However, the largest losses still occur in these latter regions. In contrast, smaller cities, those with 1 million or fewer residents, are either gaining population or suffering losses proportionately smaller than the larger cities.

Measures of Change

The interpretation of the changes occurring in central cities and suburbs depends on what measure one uses, whether income levels, racial and age composition, or family structure.

The differences in median incomes between suburban and central city families appear to be widening: Suburban median incomes were 17.8

percent higher than central city median incomes in 1970 and 23.8 percent higher in 1977. Or, to cite a different measure, over 10 percent of all central city families now receive some sort of public assistance income, but only 3.9 percent of suburban families do so.

The majority of black Americans live in central cities. However, the suburban growth rate during the 1960's and 1970's was higher for blacks than whites; still, suburban populations remain largely white.

The median age of central city residents has historically been higher than that of suburbanites, but that gap is narrowing as the general decline in fertility rates reduces the number of young children in suburbs. Finally, the proportion of all families headed by females is higher in the central cities than in the suburbs, with these differences largely due to black families, 36 percent of which are headed by females compared with 11 percent of white families.

SOME QUESTIONS

These are some of the signs of major changes since 1970 in the movement of population within metropolitan areas and between metropolitan and nonmetropolitan areas. We lack precise information on the patterns and effects of regional migrations—in and out of metropolitan areas, suburbs, and central cities, and geographical regions. Differences in the volume of such migrations and in the characteristics of the migrants are critical elements in determining the economic and social effects of migration into and out of a region, town, or city.

As noted earlier, rural areas, which historically lost population to the cities, have recently begun to keep more of their population and to attract more newcomers. Questions remain: Are former rural residents returning? Are urban-reared and educated persons opting for rural life? Are jobs, climate, recreation, or some combination of these the major attractions? Are the people left behind increasingly the disadvantaged? Do changes in family organization and in settlement patterns affect the needs for public services by specific groups, such as the elderly?

HOUSEHOLDS AND FAMILY FORMATION

Recently there has been a noticeable postponement of the age of first marriage. The median age of marriage for both men and women increased by about 1.5 years between 1956 and 1977, and the proportion of women 20–24 years old who have not married increased by one-half between 1960 and 1970.

Combining such demographic trends with other social factors is useful in analyzing social patterns. For example, the postponement of marriage is associated with higher college enrollment among women, increased number of women in the labor force, and increased unemployment among young adults. The increased number of young unmarried women of marriageable age may also be related to the fact that there are 5–10 percent more women at the usual age of first marriage (18–24) than there are young men (20–26), because the older men were born during a time of low birthrates, while the women were born during the baby boom. The same effect might work in reverse in the next five years, as the men born during the baby boom seek marriages among women born in the 1960's, when birthrates began their sharp decline.

DIVORCE

Divorce rates historically tend to rise and fall with marriage rates, with the former lagging behind the latter by seven years for first marriages and three years for remarriages. Since marriage rates peaked in 1972 and have now declined by about one-tenth from that level, it can be expected that the rate of divorce will begin to decline about 1980.

ONE-PARENT FAMILIES

In 1977, 17 percent of families with children under 18 living at home were one-parent families and 90 percent of these parents were women. Seventeen percent of all children under 18 were living with one parent, usually the mother, in 1977. Ten percent of children living with two parents live with a stepparent. Based on current trends, five years from now 20–25 percent of all households with children are likely to be one-parent families.

Although 90 percent of all families in the United States are headed by men, half of all families below the poverty level with dependent children are headed by women. Although the percentage of families headed by women is greater among blacks (33.1 percent in 1976) than among whites (11 percent), the number of white families headed by women is increasing and in absolute figures is well above that of blacks. In 1975 the Census Bureau reported 5,380,000 white families headed by women and 2,102,000 black and "other" families headed by women.

HOUSEHOLDS

Although population growth is slowing, the number of households is growing three times as fast as the total population—an increase of 20 percent between 1970 and 1978, as compared with an increase of only 7 percent in the total population. This is a result of the trend toward smaller families, more single parents, fewer or delayed marriages by young adults, more elderly couples, widows, and widowers living alone, and less doubling up because of greater affluence. The effect is that while the population of a particular city, suburb, or metropolitan area may be declining, the number of households may be increasing.[12]

The growth rate was led by one-person households, in particular those maintained by persons either divorced or never married. Half of all divorced persons who have not remarried and half of those who have never married maintain households alone. Among divorced parents, a father usually lives alone while the mother maintains a one-parent family.

The average size of households declined from 3.3 persons in 1960 to 2.8 in 1978, and may be expected to decline further to 2.6 persons by 1983.

Elderly persons will contribute substantially in the next five years to the continued growth in the number of small households. In 1977, 33 percent of women over 65 but only 9 percent of the men were living alone. If current trends continue, the number of households maintained by persons 65 years old and over is expected to increase by 10 percent, or double the expected rate of household formation for the total population.

If present trends continue, the rate of household formation in the next five years will continue to exceed by far the rate of population growth.

LABOR FORCE

CURRENT LABOR FORCE PARTICIPATION

In May 1978 the total civilian labor force exceeded 100 million persons for the first time. During the preceding five years, while the population had increased by about 8 percent, the total number of employed persons had grown by about 11 percent. Moreover, the proportion of the total population in the work force continued to grow. Unemployment, which fluctuates more rapidly than total employment rates, also increased, with the result that in 1978 there were both more employed and unemployed workers than in the previous five years.

These recent changes continue a long-term trend of increasing proportions of the population working for pay or profit. Also, the ratio of workers to nonworkers by sex and age has fluctuated widely, primarily as a reflection of the increasing proportions of women in the work force and a shorter work span for men. Reductions in the numbers of the self-employed have increased the proportions of older men who drop out of the labor force abruptly at retirement rather than continuing as self-employed persons who can adjust their work to declining ability. Also, a growing trend of basing retirement benefits on length of service, instead of on a fixed retirement age, may have led to some increase in earlier retirements. An increased number of years devoted to education has delayed entrance into the labor force. However, by age 25–29, at least 95 percent of men are in the labor force. Young women are returning to the labor market much more quickly after the birth of children, and more men are retiring earlier than the customary retirement age. A much higher proportion of men and women in the age range 25–54 are working than was true in the past. As a result, the work life patterns of men and women are becoming increasingly similar.

Earlier Retirements

Smaller families and completion of families at younger ages may contribute to earlier retirement because of reduced economic responsibility as the youngest child leaves home while the parents are relatively young. The increasing labor force participation of women may also contribute to early retirements because of additional income provided by the wife after retirement of the husband. Increased numbers of young workers may lead to continued pressures on older workers who find it more difficult to adjust to changing technologies. However, the recent change in legislation raising the age of mandatory retirement may slow these trends.

Prospective changes in the age composition of the population and of the labor force imply a substantial decline in the ratio of nonworkers to workers. (Not to be confused with "dependency ratios," which are based on working-age population, whether working or not.) This ratio has already declined from 152:100 in 1965 to 125 in 1975 and is expected to continue to decline to about 111 by 1985. By that time a growing proportion of nonworkers will be older persons.

Major changes have occurred in the kinds of jobs that are available. Significant increases have occurred in the proportion of workers employed in professional, technical, clerical, sales, and service occupations. Correspondingly jobs for unskilled workers and farm workers have declined rapidly. These changes have been shared by men and women; but women have also made major changes from household work to clerical and sales occupations.

Black workers in general have experienced the same changes in their occupational distribution, but at a much more rapid rate. The rates of labor force participation among men have been dropping somewhat more rapidly among blacks than whites. The rates for white women have been increasing more rapidly than those for black women but the rates for black women continue to be higher than those for their white counterparts.

CHANGES IN NUMBER OF WORKING WOMEN

In 1977, about half of all women in the United States were in the labor force, compared with about a third in 1950. Between 1965 and 1977 the proportion of young women (25–34) in the labor force increased from 39 to 60 percent. This increase is particularly remarkable because the majority of the women in this age-group are married and many have children at home, factors that traditionally have kept women out of the labor force. Indeed, the largest increase in labor force participation in recent years has been among mothers with preschool children.

Earnings are the sole income for most of the 16 million women in the labor force who are single, separated, divorced, or widowed. Within husband–wife families, the majority of wives now work and account for roughly 30 percent of the familys' incomes. This additional income has been an important factor in lifting some families above the poverty level. Where the husband's earnings are more substantial, the income from the wife's work may cover the children's college expenses, help maintain living standards during inflation, and permit the husband to consider other work or early retirement. It is possible that the gap between poor and well-to-do families will widen with the increase in the number of families with two high-salaried earners.

UNEMPLOYMENT

There is currently a high degree of unemployment, with an average of 6 million unemployed persons in the United States in 1978—or 6 percent of the civilian labor force. To be considered unemployed a person must be out of work, looking for work, and available for work. Persons looking for their first job are included, along with others who have been in the labor force for long periods of time. Fully retired persons are not considered unemployed.

The National Commission on Employment and Unemployment Statistics currently is evaluating the concepts and measures of labor participation and unemployment.

About half the unemployed persons lost their last job and about one-fourth were coming back into the job market after having been out for some time. The remainder were almost equally divided among those seeking their first job and those who had left a previous one. More women than men were new entrants or re-entrants, and more women than men had quit their jobs.

Unemployment rates vary considerably by age, sex, race, residence area, and occupation. They are particularly high among those born during the baby boom who are now beginning to enter the labor force (Table 6).

TABLE 6 1978 Annual Average Employment Status of the Population in Metropolitan and Nonmetropolitan Areas by Sex, Age, and Race (thousands)

Employment Status	Metropolitan Areas			Non-metropolitan
	Total	Central Cities	Suburbs	Total
Total				
Civilian noninstitutional population	107,391	45,323	62,068	51,550
Civilian labor force	68,738	28,108	40,630	31,682
Percent of population	64.0	62.0	65.5	61.5
Employed	64,529	26,029	38,499	29,844
Unemployed	4,210	2,079	2,131	1,837
Unemployment rate	6.1	7.4	5.2	5.8
Not in labor force	38,653	17,215	21,437	19,869
Males, 20 years and over				
Civilian noninstitutional population	45,158	18,660	26,499	21,848
Civilian labor force	36,459	14,484	21,975	17,005
Percent of population	80.7	77.6	82.9	77.8
Employed	34,880	13,658	21,223	16,332
Unemployed	1,579	825	753	673
Unemployment rate	4.3	5.7	3.4	4.0
Not in labor force	8,699	4,176	4,523	4,842
Females, 20 years and over				
Civilian noninstitutional population	51,219	22,267	28,953	23,269
Civilian labor force	25,897	11,268	14,629	11,518
Percent of population	50.6	50.6	50.5	47.5
Employed	24,360	10,511	13,849	10,820
Unemployed	1,537	757	781	699
Unemployment rate	5.9	6.7	5.3	6.1
Not in labor force	25,322	10,999	14,323	12,751
Both sexes, 16–19 years				
Civilian noninstitutional population	11,014	4,397	6,617	5,433
Civilian labor force	6,382	2,356	4,026	3,158
Percent of population	57.9	53.6	60.8	58.1
Employed	5,289	1,860	3,428	2,692
Unemployed	1,093	496	597	466
Unemployment rate	17.1	21.1	14.8	14.7
Not in labor force	4,632	2,039	2,591	2,275

TABLE 6 1978 Annual Average Employment Status of the Population in Metropolitan and Nonmetropolitan Areas by Sex, Age, and Race (thousands) —Continued

Employment Status	Metropolitan Areas			Non-metropolitan
	Total	Central Cities	Suburbs	Total
White				
Civilian noninstitutional population	92,782	34,856	57,926	46,798
Civilian labor force	59,566	21,732	37,834	28,890
Percent of population	64.2	62.3	65.3	61.7
Employed	56,464	20,491	35,973	27,372
Unemployed	3,102	1,241	1,861	1,518
Unemployment rate	5.2	5.7	4.9	5.3
Not in labor force	33,216	13,124	20,091	17,908
Black and other				
Civilian noninstitutional population	14,609	10,467	4,142	4,752
Civilian labor force	9,172	6,376	2,796	2,792
Percent of population	62.8	60.9	67.5	58.7
Employed	8,065	5,538	2,527	2,472
Unemployed	1,108	839	269	319
Unemployment rate	12.1	13.2	9.6	11.4
Not in labor force	5,437	4,091	1,346	1,961

SOURCE: *Employment and Earnings*, Vol. 26, No. 1, 1979. Washington, U.S. Bureau of Labor Statistics.

Unemployment by Age

One-third of the unemployed are in the prime working ages 25 to 44. In recent years 150,000 people over 65 have been reported as unemployed. One-fifth of the unemployed are between 15 and 19 years old.

Teenage Unemployment In 1977 unemployment rates for teenage blacks were more than double those for whites, with unemployment rates for black teenagers of 40 percent in many places and possibly higher in others. Youth unemployment may be undercounted, and this may be particularly true for black unemployment rates in central cities.

In the past, high unemployment rates among teenagers have been a transitory phenomenon, an inevitable consequence of the school-to-work transition, since these rates diminished to adult levels as the youths entered the adult labor market. However, there is some evidence that those who were unemployed or underemployed in their early years continue to have

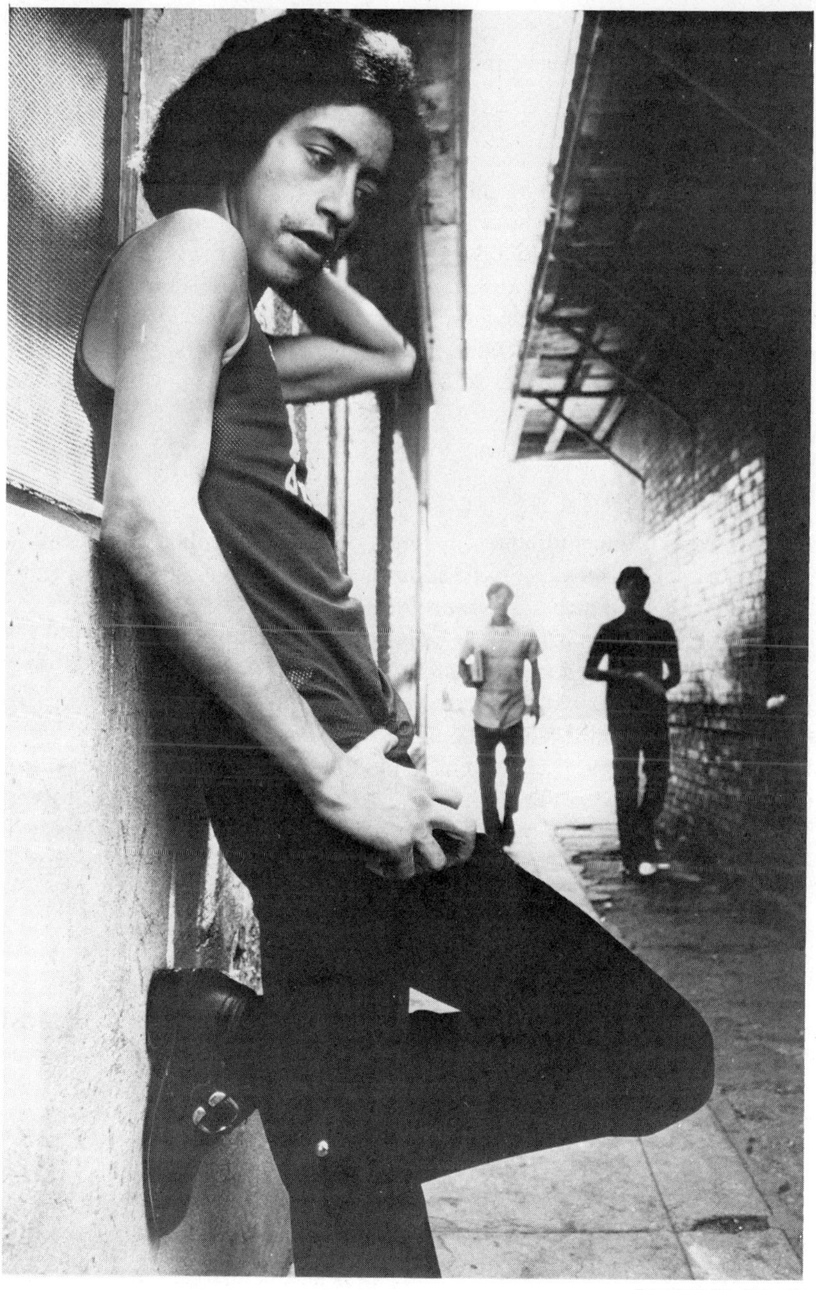

Danny Lyon, EPA—Documerica

difficulty in later years and that their earnings remain relatively low.[13] This is particularly true for black and white females and for black males.

By Race and Sex

Unemployed men exceed women in numbers in each age-group, and the number of unemployed whites exceeds the number of unemployed blacks in each age-group. But, while there are nine times more whites than blacks in the United States, the number of unemployed whites is only about four times as great as that of blacks. Women are more likely than men to experience unemployment. Unemployment rates among male family heads have been considerably below and female family heads above that for the labor force as a whole.

By Residence Area

In 1978, rates of unemployment for men over 19 years old were highest in central cities, and the rates in the suburbs were somewhat lower than those in nonmetropolitan areas. The rates for women over 19 were highest in the central cities and lowest in the suburbs. In the third quarter of 1978 unemployment rates for men and women 16–19 years old were 12.1 percent for metropolitan residents and nearly as high, 11.4 percent, for those living in nonmetropolitan areas. Unemployment rates for white teenagers were highest in central cities, but the rates for suburbs and nonmetropolitan areas were identical. Although suburban areas account for more of the population than any other, the numbers of unemployed persons were almost evenly distributed among the three residence areas—central cities, suburbs, and nonmetropolitan areas.

By Occupation

In general, blue-collar workers are about twice as likely to be unemployed as white-collar and farm workers, with unemployment rates for service workers falling between these groups. Service workers include those involved with protection, food, health support, and cleaning.

Duration of Unemployment

In recent years, about one-fifth of the unemployed have been so for 4 weeks or less. One-sixth to one-fifth have been unemployed for more than 26 weeks. The average duration of unemployment has been 14–16 weeks or between 3 and 4 months. Fewer than half the unemployed were drawing unemployment insurance benefits, some because they are new workers, others because they have exhausted their benefits, or because they were not covered in that program when working.

LABOR MARKET OUTLOOK

The large increase in the number of adults aged 14–22 due to the high fertility rates from 1957 to 1964 will probably put severe pressure on the employment market and for a time may narrow their employment opportunities. Also, some career opportunities will lag behind others. Falling birthrates have already decreased job opportunities in elementary and secondary school education and are beginning similarly to affect employment in colleges and universities.

The pressure on the youth labor market from the baby boom will continue during the next five years but will begin to ease by 1981. However, this easing, with its attendant decline in teenage unemployment, will be manifest mainly for white youths, as the number of black teenagers reaching working age will decrease only slightly. Also, other demographic forces will continue to exert pressures on the youth labor market, including shifts of population and industry out of the central cities and increased competition among the older unemployed for the low-skill jobs traditionally held by teenagers.

SOCIAL CONSEQUENCES OF CHANGED LABOR FORCE PARTICIPATION

The new working patterns for women come at a time of recent decrease in fertility, increases in marriage and divorce rates, and later age of marriage. These trends make it more likely that women will continue to work at present or higher rates. Conversely, as women are assimilated more into the job market, they are likely to marry later, have fewer children, and divorce more readily.

Employment Issues of Women

These relationships between patterns of work and changing family organization in turn highlight several social problems. For example, women are still predominantly employed in "female" occupations, and the median income from their full-time work is less than 60 percent of the median income of males.

These average lower earnings particularly affect women who live alone and must depend solely on their own earnings, even more so for women with children. Unmarried mothers may have difficulty simply providing basic needs, given both low earnings and the complications of caring for a child during working hours. Divorced women who have not remarried—

the fastest growing group of female family heads—typically also do poorly economically when they enter the labor market. Also, divorced women with children usually assume child-care responsibilities. Even though support payments from fathers are often specified in divorce and separation cases, they are frequently unpaid and uncollectable. Most divorced women eventually remarry, but usually only after a financially precarious period. Those who do not remarry often do not have the job skills needed to earn an adequate living.

Day Care of Children with Working Mothers

Half of all children under 18, and 37 percent of preschool children, have working mothers. There are obvious pressures on these mothers, many living alone, who must both keep a home and arrange for supervisory care of their children during working hours. In about two-thirds of families with working mothers, the parents, living alone or not, care for their children themselves. For the remainder, child care is provided in a variety of ways: in their own homes by relatives or others, in the homes of others, and least frequently (3 percent) in day-care centers.

Equity in Taxes and Benefits

The increase in the number of married women who work has raised problems of equity in federal income tax rates and social security benefits, both of which are oriented toward the family as the basic economic unit. Given progressive income tax rates, the two-earner married couple may pay more taxes than two single persons who live together and earn the same total income, thus incurring a "marriage tax." Social security taxes are levied on individuals, but many of the benefits are family oriented. For example, a working wife may find at retirement that her contributions to social security have earned her nothing above the retirement benefits she would have received through her husband's contribution. There probably will be increased pressure for some changes in the rate structure of income taxes and in social security benefits to allow for changes in family income patterns associated with the increased participation of women in the labor force.

SCHOOL AND COLLEGE ENROLLMENT

Sixty million persons were enrolled in schools or colleges in both 1973 and 1978. But the mix by type of school changed considerably during these years. The number of nursery school children increased by half a million from 1.3 million to 1.8 million; kindergarten and elementary school enrollments went down by 3 million from 34.5 million to 31.5 million; high-school enrollments were virtually constant at 15.5 million; and college attendees went up by 2 million from 9 million to 11 million (see p. 472). Increases at the nursery school and college level resulted largely from rising enrollment rates, whereas the other enrollment figures reflect primarily the population level of the relevant age-group at the two dates.

During the next five years elementary school enrollment may decline another million. High-school enrollment is now at its peak and is expected to decline nearly 2 million in the next five years.

College enrollment increased between 1973 and 1977 but fell by 400,000 between the fall of 1977 and the fall of 1978. The population of 18–24-year-olds will not peak until 1981. But the number who will actually be entering college is difficult to predict, and may actually decline gradually before another five years pass. In brief, the total number of college students may have come close to a plateau with little change expected in total attendance for a few years. The number of graduate-school students in 1978 (1.7 million) was somewhat larger than the number of seniors (1.4 million) and has been increasing somewhat faster than that of the rate of growth of the seniors (22 percent versus 17 percent since 1973).

White college students were 16 percent more numerous in 1978 than in 1973, whereas black college students were 49 percent more numerous. In 1978, black persons 18–24 years of age were 12 percent of the total population, and black college students were 11 percent of all college students. Corresponding proportions for persons of Hispanic origin were 6 percent and 3 percent, respectively.

In 1978 white women, white men, and black women at ages 18–21 years had somewhat similar enrollment rates—32, 35, and 29 percent, respectively. Black men, however, had a much lower rate, 22 percent.

If present trends continue, black college enrollment rates may be expected to continue to approach those of whites, but those for persons of Hispanic origin seem likely to continue to lag behind.

BIRTHRATE PROJECTIONS

Projecting the number of births for the next decade seems on its face a simple matter, since the number of potential mothers is known, their attitudes toward family size have been surveyed, and it is reasonable to assume that low fertility rates will continue. However, the assumption is tempered by the fact that the baby boom came as a surprise to most social scientists and also by the current disagreements as to what birthrates will be in the early 1980's, less than five years from now.

In 1977 the Bureau of the Census made projections based on a range of fertility rates including an intermediate rate of 2.1 births per woman (the long-term replacement rate) and a low rate of 1.7 close to the actual current rate of 1.8.

If the current fertility rate of 1.8 continues, there will be some 250 million Americans by 2000 compared with 220 million in 1979. Should the fertility rate rise to 2.1 children per woman or decline further to 1.7 children per woman, assuming no change in immigration rates, the population of the United States will rise by the year 2000 to 259 or 245 million, respectively. In all, if there were no change in the timing of childbearing and fertility were to attain exact replacement level in 1980 to 1985 and remain there indefinitely, the population of the United States would increase some 29 percent between 1975 and the year 2030, and decline slightly until the year 2100.[14]

One analysis predicts that fertility will increase in the next few years as the proportion of young men in the labor force declines. This smaller number of men will command higher incomes relative to the incomes of older men, and young men will, therefore, be able to marry at a younger age. They will also begin their families sooner and be able to support larger families. The analysis predicts increasing fertility rates about 1984 and, with increased numbers of young children, a significant change in the age structure of the population.[15]

However, another view argues that with the perfection of contraceptive methods we are fast approaching an era of few, if any, unwanted pregnancies. This, combined with transient marriage patterns, should lower fertility rates and lead to further declines in population growth rates.[16]

Regardless of the fertility rate, however, births will be the major component of the country's population growth for the next several decades. Even if the unprecedented 20 million American women now approaching childbearing ages have children merely at the replacement rate, they will add some 3.4 million babies to the population each year throughout the 1980's—an echo baby boom. Thereafter, the number of births should begin to decline as much smaller numbers in that age-group reach childbearing age and as the baby boom men and women leave it.

IMPLICATIONS OF GEOGRAPHIC REDISTRIBUTION

The changes in distribution of population will lead to changes in the need for community services—ranging from schools to highways to chronic care facilities for the elderly—since they are shaped by the size and composition of the population. Because migrants tend to be young, better educated, and therefore more able to secure well-paying jobs, they may alter the service requirements in both the areas where they settle and those they leave.

An important issue is the effect of growing and declining populations on the ability of communities to finance services. In regard to growth, an increase in population in sparsely settled rural areas may result in lower per capita costs for public services, since, in many cases, a moderate population increase enables a more efficient use of already existing facilities such as roads, schools, and hospitals.[17] However, in other cases significant increases in population may raise the per capita cost of services if existing facilities are used beyond capacity, requiring increased capital expenditures for additional facilities. Furthermore, newcomers often require services before they make commensurate contributions to tax revenues. A rapidly growing area, for example, may require an immediate expansion of school facilities, with contributions of the newcomers to taxes coming much more slowly.

Recent population changes include rapid growth in nonmetropolitan areas adjacent to existing metropolitan areas. Like the earlier rapid growth of the suburbs, this has placed additional strains on the multiplicity of local governments. There have already been a number of governmental efforts to meet the need for areawide planning.

When population declines in urban areas with established facilities, the remaining poor populations often do not generate sufficient taxes to support the social services that they need most. Also, the demand for public services in relation to changing composition of the population is unclear. Which services, such as crime protection, criminal justice, education, and income maintenance, are in greater or lesser demand as a consequence of lower population? Or does the remaining population require different services?

The costs of urban public services are difficult to control or reduce despite declines in population. Often the costs, such as municipal debt servicing, were incurred before the decline set in; and usually infrastructural systems and services, such as schools and utilities, were set up to support a much larger and wealthier population.[18]

In time, however, it is possible that declining populations may reduce the demand for many services thus reducing costs.

SCHOOLS

The falling birthrates beginning in the early 1960's have resulted in the yearly declines of about 1.5 percent of the elementary school population. Secondary school enrollments will certainly decline through the 1980's.

That population will drop from almost 16 million in 1970 to about 14 million in 1985 and 13 million in 1990, a 20 percent decline in 20 years.[19]

Declines in numbers and the migration of school-age children will have different effects in different parts of the country. While the northern school systems and cities in particular are losing population and may face greater than average declines in secondary school enrollments, systems in the new settlement areas will probably need to build more schools to meet the demands of young migrants with school-age children.

In the first half of the 1970's enrollments grew by 5 percent or more in Arizona, Florida, Alaska, Nevada, and New Hampshire. However, enrollments declined in some of the local school systems within those states. Impacts will generally parallel those of gross population movements. For example, the net migration of school-age children 5–13 years old tends to flow from central cities to suburbs to nonmetropolitan areas.

Growth and decline pose different problems for school systems.[19] Growth, unless anticipated, may result in severely crowded classrooms and hasty and inadequate financing. School systems losing population are facing personnel problems related to excess numbers of teachers and school administrators, including tenure, job protection rules, and other limits on personnel reduction. The closing of neighborhood schools on economic grounds because they serve too small a student population has also encountered strong resistance from area parents.

SOCIAL SERVICES

Social services for the poor, ill, aged, unemployed, or disabled require special facilities and trained personnel. Migration has altered the proportion of these vulnerable populations in many places, changing the level, nature, and location of pension, welfare, and rehabilitation programs and hospitals and other facilities.

Between March 1975 and March 1976, the South had a net gain and the northeastern and north-central regions a net loss in populations of the old and the very young, unemployed women, and those below the poverty line[20] (Table 7).

However, the actual fiscal consequences for the South and West of gaining populations in need of social services are problematical, because these same regions are also gaining young and relatively well-educated populations. In effect, additional costs to meet increased need for social services in these areas may be offset by increased ability to pay for them.

The future social services needs of populations in central cities, suburbs, and nonmetropolitan regions are more difficult to categorize and should be studied, analyzed, and planned for. Overall, central cities and nonmetropolitan areas in 1974 had the highest proportion of the elderly, the poor,

TABLE 7 Net Changes in Selected Characteristics of Immigrants and Outmigrants by Region, 1975–1976 (thousands)

	Northeast	North Central	South	West
Under 5 years of age	– 29	– 16	+ 21	+ 22
65 Years or over	– 18	– 18	+ 27	+ 8
Below poverty level	– 40	– 100	+ 78	+ 63
Unemployed male	+ 1	– 7	– 20	+ 30
Unemployed female	– 13	– 10	+ 19	+ 4
4 Years of college	– 42	– 18	+ 17	+ 41
Professional, technical workers				
Male	– 7	– 34	+ 16	+ 25
Female	– 4	– 15	+ 7	+ 13

SOURCE: *Current Population Reports*. P-20, No. 305, p. 108, 1977. Washington, U. S. Bureau of the Census. (From "Social Services and Population Redistribution in the 1970s," *Implications of Population Redistribution in the United States in the 1970s*, Washington, National Research Council, February 1978.)

and one-parent families led by women.[2] Countrywide, the nonmetropolitan areas had the largest drop in the proportion of the population below the poverty line. At the same time the proportion of elderly populations and female-headed families increased most rapidly outside central cities.

General inferences from these and other patterns suggest that the demand and need for social services will continue at the same or higher levels in the central cities even though they may be losing populations, that the influx of the young and the old into nonmetropolitan areas and the South will increase the demand for services related to their particular needs, with or without public financing, and that the need for programs of income support for the elderly will possibly increase in sun belt regions.

JOBS

Economic patterns tend to mirror those of populations and vice versa. As with populations, there is a continuing shift in manufacturing industries from metropolitan areas to small cities and rural areas. Since 1970 the growth in nonmetropolitan areas of wages and salaries and nonagricultural jobs has been twice that of metropolitan regions. By and large, the shifts to nonmetropolitan areas by manufacturing have been in labor-intensive industries, while capital-intensive industries have tended to stay in metropolitan regions.[21]

While there was little increase in the total number of jobs in industry between 1960 and 1975 (about 1.5 million jobs, or an 8.8 percent increase), regionally the changes were considerable. The South accounted for almost all of the job gains at the same time that the Northeast lost almost 14 percent of its 1960 total.[22]

However, while it is possible to discern relationships between regional shifts in employment and shifts in population, there is no direct quantitative link between increase in employment and population. Relationships between migration patterns and employment change with time and with the particular region. Employment in an area may be determined by the state of the national economy, the nature of the industry, or the socioeconomic status of the population.

That there is a positive but rough link between employment and population is illustrated by the fact that while population growth in the South and West has been similar only since 1970, both regions have shared a similar rate of employment growth since 1960. Growth in local employment is also coupled with the number and characteristics of migrants. Increases in population in a region generate employment through additional demands for goods and services at the same time that the migrants are employed to provide them.

MIGRATION AND RESOURCES

Land

Population density in the United States is remarkably low compared with other industrialized countries.[23] The dispersion of the population from central cities to suburbs to beyond the metropolitan zones will continue to reduce the population density in some areas of the United States while raising it in others.

The shift of people to the South and West is a shift from high to low density and from high-value to low-value land.[17] Both the low densities and low prices imply that a resident moving from Baltimore to Houston may sharply increase the amount of land he uses. The lower costs of land in settlement areas also imply future increases in low-density homes and businesses until land values rise and force more intensive use of the land.

Energy

Shifts in population between the North and the South and from high-density to low-density areas will create shifts in energy usage. Because residential heating is an important consumer of energy, the population gains by the South and Southwest from the colder parts of the country may in time result in significant national savings of fuel for heating purposes. These savings may be offset, however, by the greater use of the car and higher gasoline consumption in the South and West because of lower residential and business densities. Rural residents drive 70 percent more miles than those living in cities with populations of 100,000 or more. As rural areas continue to gain population, their consumption of gasoline will continue to mount and they may become proportionately more vulnerable to fuel shortages.

Conversely, a decline in the populations of large metropolitan areas implies a reduction in the demand for mass transit because such transportation, to be efficient and cost-effective, must rely on high population densities.

CONCLUSION

ZERO POPULATION GROWTH

A slowing rate of population growth, including a total fertility rate currently below the long-term replacement level, does not portend an early arrival at zero population growth. With 3 million births and 2 million deaths per year, the population continues to grow, even without immigration. Also fertility rates may decline further.

FUTURE AGE STRUCTURE

Whatever the overall rate of growth may be, whether increasing or leveling off, major population changes will occur in the near future. Baby boom babies are passing into adulthood, establishing new families and households, and creating new demands. As they move from teenage to adult status and into middle age, numerous adjustments in living patterns, work relations, and public policy are to be expected. The aging of the population, especially the rapid increase in the number of persons who have passed their sixty-fifth birthday, will be significant in the country's social and economic future. This will be the case even before the survivors of the baby boom become 65, sometime after 2012.

Between 1978 and 1983 the increase in the number of persons who are over 65 almost balances the expected decline in the number under 20, while the number of persons who are between 20 and 64 years of age increases by about 10 million persons. But between 1983 and 1993, the population of the elderly and those under 20 are both expected to increase by about 5 million each. This 10 million increase in these two groups is barely met by an increase of 12 million in the age-group they sandwich—the 20–64-year-olds.

Lower fertility increases the likelihood that more women will enter and thereby enlarge the labor force. Changes in retirement legislation could extend the working life of the elderly, also increasing the labor force. If both developments occur, the burden of support for persons in dependent ages would be reduced relative to the numbers in the labor force.

Consumer demand will be affected by the increase of the young middle-aged. Between 1978 and 1983 the increase in the number of persons between 25 and 44 years of age accounts for about 90 percent of the total projected population growth. This may be good news for consumer markets providing goods and services for relatively young households.

POPULATION REDISTRIBUTION

The future effects of population redistribution are immediately visible. There will be significant differences in population growth through migration among regions, states, urban and rural areas, and between metropolitan and nonmetropolitan areas. There will be differences in fertility and migration by age, sex, and color. The South and the West will be confronted with issues arising from relatively rapid growth, while the northeastern and the north-central states will face problems of managing older and declining populations.

Current information about migrants, their motives, and their adjustments to new locations is very limited, and additional information must await the 1980 census. There is evidence that trends reported through 1976 continued into 1978.

ELEMENTS OF CHANGE

Trends revealed by demographic analysis can be mechanically extrapolated to form a picture of the future of 1984 or 2000. However, such projections should not be taken as forecasts, for they do not allow for the possibility of major shifts in the underlying forces that cause demographic change.

Further understanding of these forces is essential to answer the questions raised by demographic analysis: Why are more Americans moving to smaller metropolitan areas and rural regions? Does the decision by more Americans to marry later or not at all signal a deep change in the structure of the American family? Or is the phenomenon a temporary interruption of traditional patterns just as the baby boom interrupted a long-term decline in fertility rates? What is the effect now and what will be the effects in the future of the decisions of young Americans to have fewer children than their mothers had?

Answers to such questions are complicated by the weaving in of individual motives and circumstances, but also by the fact that demographic trends themselves have changed very rapidly. Population projections into the future have had to be revised frequently. Not many years ago it seemed reasonable to project a population of about 300 million by the year 2000. The Census Bureau's preferred current projection now is 260 million, but this may also be an overestimate.

Neither the baby boom nor the speed of the recent declines in fertility were accurately predicted. It is not known what the course of fertility will be during the next five years, for changes can be and indeed are so rapid that only current observations are dependable.

Historical relationships and trends should be used cautiously as a guide to the future. It cannot be assumed, for example, that marriage and fertility rates will correlate in the future as they did in the past. To illustrate, the baby boom occurred partly because more Americans decided to marry at younger ages and thereby increased their chances of having more children. However, the fact that many Americans now are putting off marriage does not mean that when and if they do marry they will choose current patterns in family size. They might have three or four or no children instead of the "standard" one or two. Nor, if marriage rates go up, can one assume that the "usual" proportion of couples will choose to have children.

Demographic information alone is not sufficient to make intuitive judgments. For example, while the number of children three–five years old in the United States has dropped considerably in the last decade, enrollments in nursery schools and kindergartens have increased sharply. The reasons are not embedded in demography, but in other social influences, such as, perhaps, the increased number of working mothers.

It should also be kept in mind that global assessments of trends often fail to reveal highly significant developments at the subnational level. For example, while from 1970–75, 11 of the fastest growing counties were in the South, 16 of 25 counties losing the greatest percent of population were also in the South.

Whatever the difficulties in coping with the rapidity of demographic change and the uncertain insights to the future offered by past events, population trends do have long-lasting and deep effects. Since the size of current population groups is known and since mortality rates are unlikely to change drastically, barring catastrophe, changing dependency ratios can be predicted and their effects anticipated. Post-World War II increases in fertility rates will affect the American social structure well into the next century. Schools and colleges, the labor market, the types and magnitude of public services demanded, and, eventually, the needs of the elderly will all be affected by current demographic change.

OUTLOOK

The following outlook section on demography of the American population is based on information extracted from the chapter and covers trends anticipated in the near future, approximately five years.

Two significant recent demographic developments in the United States are the changes in age structure, the results of the sharp increase in birthrates after 1947 and the subsequent sharp decline after 1957; and changes in the patterns of internal migration, signified by net movement to nonmetropolitan areas.

AGE PROFILE

As a result of the 1947–64 baby boom, by 1984 the population of the United States will have an unprecedented number of young adults of working age (20–37 years old). Because of the continuing decline in the birthrate, there will be relatively fewer school-aged children and university-aged men and women. At the same time, there will be a significantly higher number of elderly people. These combined factors contribute to the aging of the population. However, the early 1980's may see the beginning of an upswing in the numbers of the very young as the large young-adult population reaches traditional marriage and childbearing ages.

GEOGRAPHIC DISTRIBUTION

Current migration and settlement patterns in the United States have changed population densities in many areas. These patterns are characterized by a decline in the populations of the urban centers of the North and Northeast and an increase in the populations of the Southwest and South. Although many young couples have moved into city centers and upgraded housing and other facilities, the overall trend is away from large cities. If current trends continue, in 1984 central cities, especially in the northern and north-central states, will have a disproportionately large number of the elderly and impoverished.

FAMILY SIZE AND STRUCTURE

Trends toward high female employment rates, high divorce rates, increased life expectancy, and later age at marriage will probably increase the number of one-person and one-parent households as well as small complete families. There will be an increased number of small households comprised of elderly people living alone and elderly couples. The number of households headed by females will increase among the urban poor.

EMPLOYMENT AND THE LABOR MARKET

The proportion of men in the labor force has declined recently because of the increased number of working women and earlier retirement among men. The age profile of the labor market in 1985 will be characterized by the large number of young adults, male and female, either just entering the market or already established. Older women will probably continue to be a significant proportion of the labor market, but there is some speculation of a significant decline in the rates of labor force participation of young women as they reach childbearing age. Because there are proportionately fewer teenagers following the baby boom, thus putting less pressure on the teenage labor market, teenage unemployment should begin to decline by 1984.

COMMERCE AND INDUSTRY

Manufacturing is expected to continue to expand in the South and Southwest along with the labor force as young adults continue to move to these areas. It is further anticipated that a demand among young adults for consumer products, housing, and services associated with marriage and family formation will stimulate economic growth nationwide.

SOCIAL AND ECONOMIC IMPLICATIONS OF CHANGING AGE STRUCTURE AND POPULATION REDISTRIBUTION

Changes in the age profile of the population and shifting patterns of migration will require major changes in infrastructural services and facilities, changes in industry, and changes in the allocation of local, state, and federal resources.

Energy demands may shift as more people move to warmer climates and require less fuel for heating homes, but at the same time the newly dispersed populations will use more automotive products, gasoline, and roads.

Nationwide there will be less demand for goods, services, and public programs for the very young and a greater demand for services for the elderly. However, in the Southwest and South there will be a greater need for youth-oriented activities. At the same time northern cities will require more services geared to the poor and elderly.

In the next five years increases in the working-age population in relation to the size of nonworking-age population should lessen the burden on individual workers and temporarily ease the pressures on funding of social security pension systems.

If migration patterns continue, there will be an increase in the tax base of the nonmetropolitan areas and small cities gaining population and a decline in the tax base of the old cities. Supplying the goods and services needed in rapidly growing areas should prove less of a burden on public resources than meeting the needs of declining cities.

REFERENCES

1. Kirk, D. Population. *International Encyclopedia of the Social Sciences* 12:342–349, 1968.

2. Alonso, W. Metropolis Without Growth. *Public Interest* 53:68–86, 1978.

3. Taeuber, C. "Some Current Population Trends in the United States." Testimony before the House Select Committee on Population (February 9, 1978), p. 2.

4. Ford, K. Contraceptive Use in the United States, 1973–1976. *Family Planning Perspectives* 10(5):264, 1978.

5. *Monthly Vital Statistics Report,* National Center for Health Statistics, DHEW Pub 79-1120, 27(11 supplement):12, 1979.

6. Baldwin, W. Adolescent Pregnancy and Childbearing—Growing Concerns for Americans. *Population Bulletin* 31(2):36, 1977.

7. Ball, R.M. *Social Security Today and Tomorrow.* New York: Columbia University Press, 1978.

8. Long, L.H., and C.G. Boertlein. *Geographical Mobility of Americans.* Washington, D.C: Bureau of the Census, 1976, p. 14–15.

9. Alonso, *op. cit.* p. 73.

10. Berry, B.J.L., and D.C. Dahmann. *Population Redistribution in the United States in the 1970's* (NRC Assembly of Behavioral and Social Sciences). Washington, D.C.: National Academy of Sciences, 1977, p. 7.

11. Morrison, P.A. "Overview of Demographic Trends Shaping the Nation's Future." Testimony before the Joint Economic Committee, (May 31, 1978), p. 7.

12. *Ibid.* p. 18.

13. Adams, A.V., and G.L. Mangum. *The Lingering Crisis of Youth Unemployment.* Kalamazoo: W.E. Upjohn Institute for Employment Research, 1978.

14. Day, L.H. What Will a ZPG Society Be Like? *Population Bulletin* 33(3):8–14, 1978.

15. Easterlin, R.A. What Will 1984 Be Like? Socioeconomic Implications of Recent Twists in Age Structure. *Demography* 15(4):397–432, 1978.

16. Westoff, C.F. Marriage and Fertility in the Developed Countries. *Scientific American* 239(6):51–57, 1978.

17. Beale, C.L. "Recent U.S. Rural Population Trends and Selected Economic Implications." Testimony before the Joint Economic Committee (May 31, 1978).

18. Bahl, R. "The Fiscal Problems of Declining Areas." Testimony prepared for the House Select Committee on Population (June 6, 1978), p. 3.

19. Katzman, M.T. Implications of Population Redistribution: Education. *The Implications of Population Redistribution in the United States in the 1970's.*(NRC Assembly of Behavioral and Social Sciences). Washington, D.C.: National Academy of Sciences (in press).

20. Perlman, R. Social Services and Population Redistribution in the 1970's. *The Implications of Population Redistribution in the United States in the 1970's.*(NRC Assembly of Behavioral and Social Sciences). Washington, D.C.: National Academy of Sciences (in press).

21. Sternlieb, G., and J.W. Hughes. New Regional and Metropolitan Realities of America. *American Institute of Planners Journal* 43(3):230, 1977.

22. Greenwood, M.J. The Employment Policy Implications of Population Redistribution in the United States. *The Implications of Population Redistribution in the United States in the 1970's* (NRC Assembly of Behavioral and Social Sciences). Washington, D.C.: National Academy of Sciences (in press).

23. Mills, E.S. Population Redistribution and the Use of Land and Energy Resources. *The Implications of Population Redistribution in the United States in the 1970's* (NRC Assembly of Behavioral and Social Sciences). Washington, D.C.: National Academy of Sciences (in press).

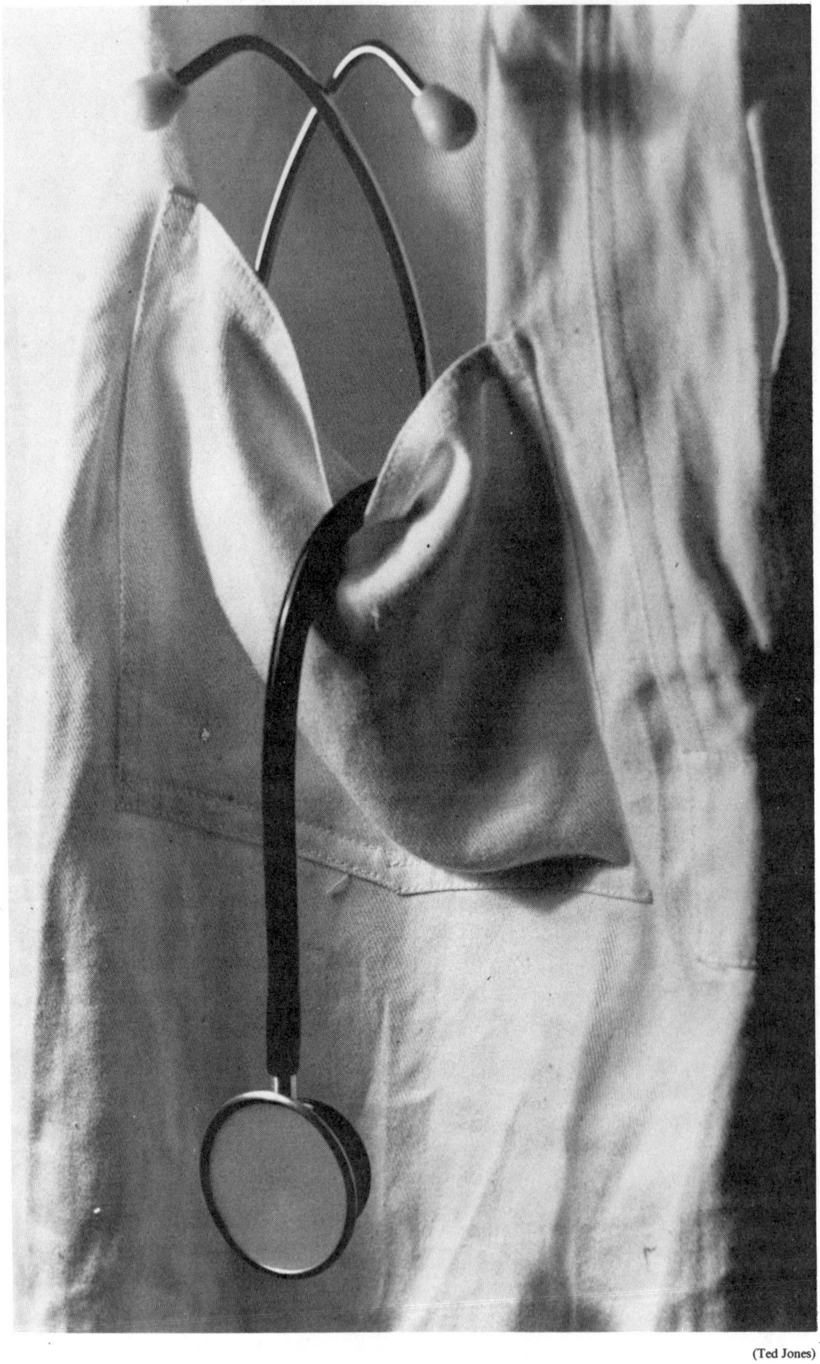

(Ted Jones)

8 Health of the American People

INTRODUCTION

Major progress has been made in public health in the twentieth century in the United States and other industrialized nations. Advances in sanitation—particularly improvement of water quality and less human contact with excreta—have improved health profoundly, especially in terms of infectious and parasitic diseases. Similarly, advances in food production and distribution have led to unprecedented quality in nutrition in the industrialized world, thus further diminishing susceptibility to infectious and other diseases.

Such improvements have been paralleled by advances in the practice and technology of medicine. Thus, the health care system can offer antibiotics for many infections, immunizations for viral and bacterial infectious diseases, and the detection of serious genetic defects in the human fetus through prenatal diagnosis. Encouraging progress has been made in developing pharmacological and other therapies for common chronic diseases such as hypertension and schizophrenia. Gross nutritional deficiency diseases have been virtually eliminated; for example, pellagra was a major cause of severe mental illness until it was found to be associated with a deficiency of niacin and protein and combated successfully on that basis. Important anemias have been brought under control. Many hormonal disorders can now be effectively treated.

ASSESSING THE NATION'S HEALTH

Progress in public health is evident in the statistics used to assess the health of the U.S. population. These vital statistics include mortality rates, life expectancy, and data on the incidence of serious illness.

Mortality Rates

Mortality rates in this country, excluding war deaths, declined steadily during the first half of this century. In 1900 the rate was almost 20/1,000 inhabitants. By 1940 the rate had fallen to 10.8/1,000 and in 1950 to 9.6/1,000. In 1975, the overall mortality rate dipped below 9/1,000 for the first time. In both 1975 and 1976, the U.S. mortality rate was 8.9/1,000. Even taking into account the continuing aging of the American population, the age-adjusted death rates declined 10 percent between 1950 and 1975.

The decline in infant mortality during the first year of life that was so marked in the first half of the century halted abruptly around mid-century:

> The relatively poor progress in reducing infant mortality since the early 1950s has been a source of increasing concern in the United States. The subject has been examined previously in the context of international, national, and local changes in pregnancy loss rates; but it is clear that continued discussion based on the analysis of old and new data is very much the order of the day. . . .
>
> In 1950 the assessment of the performance in the immediate past could well have led to an expectation of additional impressive gains in the future. Today the mood is quite different. For over a decade there has been no sizable decrease in the infant mortality rate. In fact, during the 1950s there were years in which the rate increased—a most unusual occurrence in half a century of vital statistics reporting in the United States. Events in the last few years give the definite impression that while the infant mortality rate will not remain stationary, its downward movement will be slow indeed.[1]

In 1975, the mortality rate for nonwhite male infants was 30/1,000, almost twice the rate of 15.9 for white male infants. Similarly, among female infants the rates were 25.2/1,000 for nonwhites and 12.2 for whites. The higher rates for nonwhite infants is made more telling by the fact that several industrialized nations have infant death rates even lower than those for white infants in the United States.

Major contributors to total mortality rates today are cardiovascular disease, cancer (mostly lung cancer), and accidents. Cardiovascular disease and cancer will be covered in more detail later. In recent decades, automobile accidents in the United States accounted for more than 50,000 deaths per year, a disproportionate number of them involving males between 15 and 24. In 1975, accidents, particularly automobile accidents, were by far the largest cause of death among young males. At 96.7 deaths per 100,000 young males, accidents far outpaced the second and third causes of death among this group. Accidents were also the chief cause of death among females between 15 and 24, amounting to 23.7 fatalities per 100,000.

Two final points concerning mortality rates should be mentioned. One is that they are higher for males than for females in every age group. The lower mortality rates for infant girls, as well as the lower rates of female deaths from accidents, are notable examples. There is also a sex difference in mortality from cancer and cardiovascular disease. Although sex differences in death rates are virtually universal and probably have a biological basis, they are especially evident in modern technological societies.

The second point about mortality rates is that at all ages the rates for nonwhite Americans are higher than the rates for white Americans. This is apparent in the rates for cardiovascular disease and cancer, as well as in the infant mortality figures noted earlier. Differences in mortality rates by race are strongly related to differences in socioeconomic status and can be viewed as partly reflecting the limitations of the health care system. Where socioeconomic differences are deliberately counteracted—as the U.S. Public Health Service has done through a highly organized and sustained effort in prenatal and perinatal care for American Indian mothers and their infants—both infant and maternal mortality rates among nonwhite citizens drop substantially.

Life Expectancy

Life expectancy is the average number of additional years that any one person can expect to live; typically it means life expectancy at birth. A female born in 1900 could expect to live, on the average, 51 years; a male born in the same year could expect to live 48 years. But a female born in 1975 could expect to live for almost 75 years, or 24 years more than the female born in 1900. Similarly, a male born in the United States in 1975 could expect to live to age 66, or 18 years more than the male born in 1900.

Improvements in life expectancy do not apply equally at all ages. For example, although an infant born in 1975 could expect to live about 20 years longer than an infant born in 1900, a person of 40 in 1975 could expect to live, on the average, only 6 years longer than the person who was 40 in 1900.

The increase in life expectancy in the United States during this century has been quite remarkable. By far the largest part of the increase occurred

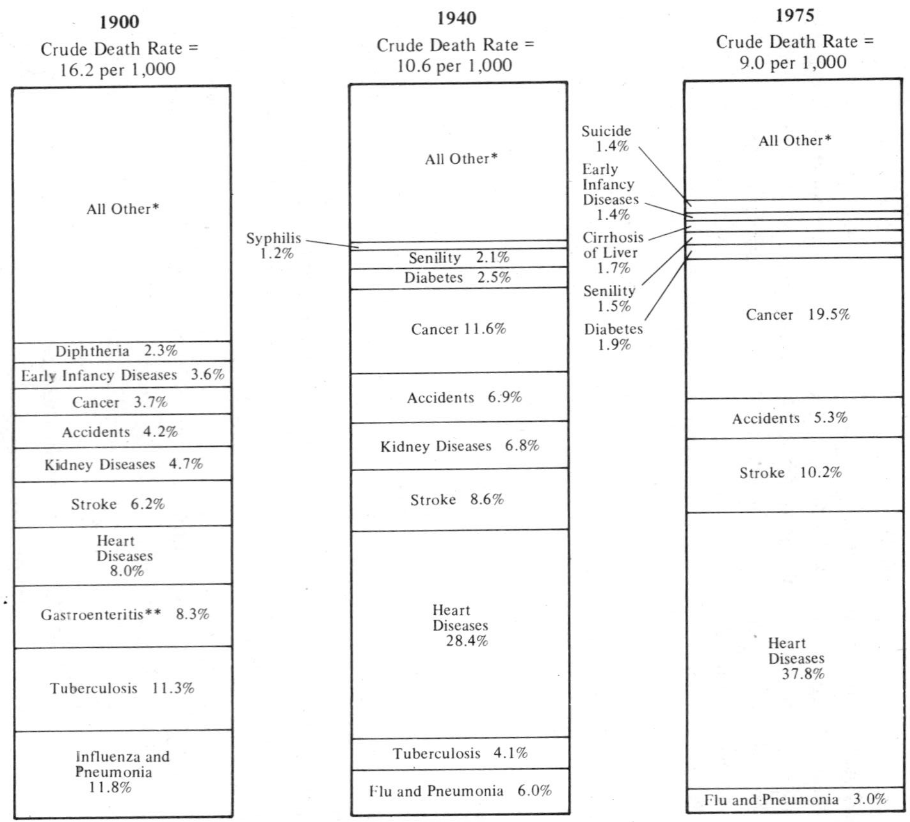

FIGURE 28 Major causes of death in the United States (1900-1975). *No disease in this category represents more than 2 percent of all deaths. **Inflammation of the stomach and intestines. (Courtesy of the Population Reference Bureau, 1337 Connecticut Ave., N.W., Washington, D.C.)

during the first half of the century. In the first two decades of this century, life expectancy at birth, regardless of race or sex, increased 4.6 years. Between 1920 and 1930 the increase was even larger, 5.6 years. Improvement during the next decade was smaller, 3.2 years by 1940; but larger by 1950, 5.3 years.

As with mortality rates, life expectancy showed relatively small improvement between 1950 and 1975. Similarly, there are significant differences in life expectancy at birth between male and female and between white and nonwhite Americans. Both white and nonwhite females can expect to live longer than white and nonwhite males. Finally, whites of both sexes can expect to live longer than nonwhites.

Male life expectancy in the United States is not outstanding compared with other countries. Male life expectancies in the Scandinavian countries (apart from Finland) surpass those in this country. As of 1975, U.S. male life expectancies also lagged behind those of other parts of Europe—including the Netherlands, Italy, Spain, Switzerland, England, Wales, Ireland, East Germany, Bulgaria, and France—and also of some countries outside Europe—Japan, Israel, Canada, New Zealand, and Cuba. However, the incidence of cardiovascular disease and its main known risk factors are declining in the United States, but are stable or increasing in most other industrialized nations. This almost unique trend promises to improve the relative standing of the United States in regard to male life expectancy.

American female life expectancies compare more favorably with those of other countries. The life expectancy for female infants born in this country in 1975 was surpassed in only eight other nations: Norway, Sweden, the Netherlands, France, Canada, Japan, Denmark, and Switzerland, in that order. One reason the United States lags behind other nations is the lower life expectancies among nonwhite Americans.

CHANGING PATTERNS OF ILLNESS

Since 1900, the pattern of fatal illnesses in this country has changed significantly (Figure 28). While this may reflect changes in the way we live, other factors can affect the ranking of specific diseases as causes of death. For example, if one disease is reduced in incidence, the next in line rises in order, without necessarily rising in incidence. Age is also a factor. If fewer people die of pneumonia, more will live to die of cancer.

Most noteworthy has been the rise of arteriosclerotic heart disease and cancer. In 1900, the leading causes of death in the United States were pneumonia, influenza, and tuberculosis—all infectious diseases. By 1940, heart disease had become the leading cause of death, with cancer second.

Other changes in the pattern during this century are also significant. Tuberculosis, once one of the most dreaded diseases and the seventh-ranking cause of death in 1940, was no longer among the 10 most important diseases by 1960. Two other leaders in 1900—diarrhea and other intestinal infections and diphtheria—were no longer among the 10 leading causes of death by 1940. Nephritis (a kidney disease), the fifth leading cause of death in 1940, had disappeared from the top 10 by 1960.[2]

However, two other causes of death—strokes and accidents—have maintained a fairly steady position among the top 10 throughout the century, although the incidence of death from strokes is decreasing. Strokes were the fifth leading cause of death in 1900 and the third in 1940, 1960, and 1970. In 1900, accidents as a single category were the seventh leading cause of death; in 1940, when motor vehicle accidents were listed separately from other accidents, the rates for the two combined made them fourth, the position maintained by accidents of all types through 1970.

Finally, certain diseases only recently appeared among the leading causes of death. These include diabetes mellitus, eighth in 1940 and seventh in 1970; arteriosclerosis, seventh in 1960 and eighth in 1970; cirrhosis of the liver, tenth in 1960 and ninth in 1970; and bronchitis, emphysema, and asthma, tenth in 1970.

In very general terms, then, the pattern among the 10 leading causes of death in the United States during the first eight decades of the twentieth century has been marked by the ascendancy of cardiovascular disease (heart disease, strokes, arteriosclerosis) and cancer, the upsurge of diabetes, cirrhosis, emphysema, and asthma, and a substantial decline in infectious diseases. The rise of cardiovascular disease and cancer to their present positions has been a long one. Even in 1900, diseases of the heart were the fourth leading cause of death in this country and cancer eighth. Moreover, while cardiovascular disease is the leading cause of death, it is now declining; this development will be considered later in the chapter. And while infectious diseases have declined in incidence, they have not disappeared from the top 10. As late as 1970, influenza and pneumonia together were still the fifth leading cause of American deaths.

The point is that the patterns of serious illness are far from immutable. Medical attacks on the principal health problems of the past have been remarkably successful. There is no inherent reason why the incidence of such current afflictions as heart disease and cancer cannot be diminished through sound basic research and the application of the resulting knowledge. For example, advances in the treatment of cardiovascular disease, combined with changing public behavior with respect to risk factors, have contributed to the average decline of 2 percent a year in the overall cardiovascular mortality rate for American men between the ages of 55 and 64 during the past decade. There are indications that this trend is extending into older age groups.

BURDEN OF ILLNESS

The burden of any illness depends on both its clinical and its economic impact.[3,4] Clinical impact is determined by how many people suffer and die; in clinical terms, heart disease or cancer are clearly more important than allergies. Economic impact has two components: the cost of providing diagnosis, treatment, and care; and the productive work lost through illness. Respiratory diseases have a particularly heavy economic impact. Ranging from pneumonia and flu to the common cold and sinusitis, they occasion more visits to doctors and more days lost from work than any other illness.

While we have long known in general terms the total cost of providing medical care for serious diseases, little effort has been made until recently to distinguish the clinical importance of diseases from their economic importance. The burden of illness is most commonly measured in mortality rates, which describe only clinical impact. In addition, mortality rates, life expectancies, and statistics on the incidence (number of new cases per unit of time, usually a year) and prevalence (number of cases in the population at a given time) of life-threatening illnesses tell us nothing of the cost and clinical impact about such common but usually nonfatal problems as dental caries, arthritis, hay fever, the common cold, blindness, deafness, schizophrenia, ulcers, and others. We must also understand the cost of these problems.

The National Center for Health Statistics, of the Department of Health, Education, and Welfare, has been paying increasing attention to standardized assessments of illness in terms of both clinical and economic burden. Specific annual indices developed for this purpose include potential years of life lost (age at death compared with average life expectancy), inpatient (hospital) days, outpatient (primary-care) visits, and work-lost days associated with various categories of illness. Such new indices of the burden of illness demonstrate the heavy impact of respiratory disease. Similarly, mental illness and emotional distress are high on several indices of burden, although they are not prominent in mortality.

The next decade is likely to see considerable improvement in the reliability of clinical and economic measures of burden of illness. These measures can help to make more rational allocations of resources for research, education, and services.

CARDIOVASCULAR DISEASES

As infectious diseases came under increasing control during the middle third of the twentieth century, diseases of the heart and blood vessels became more prominent. These cardiovascular diseases are now the leading cause of death in the United States, accounting for more than half of all deaths in 1976. They are also responsible for a heavy burden of illness and economic loss, ranking first as a cause of limited activity, Social Security disability, and hospitalization.

Cardiovascular disease and the availability of research tools have stimulated basic research and clinical investigation into the cardiovascular system and its disorders, including congenital heart disease, rheumatic heart disease, hypertension, stroke, and coronary artery disease.[5] New screening, diagnostic, and monitoring techniques have emerged, along with effective drugs, procedures for repair or replacement of diseased blood vessels, surgical repair of the heart, antibiotics for the prevention of rheumatic heart disease, and public education programs on cardiovascular risk factors and their management.

Taken together, such advances have contributed to the recent decline in mortality from cardiovascular diseases. While they are still the number one killer in this country, the rates of death from cardiovascular diseases have fallen more than 30 percent since 1950, with the decline most rapid in the past 10 years. Cardiovascular mortality has diminished for men and women of all ages and races. Between 1970 and 1976, the age-adjusted decrease was 22 percent for stroke and 16 percent for coronary artery disease. In 1975, deaths from all cardiovascular diseases fell below 1 million for the first time since 1963.

Great gains have been made in understanding the causes and treatment of these diseases, and promising medical and technological developments will be pursued for further gains. However, we know that many cardiovascular diseases are "silent." They develop slowly for many years, perhaps even from childhood, until manifested as an acute episode—a heart attack or stroke. In the long run, we need early warning signals and effective preventive measures for these diseases.

CONGENITAL HEART DISEASE

About 25,000 infants with congenital heart defects are born each year; 3,000 die before their first birthday. Each year about 6,000 persons of all ages die of congenital heart disease, a 38 percent decline in mortality since 1948. Congenital heart problems can result from either genetic or environmental factors, the latter including a preventable disease, German measles.

Until quite recently, babies born with deformed hearts were largely beyond the reach of effective treatment. But developments in biomedical engineering, physiology, and surgery, from 1950 onward, made effective operations possible. The heart–lung machine, cooling and other life-support techniques, artificial heart valves, and improvements in anesthesia all contributed to this growing competence. Moreover, sensitive diagnostic procedures gave clear guidance for corrective surgery. Follow-up studies indicate that such surgery has become steadily more effective while the risks have diminished.[6]

Nonsurgical techniques are being explored for conditions previously amenable only to surgery. One type of congenital heart defect may be modified in newborns (simplifying later surgery) by a medication that promotes normal development of the heart. This therapy works by inhibiting a prostaglandin, one of a family of biologically active organic acids that occur naturally in the body. The biochemistry of the prostaglandins has been elucidated in the past two decades, and they are emerging as a family of compounds with great functional significance for human health.

RHEUMATIC HEART DISEASE

Painstaking clinical research has clarified the sequence of events in which a childhood streptococcal infection, if not treated, may be followed shortly by rheumatic fever, which, in turn, can damage the heart valves. One mechanism involves immune reactions to streptococcal antigens. When the underlying processes are more deeply understood, another opportunity for preventive medicine may appear.

Rheumatic heart disease was a significant source of disability and death among young adults in the first half of this century. While it is far less prevalent and rarely fatal today, there are still 100,000 new cases each year, and more than 1 1/2 million adults have rheumatic heart disease. Penicillin is administered to patients who have had rheumatic fever to block recurrence of streptococcal infection; the antibiotic may be given for years or even a lifetime. This practice has sharply reduced disability and death from the disease. Also heart valves damaged by rheumatic fever can be replaced through a remarkable surgical development.

Recent studies have shown that well-organized health programs for children with rheumatic heart disease can decrease days of illness and hospital time, and prevent later complications of childhood diseases such as streptococcal infections. In one such program, prompt and appropriate use of antibiotics reduced the incidence of rheumatic fever by 60 percent.

HYPERTENSION

Hypertension, or high blood pressure, develops insidiously over many years before damage becomes apparent in such forms as stroke, heart attack, heart failure, or kidney failure. It affects an estimated 35 million adults.

Basic research in physiology, biochemistry, pharmacology, genetics, and behavior—applied to the renal, cardiovascular, endocrine, and nervous systems—has increased our knowledge of hypertension. We have learned much about its clinical signs, consequences, prognosis, and treatment. However, in more than 80 percent of the cases we still don't know its cause and call it "essential" hypertension. We need to know more, for example, of the role of dietary factors such as weight and salt intake. Moreover, hypertension is more frequent among blacks than among whites in this country, and we do not know why. Elucidation of the specific genetic and environmental factors implicated in this major public health problem remains a major scientific challenge.

Drugs for Hypertension

Drugs now available for high blood pressure can greatly reduce the risk of stroke and kidney and heart failure. However, long-term follow-up studies indicate that diuretics, the drugs used most often to control blood pressure, may also raise cholesterol and sugar levels in the blood of most individuals. Thus, while the diuretics are undoubtedly effective in lowering blood pressure and reducing its complications, especially strokes, they may increase the risk of other disorders. This preliminary finding deserves further investigation. It also demonstrates the value of long-term follow-up in clinical research generally and in clinical pharmacology specifically.[7]

The awareness of the limitations of diuretics provides a powerful stimulus for creating and testing new drugs, especially in the light of advancing knowledge pertinent to control of blood pressure. One active and promising line of inquiry involves drugs that diminish the activity of the sympathetic nervous system. This part of the nervous system can constrict the small arteries and thereby elevate blood pressure. It can be "slowed down" either by drugs that act directly on the sympathetic nervous system throughout the body or by drugs that act on its central controls in the brain. Some drugs that inhibit the sympathetic nervous system are now in regular clinical use; others are being tested experimentally. Improved drug therapy for hypertension should become available in the next five years. It will take longer to determine exactly which therapeutic regimens most effectively prevent such complications as stroke or heart failure.

Biobehavioral Approaches

Recent work on behavioral approaches to controlling blood pressure has also been promising. The Stanford three-community study, discussed below, showed that people can learn to reduce their consumption of salt, which the study found to be the best predictor of blood-pressure change among the several behavioral factors, such as weight reduction and relaxation, that were investigated. Use of biofeedback learning techniques to control blood pressure is also receiving attention.

Unknowns in Hypertension

The risk from high blood pressure occurs over a broad range of pressures—there is no sharp cutoff between "normal" and "high." More information is needed about the effects of gray-zone pressures—between the clearly elevated and those accepted as normal. Primary preventive measures are needed to stop or slow the frequent increase in blood pressure with aging. There is no reason to assume that such an increase is inevitable. Evaluation of the benefits of medication in elderly and young people is needed, as is evaluation of the benefits and costs of medication of mild hypertension. More must be known about the basic processes of hypertension to improve both pharmacological and behavioral prevention of the disease and its complications.

CORONARY ARTERY DISEASE

Coronary artery disease involves damage to blood vessels leading to the heart muscle and subsequent angina, myocardial infarction or heart attack (death of heart-muscle cells), heart failure, or arrhythmia (irregularity) of heartbeat. Coronary artery disease usually can be traced to arteriosclerosis, narrowing of the channels of the blood vessels that restricts blood flow. Coronary artery disease, usually manifested as a heart attack, accounts for 650,000 deaths annually, or two-thirds of the total deaths from cardiovascular disease.

The prevalence of coronary artery disease has stimulated much clinical innovation, apparent especially in the nationwide development of hospital coronary-care units. In such units, sophisticated electronic monitoring permits early detection of heart-rhythm abnormalities; also, drug therapies and modern equipment for electrical stimulation and resuscitation are immediately available. Artificial pacemakers are used for immediate or long-term control of heart rhythm. Techniques for assessing the extent and location of damage to heart muscle have been improved. Surgical bypass of seriously diseased coronary blood vessels has become common and in certain cases can relieve pain.

Despite the technical sophistication and intrinsic plausibility of these approaches, we are still not certain whether extended treatment in a coronary-care unit gives better overall results than careful treatment at home after initial hospital treatment. Nor do we know whether coronary bypass surgery has any effect in addition to the highly valuable one of relieving pain. In view of the gravity and cost of these procedures, we must learn to determine more precisely which coronary disease patients are appropriate candidates for given therapeutic approaches.

PREVENTION OF CARDIOVASCULAR DISEASE

Prevention of cardiovascular disease is being increasingly emphasized. The contribution of basic research to prevention probably will increase over the next decade, particularly in regard to hypertension and arteriosclerosis, the two main causes of damage to the heart and blood vessels.

More than half of the Americans with high blood pressure are unaware of their condition. And of those who are aware, about half simply do not adhere to the prescribed treatment. It is important to learn, through health-services research, why patients do not take medication that is good for their health. Such research may produce effective ways to help patients modify their behavior, not only in regard to medication, but also in regard to such health-related habits as eating, smoking, and drinking.[8]

We already know that aside from the cost of drugs, failure to take medication depends partly on such factors as waiting time for the doctor and at the clinic; inadequate follow-up by doctors and clinics; overcomplicated and confusing dosage schedules; unsatisfactory doctor-patient relationships; and inadequate explanation of possible side effects. The last problem is particularly significant to patients who had no symptoms prior to medication. Recently, there has been some careful work on the utility of nonphysicians who can counsel on medication problems. Such counselors can markedly enhance adherence to therapeutic regimens.

Public Awareness

The National High Blood Pressure Education Program was initiated in 1972 as a result of survey research in the early 1970's that showed the low public awareness of high blood pressure and its consequences.[9] A far-reaching collaborative effort between the National Institutes of Health and various private organizations, it involves screening of the public for high blood pressure and careful follow-up. From start to present, the program has sharply raised the numbers of people using effective treatment for hypertension. A survey in 1977 of 100,000 residents of Chicago showed that 9 of 10 people with elevated blood pressure were aware of their condition and that about 60 percent of hypertensives were being effectively treated. This reflects a striking increase in recent years in both awareness and treatment.

Voluntary health organizations, health professional organizations, industry, consumer groups, and the federal government have worked together effectively in this educational program. Since hypertension is especially prevalent among blacks, a special effort in minority communities has been mounted in recent years. In view of the heavy burden of illness in these communities, the lessons learned from the National High Blood Pressure Education Program may well be useful.

CARDIOVASCULAR RISK FACTORS

Sudden death is the first symptom in at least one-fourth of those with coronary disease. This fact is a powerful stimulus to research on primary prevention and on noninvasive diagnosis before overt symptoms appear. Attention is turning increasingly to the detection of cardiovascular risk factors and means of reducing them before the disease becomes clinically apparent.

Cardiovascular risk factors include age, male sex, elevated blood pressure, cigarette smoking, elevated plasma cholesterol, elevated blood glucose, obesity, sedentary way of life, water hardness, family history of heart disease before age 65, personality type, and severe stress.[10,11] Not all of these factors have been investigated with equal thoroughness, but each has some predictive power for coronary disease.[12]

Epidemiologists have done large-scale prospective studies to delineate the biological and behavioral characteristics of individuals who are likely to develop cardiovascular disease; here considering hypertension and coronary heart disease separately and also their relationship with each other. These careful, systematic studies have been conducted with populations in different parts of the United States and in other countries.

Cholesterol

Particular attention has been given to cholesterol by the public and by clinical and basic researchers. Cholesterol in the plasma is part of several more complex structures, lipoproteins, which can be separated on the basis of their density. This density analysis has improved the predictive value of cholesterol as a risk factor: Either elevated total-plasma cholesterol or elevated low-density-lipoprotein cholesterol, which correlate closely with each other, tends to be harmful; elevated high-density-lipoprotein cholesterol, which does not correlate well with elevated total-plasma cholesterol, appears to protect against coronary heart disease.

A recent report of a World Health Organization clinical study showed that a drug (clofibrate) lowered lipids such as cholesterol in the blood, and this in turn was linked to a decrease in heart attacks. However, the status of the drug in clinical therapy is not yet clear.

Genetic Effects

A relationship between dietary saturated fats, plasma cholesterol levels, and arteriosclerosis has been demonstrated by many laboratory experiments in a variety of animal species. These relationships have been confirmed in humans: The nature and extent of fat in the human diet has a bearing on arteriosclerosis. But not everyone who consistently eats a diet high in saturated fats develops serious disease.

Genetic differences affecting individual responses to different kinds of dietary fats apparently have an important bearing on the extent and severity of the disease. Such genetic influences are illustrated by a recent discovery. High concentration of fat in human blood, or hyperlipidemia, is strongly influenced by certain genes. Indeed, three distinct single-gene disorders can predispose to hyperlipidemia.[13] One is called familiar hypercholesteremia, the second hereditary hypertriglyceridemia, and the third familiar "combined" hyperlipidemia. One or another of these conditions is found in a sizable minority of patients under age 60 who have heart attacks. Evidently, individuals with these particular genetic disorders are more vulnerable to a high-fat diet.

Modifying Risk Factors

Interest is high in modifying cardiovascular risk factors, especially among those who appear to be especially susceptible genetically. The task is not easy and will require significant educational and social changes. However, there are encouraging indications that better health is possible by altering firmly established patterns of behavior, such as smoking, exercise, diet, working, and coping with stress.

Significant changes in risk factors have occurred in the United States in the past 15 years, at least in part because of public concern with health. Consumption of tobacco products, milk and cream, butter, eggs, and animal fats have all declined among adults. The decline is greater in the more educated segments of the population and is especially striking among health professionals. These declines coincide with an accelerating decline in cardiovascular mortality. While the relationship may be a coincidence, it should be investigated. One point deserves emphasis: The link between diet and cardiovascular disease is suggestive,[14] but present evidence leaves little doubt that stopping cigarette smoking will reduce the risk of cardiovascular disease, and of lung cancer.

The Stanford Program The Stanford Heart Disease Prevention Program,[15,16] a multifaceted research effort to learn how to combat heart disease, has shown that it is possible to decrease risk factors for cardiovascular disease through health education using the media. The risk factors addressed in this two-year program in the mid-1970's were cigarette smoking, high plasma cholesterol concentrations, and high blood pressure. Three communities were involved: one with only a mass-media education program, one with a mass-media program supplemented by face-to-face counseling for high-risk individuals, and one control community with no special programs. All three communities were surveyed annually. Key aspects of the experiment were:

- The mass-media materials were devised to teach specific skills—for example, preparation of a palatable low-fat diet—and to provide information on health and motivate people to use it.
- Behavioral scientists advised on mass-media approaches and face-to-face instruction.
- The campaign was suited to the intended audience; for example, it took account of the fact that part of the audience was Spanish-speaking.

Among high-risk participants who received the annual survey, mass-media education, and intensive instruction, the overall risk of cardiovascular disease was reduced 30 percent. Almost all of the reduction was achieved in the first year of the program and sustained through the second year. High-risk individuals who received only the annual survey and mass-media education reduced their risk by 8 to 10 percent the first year and by 25 percent after two years. High-risk participants in the community exposed to the survey alone did not appreciably change their risk of heart disease.

The Finnish Program A larger study of the effectiveness of reducing risk factors has been in progress for six years in North Karelia, Finland.[17] The three major components of the program are: health education through community resources, including local newspapers and radio; hypertension screening with intensive group-health education of high-risk individuals; and early diagnosis, treatment, and rehabilitation involving existing health and social services. Education and medical measures are supplemented by a law forbidding smoking in public buildings and on public transport and by the cooperation of the local dairy and food industries in reducing fat in popular foods.

As a result of these efforts, the annual rates of incidence of acute myocardial infarction, which had been increasing in Finland for many years, have begun to decline in North Karelia. There also has been an apparent decrease in the severity of heart attacks. In addition, the annual incidence of strokes had fallen by the third year of the program.

Community-Based Prevention The experiences in California and Finland will stimulate research on community-based prevention in the next five years and should be useful in planning new programs. Such investigations must explore ways of eliciting the interest and cooperation of both the public and private sectors in facilitating health-promoting behavior. If the findings thus far are duplicated in other communities, a sustained change in the methods of health education is likely.

CANCER AND RELATED PROBLEMS

Although cancer is not the most common serious disease, it is certainly the most feared. It is therefore not surprising that the National Cancer Institute (NCI) is the largest health-research organization in this country. During the 1970's NCI expanded very rapidly under the auspices of the "War on Cancer."

New leadership is now examining this enormous effort in order to achieve a balanced program sensitive to scientific opportunities, clinical needs, and social concerns. This constructive reassessment will be an important feature of science policy in the next few years.

Since cancer is also considered in other chapters, this section will be relatively brief and clinically oriented, with special attention to treatment and prevention.

NATURE OF CANCER

Cancer has many forms. All involve a defect that permits the unrestrained multiplication of cells to yield offspring cells whose growth is similarly unrestrained.[18] Some believe that the basic cell defect is the same for all cancers, but that it may be triggered in diverse ways to produce unrestrained growth.

Cancer is no longer viewed as an inevitable concomitant of aging. Environmental factors acting on genetic predispositions are now considered important in the origin of cancer. In this context, environmental factors are defined broadly and include, for example, diet and cigarette smoking. The growing recognition of environmental influences is based in part on large geographic variations in the incidence of specific cancers, linked with evidence that migrant populations tend to shift to the patterns of incidence of their new region.

Two lines of inquiry, which may have long-term clinical significance, have not yet been sufficiently resolved to assist in diagnoses, prevention, or therapy. These involve immune responses in carcinogenesis and the role of viruses in human cancer. One hypothesis is that carcinogenesis, or production of a cancer, occurs normally throughout life (perhaps in part owing to background radiation), but that it is kept under control by the immune system. Viruses are known to cause some cancers in animals—leukemia in cats is one example. But despite considerable research, it has not been proved that viruses cause cancer in humans. In the long run, research in both immunology and virology will clarify mechanisms of carcinogenic transformation and provide clues to how this transformation might be inhibited by pharmacological or immunological means.

Thresholds and Synergism

An important question is the level of exposure to carcinogenic substances that initiates carcinogenesis (see p. 449). Most toxic substances fed to test animals produce no observable effect or response below a characteristic threshold dose. However, a threshold dose for carcinogenesis has not been demonstrated for any substance; if threshold doses exist, they cannot be detected by the available toxicological methods. If carcinogens do not in fact have threshold doses, exposure to a very small amount of a carcinogen could initiate carcinogenesis, although the probability that it would do so would be slight; known carcinogens must be fed to 50 or 100 test animals in quite large amounts for months to induce cancer in only a few of them. Quantitative dose–response data are available for some agents that affect the incidence of cancer in humans. These agents include ionizing radiation and cigarettes. For each, the data are consistent with the absence of a threshold dose; if one exists, it is undetectable by current toxicological methods.

Aside from the probable absence of a threshold for carcinogens, synergism must also be considered; that is, the combined effect of two agents may be far greater than the sum of their individual effects. An example of this compounding of risk is the combined effect of cigarette smoking and exposure to asbestos. Smoking alone increases the risk of lung cancer about 10-fold—the actual increase depends on the number of cigarettes smoked. For asbestos workers who do not smoke, the risk of lung cancer is 7 times that for nonsmoking, nonasbestos workers. However, for an asbestos worker who also smokes, the risk of lung cancer is more than 12 times that for a nonsmoking, asbestos worker and 90 times that for a nonsmoking, nonasbestos worker.

Thus, current evidence provides no basis for assuming that there is a totally safe level of any carcinogen. In real life, however, it is often impossible to avoid completely a carcinogen. The establishment of socially acceptable exposure levels will require dependable, quantitative risk estimates based on better data than now available. It should also be kept in mind that cancer can develop in the absence of any known external carcinogens.

EPIDEMIOLOGY AND CANCER

Epidemiology has a long history of accomplishment in the understanding and control of infectious diseases through public-health measures. Today it is being applied increasingly to cancer, cardiovascular disease, and mental illness. The epidemiologist attempts to determine the cause of a disease by comparing possible causes statistically with the incidence and distribution of the disease in a population until a correlation emerges. With cancer the task is difficult for several reasons: physical, chemical, biological, and social environments are very complex; many presumed carcinogens are present; exposure conditions are highly variable; and latent periods after exposure are very long. However, high exposure simplifies the problem. Epidemiological studies of groups of people exposed to unusually high doses, as in an industrial setting, have led to the identification of several dozen human carcinogens.

Large-scale epidemiological research will be required to correlate the changing patterns of environmental agents with parallel changes in the incidence of cancer. Special attention should be devoted to those cancers that occur most frequently and with large variations in geographic incidence (for example, gastrointestinal cancers).

Although some man-made chemicals have been identified as potential carcinogens, the firmly established incidence of cancer from these compounds accounts for only a tiny fraction of all cancers.[19] Cigarette smoking is the one environmental factor for which firm data demonstrate a strong association; occupational exposures are also an important source of environmental carcinogenesis. Other environmental factors, including those of natural origin, that are probably responsible for many and perhaps most cancers remain unknown. Research on these problems will intensify during the next five years.

DETECTION OF CANCER

Skin cancers are easily detected and therefore quickly treated. Deep-lying cancers, such as those in the lung, typically are detected late, and treatment is less effective. Special effort is being devoted to earlier detection of cancer in various deep-lying locations. One of the most successful screening techniques has been the "Pap smear," in which a few cells from the uterine cervix are examined. This test has contributed to a significant decrease in the death rate from cancer of the cervix among American women. The mortality rate dropped over 50 percent between 1948 and 1971, falling from 38 to 15 per 100,000 for white women and from 75 to 31 per 100,000 for nonwhite women.

The most efficient timing of screening tests is currently under study. Timing is important, for example, in mammography, a radiologic diagnosis for breast cancer. This disease affects nearly 1 woman in 13 and kills 30,000 annually. Regular screening of women by mammography during childbearing years probably exposes them to more radiation than the yield of discovered cancer is worth—the risks outweigh the benefits. But, the risks of mammography after age 50 may well be outweighed by the benefits of early detection of breast cancer. Such concern has led to technological improvement: dosage is about half what it was at first.

The new fiber-optic endoscope—an instrument for looking at the inside of a hollow organ—is much more flexible than instruments previously available, and therefore facilitates exploration for cancer of the lung and colon—two of the most common cancers. Similarly, the relatively noninvasive radiologic technique of computed tomographic scanning can detect tumors early, especially in the head. Other techniques of medical imaging for noninvasive diagnosis are under intensive development.

The immunology of tumors may become clinically useful in the next 5–10 years. Malignant tumors synthesize distinctive antigens, some of which are released into the circulating blood.[20] These antigens now are detectable in the blood only at an advanced stage of cancer. But the immensely sensitive and specific technique called radioimmunoassay is becoming increasingly useful for such analyses.[21] The method combines immunologic techniques with the measurement of substances labeled with radioisotopes. Its use to detect antigens in blood is especially promising for early diagnosis of gastrointestinal cancer.

TREATMENT OF CANCER

Surgical removal is still the best treatment for most cancers. If all the malignant cells can be excised, the cancer may not recur. Unfortunately, these abnormal cells commonly can invade surrounding tissues and, worse yet, spread through the blood or lymph systems to form new growths at distant locations. If this has occurred even microscopically by the time of initial diagnosis, removing the primary tumor will only buy time. A recurrence elsewhere is likely. Therefore, surgery is often supplemented by radiation therapy or chemotherapy, or both.

Surgery has a sharply focused target, radiation is usually directed at a specified region near the site of the tumor, and chemotherapeutic agents reach cells throughout the body. The well-established combination of surgery and radiation therapy is capable of curing (in terms of five-year survival) about one-third of all cancer patients. This fraction does not include patients with skin cancer, which is an easier problem.

Chemotherapy increasingly is being used after surgery when recurrence is especially likely. It works best when relatively few cancer cells have spread through the body. For the most part, chemotherapy is relatively new, and its long-term effectiveness is not yet clear. The technique does seem to lessen the risk of relapse in one form of bone cancer and may be helpful in some forms of breast cancer. Current research involves combined use of two or more chemotherapeutic agents, each of which has shown anticancer potency, at least in the short run. These chemotherapeutic combinations are also being tested in combination with established and newer forms of radiation therapy. Radiotherapists not only are trying new types of beams but also are combining them with radiosensitizers—chemical agents that selectively enhance the sensitivity of tumor cells to radiation.

A notable success of chemotherapy, particularly when combined with radiation therapy, is the treatment of Hodgkin's disease, which in the past was uniformly fatal; similar gains are being made in the treatment of acute lymphatic leukemia in children.[22] In both cases about half the patients can be cured. While these are certainly encouraging, indeed ground-breaking advances, these cancers account for less than 1.5 percent of all cancers.

PREVENTION OF CANCER

Given the gravity of cancer, the difficulty of early detection, and the limitations of treatment, it is crucial to ask what science can contribute to prevention. The past quarter century has seen unprecedented progress in the life sciences, and medical scientists should examine systematically the implications of these extraordinary advances for prevention of disease in general and of cancer in particular. A significant amount of cancer may be preventable by use of existing knowledge. Much can be accomplished by identifying risk factors and learning to modify exposure to them—as the record abundantly demonstrates. As we learn more and more of the mechanisms that underlie risk factors and their modes of action, systematic efforts can be made to use that knowledge for prevention.[23]

Identifying Carcinogens

A multifaceted approach will be necessary to reduce the cancers influenced strongly by environmental factors. This includes the use and further refinement of methods for identifying carcinogens, including *in vitro* screening tests, tests in mammals, and epidemiological studies. *In vitro* tests in particular warrant rapid development because they cost far less in time and money than do tests in laboratory animals. Chemicals that test positively for carcinogenicity should be seriously suspected of being carcinogenic in humans. However, it is still considered necessary to confirm the results of *in vitro* tests by tests in animals.

Chemicals shown to be carcinogenic in laboratory animals are likely to be carcinogenic for humans; in several well-documented cases, a carcinogen was identified in animal tests before its carcinogenicity for humans became clear. Conversely, with few exceptions, such as arsenic, substances that cause cancer in humans also do so in animals. The correlation is good but not perfect from animals to man and from man to animals. Animal tests, therefore, are very good but not infallible predictors of carcinogenic potential for humans. Also, animal species may vary widely in their sensitivity to a particular carcinogen, so that a quantitative estimate of degree of risk (or potency) can be extrapolated from animals to man only with considerable caution.

The several hundred compounds proven carcinogenic in animals should be viewed as potential carcinogens in man unless proven otherwise. Research on them is well justified; also the populations exposed to suspected carcinogens should be studied to provide better data for assessing risk to humans.

CIGARETTES AND HEALTH

Of all the opportunities for preventing cancer in the foreseeable future, the most important by far lies in cigarette smoking.[24] If research can help to markedly lower the number of cigarette smokers, gains will be made not only in prevention of lung and bladder cancer, but also of serious respiratory diseases and major cardiovascular diseases. Here then is a truly critical problem in health.

Cancer of the lung, the cause of 92,000 deaths annually, was uncommon in the early part of this century. It began to rise sharply around 1935 in men and 1965 in women. Each increase began about 20 years after

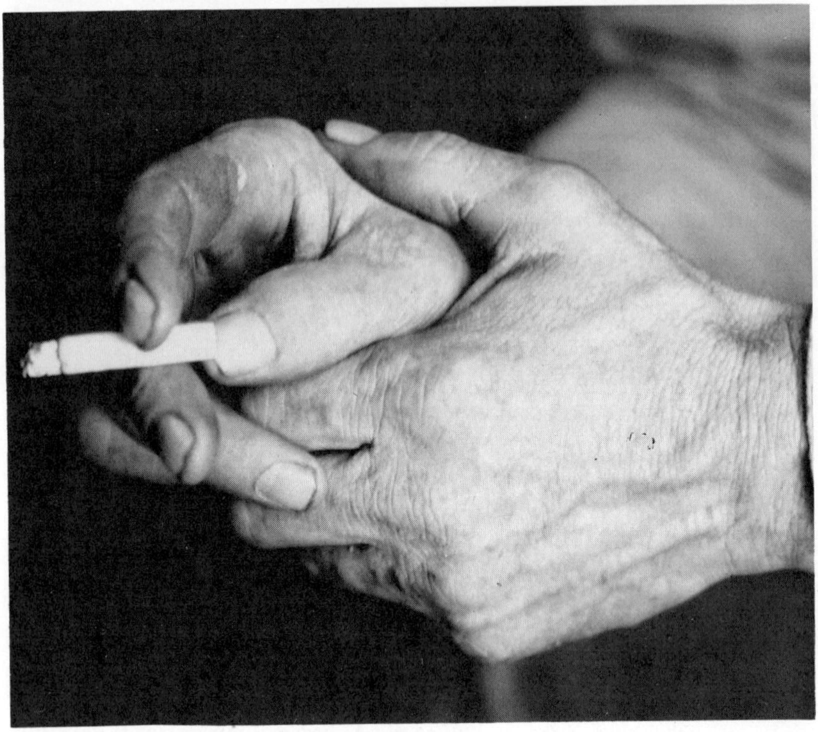

(Bill Gillette, EPA Documerica)

cigarette smoking became widespread among the sex. The increase in deaths from lung and other respiratory cancers has been so large that, if these deaths were excluded, total cancer rates among persons aged 55–64 would drop approximately 40 percent for men and 15 percent for women.

A 20-year epidemiological study of mortality in relation to smoking among British physicians casts light on both the consequences of smoking and the benefits of stopping. The data indicate that between a third and a half of all cigarette smokers die because of their smoking. Deaths in the study were due chiefly to heart disease among middle-aged men, but lung cancer was also a prominent cause. As a whole, the population reduced its cigarette consumption substantially during the 20 years of thorough study. At the same time, the incidence of lung cancer became less common among the population. This and other studies indicate that pathology related to cigarettes is at least partially reversible, even after years of smoking. But here, as elsewhere, prevention is likely to be most beneficial to long-term health.

Children and nonsmoking adults may be exposed to the smoke of others. The unborn infant particularly needs protection, since it may suffer increased risk of death or retarded growth with long-term effects if the mother smokes. Increased risk of respiratory infections occurs among infants whose parents smoke, and these infections may predispose to adult chest disease.

INCIDENCE OF SMOKING

Knowledge of the risks of smoking has resulted in a decline in smoking among adult Americans. More than 30 million smokers have quit since the Surgeon General's report, *Smoking and Health*, was published in 1964. Sharper declines in smoking than in the general population have occurred among doctors, dentists, and pharmacists—professionals who daily observe the long-term effects of cigarettes on patients and customers.

Thirty-five percent of American adults smoke cigarettes. The proportion of teenagers who smoke has been increasing, especially in the 10–15 age-group and above all in females in this age-group; but there are indications of a very recent decline in this trend. Surveys show that Americans know of the relationships between smoking and lung cancer, but not that tobacco-related cardiovascular diseases take a heavier toll.

During the past 20 years, many clinics, techniques, and devices have been developed to help smokers quit. While several methods are effective for a short time, most smokers eventually resume smoking. The prevention of such relapses has been tackled recently as a problem that requires solutions different from those that help smokers first quit the habit, and will likely be given high research priority in the next five years. Techniques based on learning principles show promise for stopping smoking permanently, but are still at an early stage of experimentation.

LOW TAR AND NICOTINE

There has been considerable effort in recent years to devise less hazardous cigarettes. Usually, lower hazard is attributed to "low tar and nicotine" in the inhaled smoke. While the low-tar-and-nicotine approach deserves systematic investigation for both risks and benefits, the facts are far from clear since cigarette smoke contains a variety of toxic substances. For example, carbon monoxide may be much more important in causing cardiovascular disease than in causing cancer.

SMOKING AND ADOLESCENCE

Adolescence is a critical period of biological and psychological change, and also of drastic change in social environment. And it is a time when lifelong behavioral patterns are formed involving cigarette smoking, use of alcohol and other drugs, automobile driving, habits of diet and exercise, and patterns of human relationships. Adolescence has been neglected until recently in biomedical and behavioral research.

The links between puberty and the endocrine hormones can now be studied with high precision as a result of recent advances. These include especially the discovery of the brain hormone that controls the reproductive system in both sexes and the development of radioimmunoassay methods for measuring hormones and their derivatives. Knowledge is needed relating endocrine and bodily changes of puberty to parallel emotional and behavioral changes.

Peer Counseling

The impact of peers, influential adults, and the mass media on adolescents needs careful, systematic study to clarify the determinants of behavior harmful to health. Peer counseling is one promising approach to guidance during adolescence. Such programs rely on the credibility of peers during adolescence in training students to help other students. The approach has been used in junior and senior high schools, and the findings suggest that such programs can be useful for both students and counselors.

Peer counseling is now being tested as a means of discouraging the onset of smoking. The test[25] is a field experiment, based on a model curriculum, in which 16-year-olds teach 12-year-olds how to resist peer pressure to smoke. Hundreds of students in matched experimental and control schools are compared for cigarette-smoking rates at the start of the test and during follow-up periods of up to two years. So far, the results indicate much less smoking among the adolescents in the peer-counseling program than among the controls.[26]

Since smoking rates have been rising more rapidly among early adolescents than in any other segment of the American population, this age group will be a critical focus in the next five years for research aimed at prevention of diseases related to smoking.

MENTAL ILLNESS AND BIOBEHAVIORAL SCIENCES

During the past quarter century, drug therapy for major psychiatric disorders has drastically decreased the numbers of patients in public mental hospitals. Research on drug therapy has led to more specific methods for diagnosis, assessment of severity, and criteria for improvement of mental illness. The new psychopharmacology also drew psychiatric research into modern biological science, stimulating investigation of possible biological mechanisms underlying mental disorders.[27] For the first time, a substantial cadre of scientifically well-trained people emerged to do research on mental illness.

MAJOR MENTAL DISORDERS

Schizophrenia is a profound disorder of behavior, emotional responses, thinking, and perception. About 200,000 schizophrenic patients are now hospitalized in the United States, occupying one-fourth of all hospital beds; and another 400,000 are in outpatient clinics or out of treatment altogether.

Schizophrenia can be disabling. Its symptoms, which usually appear in late adolescence or early adulthood, include altered movements ranging from total immobilization to frenetic and purposeless activity with peculiar mannerisms. Perceptual disorders in schizophrenia often include hallucinations such as hearing voices when no one is present. Disturbances in thinking are common and often lead to distorted concepts, bizarre speech, and grossly illogical beliefs—most vividly expressed in paranoid delusions involving profound and pervasive distrust of others. Emotional expression is sometimes completely absent, sometimes highly inappropriate to the words and actions of the individual. During relapses, most schizophrenics seem incapable of experiencing genuine satisfaction of any kind.

Manic psychosis is another severe mental disorder. Its victims usually are intensely excited and pseudoelated and speak and move very rapidly. They often seem enormously energetic, relentlessly driven, and bold to the point of grandiosity. They tend to be easily distracted, easily irritated, impatient, and rarely able to complete a task. Words and fragmentary ideas may flow so rapidly as to be incoherent.

Psychotic depression is another disabling condition that frequently leads to hospitalization. During the 1970's, about 1,500,000 people have been treated annually in the United States for depression—some in hospitals but most outside. This condition goes far beyond the ordinary sadness of everyday life, even beyond the grief of bereavement. Psychotic depression involves deep and persistent feelings of hopelessness, helplessness, and worthlessness. These feelings are accompanied by severe loss of appetite and weight, disturbed sleep, and loss of interest in familiar activities. Some depressed patients also have delusions—for example, that one has committed hideous crimes or that one's internal organs are disintegrating. Patients think of death and suicide. Indeed, severe depression is the principal context in which suicide is undertaken, although it is associated with other disorders, including schizophrenia. Suicide is the tenth leading cause of death in the United States: about 24,000 suicides are reported each year, and many go unreported.

Treatment of Major Mental Disorders

Before drug therapies, there were few opportunities for treating outpatient mental cases, and most general medical hospitals did not admit severely ill psychiatric patients. Those who were admitted were quickly transferred as a rule to a public mental hospital, usually one under state auspices. Such hospitals were largely custodial institutions, not medical facilities oriented to active treatment. Pessimism prevailed. Each year the number of patients in mental hospitals increased; very few patients were discharged. The patients' living areas typically were crowded and dilapidated. Though some such hospitals were humane and compassionate, others were of very poor quality.

Psychopharmacological Therapy

Drugs called phenothiazines were introduced in the United States in 1955. They quickly proved valuable for schizophrenic patients and moderately helpful for patients suffering from manic psychosis. Inhibitors of the brain enzyme monoamine oxidase (MAO) and the tricyclic antidepressants, two different classes of drugs, were first used in the United States in 1956 and 1957. They were reasonably effective in relieving severely depressed patients. Finally, the use of lithium carbonate to treat manic psychosis was discovered in Australia in the late 1940's and tested extensively in Europe in the 1950's. The compound was studied further in this country and found to be a remarkably effective treatment for manic-depressive illness. But dramatic as the treatment is in most cases, a significant minority of manic-depressive patients do not respond well to it. Research is under way to clarify the mechanism of action of lithium to enable still more effective treatment.

In the climate of optimism that the new drugs fostered, imaginative innovations in psychosocial therapies were undertaken. These gradually led to a set of useful therapies: crisis intervention, brief psychotherapy, group therapy, family therapy, the therapeutic community, and behavior-modification therapies that apply the principles of learning theory elucidated in psychology laboratories.

Effect on Patient Population

The patient population of public mental hospitals in the United States reached its peak of 559,000 in 1955. In 1956, for the first time since records have been kept, the number declined. This population has declined steadily since then despite a further increase in the national population and a steady increase in the admission rate. By 1967, the patient population had fallen to 426,000 and by 1973 to 249,000. If the pre-1956 trend had continued, the number of patients in public mental hospitals in 1979 would be almost a million.

Most of these patients are back in the communities from which they came. Many were prepared for reentry by teaching them new interpersonal and occupational skills and by the use of transitional facilities such as halfway houses or community lodges.

Need for Aftercare

A large proportion of the patients, especially those who have not been chronically hospitalized, are able to go to work, form attachments with family and friends, and take part in community life. Nevertheless, many discharged mental patients would benefit from moderate and continuing outpatient treatment and involvement in a community network of mutual assistance. Some do get such adequate aftercare; many do not. Providing such care is especially difficult in vast, disorganized, poor urban areas, but some encouraging prototypes exist.

In some cases, patients were moved too rapidly out of the large state and county mental hospitals and communities were not able to absorb them constructively. Many chronically ill patients discharged from mental hospitals are living in cheap hotels and deteriorated single-room occupancy dwellings. In these neighborhoods, former patients are often abused by people who take advantage of their helplessness. Sometimes, they come to public attention through violent crimes, either as victim or as perpetrator.

PSYCHOPHARMACOLOGY AS A STIMULUS TO RESEARCH

In addition to benefitting individual patients, psychopharmacological agents also have powerfully stimulated clinical and basic research. Clinical research was needed to check the effectiveness of medications against mental illness. This work led to efforts to define mental disorders and their subcategories more precisely and to the development of more rigorous methods for assessing improvement or worsening of psychiatric symptoms. The work brought into this field of medicine a crucial method of pharmacological research: the random-assignment, double-blind design of experiments. This research method eliminates much of the bias of the placebo effect—spontaneous recovery from symptoms—and the bias of the staff either for or against a new medication, which might influence staff ratings.

Research on brain function has also been stimulated by drug treatments for mental disorders, principally in initial attempts to explain the mechanism of action of psychopharmacological agents. The work brought a generation of young investigators into the emerging science of neurobiology and exposed them to the awesome, fascinating complexity of the human brain.

NEUROTRANSMITTERS AND PSYCHOPHARMACOLOGY

The human brain is composed of as many as 10 billion nerve cells or neurons, whose long processes or axons conduct electrical impulses (see p. 95). The axons connect to other neurons in elaborate circuitry. One axon does not actually touch the next neuron, but rather is connected to it by a synaptic cleft, a microscopic gap chemically bridged by a small molecule that transmits nerve impulses across the gap. When the neuron fires, it releases this molecule, a neurotransmitter, which then diffuses across the gap and attaches to a receptor on the second neuron. There the neurotransmitter, also called a neuroregulator, activates a mechanism that may stimulate, inhibit, or modify the firing of the second neuron. Several neurons—some stimulating, some inhibiting—may be connected to still another neuron by these neurotransmitters. Neurotransmitters include four compounds called biogenic amines: dopamine, norepinephrine, serotonin, and acetylcholine. Psychoactive drugs probably act by changing the functional activity of one or more neurotransmitters.

Two hypotheses of special interest relate neurotransmitter malfunction to severe mental illness. The biogenic-amine hypothesis states that vulnerability to depression is enhanced by a functional deficit in two neurotransmitters, norephinephrine or serotonin or both, in particular parts of the brain, and that a functional excess of these same neurotransmitters at crucial brain sites increases vulnerability to mania. The dopamine hypothesis states that excess activity of the brain neurons using the neurotransmitter dopamine predisposes to schizophrenia.

The Biogenic-Amine Hypothesis

The biogenic-amine hypothesis of depression and mania originated in observations on the effects of two psychoactive drugs, reserpine and iproniazid. Reserpine was used in the 1950's to treat manic psychosis. The drug was also useful for hypertension, but some patients treated chronically with it for hypertension developed a depressive disorder. Similarly, and also in the 1950's, some patients given iproniazid for tuberculosis developed feelings of intense well-being. In 1956, this observation led clinical investigators to treat depressed patients with iproniazid, with encouraging results.

Research soon revealed that reserpine depletes serotonin and norepinephrine in the brain, while iproniazid inhibits MAO, which is partly responsible for inactivating the biogenic amines. When MAO activity is blocked by iproniazid, the level of biogenic amines in the brain increases. So reserpine causes depression and lowers the level of brain biogenic amines, while iproniazid raises the level of biogenic amines and relieves depression. Subsequently, numerous other drugs were developed that also inhibited MAO and, like iproniazid, tended to decrease depressive symptoms.

The findings with reserpine and iproniazid suggested that chemical changes in the brain could alter emotional distress in predictable ways. Two parallel and important research strategies followed. One was to look for new agents that might alter the mood of severely depressed patients; the other was to study the chemical mechanisms by which such substances might alter mood.

The tricyclic antidepressants proved to be even more effective and considerably safer than the inhibitors of MAO in treating depression.[27] The effects of both lithium and the tricyclic antidepressants on brain chemistry are consistent with the biogenic-amine hypothesis. Lithium slows the release of biogenic amines from neurons and speeds the removal of biogenic amines from the synaptic cleft, thus decreasing their functional activity. Tricyclic antidepressants, on the other hand, slow the removal of biogenic amines, presumably increasing their functional activity. The biogenic-amine hypothesis is, therefore, supported by the action of four pharmacological agents: reserpine, iproniazid, lithium, and the tricyclics.

But the tests of the biogenic-amine hypothesis of depression and mania have not been truly decisive. Drugs with effects similar to those of the tricyclic antidepressants on biogenic amines are not effective antidepressants. Moreover, the biogenic-amine hypothesis has not been confirmed by the precursor-loading strategy. The precursors are dietary substances converted to biogenic amines in the brain and the strategy is straightforward: If depression is caused by decreased functional activity of biogenic amines, such activity might be restored by feeding patients large amounts of precursors of biogenic amines. Several of these precursors have produced disappointing results in depressed patients. They have some effect on mood, but are not effective against depression. Also, direct attempts to find altered levels of biogenic amines in patients with mania and depression have produced suggestive but inconclusive results. Thus, the biogenic-amine hypothesis of depression and mania is stimulating and useful, but may well be an oversimplification. Nevertheless, the research has yielded significant new insights into brain function and has improved the quality of treatment.

The Dopamine Hypothesis

The dopamine hypothesis—that excess functional activity of the neurotransmitter dopamine predisposes to schizophrenic symptoms—also originated in both biochemical and clinical studies of psychoactive compounds (see p. 116). Drugs effective against schizophrenia seem to work by blocking the receptor for dopamine in brain neurons since their antipsychotic potency correlates fairly closely with their potency in blocking the dopamine receptor. If dopamine blocking and hence a decrease in dopamine neurotransmission helps to reduce schizophrenic symptoms, increased dopamine activity might cause or worsen schizophrenia. There is some evidence that it does. Some chronic users of the psychoactive drug amphetamine, which increases the activity of circuits in the brain that respond to dopamine, develop a syndrome remarkably like paranoid schizophrenia. Many of these paranoid amphetamine users have no history of mental illness.

Still, the dopamine hypothesis remains unproved. Direct attempts to find increased dopamine activity in schizophrenic patients have been inconclusive. Nonetheless, like the biogenic-amine hypothesis of depression and mania, the dopamine hypothesis of schizophrenia is an important stimulus to clinical and neurochemical research. It has led to the discovery of a number of pharmacological agents now in use that have improved the treatment of schizophrenia.

DRUG SIDE EFFECTS

Drugs that relieve severe mental illness can have important side effects. The antipsychotic medications—those used to treat schizophrenia, for example—are remarkably safe, considering their long-term use. However, a few patients who take them for many years develop involuntary twitches of muscles of the jaws, cheeks, and tongue. The disorder sometimes subsides gradually when the drug is stopped, but it is disturbing and sometimes continues long after the drug is stopped. Recently, it has been discovered that this unfortunate side effect can be alleviated by administering choline, one of the B vitamins.

The antidepressant drugs so far have not shown adverse long-term side effects, but they do have an unfortunate property. Unlike the phenothiazines, the tricyclic antidepressants can be taken in fatal overdoses by patients attempting suicide. The problem is particularly poignant, because, while the tricyclic antidepressants are quite useful for many depressed patients who may kill themselves, the drugs usually require about two weeks to work. The search for safer, more rapidly acting and effective antidepressants, with fewer side effects, will be a dynamic area in the next five years, one relying on expanding knowledge of neurobiology.[28–30] As in the past, the psychopharmacological efforts will be paralleled by work on the psychotherapy of depression.

Abuse of Psychopharmacological Drugs and Alcohol

Psychopharmacological agents have also been abused by some for their effects on mood, thinking, and behavior. On balance, however, these drugs are much less serious in this respect than is alcohol. Recent epidemiological, toxicological, and clinical research has shown that alcohol-related disorders constitute a major portion of the burden of illness. Only now—and quite belatedly—is alcohol abuse becoming a truly major focus for research in the biomedical and behavioral sciences.[31] This trend is likely to accelerate in the next five years.

ANIMAL MODELS

However new drugs are developed, they must be tested in animals before they are given to humans. Tests in animals, besides revealing toxic effects, help investigators to predict the pharmacological actions of drugs in humans. Thus, animal pharmacologists have developed a number of tests in rodents that predict antidepressant and antipsychotic activity of new medications in man. These tests were developed empirically by examining the actions of drugs known to be effective in humans on numerous animals.

A promising but expensive recent approach to animal testing is the development of animal models that more closely approximate human mental illness. Models of behavioral disorders using nonhuman primates are being created and evaluated. Several of them are useful not only in tests of individual animals, but also in studies of nonhuman primates in social groups. For example, separating a mother monkey from her infant produces a monkey model of human depression. Chronic administration of amphetamine to nonhuman primates produces a disorder resembling a simplified form of paranoid behavior.

The principle of a relatively close biological relationship of nonhuman primates to humans that is helpful in this work can also be applied in other contexts. For example, further research on immunization against a form of viral hepatitis and on reproductive biology pertinent to contraception relies heavily on the use of higher primates, especially chimpanzees. However, the availability of primates has been sharply curtailed by countries that traditionally have supplied them, and the United States does not have adequate breeding colonies. This situation could seriously inhibit progress on problems of biobehavior, reproduction, and infectious disease.[32]

AGING AND HEALTH

The number of Americans aged 65 and over has risen from 4 percent of the population in 1900 to 11 percent today. Their personal health-care expenditures in fiscal 1976, when they comprised about 10 percent of the population, were nearly 30 percent of the total for all Americans.

The National Center for Health Statistics (NCHS) has recently projected health trends for the next 25 years. Especially significant for health services is the NCHS projection for the postretirement portion of the population. The population aged 65 years and older grew 76 percent from 1953–78, compared with 38 percent for the entire population. If mortality rates decline as expected, the population aged 65 and older will continue to grow about twice as fast as the entire population during the next quarter century and will be the fastest growing segment of our population.

Chronic disorders tend to increase with age. Their effect on limitation of activity is a measure of health status reported regularly in NCHS Health Interview Survey. The current rates of activity-limitation, applied to the populations projected for 25 years hence, indicate that the number of persons with activity-limitation will rise from 31 million in 1978 to 42–46 million in 2003, an increase of 35–48 percent among all ages. We can reasonably expect a corresponding increase in demand for health services.

The main causes of limited activity today in persons over age 65 are cardiovascular diseases, arthritis, and mental and neurological disorders.[33] Impaired vision and hearing are also especially important among the aged. In recent years, the number of elderly people in mental hospitals has decreased substantially. However, nursing homes—varying greatly in nature and quality of care—are becoming increasingly important in the care of the elderly. A very large increase in the number of nursing-home residents is likely in the next 25 years. The special problems of health care for the elderly, such as transportation to the source of care, will also have to be dealt with.

RESEARCH ON AGING

The maximum life span for humans is about 100 years, but few people live that long because of the interaction of aging and disease. The distinction between aging and the diseases associated with aging is important. Atherosclerosis, cancer, and senile dementia, for example, are not inevitable consequences of aging and can be considered separately from normal aging and each of these conditions investigated in its own right.

Cellular Biology

Research in cellular biology is clarifying the basic processes of aging. Cultivation of human cells *in vitro* provides a model for studying cellular changes in aging, since such cells have a finite life. Using cell fusion, investigators are beginning to determine the extent to which longevity is programmed in the nucleus or in the cytoplasm. This work is an attempt to find the basis for the difference in longevity between cell types. Cells from patients with conditions associated with premature aging may provide important clues to the genetic basis of aging.

Immunological Studies

The immune system, which produces antibodies in response to invading antigens, deteriorates with age; the decline can lead to increased susceptibility to infection and possibly to some tumors. In one study, 80 percent of individuals over age 80 who had no or minimal immune response to specific antigens were dead within two years, as contrasted with 35 percent of those who had a vigorous immune response.

Autoimmunity, or immune responses to one's own tissues, generally increases with age in experimental animals and in humans. Animal studies have shown that increases in autoantibodies and abnormal immunoglobulins, or antibody proteins, can be slowed by manipulating diet and other environmental factors. The potential reversibility of these immunologic changes suggests that research may be able eventually to ameliorate the effects of aging or at least the susceptibility with age to specific diseases.

Neurological Studies

Senile dementia is pathologically distinct from normal aging. This condition, severe in 5 percent and moderate in 10 percent of those 65 and over, is characterized by loss of initiative, decrease in judgment, difficulty in selecting appropriate words, severe loss of recent memory, difficulty in performing calculations, disorientation, and personality deterioration.

Recent research on senile dementia has focused on changes in choline acetyltransferase, an enzyme responsible for production of the neurotransmitter acetylcholine in certain nerve cells. Although the level of the enzyme declines in normal aging, it declines much more in patients with senile dementia (of the important Alzheimer type) than in age-matched controls. Acetylcholine is one of the major neurotransmitters in a part of the brain having a vital role in memory. Receptors for acetylcholine are present in normal quantity even when the enzyme responsible for producing the neurotransmitter is not. It may thus be possible to find a substance that activates the receptors and so alleviates Alzheimer senile dementia. Very recent evidence indicates that feeding extra choline to patients with senile dementia can improve memory.

Other neurologic research has demonstrated age-related changes in functional activity of neurotransmitters. The finding may explain the higher incidence of Parkinsonism in older patients. The ailment is associated with inadequate functioning of brain cells that normally function in response to the neurotransmitter dopamine. The life of rats has been prolonged by feeding them the amino acid L-dopa, the precursor of dopamine. The compound presumably activates pathways in the brain that respond to dopamine and has proved useful in relieving the symptoms of Parkinsonism in humans.

Other studies have suggested that pituitary glands in older people may secrete a compound that diminishes the response of peripheral tissues, such as toes and fingers, to thyroid hormone. Individual differences in secretion of this compound could affect activity and vulnerability to cold. Also, some investigators are seeking a pacemaker for the aging process in the brain's pituitary–endocrine system.

Various hormones and hormone fragments are being studied with respect to brain function, including learning and memory. A new approach to the treatment of senile brain disease involves the use of a hormone fragment that seems to improve memory in certain behavioral tests in animals. This medication, ACTH (4-10), is a fragment of a stress-related human pituitary hormone. It is being studied in patients with defective memory.

Pharmacology and the Aged

Pharmacological treatments currently show some promise for certain ailments of the aged. Drugs that dilate blood vessels may be useful in treating senile patients whose symptoms are due to decreased blood flow to the brain. Such medication is already in clinical use. However, further clinical trials are needed to determine how effective the drugs are and how to identify patients most likely to benefit from them. The same is true of anticoagulant therapy, such as aspirin on a long-term basis, for prevention of stroke.

The clinical pharmacology of elderly patients presents distinctive problems. We are only beginning to understand the changes in enzyme and organ function, the effects of drug interactions, and the difficulties in adhering to a therapeutic regimen because of poor memory and altered eating habits. For practical reasons, most studies are performed on young people, and their applicability to the older population is limited. Research in clinical pharmacology will benefit from a perspective that considers the entire life span.

DEPRESSION IN THE ELDERLY

Depression is common among the elderly, who account for almost one-quarter of all reported suicides. Some patients with the personality changes, thinking, and memory defects of senility seem to improve markedly if treated with the same medications given younger patients suffering from depression. There is no reliable way so far to predict which elderly patients are most likely to benefit from antidepressants. In this respect, as in others, the aged have been neglected both in current services and research.

Of the mental illnesses of the aged, depression is more common than is serious loss of awareness and judgment resulting from pathological changes in the brain. The aged are subject to multiple stresses, including loss of family members, friends, jobs, economic assets, all of which contribute to their depressions and other psychiatric disorders. Social isolation is significant in these difficulties.

SOCIAL EFFECTS

Clinical research has shown that senile, incompetent behaviors of elderly people are often related to social factors. Many persons can function effectively at advanced ages when placed in a network of mutual aid and social support and given meaningful tasks and a basis for self-respect.

Retirement is one of the critical changes of later life. We know little about biological and physiological changes in response to retirement. Preparation for retirement can be studied in field experiments, by systematically comparing different approaches in similar populations and examining their outcomes.

The family remains the major social, economic, and emotional resource for dependent older persons. However, many problems related to dependency require study, especially because they are so pertinent to the individual's ability to function in society. Special attention should be directed to the socially isolated, who make up most of the 20 percent of those people over 65 who will spend some time in a long-term care facility and who constitute a vulnerable subset of the elderly population.

CHANGES IN SLEEP/WAKE MECHANISMS

While the changes in sleep/wake mechanisms with aging are poorly understood, they are associated with many complaints of insomnia among older people.

There is a resultant medical problem because the elderly are particularly vulnerable to the hazards of sleeping medications, since they are more likely to already have the problems associated with those same hazards, such as respiratory disorders and impaired kidney function. Also, they are more likely to suffer from disorders that require the use of other prescription drugs and so are exposed to the risk of toxic interactions among several different medications.

There is little data on the efficacy of sleeping medication for older people. The few available studies show no drug more effective than another. Given the paucity of data on a clinical problem that is truly difficult for so many elderly people, a serious effort to learn more of the physiology and pharmacology of sleep and its disorders in the elderly is indicated in the years ahead.

HEALTH SERVICES FOR OLDER PEOPLE

There is a growing consensus in clinical medicine and public health on two principles regarding health, disease, and adaptation in older people. First, the care of the elderly should be designed to maintain maximum possible functional and social independence. And, second, health care and social services should be provided in a manner that preserves the dignity of the elderly individual and provides opportunities for personal choice.

Although aging is likely to bring some decline, many elderly people function well until shortly before death. For those who are functionally disabled, the rate of decline may be decreased substantially or minimized through early detection and appropriate assistance. To promote independence among the elderly, we must improve our ability to identify environmental, genetic, and social risk factors in the development of functional dependency.

Current practice emphasizes skilled institutional care for the elderly. Many individuals require and benefit from intensive institutional care, but the lack of alternative types of care can result in inappropriate use of high-intensity services. Indeed, studies of older people have demonstrated the enhancement of dependency by unnecessary services. If individuals are blocked from taking responsibility for certain daily tasks, such as preparing meals and personal care, their abilities tend to diminish rapidly.

Public concern for the functionally dependent elderly is warranted for economic as well as humanitarian reasons. It is in the interest of both society and the elderly individual to forestall dependency or to minimize its impact once functional capacity begins to decline. It is important to consider those who are at high risk of becoming dependent, such as the recently widowed, in addition to individuals who already depend on others for care.

We need better ways to compare systematically the different arrangements for minimizing functional dependency in the elderly. We must also seek more effective means of linking medical and social services. And we must examine ways of improving both long-term institutional care (as Sweden has done) and home-care services (as Great Britain has done).

A sustained effort will be required to meet the health needs of the older people of this country. Their rising numbers and special problems offer real challenges in basic science, in clinical investigation, and in health-services research.

GENETIC FACTORS IN DISEASE

An emerging line of inquiry that will become more visible in the next five years is the influence of genetic factors on the response to environmental agents. The combination of biomedical research with family and population studies is likely to bring new understanding to the prevention and treatment of diseases that are so burdensome in industrialized countries—for example, cancer, arteriosclerosis, depression. This work will require identification of genes involved in susceptibility and resistance to these diseases. Greater understanding of the interaction of such genes with specific environmental factors may point the way to sharply focused preventive techniques.[33–37]

A special target of such research is likely to be substances to which a relatively large fraction of the population is genetically susceptible and to which it is widely exposed—substances found in diet, medication, and occupation.

INNOVATION IN HEALTH CARE DELIVERY

The importance that American society attaches to health has provoked in recent decades a continuing debate about how health care should be provided. Likely to intensify in the coming five years, this debate has been accompanied by new arrangements for providing health care. We seem to be moving gradually away from the traditional arrangement, whose dominant characteristics include the delivery of medical services by physicians working independently, greater emphasis on curing than on preventing illness, and the payment of fees after the delivery of services. This mode remains strong, but others are emerging.[38]

Innovation in providing medical services is hardly surprising, given the increase in the nation's population, the intensive development of medical technology, the rise of the medical specialist, the growth of complexity and cost in health care, and the social and technical inventiveness that characterizes American society.

The earliest questions about the effectiveness of the traditional method of delivering health care in the United States were essentially technical ones of a kind familiar to those in many other occupations. For example, exponential growth of knowledge in this century, combined with unprecedented technological advances, have made specialization imperative, particularly in science and the science-based professions. In medicine, specialization has meant a declining role for the general practitioner, the individual physician able to deal with every health problem from measles and heart disease to broken bones and depression.

GROUP PRACTICE

As diagnosis and treatment became the joint effort of two or more physicians, a part of the medical profession concluded that technical efficiency in curing disease could be improved if physicians became partners or members of larger groups. By the early 1960's, physicians were tending increasingly to enter private group practice, with continued emphasis on curative medicine and no change in the standard fee-for-service arrangement. This trend toward multispecialty group practice is stronger and more varied in this country than elsewhere.

Service innovations must be as much concerned with the quality of care as with its cost.[39,40] The various kinds of group arrangements bring to clinical practice the kind of peer review that is effective in judging scientific quality. Peer review is quite formal in the clinical context, but it also includes informal peer contact, particularly among people of different clinical disciplines. Such interactions tend to maximize the diffusion of information, to keep practioners abreast of new developments, and to ensure that procedures are executed in accordance with well-recognized standards based on the best-available evidence.

The basic asset of group practice in organized settings is the pooling of resources, including ideas, techniques, instruments, facilities, and clinical judgments. Within this framework, it is quite possible to have an enduring doctor–patient relationship with a primary-care physician, specialists being brought in when needed. Group ventures can be organized in many ways, and we are now seeing traditional American ingenuity and pluralism being applied to the challenges of the new science base and changing social context of health care.

ACCESS TO HEALTH CARE

In the United States in the middle and late 1960's, public attention shifted to another problem: how to assure that all Americans—regardless of race, age, income, or where they live—have access to at least a decent minimum of health care.[41] This goal has been achieved in almost all advanced countries. In the United States, it became apparent that millions of citizens could not obtain such care—or were severely restricted in their efforts to do so—because of inadequate income, racial discrimination, or residence in areas where doctors were reluctant to live, particularly inner-city neighborhoods and rural areas. The national response to these problems included the creation of the multibillion-dollar Medicare and Medicaid programs, efforts to increase the numbers of minority students in medical schools, publicly funded clinics in poor areas of the large cities, and continuing nationwide effort by certain groups (particularly those representing labor unions, poor people, and minorities) to establish through federal legislation a national health insurance program that would pay most health care bills of all Americans.

More recently, concern over costs, combined with growing appreciation of the value of preventing illness, has led to increasing interest in alternatives to both the traditional private method for delivering health care and to a federally controlled and financed system. Several alternative delivery systems have become reasonably well established in the 1970's. Indeed, there is now a ferment of innovation in the organization and financing of health care delivery systems. These pluralistic developments are likely to ramify in the next five years, and they deserve careful evaluation.

ALTERNATIVE DELIVERY SYSTEMS

Several kinds of alternative delivery systems now exist; the most prevalent is the Health Maintenance Organization, or HMO. As of August 1978, 199 HMO's were providing health care to more than 7 million Americans, or over 1 million more than a year earlier.

Prepaid Group Practice

The predominant type of HMO is the Prepaid Group Practice (PGP) plan. More than 90 percent of those enrolled in HMO's in August 1978 were members of either a PGP plan or of its principal variation. PGP plans have two essential characteristics. The first is that families or individuals enrolled in them agree to pay a set monthly premium to the HMO, whether or not they need medical care. That fee covers all care, including hospitalization. In many cases the monthly premium is paid by an employer or by government; the federal government, for example, pays monthly premiums for Medicare recipients who belong to HMO's. The second characteristic of a PGP plan is that those enrolled are not entirely free to choose their own physicians or other health personnel. In most cases, the HMO's staff physicians provide primary care and also specialized care to the extent that they can. Where additional specialized care is needed, some HMO's require, and others recommend, that the patients consult specific specialists.

Independent Practice Associations

The other principal type of HMO is the Independent Practice Association (IPA); by August 1978, IPA's were providing service to 9.5 percent of the people enrolled in HMO's. In essence, the goal of IPA's is to give their members the advantage of fixed monthly payments and at the same time let them choose their own doctors. Participating physicians practice as they always have, either alone or as members of a private partnership or other group. The IPA customarily reimburses them out of the monthly premiums on a fee-for-service basis. The appeal of the IPA depends in part on how many physicians in a community agree to participate. Giving patients the opportunity to choose their own physicians has little meaning unless most of the community's physicians participate in the association.

Assessments of HMO's

HMO's must be managed carefully and must deliver services efficiently. Since HMO's agree to provide their stated benefits for a fixed payment in advance, they must do so or risk a loss. It is this fixed payment in advance which is largely responsible for a HMO's ability to hold costs down, since it forces the staff to provide care within an annual budgetary limit. In contrast, the traditional method requires payment for costs already incurred, regardless of amount.

Hospitalization costs are the highest of all types of medical care. They were also most prone to inflation between 1972 and 1977, when hospital service charges rose 60.1 percent, or an average of 12 percent a year. HMO's have been able to reduce the cost of health care chiefly by reducing the time that their members spend in hospitals. A comparison of hospital use in seven large states between HMO members and persons with Blue

Cross/Blue Shield coverage, for example, showed that the HMO members spent 782,000 fewer days in the hospital in 1976 and had less surgery.

Such examples are encouraging but not compelling. It is possible that the savings are achieved by reducing medical services to the patient below the level necessary for good health. It is also known that some patients go outside the system for surgery; if such patients were to become a large proportion of HMO members, the cost savings would be more apparent than real.

One substantial systematic study compared the quality of care under an HMO, the Health Insurance Plan of Greater New York, with other forms of delivery. The results indicated that HMO members experienced fewer premature births and fewer newborn deaths than did a New York City sample under the care of private physicians. The observations on premature births are of special interest because these babies are at high risk of mental retardation and long-term disability. Other studies of maternal and child health care in organized settings, not necessarily HMO's, indicate that such approaches can be remarkably effective in reducing death and illness among mothers and children. However, such experiments are complicated by uncertainties about the populations studied. For example, members of HMO's may be habitually more careful of their health than nonmembers. We need a much better understanding of the true effectiveness of HMO's and similar organizations. In fact, few comparisons have been made of the health outcomes of different service-delivery systems. Such comparisons require greater attention in the next 5–10 years.

Scholars and clinicians concerned with innovation in health care are increasingly interested in creating efficient and high-quality systems that can compete effectively in the health care marketplace on a nationwide basis.[42] A variety of useful models are already functioning in several regions. Whether they spread widely will depend on the professional and social context. The following characteristics are desired in such systems:

- A group or association of physicians accepts responsibility for providing comprehensive health services to a defined population.
- People join through free choice among this and other health plans, thus providing a competitive incentive for the health plan to see to it that its personnel give patients judicious, caring, willing service.
- The system provides health maintenance services to the extent that they are effective; this includes such preventive measures as immunizations, prenatal and perinatal care, and counseling on health-protective behavior.
- The health plan has built-in cost constraints. The physicians accept responsibility for the total per capita cost of the care of their patients, and

they use their best judgment of how to give appropriate care. They systematically consider such trade-offs as substituting ambulatory care or home care for more costly inpatient care—for example, to make services more accessible to patients who are poor or who live far away.

- The mix of specialists, facilities, and other resources are matched to the needs of the enrolled population.
- There is built-in quality control through peer review, ease of consultation, and follow-up on patient satisfaction and health outcome. Specialized procedures are performed by specialists whose annual volume of such cases is sufficient to maintain their proficiency.
- There is continuity of health plan membership.
- The health plan keeps a unit medical record for each patient, so that a new doctor can quickly and reliably ascertain what has been and is being done for the patient and so that unnecessary duplicate tests and conflicting prescriptions can be avoided.
- Primary care physicians and specialists work closely together in the same system.
- General and mental health professionals interact freely within the system, so that mentally ill patients are not excluded or stigmatized and so that the behavioral aspects of general health, such as cigarette smoking, can be handled effectively.

How and to what extent these characteristics can be achieved in actual practice remains to be seen.

Other Alternatives

Efforts are being made to extend primary, or "front-line," health care to rural areas and low-income urban areas. One such effort is the National Health Service Corps—young physicians whose medical education is subsidized in return for postgraduate service in underserved areas. Also emerging is a cadre of specially prepared nurses—nurse practitioners—who provide primary care in underserved areas.

Innovations in health insurance, both public and private, include a trend toward providing a reasonable "floor" of health care for those who can afford very little or no health insurance at present, and also increasing coverage among the general population for preventive and ambulatory services.

Efforts also are being made to build close functional links between mental-health and general-health services. Also under way are analyses of organizational changes aimed at more effective health services for children and for the elderly.

HEALTH-SERVICES RESEARCH

The evolution and assessment of useful innovations in health care will depend on continuing advances in biomedical research, including clinical investigation and to an increasing extent on the young field of health-services research.

The need for accurate, dependable information about health services in the United States is becoming increasingly apparent. Management of the personnel, facilities, and technologies that comprise modern health care institutions requires information similar to that needed to manage other complex enterprises. The growing involvement of the public sector in the financing and provision of health services, combined with the need to address increasingly complex issues of resource allocation, requires more knowledge than ever before. The systematic comparison of alternative delivery systems will be a great challenge to the health sciences in the years ahead.

PERSPECTIVES ON HEALTH

At the end of World War II, medical care in the United States was provided largely by physicians working as independent general practitioners and by a modest network of community hospitals with limited diagnostic and therapeutic capabilities. Penicillin—the first wonder drug—had recently become available; the chief diagnostic tools of the practicing physician were the stethoscope and the X-ray. Few could have foreseen broad-spectrum antibiotics, polio immunization, open-heart surgery, organ transplants, and computed tomographic scanning.[43]

The American system of health care today bears only a family resemblance to that cottage industry of 35 years ago. Health professionals use a wide array of drugs and diagnostic and treatment equipment requiring a high level of technical competence. The nation has hundreds of thousands of physicians and other health professionals in many specialties and subspecialties—surgery, internal medicine, gynecology, urology, psychiatry, pediatrics, dentistry, nursing, and many more—as well as thousands of laboratory and equipment technicians, physical therapists, physicians' assistants, and other support personnel. Furthermore, hospital operation is now so complex that a cadre of specifically trained hospital administrators is required. The American system of health care, in short, has become a large and complex enterprise, accounting in 1978 for almost 9 percent of the Gross National Product.

Recent years have brought profound insight into the causal processes of many diseases, including nephritis, endocrine disorders, arthritis, gout, ulcer, and hepatitis; and of such genetic disorders as cystic fibrosis, muscular dystrophy, and lipidoses. Numerous surgical, pharmaceutical, and palliative measures prolong life in some cases and make it tolerable in many others. There have in fact been clear advances on many fronts in the vast field of health.

NEW PROBLEMS IN HEALTH CARE

The recent growth of American medicine has been accompanied by significant new problems: doubts about the wisdom of some of the newer medical techniques, uncertainties about the proper relationship between patient and physician, and continuing national concern about those—such as the poor and the elderly—still unable to obtain the full benefits of current medical knowledge. There is also reason for concern that the needs of children and adolescents, regardless of family income or social status, are not being met adequately. Within the past five years, these problems have been augmented by serious concern with the rising costs of medical care—recently increasing at about twice the general inflation rate. Furthermore, there is a realization that a continued rise in spending on personal health services may not yield commensurate benefits; greater attention to hazards to health in the environment and in individual behavior might be a better investment.

Means of containing health care costs that serve both the public health and our personal and national financial imperatives will not be found easily. In the long run, we may reduce costs by acquiring deeper understanding of human biology and behavior and applying this understanding to prevention of disease.

Research and Health

In principle, all basic research in the life sciences has implications for the prevention of disease.[44] The more we know of the human organism, the more we should be able to affect basic life processes—such as the reproduction of cells or the secretions of glands—so as to prevent disease. Research in the life sciences already has contributed much, but scientists still do not understand most of the bodily mechanisms to a degree that would allow prevention of disease at its primary site. "The Living State" chapter illustrates progress being made in this direction.

IMPROVEMENT OF HEALTH

The health of the public may be improved by preventing disease in various ways: personal health services, environmental measures, and changes in individual behavior.

Personal health services include the procedures and counseling by physicians and their collaborators in private offices, clinics, hospitals, and other organized settings—what is usually called medical care. The most striking example of prevention in medical care is immunization against communicable disease. The effectiveness of immunization against such diseases as poliomyelitis, diphtheria, and measles has been demonstrated beyond all reasonable doubt. A vaccine against pneumococcal pneumonia is now coming into use, and work is proceeding on vaccines against gonorrhea and malaria, although both are some years away from use. Yet millions of Americans remain unimmunized against diseases for which effective vaccines are available.[45] Increasing the immunization rates for polio and measles to the levels of Western Europe is a challenge that can be met in the next five years. Efforts in this direction in the U.S. Public Health Service are gaining momentum.

Environmental measures for improving health are well illustrated by the effectiveness of fluoridated drinking water in reducing dental caries. For example, 12–15 years of communal water fluoridation in five cities reduced the incidence of caries in children 41–70 percent. The benefits for adults are more difficult to assess. However, a comparative study of two communities—one fluoridated, the other not—indicated that people over 45 had at least 25 percent fewer caries with fluoridation.

In recent years, individual behavior has been viewed increasingly as a leading key to good health. The health sciences now are extending their scope to individual behavior patterns and their relationships to the illnesses most prevalent in American society. For example, deaths from lung cancer, coronary heart disease, chronic respiratory disease, and other conditions attributable to cigarette smoking constitute a major part of the burden of illness in this country. Yet surveys indicate that more than one-third of American adults still smoke cigarettes and that, until very recently, an increasing proportion of adolescents aged 10–15, especially girls, were taking up the habit. Smoking in the 10–15 age-group is at a very high level. Should these patterns of behavior persist, the consequences for the health of the American people could be serious indeed.

Building public understanding to the point where Americans take these behavioral risks to health seriously will be a long-term and difficult process. We badly need to find more effective methods of health education in the next 5–10 years. The smoking problem in particular is likely to receive much more public and scientific attention than in the past.

Because health depends on medical care, on environmental conditions, and on individual behavior, we must take a variety of approaches to maintaining and improving the health of the American people. The circumstances call for a broad spectrum of sciences working together over the long term.[46,47]

GOALS FOR THE HEALTH SCIENCES

The burden of illness in this country shows clearly that science still has much to do in reaching a fundamental understanding of how the body functions and why it malfunctions. Achievement of deeper understanding of the body's vital processes will require sustained effort in every part of the health-sciences spectrum, from basic science through clinical investigation to health-services research. One can observe today a revival of interest in older disciplines, such as epidemiology and biostatistics, the strengthening of such relatively new disciplines as biomedical engineering and behavioral sciences, and novel combinations of disciplines for doing research on prevention of disease and delivery of medical services.

As new scientific opportunities arise, their relevance to disease must be determined. Not so long ago the then-emerging discipline of biochemistry was viewed with suspicion by chemists as weak chemistry and by biologists as weak biology. Today, this hybrid discipline plays a central role in biomedical research. Similarly, not so long ago most medical scientists doubted that genetics would have any practical significance for health in the twentieth century. Today, genetics is one of the most dynamic fields in medicine.

At one end of the health-sciences spectrum is basic research, usually initiated by a laboratory investigator with no particular prevention or treatment goal in mind. Next in the spectrum are small-scale clinical investigations designed to determine the significance of new knowledge to humans. Such clinical research often stimulates basic science. Small-scale investigations sometimes are followed by medium-scale experiments, usually involving several hundred people, and then by large-scale, controlled field trials involving thousands of persons. Such large-scale experiments typically seek to delineate the effects of particular risk factors and the results of specific kinds of interventions on human health.

At the other end of the research spectrum is health-services research, which is concerned with organization of medical services and their quality, availability, and cost. Careful and systematic investigation of these matters is a recent development. For that reason, such research still lacks a secure institutional base, either in government or among the nation's leading universities.

The prevention of rheumatic heart disease illustrates the value of linkages across the health-research spectrum. Basic research on infectious diseases led to clinical investigation of the link between streptococcal infection and rheumatic fever. This resulted in clinical work that showed the efficacy of penicillin for preventing rheumatic fever and thus rheumatic heart disease. These interrelated investigations took years to complete, and work is still progressing on some aspects of the problem—for example, health-services research on effective implementation of methods of preventing rheumatic fever in ordinary and, especially, poor communities. But the example is clear: The sequential linkage of basic research, clinical investigation, and health-services research drastically reduced the toll of a serious disease in a few decades.

Assessing New Techniques

An important aspect of health-services research is assessment of the benefits, costs, and risks of new interventions and treatments. These investigations are difficult and sometimes expensive. In some cases, we are still hard pressed to measure benefits; the difficulties are compounded when a long time passes between treatment and a measurable effect on health.

Despite the difficulty and added expense, these assessments of new technology and treatments are crucial because of escalating costs, intensified public scrutiny, and the potential of great harm that accompanies the potential of great benefit. More and more in recent years we have seen new kinds of medical techniques widely used and accepted without a thorough, objective assessment of potential clinical utility or probable impact on the existing health care system. Computed tomographic scanning is one recent example.[6] In the next five years, serious efforts will be made to strengthen clinical systems for determining the proper applications of promising technology.

BROAD APPROACH TO RESEARCH

Each part of the spectrum of research in the health sciences must be fostered if the task of improving health is to be pursued effectively. Overriding emphasis in any one part of the spectrum may act against vigorous cross-stimulation among the sciences and continuing improvement in the health of our population. Communication and cooperation among various parts of the spectrum must be strong, so that knowledge developed through basic research can be translated into practical applications at the clinical level and so that the clinical problems uncovered by medical practitioners can be relayed back to those conducting research. Mechanisms are being developed—for example, by the National Institutes of Health, by the Congressional Office of Technology Assessment, and by professional societies—for arriving at a technical consensus on the potentials of new information for providing specific diagnostic, therapeutic, and preventive advances.

As the national prevalence of various illnesses changes, the scientific efforts needed to reduce illness also change. During the first half of this century, the chief goal of American medicine and public health was to develop ways to control the most serious contagious diseases, which were then the principal causes of death in the United States. As that goal has been approached, with remarkable though not total success, the health sciences have gradually shifted their attention to the types of illness and injury most characteristic of American life today: cardiovascular diseases, cancer, respiratory diseases, mental illness, major vehicle accidents, abuse of alcohol and other drugs, and chronic disorders associated with aging. Most of these problems have biomedical, environmental, and behavioral components, so that multifaceted approaches are needed to make progress against them.

The nation is beginning to test the extent to which the methods of science can be brought to bear on all factors that determine the health of the American people—building on the firm base of biomedical research to include behavioral and environmental influences and the effects of health care per se. Research now extends from the laboratory bench to the patient's bedside to daily community activities. To further improve health, we must have deeper insight into molecules, tissues, organisms, populations, and health care systems.

OUTLOOK

The following outlook section on health of the American people is based on information extracted from the chapter and covers trends anticipated in the near future, approximately five years.

The nation is learning that health depends only partly on the traditional methods of health care. Environment and behavior are also crucial, especially in preventing disease. The scope of research in the health sciences has been expanding steadily and today extends from traditional biomedical investigations to the patient bedside to environmental and behavioral factors in everyday life to the organization of health services in the community.

In recent decades, new kinds of prevention and treatment have been applied successfully to a wide range of disorders, including infectious diseases, vitamin deficiencies, damaged organs, mental illness, cancer, and cardiovascular diseases. But a heavy burden of illness remains.

BURDEN OF ILLNESS

The burden of illness can be assessed both clinically and economically; clinically—how many people suffer and die—cardiovascular disease and cancer predominate; economically, respiratory diseases have particularly heavy impact in terms of work days lost and other indices.

Mental illness, although not a leading cause of death, takes a heavy toll in suffering, disability, and economic costs. Diabetes and arthritis also are large components of the national burden of illness. Recent research has shown clearly that alcohol-related disorders comprise an important part of the burden of illness. Only now—and quite belatedly—is alcohol abuse becoming a truly major focus for research in the biomedical and behavioral sciences. This trend in research is likely to accelerate in the next five years.

The next decade is likely to see considerable improvement in the reliability of clinical and economic measures of burden of illness. These measures can enable a more rational allocation of resources for research, education, and services. For instance, the population 65 years and over is increasing steadily. Since chronic disorders tend to increase with age, it is reasonable to expect a corresponding increase in demand for health services. We can also anticipate an increase in the special problems in delivery of health care to the elderly, such as tranportation to the source of care.

HEALTH AND BASIC SCIENCE

As the prevalence of various kinds of illness changes, the scientific efforts needed to reduce illness will also change. One can observe, for example, a revival of interest in older disciplines, such as epidemiology and biostatistics, the strengthening of such relatively new disciplines as biomedical engineering and the behavioral sciences, and an upsurge of interdisciplinary research on prevention of disease and delivery of health services. Advances in these fields—and in such fields as biochemistry, genetics, neurobiology, immunology, and pharmacology—potentially could yield marked gains in health over the next few decades.

Major strides in basic science can be expected in the next five years. Still, we know too little to be able to relieve much of the burden of illness in the near future. The interplay of basic research with clinical investigation can provide deeper insight into molecules, tissues, organisms, populations, and health care systems.

CARDIOVASCULAR DISEASE

Large-scale experiments on the prevention of cardiovascular diseases have highlighted the need to strengthen the linkages of basic research, clinical investigation, and health-services research. Such linkages can greatly reduce the toll of these serious diseases. Efforts will be focused on prevention and treatment of arteriosclerosis and hypertension.

More must be known about the basic processes of hypertension and arteriosclerosis to enhance our ability to prevent the diseases and their complications by both pharmacological and behavioral means. Also, we must learn to identify cardiovascular disease patients more precisely to better fit appropriate therapy.

Improved drug therapy for hypertension should become available in the next five years. It will take longer, however, to determine which therapeutic regimens will work best for which patients in preventing such complications of the disease as stroke or heart failure.

Hypertension is more frequent among blacks than among whites in this country. We do not know why. Identification of specific genetic and environmental factors in this public health problem remains a major scientific challenge.

Basic research on arteriosclerosis is likely to advance rapidly in the next five years. Progress should be made in clarifying the nature and formation of the basic lesion causing the disease, in developing technical advances that might lead to noninvasive diagnostic techniques, and in providing a firmer foundation for preventing the disorder.

CANCER

Environmental factors acting upon genetic predispositions are now considered important in the origin of cancer. Where such causal factors can be identified, they can often be minimized in the environment so as to prevent cancer. A goal for epidemiology is improved understanding of the relationship of changing patterns of environmental agents to the incidence of cancer. Special attention should be given to those cancers that occur most frequently and with large geographic variations in their incidence.

Even after a carcinogen has been identified, it may not be possible to remove it from the environment or to avoid it totally. Thus, we will need to establish socially acceptable exposure levels. Such levels must be based on dependable, quantitative risk estimates derived from data much better than most of that now available.

Cigarette smoking is responsible for 80 percent of the incidence of lung cancer, so that the advantages and disadvantages of low-tar-and-nicotine cigarettes deserve systematic investigation. Careful attention should be paid to the effects of cigarette smoking, not only on lung cancer, but also on cardiovascular and other cigarette-related diseases.

The search for more effective combinations of surgery, radiotherapy, and chemotherapy for treatment of various cancers will proceed intensively during the next five years.

BEHAVIOR AND DISEASE

Modifying risk factors for various diseases through changes in behavior is important but not easy. Significant educational and social changes will be required. However, there are encouraging indications that firmly established patterns of behavior—such as smoking, exercise, diet, working, and coping with stress—can be changed. Work in the area will intensify during the next five years, and will involve combined efforts in biomedical and behavioral research.

To the extent that behavior harmful to health can be reduced, the unborn infant also will benefit. The unborn infant is at increased risk of death or abnormality if the mother smokes or drinks alcohol during pregnancy. The vulnerability of the fetus to maternal intake will be investigated intensively in the next several years.

Because cigarette smoking contributes so heavily to cancer and cardiovascular disease, research on stopping smoking is highly pertinent. The focus will be not so much on stopping initially as on preventing relapse. This is likely to be an important area of study in the next five years.

SMOKING AND ADOLESCENCE

Smoking rates in the 1970's have been rising more rapidly among early adolescents, but have stabilized or even decreased in other parts of the American population. Early adolescence, therefore, will be a critical focus in the next five years for research on prevention of diseases related to smoking. Adolescence is a critical period of biological and psychological change, and of drastic social change. It is a time when lifelong behavior patterns crucial to health are formed. These patterns involve not only cigarette smoking, but also use of alcohol and other drugs, automobile driving, diet and exercise, and human relationships. The impact of peers, influential adults, and the mass media on adolescents needs careful, systematic study to clarify the bases of health-related behavior. Research on prevention of diseases due to smoking will focus on the adolescent age group in the next five years.

RESEARCH ON ANIMALS

Some areas of health research, including behavioral research, rely heavily on the use of higher primates, because of their biological similarity to humans. It will be essential in the next five years to develop adequate breeding colonies and research facilities for such primates. They are needed to make critical progress on problems of behavioral biology, of reproduction (e.g., contraception), and infectious disease (e.g., viral hepatitis vaccine).

MENTAL ILLNESS

Drugs for treating mental disorders have more than halved the population of public mental hospitals in this country since about 1955. The drugs also have spurred rapid growth in research on the chemistry and biology of the brain and particularly on the biochemical bases of mental disorders. This research in part has led to more effective drugs for mental illness and is likely to continue to do so. Better drugs also can be expected to result from systematic modification of existing compounds.

AGING

The rising proportion of elderly people in the American population is leading to a new era in basic, clinical, and health-services investigations of the problems of aging and health. These efforts include work on immune responses as protective mechanisms; changes in patterns of hormonal secretion with age; neurobiology in relation to memory, sleep, and other brain functions; the distinctive problems of drug therapy in the elderly; maintenance of functional independence in coping with chronic diseases; and linkage of health and social services in maintaining health into later life. Such efforts will intensify in the next several years.

GENES AND ENVIRONMENT

Studies of the influence of genetic factors on responses to environmental agents will expand in the next five years. The work will require the identification of genes involved in susceptibility and resistance to particular diseases. The interaction of such genes with specific environmental factors may point the way to sharply focused preventive techniques. These investigations are likely to concentrate on substances that are widespread—as in diet, medication, and occupation—and to which a relatively large fraction of the population is genetically susceptible.

HEALTH CARE DELIVERY

Innovations in health care delivery are likely to be prominent in the next five years. Various organized settings, such as group practice and health maintenance organizations, are being used increasingly. Efforts to extend primary health care to rural areas and low-income urban areas include the National Health Service Corps—young physicians whose medical education is subsidized in return for postgraduate service in underserved areas. Also emerging are specially trained nurse practitioners to provide primary care in underserved areas.

Efforts are being made to build close functional links between mental-health and general-health services. Also under way are analyses of organizational changes aimed at more effective health services for children and for the elderly.

HEALTH-SERVICES RESEARCH

The search for improvement in health services will require not only carefully designed innovations, but also systematic assessments—the function of health-services research.

The need for accurate, dependable information on the organization, quality, availability, and cost of health services in the United States is becoming increasingly apparent to health care professionals, government officials, and the public. Management of the personnel, facilities, and technologies that comprise modern health care institutions requires information similar to that needed to manage other complex enterprises. The systematic comparison of alternative health care delivery systems will be a great challenge to the health sciences in the years ahead. However, such research still lacks a secure institutional base, either in government or among the nation's leading universities.

REFERENCES

1. Shapiro, S., E.K. Schlesinger, and R.E.L. Nesbitt. *Vital and Health Statistics* (National Center for Health Statistics) 3(4):1, 1965.

2. *Research Needs in Nephrology and Urology, Vol. 1.* Report of the Coordinating Committee, National Institutes of Arthritis, Metabolism and Digestive Diseases (DHEW Pub. No. [NIH] 78-1481). Washington, D.C., 1978.

3. MacMahon, B., and J.E. Berlin. Health of the United States Population. In: *Horizons of Health,* H. Wechsler *et al.* (eds.). Cambridge, Mass.: Harvard University Press, 1977, pp. 13–21.

4. Rice, D.P., J.J. Feldman, and K.L. White. *The Current Burden of Illness in the United States* (Occasional Paper of the Institute of Medicine) Washington, D.C.: National Academy of Sciences, 1976, p. 26.

5. Smith, T.W. The Heart and the Vascular System. In: Horizons of Health, H. Wechsler *et al.* (eds.). Cambridge, Mass.:Harvard University Press, 1977, pp. 179–197.

6. Moore, F.D. Surgical Care. In: Horizons of Health, H. Wechsler *et al.* (eds.). Cambridge, Mass.: Harvard University Press, 1977, pp. 343–353.

7. Melmon, K., and H. Morelli (eds.). *Clinical Pharmacology,* 2nd ed. New York: Macmillan, 1978.

8. Sackett, D.L., and R.B. Haynes. *Compliance with Therapeutic Regimens.* Baltimore: The Johns Hopkins University Press, 1976.

9. Shapiro, A.P. Behavioral and Environmental Aspects of Hypertension. *Journal of Human Stress* 4(4):9–17, 1978.

10. Levi, L. (ed.). *Society, Stress and Disease, Vol 1., The Psychosocial Environment and Psychosomatic Diseases.* London: Oxford University Press, 1974.

11. Rose, R.M., C.D. Jenkins, and M.W. Hurst. *Air Traffic Controller Health Change Study: A Prospective Investigation of Physical, Psychological and Work-Related Changes.* Boston: Boston University School of Medicine, 1978.

12. Levy, R.I., and M. Feinleib. Coronary Artery Disease: Risk Factors and Their Management. In *Heart Disease* (in press).

13. Havel, R.J. Classification of Hyperlipidemias. *Annual Review of Medicine* 28:195–209, 1977.

14. Truswell, A.S. Diet and Plasma Lipids—A Reappraisal. *The American Journal of Clinical Nutrition* 31:977–989, 1978.

15. Farquhar, J.W., *et al.* Community Education for Cardiovascular Health. *The Lancet,* 2:1192–1195, 1977.

16. Maccoby, N., *et al.* Reducing the Risk of Cardiovascular Disease: Effects of a Community-Based Campaign on Knowledge and Behavior. *Journal of Community Health* 3(2):100–114, 1977.

17. Puska, P. High Risk Hearts. *World Health.* Geneva: The World Health Organization, 1976, pp. 12–15.

18. Frei, E. Cancer. In: *Horizons of Health,* H. Wechsler *et al.* (eds.). Cambridge, Mass.: Harvard University Press, 1977, pp. 25–39.

19. Tomatis, L., *et al.* Evaluation of the Carcinogenicity of Chemicals: A Review of the Monograph Program of the International Agency for Research on Cancer (1971–1977). *Cancer Research* 38:877–885, 1978.

20. Isselbacher, K.J. The Gastrointestinal System. In: Horizons of Health, H. Wechsler *et al.* (eds). Cambridge, Mass.: Harvard University Press, 1977, pp. 243–251.

21. Yalow, R.S. Radioimmunoassay: A Probe for the Fine Structure of Biologic Systems. *Science* 200:1236–1250, 1978.

22. Kaplin, H.S. *Hodgkin's Disease.* Cambridge, Mass.: Harvard University Press, 1972.

23. Upton, A.C. "Prevention—The Ultimate Goal." Presented at Conference on Cancer Prevention—Quantitative Aspects sponsored by National Cancer Institute, NIH, held in Reston, Va., September 1978.

24. *Smoking and Health: A Report of the Surgeon General.* Prepublication copy, Office of the Assistant Secretary for Health, DHEW, January 1979.

25. Kety, S.S. The Biological Bases of Mental Illness. In: *Horizons of Health,* H. Wechsler *et al.* (eds.). Cambridge, Mass.: Harvard University Press, 1977, pp. 111–123.

26. McAlister, A.L. Tobacco, Alcohol and Drug Abuse: Onset and Prevention. In: *Disease Prevention: Report to the Surgeon General on Health Promotion and Disease Prevention, Part II.* Washington, D.C.: Institute of Medicine, 1979.

27. Hollister, L.E. Tricyclic Antidepressants. *The New England Journal of Medicine* 299:1106–1109, 1978.

28. Bloom, F., *et al.* Endorphins: Profound Behavioral Effects in Rats Suggest New Etiological Factors in Mental Illness. *Science* 194(4265):630–632, 1976.

29. Guillemin, R. Peptides in the Brain: The New Endocrinology of the Neuron. *Science* 202(4366):390–402, 1978.

30. Schally, A.V. Aspects of Hypothalamic Regulation of the Pituitary Gland. *Science* 202(4363):18–28, 1978.

31. Mendelson, J.H., and N.K. Mello. Alchohol and Drug Abuse. In: Horizons of Health, H. Wechsler *et al.* (eds.). Cambridge, Mass.: Harvard University Press, 1973, pp. 124–140.

32. *Report of the Task Force on the Use and Need for Chimpanzees of the Interagency Primate Steering Committee.* Bethesda, Md.: National Institutes of Health, 1978.

33. Geschwind, N. Neurological Disorders. In: *Horizons of Health,* H. Wechsler *et al.* (eds.). Cambridge, Mass.: Harvard University Press, 1977, pp. 161–175.

34. Motulsky, A.G. Family Detection of Genetic Diseases. In: *Early Diagnosis and Prevention of Genetic Diseases,* L.N. Went *et al.* (eds.). Leiden: Leiden University Press, 1975, pp. 101–109.

35. Motulsky, A.G. "Genetic Approaches to Chronic Common Diseases." *Centenary Lecture Series. I. Medical Genetics.* University of Western Ontario, London, Canada, March 1978, in press.

36. Omenn, G.S., and A.G. Motulsky. "Eco-Genetics: Genetic Variation in Susceptibility to Environmental Agents." *Genetic Issues in Public Health and Medicine.* Charles C Thomas: Springfield, Ill., 1978, pp. 83–111.

37. Rotter, J.I., and D.L. Rimoin. Heterogeneity in Diabetes Mellitus—Update 1978. Evidence for Further Genetic Heterogeneity Within Juvenile-Onset Insulin-Dependent. *The Journal of the American Diabetic Association* 27:599–608, 1978.

38. Rutstein, D.D. *Blueprint for Medical Care.* Cambridge, Mass.: The Massachusetts Institute of Technology, 1974.

39. Fine, J. Proposal for a National System of Peer Review. *Bulletin of Atomic Scientists* 33(7):38–43, 1977.

40. Rutstein, D.D., *et al.* Measuring the Quality of Medical Care. *The New England Journal of Medicine* 294:582–588, 1976.

41. Lewis, C.E., R. Fein, and D. Mechanic. *A Right to Health: The Problem of Access to Primary Medical Care*. New York: John Wiley and Sons, 1976.

42. Enthoven, A. "Incentives and Innovation in Health Services Organization." Paper delivered at the Annual Meeting of the Institute of Medicine, October 26, 1978.

43. *Computed Tomographic Scanning* (IOM Pub. No. 77-02, Washington, D.C.), April 1977.

44. Frederickson, D.S. Health and the Search for New Knowledge. *Daedalus*106(1):159–170, 1977.

45. *Evaluation of Poliomyelitis Vaccines* (IOM Pub. No. 77-02, Washington, D.C.), April 1977, p. 15.

46. Comroe, J.H. The Evolution of Biomedical Science: Past, Present, Future Perspectives. *Journal of Medical Education* 52:3–10, 1977.

47. Hamburg, D.A., and S.S. Brown. The Science Base and Social Context of Health Maintenance: An Overview. *Science* 200(4344):847–849, 1978.

BIBLIOGRAPHY

Aging and Medical Education (IOM Pub. No. 78-04, Washington, D.C.), September 1978.

The Application of Advances in Neurosciences for the Control of Neurological Disorders. Geneva: World Health Organization, 1978.

Arteriosclerosis: The Report of the 1977 Working Group to Review the 1971 Report of the National Heart and Lung Institute Task Force on Arteriosclerosis, December 1977 (DHEW Pub. [NIH]No. 78-1526), 1978.

Assessing Quality in Health Care: An Evaluation (IOM Pub. No. 76-04, Washington, D.C.), November 1976.

Barchas, J. *et al.* (eds.). *Psycho-Pharmacology: From Theory to Practice.* New York: Oxford University Press, 1977.

Baselines for Setting Health Goals and Standards. Papers on the National Health Guidelines (DHEW Pub. No. [HRA] 76-640), September 1976.

Berger, P., B. Hamburg, and D. Hamburg. Mental Health: Progress and Problems. *Daedalus* 106:261–276, 1977.

Besser, M. (ed.). *Medicine, 1977.* New York: John Wiley and Sons, 1977.

Beyond Tomorrow: Trends and Prospects in Medical Science, A Seventy-Fifth Anniversary Conference. New York: The Rockefeller University, 1977.

Biomedical Research in the Veterans Administration (NRC Committee on Biomedical Research in the Veterans Administration). Washington, D.C.: National Academy of Sciences, 1977.

Breslow, L., and A.R. Somers. The Lifetime Health Monitoring Program: A Practical Approach to Preventive Medicine. *The New England Journal of Medicine* 296(11):601–608, 1977.

Brown, S.S. *Policy Issues in the Health Sciences: A Staff Paper* (IOM Pub. No. 77-002, Washington, D.C.), October 1977.

Bunker, J.P., B.A. Barnes, F. Mosteller (eds.). *Costs, Risks, and Benefits of Surgery.* New York: Oxford University Press, 1977.

Cairns, J. *Cancer: Science and Society.* San Francisco: W.H. Freeman, 1978.

Carlson, R.J., and R. Cunningham (eds.). *Future Direction in Health Care: A New Public Policy.* Cambridge, Mass.: Ballinger Publishing Co., 1978.

Comroe, J.H., and R.D. Dripps. Scientific Basis for the Support of Biomedical Science. *Science* 192(4235):105–111, 1976.

Conference on Health Promotion and Disease Prevention (Vol. I, Themes) (IOM Pub. No. 78-002, Washington, D.C.), June 1978.

Conference on Health Promotion and Disease Prevention (Vol. II Summaries). (IOM Pub. No. 78-003, Washington, D.C.), June 1978.

Cooper, B.S., and D.P. Rice. The Economic Cost of Illness Revisited. *Social Security Bulletin* (DHEW) February: Vol. 39, pp. 21–36, 1976.

Davis, K., and C. Schoen. *Health and the War on Poverty: A Ten-Year Appraisal.* Washington, D.C.: The Brookings Institution, 1978.

The Elderly and Functional Dependency (IOM Pub. No. 77-04, Washington, D.C.), June 1977.

Fisher, K.D., and A.U. Nixon (eds.). *The Science of Life: Contributions of Biology to Human Welfare.* New York: Plenum Press, 1972.

Forward Plan for Health, FY 1978–82. U.S. Department of Health, Education and Welfare, Public Health Service (DHEW Pub. No. [OS] 76-50046), 1976.

Freinkel, N. (ed.). *The Year in Metabolism: 1977.* New York: Plenum Publishing Corporation, 1978.

Hamburg, D.A., and H.K.H. Brodie (eds.). *American Handbook of Psychiatry.* New York: Basic Books, Inc., 1975.

Health in America: 1776–1976. U.S. Department of Health, Education, and Welfare, Public Health Service, Health Resources Administration (DHEW Pub. No. [HRA] 76-616), 1976.

Health in the United States, 1975 (DHEW Pub. No. [HRA] 76-1232), 1976.

Health in the United States 1976–1977 (DHEW Pub. No. [HRA] 77-1232), 1977.

Lalonde, M., *A New Perspective on the Health of Canadians.* (Information Canada, Ottowa), 1975.

Luria, S.E. *Life: The Unfinished Experiment.* New York: Charles Scribner's Sons, 1973.

McLachlan, G. (ed.) *A Question of Quality?: Roads to Assurance in Medical Care.* New York: Oxford University Press, 1976.

Papers on the National Health Guidelines: The Priorities of Section 1502 (DHEW Pub. No. [HRA] 77-641, Washington, D.C.), 1977.

Perspectives on Health Promotion and Disease Prevention in the United States (IOM Pub. No. 78-001, Washington, D.C.), January 1978.

Preventive Medicine USA. Task Force reports sponsored by The John E. Fogarty International Center for Advanced Study in the Health Sciences, National Institutes of Health and the American College of Preventive Medicine. New York: Prodist, 1976.

Priorities for the Use of Resources in Medicine (DHEW Pub. No. [NIH] 77-1288, Bethesda, Md.), 1976.

Recommendations for a National Strategy for Disease Prevention. Atlanta, Ga.: U.S. Department of Health, Education, and Welfare, Center for Disease Control, June 30, 1978.

Report of the President's Biomedical Research Panel (DHEW Pub. No. [OS] 76-500, Washington, D.C.), 1976.

Report of the President's Biomedical Research Panel, Appendix A: The Place of Biomedical Science in Medicine, and the State of the Science (DHEW Pub. No. [OS] 76-501), 1976.

Respiratory Diseases: Task Force Report on Prevention, Control, Education (DHEW Pub. No. [NIH] 77-1248, Washington, D.C.), March 1977.

Rogers, D.E. *American Medicine: Challenge for the 1980s.* Cambridge, Mass.: Ballinger Publishing Company, 1978.

Segal, J. (ed.). *Research in the Service of Mental Health: Report of the Research Task Force of the National Institute of Mental Health* (DHEW Pub. No. [ADM] 75-236, Rockville, Md.), 1975.

Shannon, J. Federal Support of Biomedical Sciences: Development and Academic Impact. *Journal of Medical Education.* 51(7):1–98, 1976.

Smith-Exton, A.N., and J.G. Evans (eds.). *Care of the Elderly: Meeting the Challenge of Dependency*. New York: Grune & Stratton, 1977.

Somers, A.R., and H.M. Somers. *Health and Health Care*. Germantown, Md.: Aspen Systems, 1977.

Surgery in the United States: A Summary Report of the Study on Surgical Services for the United States. The American College of Surgeons and The American Surgical Association, 1975.

Tanner, J.M. (ed.). *Developments in Psychiatric Research*. London: Hodder and Stoughton, 1977.

Usdin, E., D.A. Hamburg, and J.D. Barchas (eds.). *Neuroregulators and Psychiatric Disorders*. New York: Oxford University Press, 1977.

Walsh, J. *The Biomedical Sciences—1975: Report of a Macy Conference*. New York: Josiah Macy Jr. Foundation, 1975.

Wechsler, H., J. Gurin, and G.F. Cahill (eds.). *The Horizons of Health*. Cambridge, Mass.: Harvard University Press, 1977.

White, A. *et al. Principles of Biochemistry*, 6th ed. New York: McGraw-Hill, 1978.

White, K., and M.M. Henderson. (eds.). *Epidemiology as a Fundamental Science: Its Uses in Health Services Planning, Administration, and Evaluation*. New York: Oxford University Press, Inc., 1976.

(Jim Holland. Stock, Boston, Inc.)

9 Toxic Substances in the Environment

INTRODUCTION

A principal effect of technological development during recent decades has been the accelerating rate of human alterations of land, vegetation, water, and air. Most visible is the transformation of the American landscape. Cities spread outward, carrying residences, factories, and roads, while inner sections are rebuilt. Streams are dammed and channeled; wetlands drained or restored; forests cleared, reserved, thinned, and replanted. Farmlands are affected by changes in patterns of crops, cultivation practices, and uses of fertilizer, pesticides, herbicides, and water. Mineral lands are stripped and sometimes reclaimed.

Each transformation affects in some fashion the basic processes of the interlocking systems of land, water, and air that support life. Each is intended to enhance the capacity of parts of these systems to serve human needs—for food, fiber, energy, transport, recreation, and solitude. Each, however, may also entail hazard to other parts of the system; each may reduce the productive capacity of the earth and perhaps impair the health and welfare of the people whom the changes are expected to benefit. The hazard may result from inadvertently altering a system, as when land cultivation disturbs stream flow, or from exposing people to new risks, as when a hospital is constructed in an active seismic zone.

Some of these linkages are direct and relatively measurable—such as the obliteration of farmland by the asphalt parking lot of an industrial plant, thus impeding groundwater infiltration in exchange for new manufacture. Others are far more complex and difficult to quantify: For example, the cutting of a forest reduces the standing stock of carbon and possibly renders global climate more vulnerable to carbon dioxide (CO_2) buildup, as noted in Chapters 1 and 5.

In this chapter we review only one facet of the massive changes under way: the issues raised by the presence of toxic substances in the environment—in particular those related to the effects of man-made chemicals on human health. An immense array of substances in the natural environment may also be toxic to plants, animals, and humans. Indeed, some of the chemicals that are necessary for life are nevertheless toxic or even lethal in high doses. Ions of zinc and copper, for example, are essential nutrients, but fatal at high concentrations. Even salt and oxygen are toxic in large enough amounts. Natural selenium in certain soils causes disease in grazing animals and tooth deformation in humans.[1] Marine plants and sponges generate a great variety of halogenated organics and have the ability to tolerate and detoxify such materials.[2] Many chemical interactions in undisturbed natural systems are imperfectly understood or barely suspected. For example, natural sources appear to account for the methyl mercury found in the livers of some marine fish and mammals. However, the adverse effects of methyl mercury on these animals may be ameliorated when selenium is also present.[3]

Interactions of this sort among humans are extraordinarily difficult to study. Epidemiologists find that their subjects move from place to place, drawing upon different soils and groundwater. In any given area, new substances are being introduced, thus changing the background. Nevertheless, there is great pressure on science to understand the consequences of the increasing introduction of man-made substances into the environment and to consider possible ways of dealing with them. Of most public concern has been the diffusion of man-made chemicals for household, agricultural, and industrial purposes. They attract attention because of dramatic instances of local contamination and because they are suspected of contributing to broader injury both to the health of humans and of ecosystems.

There is a linkage between the effects of toxic chemicals on human health and on the environment. For example, air pollutants such as ozone and sulfur dioxide, which can damage crops, are also associated with human respiratory ailments.

Much remains to be learned. Some toxic chemical hazards are well known. Other chemicals are suspect. They arouse concern but their dangers have not been clearly identified. Furthermore, an earlier viewpoint, that there are low levels of exposures to chemicals that are totally without effect (totally safe levels), is currently being challenged in some areas, especially in regard to cancers. At the same time, rapid advances in analytical techniques make it possible to detect smaller and smaller quantities of hazardous chemicals. Thus, for a number of reasons we are moving towards a position where quantitative assessment of both risks and benefits must be increasingly used. Since it is impossible to avoid all risk, society will have to decide in each case if the benefit from a hazardous material justifies the risk.

Environment is emerging as a factor in today's major diseases. The preceding chapter has shown that since 1900 cardiovascular disease and cancer have replaced microbial infection as the major health threats to Americans. Among the cardiovascular diseases "the elucidation of specific genetic and environmental factors" implicated in hypertension remains a scientific challenge (see p. 389). Water hardness is one of the dozen or so factors deserving investigation as a basis for preventing coronary disease (see p. 391).

Environmental factors, defined broadly to include diet and cigarette smoking, are now considered important in the origin of cancer (see p. 401). This recognition is based in part on the large geographic variations in incidence of specific cancers, and it encourages speculation as to what kind of background or anthropogenic factors may account for the differences in incidence among people in different regions of the United States. The large number of presumed carcinogens, the highly variable conditions of exposure to them, the long latent periods after exposure (see p. 453), and the high mobility of population make it extremely difficult to work out the epidemiology of cancer. The five years ahead will see major efforts by epidemiologists to refine their understanding of the large variations in geographic incidence.

The situation at present is summed up in the preceding chapter:

> Although some man-made chemicals have been identified as potential carcinogens, the firmly established incidence of cancer from these compounds accounts for only a tiny fraction of all cancers. Cigarette smoking is the one environmental factor for which firm data demonstrate a strong association; occupational exposures are also an important source of environmental carcinogenesis. Other environmental factors, including those of natural origin, that are probably responsible for many and perhaps most cancers remain unknown. Research on these problems will intensify during the next five years.

Interest in the outcome and application of that research will expand even more rapidly as more refined measurements reveal the presence of substances whose presence was unsuspected, as additional carcinogenic factors are identified, and as more toxic chemicals are found. Even if there were no expansion in the production and number of chemical compounds, the question of the hazard of those already circulating in the environment would still be important. To the extent that production enlarges, that question will command even more public attention.

GROWTH OF CHEMICAL TECHNOLOGY

Over the past three decades, the production and use of industrial chemicals expanded a hundredfold. The welfare of our nation has come to depend in no small measure upon these and other chemical products. Synthetic fertilizer is essential to food production at current levels in many parts of the world. Reliability and output of crop yields would be substantially diminished without agricultural chemicals such as pesticides and weed killers. Almost all pharmaceuticals are synthetic chemicals. The smelting and purification of metals are chemical processes. Most of the gasoline, rubber, plastic, adhesives, detergents, textiles, antifreezes, disinfectants, cosmetics, solid-state devices, films, paints, and much of the building material we use are synthetic.

For many years the American people looked upon expansion in the manufacture and use of synthetic chemicals as a major area of national progress. Many of the newly generated substances apparently have no harmful effects at current levels. Some of them, however, are toxic, and over the past decade we have come increasingly to realize that this advance in productivity and convenience has been accompanied by both real and potential impacts on human and environmental health now and in the future. This focuses much public attention on the dangers inherent in the uncontrolled manufacture and use of synthetic chemicals and on the ways in which we can in the future improve the safety of the products and processes we use.

Some 80 generally distributed chemicals (not immediately used in the synthesis of other chemicals) are each produced in amounts greater than 100 million pounds annually. Over 25 million pounds each of 25 other such chemicals are manufactured yearly.[4]

Figure 29 shows the growth of the synthetic organic chemical industry in the United States since 1917. Reported annual production of manufactured chemicals includes some 10,000 individual chemicals. (However, about 70 percent of the total output is made up of some 500 large-volume chemicals.[5]) Many more, produced in small volume, are not reported. It has been estimated that approximately 70,000 chemicals are in some degree of current use, with perhaps 1,000 new chemicals being introduced each year.[6] Growth rates for six of the major basic compounds used in the manufacture of large volume chemicals during the past two decades are shown in Figure 30. Some of these starting chemicals are utilized in a wide variety of ways. As an example, the many uses of ethylene, the largest volume starting chemical, are shown in Figure 31. Whether or not these production rates continue, the number and volume of substances released into the environment and their uses are large and promise to remain so.

In the face of increasing demand and production over the past several

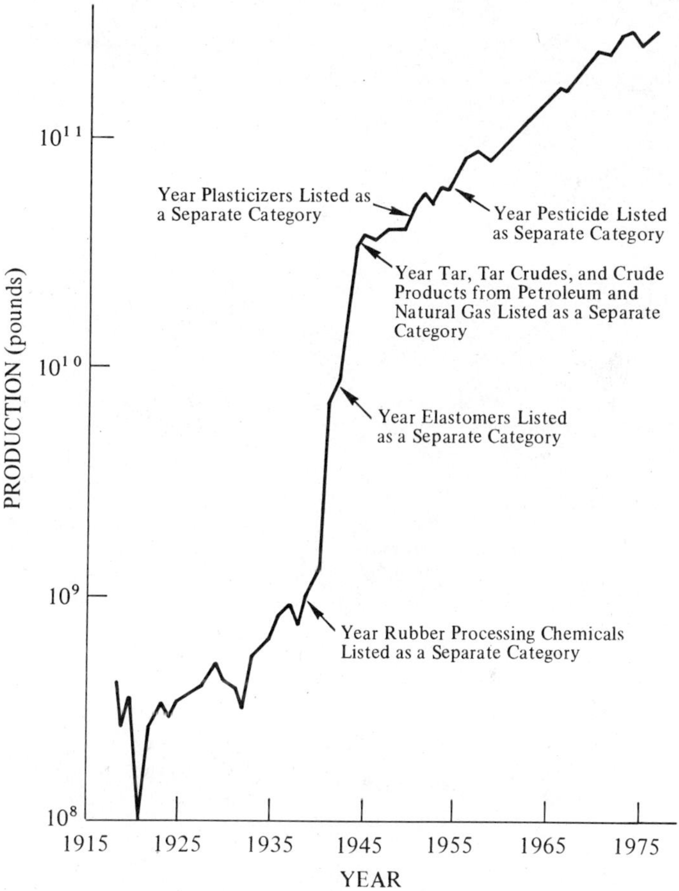

FIGURE 29 Synthetic organic chemical production. (Synthetic Organic Chemicals: U.S. Production and Sales, 1918-1976, Washington, D.C., U.S. International Trade Commission)

years, considerable advance has been made in curbing pollution and controlling the spread of toxic chemicals. Federal regulations have brought substantial reductions in air pollutants such as sulfur dioxide, workplace contaminants such as vinyl chloride, and industrial-waste discharges into rivers and lakes. Some of this progress has resulted from the voluntary cooperation of industry with federal agencies in restricting uses and production of PCB's, and in early reduction in the use of halocarbons as an aerosol propellant.

Much remains to be done. The pressing need now is for better knowledge—both to determine the dimensions and importance of suspect-

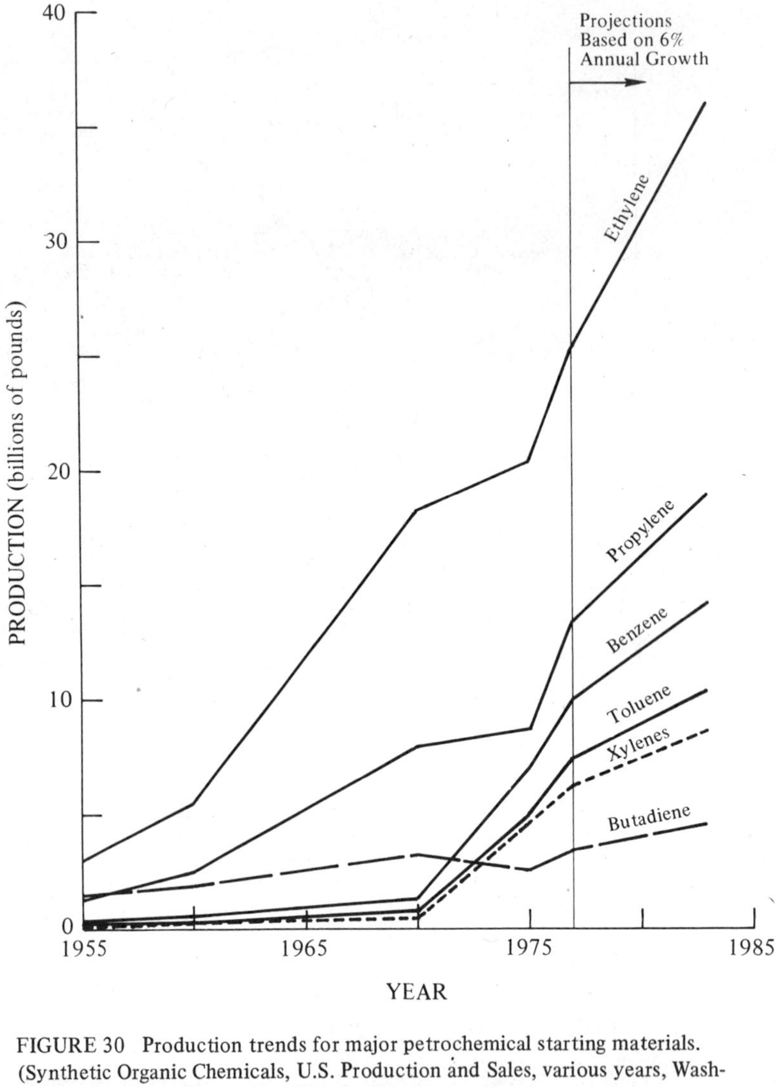

FIGURE 30 Production trends for major petrochemical starting materials. (Synthetic Organic Chemicals, U.S. Production and Sales, various years, Washington, D.C., U.S. International Trade Commission [Benzene production estimates for 1955, 1960, and 1970 courtesy of *Chemical Economics Handbook,* SRI International, Menlo Park, Calif.])

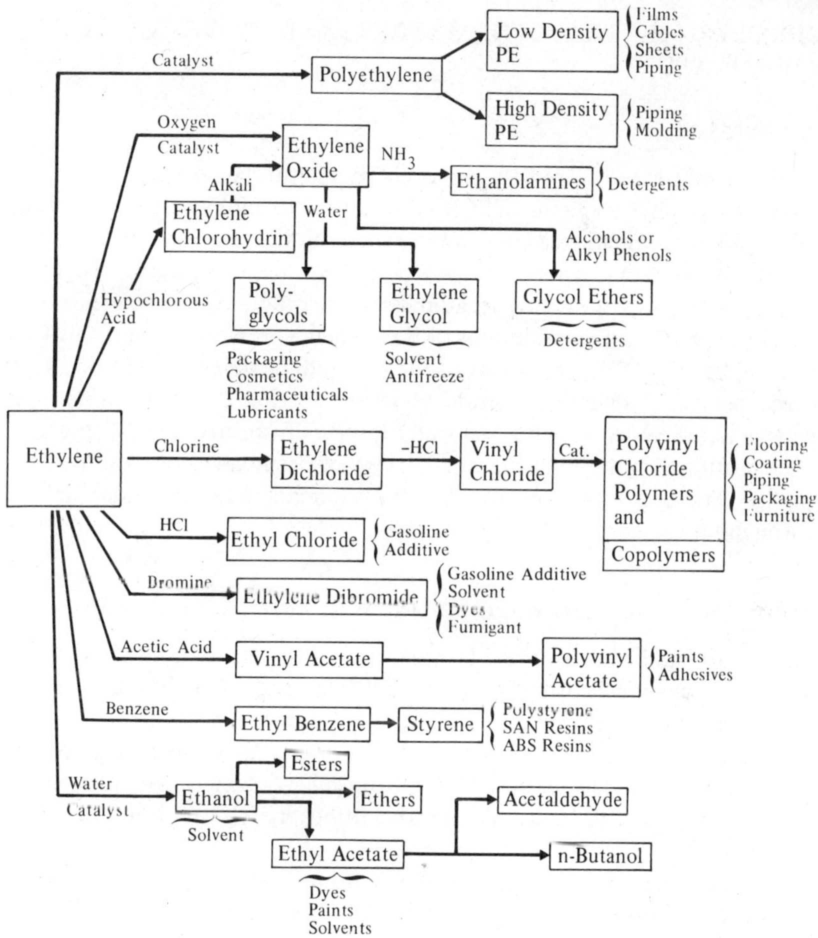

FIGURE 31 Uses of ethylene.

ed chemical hazards and to improve regulatory efforts to control recognized problems. Only thus can we separate large from small problems in order to set our priorities. In many cases the necessary knowledge will be slow in coming or never completely available.

The following discussion of a few known and suspected chemical hazards (both man-made and natural) is intended to illustrate the pervasiveness of the problems, gaps in our knowledge of causes and effects, and ways that scientific information can be brought to bear in dealing with issues that threaten to become yet more troublesome.

PERSISTENCE, TRANSFORMATION, AND MOVEMENT OF CHEMICALS

PERSISTENCE

Some chemicals are extremely persistent in a natural environment—highly resistant to physicochemical breakdown and to biological degradation. Such persistent chemicals collect and remain in the sediments of lakes and rivers. Some move up the food chain from microorganisms to other creatures, to fish, and finally to humans, increasing in tissue concentration at each stage. The accumulations of these chemicals in the tissue of fish are thus higher than the average levels in surrounding waters.[7] The stability of these persistent chemicals tends to make possible their diffusion over substantial areas by both air and water (especially air over long distances).[8] The resulting contamination can be very widespread. For example, the highly persistent PCB's are found in both humans and nonhuman species throughout the world.[9]

CHEMICALLY TRANSFORMED, SECONDARY, AND TRACE CHEMICALS

Chemicals can be transformed by various means—for example, by reaction with other chemicals or by the action of bacteria. Bacteria in the sediments of streams and lakes can act on inorganic mercury and produce the more toxic methyl mercury. Because methyl mercury derivatives are volatile, they can contribute mercury to the atmosphere, which can then spread over long distances and add to the atmosphere's natural burden of mercury.[10]

Secondary chemical pollutants derive from complex processes in which a totally new toxic agent is generated. For example, ozone and peroxyacetyl nitrate (PAN) are produced in the lower atmosphere through the photochemical action of sunlight on nitrogen oxides and hydrocarbons in the air (from auto exhausts, power plants, and so on).[11] Ozone and PAN can damage agricultural crops. (It is of interest here to note the relationship between site and effect. In the upper atmosphere, ozone filters some of the potentially carcinogenic ultraviolet rays from sunlight and is beneficial rather than harmful.)

Sometimes a chemical product presents a relatively minor hazard compared to that posed by trace amounts of contaminants that it contains. An example is dioxin (tetrachlorodibenzedioxin or TCDD), a substance found in the defoliant 2,4,5-T. Although dioxin is present in only minute quantities, it has been a major cause of concern about the defoliant's possible toxicity.[12–14] We are now aware of the importance of some trace contaminants even at concentrations of a few parts per billion.

Deposits of Chemical Wastes

One of the most perplexing aspects of chemical pollution is the accumulation of chemical wastes in landfills and sediments. Hudson River sediments, for instance, contain the accumulation of many years of PCB discharges. Dredging is being proposed to remove the PCB's from the river. But it has not been shown that this approach would be successful. Indeed, it could make the situation worse by simply spreading the contamination. Moreover, there would be the problem of where and how such a vast volume of contaminated material (an estimated 1.7 million cubic yards)[15] could be safely disposed of with no environmental or health impact.

Other approaches have been suggested: overlaying the sediment with impervious material, sequestration, and bacterial or chemical degradation. A key question in each case is whether it is possible to improve on natural degradation and sequestration, slow and inadequate though they may be.

Widespread chemical dumping has been a pattern of the past, and some dump sites have been later used for landfill. The Love Canal near Niagara Falls, New York, is one such site. Thus far, the human health impacts of these abandoned and refilled chemical dumps have not been fully evaluated.[16]

MAJOR PATHWAYS OF CHEMICAL POLLUTION

Air Pollutants

The nation has been concerned about air pollution for some years. The general dimensions of the health impacts of pollutants such as the sulfur oxides, ozone, nitrogen oxides, carbon monoxide, and particulates are only now being defined.[17] Current estimates vary widely. Major gaps in our knowledge of these health effects persist—especially detailed information on the health effects of actual concentrations of specific pollutants on specific segments of the population.

It is generally believed, for example, that acid particulates alone or in combination with irritant gases contribute to the respiratory effects of air pollution. If the responsible chemical species and various combinations among a constellation of chemical agents could be identified, it should be possible to design more efficient and selective controls.[18]

More refined knowledge of the quantitative impacts of air pollutants, in terms of specific agents and disease patterns, is also needed. Irritant air pollutants such as sulfur oxides, nitrogen oxides, acid particulates, and ozone are thought to contribute to morbidity and mortality. But not all parts of the population are equally endangered by these pollutants. The young and the old seem to be more susceptible, especially those elderly persons who already suffer from cardiorespiratory disease.[11,19,20] Knowledge is needed of the role of host factors, such as age, sex, prior or concurrent disease, and genetic makeup. Also, a better understanding of the interaction and role of the various components of the acid particulate complex could aid in refining and improving control procedures.

The role of air pollution in contributing to lung cancer remains controversial. It is generally agreed that there is an excess of lung cancer in urban as compared with rural areas and that carcinogenic and cancer-enhancing air pollutants are in higher (but not massive) concentrations in urban than in rural areas. However, in view of other differences in the two settings, there is no general consensus that the urban–rural differences in lung cancer rates can be confidently attributed to air pollution. One still-controversial estimate suggests that general air pollutants can increase the incidence of lung cancer in males who smoke by perhaps 10 percent beyond the average effect of cigarette smoking alone.[21] This estimate does not apply to nonsmokers.

Water Pollutants

There are two well-known pathways for the chemical contamination of lakes, rivers, and groundwaters: current discharges or runoffs, and gradual releases from accumulated deposits of chemical wastes. During recent years, persistent pesticides (such as Mirex, Kepone, and DDT), mercury, PCB's, and other chemicals have been found in many of our groundwaters and freshwater lakes and streams.

A third, perhaps significant pathway derives from the internationally employed chlorination of water supplies. Because of chemical treatment, microbial and viral contamination of water bodies, the classical potential hazard to human health, is no longer an active threat of massive dimensions. The control of intestinal infections by the chlorination of water supplies has been one of the most dramatic public-health advances of the past century. However, there is now concern that this may produce chemicals such as chloroform by interaction with otherwise harmless organic substances in the water. Laboratory data show that chloroform and several similar compounds cause liver cancer in rodents.[22] There is also debated epidemiological evidence of increases of cancer among populations using chlorine-treated water.[23]

Any change in this extremely successful chlorination process should be approached carefully. To remove those organic chemicals that can react with chlorine to produce chloroform or similar compounds, the Environmental Protection Agency has proposed charcoal treatment of water prior to chlorination in cities with populations above 75,000.[24] This additional treatment, however, may have undesirable side effects, and would be costly and an additional burden on public health authorities. Hence, alternative methods for the control of waterborne pathogens, such as ozone treatment, are being investigated. The magnitude of the cancer risks involved in chlorination should, in addition, be more thoroughly assessed.

Trace quantities of many other organic compounds are present in water. Some of these compounds have been identified as weak carcinogens or mutagens in experimental animals or other biological systems.[25] There are currently no fully acceptable means for evaluating the health risks (if any) from drinking water containing trace amounts of such compounds; appropriate techniques should be sought.

Exposure in the Workplace

The workplace has been a tragic but revealing source of information about chemical carcinogens. Some 20 chemicals or processes have been linked to the incidence of cancer in exposed workers. Some, such as 2-naphthylamine, chromium metal production, and asbestos, have been well known for years. Others, such as vinyl chloride and the chloromethyl ethers, have been recognized more recently.[26] The contribution of occupational exposure to chemicals to overall incidence of cancer in the United States is uncertain; published estimates have ranged from 1 to 20 percent.[27]

Cancer-causing chemicals in the workplace are presently regulated through a number of procedures, such as allowable exposure concentrations for asbestos and vinyl chloride and stipulated work practices for certain chemical carcinogens. Several new approaches have been suggested, among them a permit system requiring the stipulation of control procedures and a total-enclosure approach, under which carcinogenic compounds would be isolated from workers and monitoring would involve the testing of tightness rather than allowable concentrations. An evaluation of new approaches might yield a reduction in risk for workers and better, more cost-effective controls.

Some occupational diseases, such as silicosis, coal miners' pneumoconiosis (black lung disease), and byssinosis (brown lung disease, associated with the vegetable fiber industries) have been with us for many years. Recently, attention has been given to the effects of chemicals on the reproductive and nervous systems. The soil fumigant, 1,2-dibromo-3-chloropropane, has been linked to sterility in production workers[28] and lowered sperm counts in field workers.[29] Some female anesthetists have suffered abortions and congenital defects in offspring.[30] The solvent methyl-*n*-butyl ketone has been associated with neurological disorders.[31]

What is urgently needed is an assessment of the prevalence and seriousness of the various occupational diseases to aid in setting control priorities. Also needed are evaluations in the laboratory of many of the workplace chemicals now in use and or to be introduced.

The Marketplace

Among consumer products, both the degree of control and appreciation of hazards remain very limited. A case in point was the rapid introduction of the flame retardant chemical Tris (2,3-dibromopropylphosphate) for use in treating children's sleepwear. The chemical is now restricted after being found in laboratory tests to be mutagenic and carcinogenic.[32] The Tris incident clearly suggests that more rigorous pretesting is needed before product introduction.

Many household products (so-called under-the-sink chemicals and agents) have not been adequately tested, especially for chronic effects. There is a clear need for better understanding of the effects of chemicals used in consumer products, with particular emphasis on chronic effects and chemical interactions.

Toxins in food may be introduced by man or occur naturally. Aflatoxin, for example, is a product of mold growth that can contaminate peanuts, cotton seed, and other agricultural products. There is strong suspicion of an association of increases in liver cancer in Asia and Africa with the consumption of food contaminated with aflatoxins.[33] Even more important is the epidemiological evidence that general dietary choices may substantially influence the incidence of human cancer (see p. 453).

Man-made chemicals in food have aroused widespread but sometimes inappropriate concern. Such chemicals include food additives that have been used for decades under the so-called GRAS (generally regarded as safe) concept. The GRAS list is being reexamined, a process that may raise many problems of hazard evaluation.

An important legal element is the Delaney Clause, which forbids the deliberate use, in any amount, of any food additive found to cause cancer in man or animal. The wisdom of the clause has been much debated, and the issues will be clarified as valid test cases are examined.

Two recent cases have attracted much public interest. Recent laboratory tests showed that saccharin is a definite but weak carcinogen.[34] These data would seem to require action under the Delaney Clause, even though the risks to humans from saccharin seem very modest.

Nitrite, a preservative used in meats such as salami, wieners, bacon, ham, and so on, is the focus of another controversy. It can combine with secondary amines in the stomach to form nitrosamines, a class of chemicals known to be carcinogenic.[35] Although nitrite does provide protection against botulism, the same result might be achieved with smaller amounts of the chemical than are now being used. However, higher levels are presumably favored by consumers (and meat processors) because of the red color they bring to meats. A further complication is that nitrates—found in most diets—are reduced naturally in the body to nitrites in significant quantities.

The principle underlying the Delaney Clause is clearly acceptable for compounds with minimal benefits or for which noncarcinogenic substitutes are available. But quantitative risk assessment becomes necessary when it is asserted that there are benefits (including benefits to health) and no ready substitutes, or when an agent is of natural origin and its complete removal impractical. Even though assessments of both risks and benefits have substantial uncertainties at this time, they should under these circumstances be taken into account in regulatory decisions.

Cigarette smoking and alcohol and drug abuse are also major health problems (see p. 401). It is useful to note, however, that smoking can substantially increase the adverse health effects of toxic substances in the environment. The most vivid example is the interaction of smoking with certain inhaled asbestos fibers—raising the risk of lung cancer some eight times.[36] As discussed earlier in this chapter, there also seems to be an interaction between smoking and air pollution.

COMPOUNDS OF SPECIAL CONCERN

SOME PROBLEM METALS

Compounds of some of the old, familiar metals such as arsenic, mercury, cadmium, and lead are still potentially serious hazards to human health. There appears to be clear evidence that some arsenic compounds are carcinogenic in humans, especially in association with pollutants such as sulfur dioxide. There is also evidence that high natural levels of arsenic compounds in water supplies are associated with cancer in certain population groups such as in Taiwan. Strangely, tests of arsenic carcinogenicity in animals have all proved negative.[37]

Much public attention was given to methyl-mercury poisoning after the discovery that many Japanese families eating contaminated fish caught in Minimata Bay were paying a heavy price. Clinical studies revealed damage to the central nervous system and congenital defects in the offspring of exposed mothers.[10] Sale of fish containing mercury is now regulated in the United States, although the actual health effects of exposure to low levels of methyl mercury are unknown. Since methyl mercury is found in fish from many U.S. waters, including the sport fishing waters of the Northeast, health and wildlife-management officials face difficult decisions.[38]

Exposure to cadmium in the workplace is associated with respiratory and renal disease. More significant to the general population is the fact that cadmium largely remains in the body once it is absorbed. Thus, a newborn infant starts life with only a few micrograms of cadmium in his body. But by the age of 60 or 70, the cadmium burden, mainly from food, in his kidneys and liver is many milligrams—a 1,000-fold increase. The level in the kidneys by this age is approximately one-fourth of the average required to produce renal disease.[39] It is not known whether and to what extent kidney deterioration normally associated with the aging process is due to the natural accumulation of cadmium in normal kidneys.

Cigarette smoke is another source of cadmium. Absorption of cadmium through the lungs from cigarette smoke is, in contrast to absorption from food, very efficient—up to 40 percent.[39,40] As a result, smokers generally have higher cadmium burdens than nonsmokers.

The ancient problem of lead poisoning still persists today, sometimes in acute forms. There are poorly maintained buildings in ghetto neighborhoods of older cities whose interior walls are covered by many layers of paint with high lead content. Over the years, significant levels of lead have been found in children from such areas, with noticeable adverse effects on mental development and neurological health.

There is also concern over so-called normal levels of lead in the body. Lead levels in the blood of city dwellers (heightened by the lead from auto exhausts) are higher than those of rural residents by some 25–30 percent. Our understanding of biochemical disturbances stemming from increased lead burdens in the body is beginning to suggest that moderately elevated levels may be associated with biochemical changes related to synthesis of hemoglobin in the oxygen-carrying pigment of red blood cells.[41] The practical implications of these findings are unknown.

Again, there is a lack of knowledge as to whether, how, and to what extent chronic low-level exposures to toxic metals contribute to adverse health effects.

SOME HALOGENATED HYDROCARBONS

Three halogenated hydrocarbons—DDT, PCB's, and PBB's (polybrominated biphenyls)—illustrate different patterns of distribution, exposure, and sources of exposure. All are fat-soluble chemicals that are slowly metabolized. They tend to accumulate in the fat where they remain as a source of continuing exposure to the rest of the body.

DDT saved millions of lives through its contribution to the control of malaria, but its wide use as an insecticide spread the chemical into the atmosphere, into lakes and rivers, and, through concentration in the food chain, into fish and humans.

PCB's were originally developed for use in electrical equipment, especially condensers and transformers. The special properties of these compounds, such as low flammability, low freezing point, and low volatility, also led to their extensive use as heat-exchange liquids, plasticizers, and so on. PCB's have now spread throughout the environment, reaching many parts of the biosphere, including humans.[9]

The carcinogenic potential of PCB's has caused concern. But the more serious problem may be the effects of PCB's on nursing infants. They are the most intensely exposed to the chemicals, since they are dependent on mothers' milk, which, in many parts of the country, contains PCB's.[42] Laboratory studies on nonhuman primates have shown that a dietary intake of PCB's has produced abortions and underweight offspring.[43] The application of these findings to humans is uncertain.

The problem of PBB's appears to be mainly restricted to Michigan where, through a packaging error, the chemicals were added to cattle feed supplement. This resulted in contamination of some cattle and the exposure of some people in the area through the consumption of milk and beef.[44] The extent of the human impact is still being evaluated.

Chlorinated pesticides, such as Chlordane, Mirex, Kepone, and so on, present dissemination problems similar to those of DDT, PCB's, and other such chemicals. In addition, laboratory experiments have shown that the chlorinated pesticides DDE, Mirex, Lindane, and Chlordane produce liver tumors in rodents.[45]

TCDD (dioxin) first came into prominence as an impurity in the herbicide 2,4,5-T. It has since been identified as being involved in several episodes of an accidental nature, the most recent of which was an explosion in Seveso, Italy.[46] Health effects of TCDD on humans are not well established. However, TCDD has been extensively studied in animals and has been found to produce skin disorders, to be teratogenic (producing congenital defects), and injurious to the immune system.[13]

ENVIRONMENTAL FACTORS IN DISEASE

The widespread publicity accorded carcinogenic chemicals has created apprehension in many people. The identification of a significant number of new chemical carcinogens has led to the mistaken impression that nearly all chemicals can cause cancer. In fact, only a fraction of the chemicals tested are carcinogenic.

It is also important to recognize that the modest increase in the total cancer rate over the past few years is due to the prevalence of lung cancer associated with smoking; otherwise, cancer incidence has actually somewhat declined.[47] However, in terms of the impact of chemicals in the environment, these total incidence rates mean less than statistics on cancer rates by organ site, since different chemicals affect different organs. Lung cancer, mainly attributable to cigarette smoking, has gone up dramatically and is still increasing, especially among women. Stomach cancer has dropped sharply. Pancreatic cancer is rising.[48] Thus, the pattern is mixed and requires detailed analysis. Except for lung cancer, the other trends remain largely unexplained.

An important concept involving the role of external, or environmental, factors in carcinogenesis has developed over the past 20 years. Different ethnic groups have cancer patterns that seem to be related to their cultural patterns—such as diet, occupation, housing, etc.—rather than to their genetic makeup. Thus, native Japanese have a high stomach cancer rate. Their descendants in Hawaii have a significantly lower rate, and their descendants who live in the continental United States, a still lower incidence.[49] Ethnic groups tend to assume the cancer rate (by organ site) of the region to which they have migrated.

Thus cultural patterns, especially diet, may be the basis for these different rates. This does not mean that genetic factors are unimportant, since individual susceptibility, within specific populations, almost certainly must vary because of individual genetic variation. The range of such genetic variation may be more or less the same in different population groups.

The strong implication of these epidemiological findings is that cultural factors may be the source of major differences in cancer rates in various ethnic groups. If this inference is correct, then the excesses over the minimal background rate of cancer suggest that the additional cancers attributable to cultural factors are substantial.

This observation, together with data on such known cancer sources as smoking and occupational exposures, has led to the estimate that a large proportion (possibly high as 80–90 percent) of human cancer may stem from cultural factors,[49] including foods that have been part of the human diet for thousands of years.

There is a need to refine our knowledge of the contributory causes of cancer, whether from diet, occupation, or other sources. A number of approaches are required to develop this knowledge, including epidemiology, the continued testing of agents (alone, and together with other carcinogens and enhancing agents), and studies of mechanisms of action. Risk-assessment procedures can be improved by refining methods of translating laboratory data into human risk evaluations, especially in extrapolating from high doses to low doses over the long term.

In the long run, critical to all such evaluations will be knowledge of the minimal rate of carcinogenesis in man, inherent in our own biology, and the mechanism of such carcinogenesis. And it seems equally important that understanding be sought of the fact that so seemingly bizarre a diversity of chemical structures all engender cancerous transformation of animal cells.

Cardiovascular Disease

Evidence continues to accumulate showing lower rates of cardiovascular disease in hard-water areas than in areas where the water is soft.[21] The issue is still unresolved. We need more understanding of the role of water hardness in cardiovascular disease to determine whether hard water is beneficial and, if so, what compounds are responsible.

Germ-Cell Mutations and Injury to the Reproductive Process

There is a strong possibility that mutations of germ cells could result in damage that would become apparent in future generations. Mutations arise naturally from imperfections in the biochemical process of replication. In fact, without such mutations organisms could not evolve; but additional mutations are caused by chemicals and radiation, and the majority of these are deleterious. Indeed, the Environmental Protection Agency has already moved ahead on a proposal that the potential of pesticides to produce heritable mutations should be systematically tested as part of the normal registration procedure.[50]

Intense attention is currently being given to the potential of chemicals to produce birth defects. Well-developed methods of testing in this important area are now available, but there is a further need to focus more generally on the impact of chemicals on reproductive physiology and on the reproductive process in general. There are instances of sterility and other alterations of reproductive physiology that were caused by certain chemicals (see p. 449). This field has been relatively neglected.

ADVERSE EFFECTS ON NONHUMAN SPECIES

There is a large variety of materials and processes in natural systems that need to be understood for purposes of resource management and as background for assessing the effects of man-made additions to the natural background. Birds and fish are widely contaminated by such persistent chemicals as PCB's[9] and methyl mercury.[51] Acid rain from sulfur oxides and nitrogen oxides in the air appears to be affecting aquatic life, especially in lakes in the Northeast.[18] The movement of sewage effluent, agricultural wastes, and fertilizer residues is also a major problem. Such pollution changes aquatic ecosystems by increasing the volume of nitrates in lakes and streams.[52] As noted earlier, interactions among a pollutant's impacts on the environment, nonhuman species, and human beings are not infrequent. When a lake is contaminated by acid rain, for instance, acidification can alter the leaching of metallic compounds from the lake bed,[53] thus increasing the exposure of aquatic biota (and people, if the lake is a source of drinking water) to toxic substances that are a part of the natural sediments. We need better understanding of the extent, nature, and importance of the effects of acid rain. The damaging effects of individual air pollutants on some agricultural crops have been fairly well described. However, knowledge of the impacts of some chemicals remains inadequate, as does understanding of the role of synergism between air pollutants.

The impacts of chemicals on nonhuman species involve a rather different set of considerations than is the case with human beings. When assessing the effects of chemical contamination on a natural ecosystem, there is less concern over injury to a single individual (or small numbers of individuals) than with the survival of populations, communities, or even entire species. Weakening of the hold of a particular species in an ecosystem could lead to the proliferation of other species; for example, the emergence of the red spider mite after an area has been sprayed with DDT.[54] The consequences for a natural ecosystem could vary, depending on the circumstances. We need more and better information on the effects of pollutants on natural systems. Efficient pretesting methods are needed to predict these effects as well as the degree of natural resilience that such systems might have.

SCIENTIFIC, TECHNICAL, AND POLICY RESPONSES

REDUCING OUR RISKS

Safety evaluation of chemicals was relatively simple in the past. It was usually assumed that there is a safe level for every chemical and that this level can be determined through modest laboratory tests (to which a safety factor was applied). This concept was based in part on the assumptiom that a threshold exposure level existed below which no adverse effects would occur. Paracelsus (1493–1541) wrote: "All things are poisons, for there is nothing without poisonous qualities. It is only the dose that makes a thing a poison." For a wide range of chemicals that statement holds true today.

The situation may, however, prove different for mutagens and carcinogens as compared to other poisons.* Quite possibly for some carcinogens there is no "threshold," no concentration below which these compounds are safe; it may turn out that, with respect to these materials, health problems can be expressed only in quantitative, statistical terms. A trace of some carcinogens may inevitably occasion a minute but nonzero increase in the probability of cancer, whereas a larger dose will lead to a much greater probability. The situation is like that of crossing a street; some crossings are more dangerous than others, but none is absolutely safe. Devising proper procedures under these circumstances involves difficult social and political as well as medical questions.

The concept of a threshold for carcinogens has been intensely debated over the past few years. At this time, it is fair to say that, although there may be cases in which such thresholds appear to exist, they have not yet been reliably demonstrated.

Several instances are now known—for example, cigarette smoking and aflatoxin—in which the human dose–response curves for cancer appear to pass through or near zero and can be regarded as linear over the lower dose regions. In other words, even a very small dose would have an effect, and the effect is directly proportional to the size of the dose.

Although there are as yet no reliable examples, a true threshold may exist for some chemicals as cancer-causing agents. The probability that such thresholds exist is suggested by the presence of mechanisms that are known to inactivate harmful chemicals in varying degrees.†

The Congress has set up a number of federal agencies designed to protect the public against poisonous chemicals and advance the cause of safety. The Environmental Protection Agency has primary authority over much of this area. The Toxic Substances Control Act (TSCA), the Clean Air Act, the Safe Drinking Water Act, and the Occupational Health and Safety Act have been added to earlier legislation on pure food and drugs. Since many of these laws are new and since experience with their administration is limited, Congress will likely find ways to provide even

*There is a qualitative difference between carcinogens and other toxins. A low dose of a kidney poison, for example, may irreversibly damage a small fraction of kidney cells, but not constitute a threat to life. A very low dose of a carcinogen, in theory, could occasion the neoplastic transformation in only one or a few cells; yet, should they escape the immunologic surveillance mechanisms and reproduce sufficiently, lethal cancer might result.

†Generalizations concerning carcinogens are fraught with risk, however. The specific change(s) in the cell genetic apparatus are unknown as is the rate of spontaneous carcinogenesis. To the extent that the alterations in DNA occasioned by mutagens are repairable by the known efficient repair enzymes of normal cells, such effects may find no biological expression. Some potential carcinogens are active only after they have been metabolically altered by the liver or other cells; other carcinogens are excreted or inactivated by the body. Hence, it seems unreasonable to consider that there should be a single form of the dose response curve, differing only in slope (potency).

more effective protection. We must learn to deal more rapidly with the large numbers of toxic and potentially toxic substances in commerce and in our environment and at the same time to encourage development of safer products, such as less harmful pesticides and new pharmaceuticals. Although much has been and is being accomplished, experience with the new laws has quite naturally illuminated some areas of difficulty.

A major uncertainty today relates to the concept embodied in the legislation established in TSCA of "unreasonable risk." Each of us voluntarily accepts certain risks. The dangers of smoking are well advertised. A cigarette smoker is essentially making a personal risk–benefit assessment. But there is no practical alternative to using the community water supply or breathing air. In many instances, as in choosing modes of travel or buying paint, the individual may not be fully aware of the possible risks involved. The administrator of TSCA is asked to evaluate whether the benefits from the manufacture of a particular chemical justify the imposition of involuntary risks. The situation is still more complex when the benefits accrue to a different group from those who suffer the risks.

The problems are not easy ones, and the assignments given by law to federal regulators are difficult in the extreme. The most difficult decisions, perhaps, are those of the Commissioner of Food and Drugs, who puts some patients at risk if he grants approval for a new and possibly useful drug and others at risk if he refuses it.

It is becoming recognized that total safety is unattainable and that benefits from the use of chemicals range from trivial (and not worth even a minor risk) to vitally significant (and worth a substantial risk). It is clear that quantitative assessments are desirable to arrive at balanced judgments as to the full consequences—both social benefits and social costs—of using or not using a particular suspect chemical. What is much less clear is how to quantify benefits when they involve health, environmental quality, and other value-laden areas.

In part because of the paucity of basic data and of experience in evaluation methods, society has tended to respond erratically to assertions of chemical threats. In the case of the 1973 restriction on the use of spray adhesives, federal response proved hasty and overreactive. On the other hand, the action to curb the exposure of workers to asbestos after health effects had been demonstrated came at a rate that many observers considered unduly slow. Indications are that we are now entering an era in which there will be increasing demand for quantitative risk–benefit assessments. This balancing process could lead to a new formulation of acceptable risk. Acceptability is a social, not a scientific, criterion.

What is the best procedure for those most affected (or their representatives) to balance risks and benefits? In most current situations, a regulatory agency makes the judgments. In others, the Congress has laid down specific guidelines for acceptable risks—air-pollution laws, for example.

In some countries, tenured boards have been set up, with representations from the various constituencies affected by benefits and risks. Although this approach may be more cumbersome, it provides for open public debate on an issue and its consequences prior to the final decision.

Sociopolitical balances are mainly shaped by current value judgments, but they may also involve technical considerations. One is the development of a quantitative risk-assessment procedure. Others involve the procedures for estimates of benefits and costs of control. The development of quantitative assessments and their accurate interpretation to the people at risk will be difficult. The applicable science is still very much in its infancy, and precise estimates will rarely be possible; yet, such estimates, even crude ones, are better than none.

The public response to new scientific findings concerning chemical risks may be influenced by a variety of political and social factors not directly related to the evidence. For example, it is known that municipal referendums on adoption of water-supply fluoridation may be affected by such considerations as administrative methods and attitudes toward government.

RESEARCH RESOURCES

There are several resources available to develop the needed knowledge on environmental and health effects of chemicals: the academic community; the National Institutes of Health, especially the National Cancer Institute (NCI) and the National Institute of Environmental Health Science (NIEHS); the National Institute of Occupational Safety and Health (NIOSH); the National Center for Toxicological Research (NCTR); the Environmental Protection Agency (EPA); the Department of Energy; and, to a lesser extent, other federal agencies.

Nevertheless, a number of shortcomings remain. The greatest need lies in the area of basic research. We are only now beginning to understand the fundamental nature of cancer. Just as medieval communities were helpless in the face of bubonic plague because they did not realize that it was carried by rats and fleas, so we may be unable to check the inroads of cancer until we understand its etiology. Meanwhile, we must do the best we can. There is clearly a need for better coordination and especially better linkage between the more research-oriented units and the regulatory agencies. The new National Toxicology Program, bringing together the NCI Bioassay activity and parts of NCTR and NIEHS under the direction of the director of NIEHS, and the oversight of an executive committee that includes the heads of the regulatory agencies, could improve this linkage.

A present limitation in this field is the shortage of trained toxicologists and epidemiologists.

The impacts of chemicals on nonhuman species are not receiving sufficient attention at the national level. As a result, there are difficulties in making judgments on the risks many chemicals pose for them.

A field that has been seriously neglected, but which now has been recognized by both EPA and the Congress, is research to anticipate future problems. It is desirable to anticipate potential problems from chemicals as early as possible, perhaps at the early industrial development stage. An early-warning research program could establish a basis for developing low-risk alternatives or needed control procedures; such measures would be superior to the present tendency to wait for problems to emerge as full-blown crises. Such anticipatory research must, of course, be based on carefully reasoned and highly effective technological forecasting.

STANDARDIZATION OF TESTS

Both administrators and manufacturers understandably seek as much simplification and standardization of safety testing as possible. While some standardization is desirable (such as minimum standards of good laboratory practices), there is a danger in carrying it too far too soon. Premature freezing of techniques can stifle the development of what is still an infant science. It would be unfortunate to lock in procedures that may not be adequately informative or efficient. In a rapidly evolving field such as this, a best-available-procedures approach to safety evaluation may be best.

INTERNATIONAL COLLABORATION

Ongoing research in the United States needs to be closely linked with parallel activities in other nations, not only because of what can be learned from their experience, but also because of the influence the United States has on standards elsewhere. To these ends there is also a need for mechanisms to insure that international resources for such evaluations are used effectively, without unnecessary duplication or overlap, and that they take into account differences in physical and cultural environments.

The World Health Organization has authorized a major expansion of its program for evaluating chemical safety for humans through a greater (and to some degree decentralized) reliance on cooperative centers throughout the world. Similar resources for evaluating environmental effects of chemicals are needed but are not now available.

OUTLOOK

The following outlook section on toxic substances in the environment is based on information extracted from the chapter and covers trends anticipated in the near future, approximately five years.

Some toxic chemical hazards are well known; others remain poorly defined. Gaps in our knowledge of causes and effects are serious, and the need to fill them is urgent. More knowledge will help to reduce or eliminate the hazards from toxic agents while permitting us to enjoy the benefits of the great majority of chemicals without fear of harm to human health or the environment.

Manufactured chemicals have become ubiquitous. Their production is steadily rising and their uses multiplying. Over the next five years, a great number of new chemicals will be introduced—perhaps as many as 1,000 annually.

But we are still struggling with problems from the past. Chemical wastes were often dumped haphazardly in disposal sites, and some of these sites (e.g., the Love Canal near Niagara Falls, New York) were later used for landfill. Estimates are that hundreds of these abandoned chemical dumps exist, most still undiscovered. Many will be found in the coming years, and the government will have to establish procedures for dealing with them. Thus far, the human health hazards posed by these discarded and refilled dumps have not been fully evaluated.

Highly persistent chemicals, such as PCB's, are continuing to build up in the sediments of our freshwater lakes and streams. These chemicals move up the food chain from microorganisms and other creatures to fish and finally to humans. Through this biomagnification, such chemicals are further accumulating in humans and nonhuman species. No effective means have yet been developed to sequester, degrade, contain, or remove chemically laden sediments. Moreover, there are fears that some of the proposed solutions to the dilemma may actually worsen the situation.

A rising proportion of the nation's electrical power will probably be generated by coal. Although power-plant pollution controls will be maintained, the total environmental burden of such air contaminants as acid particulates may well increase. While the general dimensions of the health impacts of these and other pollutants are being studied, detailed knowledge on the impacts of actual concentrations of specific pollutants on specific segments of the population is still lacking.

Rapid advances in analytical techniques are making it possible to detect smaller and smaller amounts of toxic chemicals—as small as parts per billion. In consequence, trace chemicals will be discovered in previously unsuspected places. Already trace amounts of organic compounds, some identified as weak carcinogens or mutagens in experimental animals, have been found in drinking water. Currently there is no fully acceptable way of evaluating health risks (if any) from drinking water containing traces of toxic chemicals. A related concern is that the chlorination of drinking water may, through interaction with otherwise harmless organic substances, produce potential carcinogens.

A great deal more attention is being focused on workplace exposures to chemicals following the passage of the Occupational Safety and Health Act. Some 20 chemicals or processes have already been linked to cancer in exposed workers. More such associations, involving mutagens as well as carcinogens, may be found in the next few years as new evidence of long-term exposure effects is discovered.

IMPLICATIONS

Known and potential problems stemming from toxic chemicals raise a range of implications, some general and others specific. Underlying a number of them is the challenge to the threshold concept, which holds that certain low levels of exposures to chemicals are totally without effect. The threshold theory has been intensely debated over the past few years, particularly with regard to cancer. At this time, it is fair to say that although such thresholds may exist for some toxic chemicals, none has yet been reliably demonstrated for any carcinogen.

As a result of the growing conviction that there is no completely safe level of exposure to any carcinogen, we are entering an era of quantitative risk–benefit assessment. If complete safety is unattainable, we must be able to evaluate the benefits of individual chemicals: Are they vitally significant (and worth a significant risk) or trivial (and not worth even a minor risk)?

The techniques of quantitative risk assessment are still in their infancy. Over the next five years, there will be increasing efforts to evolve better methods of extrapolating the results of tests on laboratory animals to humans—in particular extrapolating from high doses given to test animals to long-term low doses to which humans are generally exposed—and to gain better understanding of the biological processes involved. The goal is to move, as rapidly as possible, to quantitative descriptions of dose–response curves at low-dose levels.

Even more elusive than quantitative risk assessment is the quantitative evaluation of benefits. There will be efforts in the near future to develop more sophisticated means of both presenting and interpreting quantitative risk–benefit evidence so it can be understood by affected groups and responsible government officials.

This approach to safety evaluation will increasingly be applied to new chemicals, chemicals in the workplace, chemicals in the environment, and the reassessment of existing household chemicals. There will be no single definition of acceptable risk. Each decision will depend on the uses of a particular chemical, its potential substitutes, the risk–benefit ratio, and yet other factors—all within a context of prevailing social values. Such decisions will be as important in determining control strategies as in arriving at acceptable risks.

Continuing efforts will be required to develop new technology for effectively sequestering, degrading, containing, or removing chemically laden wastes in freshwater lakes and rivers as well as in abandoned dumps. The challenge is to isolate or neutralize such wastes.

More knowledge will be gained of the specific effects and components of air pollutants, such as acid particulates, on specific segments of the population. The results could make possible the design of more efficient and selective controls.

Epidemiological studies will further refine knowledge of the effects of chemical exposures in the workplace. There will be more laboratory evaluations of those chemicals now in use and new chemicals being introduced. There will also be increased efforts to uncover presently unrecognized chemical hazards in the workplace. New and better control methods to protect workers from dangerous chemicals will be stressed, including education in their handling. What is further needed is an evaluation of the prevalence and seriousness of various occupational diseases as a basis for setting control priorities.

The cancer risk involved in chlorination of drinking water will be more thoroughly assessed and, if necessary, alternative methods of controlling waterborne pathogens investigated. Also, better methods will be sought to evaluate the hazards of trace chemicals in drinking water.

The growing awareness of potential chemical hazards is a positive development. During the next five years, we will come to know more about the nature, extent, and seriousness of the problems posed by toxic chemicals. Indeed, new problems may be uncovered. But as more knowledge is gained, we will learn how to better control and reduce toxic chemical hazards.

Increased training now getting underway will add significantly to the ranks of qualified toxicologists and epidemiologists within the next five years but will only begin to meet major needs toward the end of that period.

REFERENCES

1. *Selenium* (NRC Committee on Medical and Biologic Effects of Environmental Pollutants). Washington, D.C.: National Academy of Sciences, 1976.
2. Fenical, W. Halogenation in the Rhodophyta. *Journal of Phycology* 11(3):245–259, 1975.
3. Deijer, J., and A. Jernelov. *Environmental Health Perspectives* 25:43–45, 1978.
4. Report of Workshop Panel to Select Organic Compounds Hazardous to the Environment (National Science Foundation). Springfield, Va.: National Technical Information Service, No. PB-287-996, 1975. Research Program on Hazard Priority Ranking of Manufactured Chemicals: Phase II Chemical Report (prepared for the National Science Foundation by the Stanford Research Institute). Springfield, Va.: National Technical Information Service, Nos. PB-263-161 (Chemical Nos. 1-20); PB-263-162 (Chemical Nos. 21-40); PB-263-163 (Chemical Nos. 41-60); PB-263-164 (Chemical Nos. 61-79); PB-263-165 (Appendix and References), 1975.
5. *Annual Reports on U.S. Production and Sales of Synthetic Organic Chemicals, 1917–1976.* Washington, D.C.: U.S. International Trade Commission.
6. *Implementing the Toxic Substances Control Act: Where We Stand.* Washington, D.C.: EPA Office of Toxic Substances, 1978.
7. Metcalf, R.L. Biological Fate and Transformation of Pollutants in Water. *Chemical and Biological Fate of Pollutants in the Environment*, Part 2, I.H. Suffet, ed. New York: John Wiley & Sons, 1975.

8. *Principles for Evaluating Chemicals in the Environment.* (NRC Committee for the Working Conference on Principles of Protocols for Evaluating Chemicals in the Environment). Washington, D.C.: National Academy of Sciences, 1975.

9. *Polychlorinated Biphenyls and Terphenyls*(U.N. Environmental Program and WHO). Geneva: World Health Organization, 1976, p. 65.

10. *Environmental Resources* 4:1–69, 1971.

11. *Ozone and Other Photochemical Oxidants* (NRC Committee on Medical and Biologic Effects of Environmental Pollutants). Washington, D.C.: National Academy of Sciences, 1977.

12. MacLeod, C.M. A Report of the Panel on Herbicides of the President's Science Advisory Committee. Washington, D.C.: Office of Science and Technology, 1971.

13. Firestone, D. *Ecological Bulletin* (Stockholm) 27 (in press).

14. Huff, J.E., and J.S. Wassom. Chlorinated Dibenzodioxins and Dibenzofurans, A Bibliography. *Environmental Health Perspectives*, Experimental Issue 5, September 1973.

15. Axelrod, David, personal communication.

16. *Love Canal: Public Health Time Bomb.* Albany: New York State Department of Health, 1978.

17. *Air Quality and Automobile Emission Control: Summary Rept.* (NRC Coordinating Committee on Air Quality Standards), Vol. 1. 1974, p. 11.

18. *Federal Register*, 43:2229–2240, 1978.

19. *Airborne Particles* (NRC Committee on Medical and Biologic Effects of Environmental Pollutants), 1977.

20. *Sulfur Oxides* (NRC Committee on Sulphur Oxides). National Academy of Sciences, 1978.

21. Cederlof, R., *et al.* Air Pollution and Cancer: Risk Assessment Methodology and Epidemiological Evidence. *Environmental Health Perspectives* 22:1–13, February 1978.

22. *Drinking Water and Health* (NRC Committee on Safe Drinking Water). Washington, D.C.: National Academy of Sciences, 1977, pp. 439 and 715.

23. *Epidemiological Studies of Cancer Frequency and Certain Organic Constituents of Drinking Water—A Review of Recent Literature Published and Unpublished* (NRC Safe Drinking Water Committee). Washington, D.C.: National Academy of Sciences, 1978.

24. *Federal Register*43:5756–5780, 1978.

25. *Annals of the New York Academy of Sciences*, Vol. 298, 1977.

26. Cole, P., and M.B. Goldman. Environmental Factors, Occupation. *Persons at High Risk of Cancer: An Approach to Cancer Etiology and Control*, Joseph F. Fraumeni, Jr., ed. New York: Academic Press, 1975, p. 167.

27. *Estimates of the Fraction of Cancer in the U.S. Related to Occupational Factors.* Washington, D.C.: DHEW, September 15, 1978.

28. Whorton, D., *et al.* Infertility in Male Pesticide Workers. *Lancet* 2:1259–1261, 1977.

29. Glass, R.I., *et al.* Sperm Count Depression in Pesticide Applicators Exposed to Dibromochloropropane. *American Journal of Epidemiology*, Vol. 9, 1979 (in press).

30. Corbett, T.H., *et al.* Birth Defects Among Children of Nurse Anesthetists. *Anesthesiology* 41:341–344, 1974.

31. Allen, N. Chemical Neurotoxins in Industry and the Environment. In: *The Nervous System*. New York: Raven Press, 1975, pp. 235–248.

32. Blum, A., *et al.* Children Absorb Tris-BP Flame Retardant from Sleepwear: Urine Contains the Mutagenic Metabolite, 2,3-Dibromopropanol. *Science* 201(4360):1020–1023, 1978.

33. Linsell, C.A., and F.G. Peers. Aflatoxin and Liver Cell Cancer. Symposium on Liver Carcinoma. *Transactions of the Royal Society of Tropical Medicine and Hygiene* 71(6), 471–473, 1977.

34. *Saccharin: Technical Assessment of Risks and Benefits,* Part 1 (IOM and NRC Committee for a Study on Saccharin and Food Safety Policy). Washington, D.C.: National Academy of Sciences, 1978, p. 16. Also, *Food Safety Policy: Scientific and Societal Considerations* (IOM and NRC Committee for a Study on Saccharin and Food Safety Policy). Washington, D.C.: National Academy of Sciences, 1979.

35. *Human Health and the Environment—Some Research Needs—2* (Second Task Force for Research Planning in Environmental Health Science) (DHEW Pub. No. [NIH] 77-1277). Washington, D.C.: U.S. Government Printing Office, 1977.

36. *Asbestos* (IARC Monographs on the Evaluation of the Carcinogenic Risk of Chemicals to Man, Vol. 14). Lyon, France: International Agency for Research on Cancer, 1977, p. 102

37. *Arsenic* (NRC Committee on Medical and Biological Effects of Environmental Pollutants). Washington, D.C.: National Academy of Sciences, 1976.

38. Eisenbud, M. *The Health Implications of Methyl Mercury in Adirondacks Lakes.* Albany: New York State Department of Health, 1978.

39. Friberg, L., *et al. Cadmium in the Environment.* Cleveland, Ohio: CRC Press, 1974.

40. Elinder, C., *et al.* Cadmium in Kidney Cortex, Livers, and Pancreas from Swedish Autopsies. *Archives of Environmental Health* 31(6):292–302, 1976.

41. *Air Quality Criteria for Lead* (EPA Pub. No. 600/8-77-017). Washington, D.C.: Office of Research and Development, 1977, pp. 1-9.

42. Savage, E.P. *National Study to Determine Levels of Chlorinated Hydrocarbon Insecticides in Human Milk,* 1975–76, and supplementary report, 1977. Springfield, Va.: National Technical Information Service, No. PB-284-393.

43. Allen, J.R., and D.A. Barsotti. The Effects of Transplacental and Mammary Movement of PCBs in Infant Rhesus Monkeys. *Toxicology* 6:331–340, 1976.

44. Proceedings of a Workshop on Scientific Aspects of Polybrominated Biphenyls, Oct. 24–25, 1977 at Michigan State Univ., *Environmental Health Perspectives,* Vol. 23, April 1978.

45. International Agency for Research on Cancer. *The Evaluation of the Carcinogenic Risk of Chemicals to Humans: Some Organochlorine Pesticides, 5, 1974. Some Halogenated Hydrocarbons* (in press).

46. Resoconto, D. *Riunione Di Esperti Sul Problemi Determinati Dall-Inquinamento da Diossina,* Proceedings of the Expert Meeting on the Problems Raised by TCDD Pollution. Milan: Commission on the European Communities, 30 September–1 October, 1976.

47. Chiazze, L., D. Levin, and T. Silverman. Recent Changes in Estimated Cancer Mortality. *Incidence of Cancer in Humans.* Cold Spring Harbor, N.Y.: Cold Spring Harbor Laboratories, 1977, pp. 33–44.

48. Levin, D.L., *et al.* In: *Cancer Rates and Risks,* 2nd ed. Bethesda, Md.: National Institutes of Health, 1974.

49. Higginson, J., and C.S. Muir. The Role of Epidemiology in Elucidating the Importance of Environmental Factors in Human Cancer. *Cancer Detection and Prevention* 1:79–105, 1976.

50. EPA Proposed Rules for Pesticide Programs: "Mutagenicity Testing." *Federal Register* 43:37388–37394, 1978.

51. *Mercury* (U.N. Environment Program and WHO). Geneva: World Health Organization, 1976, p. 60.

52. Hutchinson, G. E. *A Treatise on Limnology,* 1. Chemistry of Lakes. New York: John Wiley & Sons, 1957, p. 386.

53. Likens, Eugene, personal communication.

54. Nisbet, Ian, personal communication.

IV INSTITUTIONS

(Scott Tobias. © 1978 MIT *Technique*)

10 Academic Science and Graduate Education

INTRODUCTION

Preceding chapters of this report have presented some of the more recent accomplishments of American basic and applied science. Much of this work has been carried out in U.S. universities by their faculties and their graduate and postdoctoral students. With few exceptions, the individuals responsible for these advances have been educated in the public and private universities and colleges distributed throughout the nation.

Thus the universities are a major force in American science. Today, their faculties, students, and support staff are responsible for more than half the basic scientific research performed in this country. Their graduates in science and engineering populate private and government laboratories and engineering centers throughout the United States.

The conduct of basic research in U.S. universities historically has been closely linked to the graduate education of scientists and engineers. One feeds the other, in contrast to the pattern in Europe and the Soviet Union, where centralized research institutes are essentially separated from education systems. Basic science in the United States is dependent upon the coupling of research and education at the postgraduate level, so that the vitality of American science is a function of the vitality of academic institutions, particularly of the major U.S. universities that carry out most academic science and award most of the advanced degrees in the sciences and engineering.

Although federal support lessened in the early 1970's, the strong national commitment to fundamental scientific research continues—perhaps best symbolized by the provision of $3 billion federal in 1978 to fund research and graduate education in U.S. universities. As a result of this strong support, particularly during the growth years of the 1950's and 1960's, the United States has developed a broad, diversified, and highly competent academic research and related postgraduate capability. The universities are now and will continue to be remarkably productive both in increasing our understanding of nature and man and in educating new generations of scientists and engineers.

Nevertheless, the institutions of academic R&D have been facing serious problems that threaten their productivity and may well slow their present momentum. The earlier rate of growth in financial support for academic R&D has slowed to relatively level funding in constant dollars over the past decade. Averaged annual growth (in constant 1972 dollars) from 1953 to 1960 was 12 percent; from 1960 to 1964, 14 percent; from 1968 to 1974, essentially zero percent; and from 1974 to 1978, about 4 percent.[10] Yet over this period the academic research population has continued to grow, increasing the demand for research support. As research becomes more sophisticated, the required instrumentation, facilities, and supporting services become more expensive, adding further costs beyond those due to inflation. Federal agencies can now fund a smaller fraction of worthy research proposals; and other demands on university resources have sharply increased, causing some private institutions to draw upon their basic endowment capital.

The increased competition for research funding has led to longer, more detailed proposals, prepared and reviewed by scientists on time formerly spent on research. Funding agencies, in an effort to meet increased demand, have tended to reduce both the amounts and the time period of support. Funding available for direct support of research has also been eroded by rising indirect cost rates resulting in part from the need to satisfy regulations of federal and local agencies. These requirements often make serious demands on the time of the research investigator as well as on university administrators.

Given these problems, it is appropriate to ask whether academic science in the United States can maintain its high rank. There is concern that academic science is on the verge of a decline, and that it may be a decade before its decreased vitality and momentum are fully apparent, by which time the direction may be very difficult and expensive to reverse.

The period immediately ahead will require adjustments in the policies of universities and in their relationships with the federal government as the principal external sponsor of academic research. Some of these problems are now being addressed by *ad hoc* groups, notably the Sloan Commission on Government and Higher Education and the National Commission on Research. The general resolution of the problem, however, will require the thoughtful attention and best effort of all partners—universities, government, and industry—if the U.S. research enterprise is not to suffer.

SCIENTIFIC RESEARCH AND GRADUATE EDUCATION

American scientists and engineers begin their professional education as college undergraduates. Through introductory and intermediate courses in mathematics and the various natural sciences, they learn scientific principles and gain an overall view of the history and subject matter of the various disciplines. Inadequate education at this level is difficult for students to overcome later.

Graduate education in the sciences and engineering consists of advanced instruction in the current body of knowledge, methods of thought, and research techniques, all of which must be assimilated by those gaining expertise in these professions. However, in addition to receiving highly specialized instruction, graduate students participate in and contribute to research, under the guidance of senior faculty members, often as paid assistants. This firsthand experience in the laboratory starts them on the path to maturity as independent investigators. Formal graduate education concludes with the writing of a dissertation based on the student's research and intended to be an original contribution to knowledge.

Many of those who receive doctoral degrees in the sciences subsequently accept postdoctoral research appointments at academic institutions. Those who hold these positions are given the opportunity to engage in one or two years of advanced research, often of their own choosing and design, at leading scientific centers having specialized facilities and faculties. Postdoctoral appointments, which usually place few other responsibilities on their holders, give younger scientists the variety and intensity of training they need to become fully qualified investigators.

Active research, then, permeates academic science and is interwoven with the education of both undergraduates and graduate students. This interrelated arrangement not only gives students a better understanding of the current status of their science but also provides members of the faculty and their graduate assistants with the intellectual stimulus that strengthens and enlivens teaching.

ENROLLMENT, DEGREES, AND JOBS

THE 1960'S

The 1960's were a period of expansion for higher education, including academic science, in the United States. Undergraduate enrollment increased by an average 7 percent a year, which meant that colleges and universities everywhere needed more teachers, administrators, buildings, and facilities. During this period, the federal government substantially increased its funding for scientific investigations sponsored by federal agencies and for graduate fellowships and traineeships, particularly in the physical and biomedical sciences.

The natural consequence of these trends was a substantial increase in the number of graduate students and in the number of persons gaining Ph.D. or Sc.D. degrees. Between 1965 and 1971, the awarding of doctorates in all fields increased at an annual rate of 11.7 percent; the number of those with doctorates in science or engineering rose at a rate of 10.3 percent.[1] By 1971, the nation's universities were awarding more than 30,000 doctorates annually, both scientific and nonscientific, and many of those earning doctorates were able to secure academic jobs.

THE 1970'S

However, the early 1970's marked the beginning of a difficult period of readjustment for higher education, particularly in the natural sciences. Large annual increases in earlier years of funds for scientific research by academic investigators were replaced by more modest increments or by actual declines measured in dollars of constant purchasing power. The number of doctorates awarded annually in the physical sciences, engineering, and the life sciences began to fall, as did graduate enrollments in the physical sciences, mathematics, and engineering. After reaching a peak of 33,755 in 1973, awards of doctoral degrees in all fields, both scientific and nonscientific, started to decline. As shown in Table 8, by 1977, doctoral production in the sciences and engineering had dropped from the 1971 figure of 14,311 to 11,777. Undergraduate enrollments, however, continued to increase during the decade, but at an average rate of only 4 percent.

General financial pressures within the universities to maintain their expanded establishments, combined with a slower rate of growth in undergraduate enrollment, also reduced opportunities for those seeking academic employment. Between 1969 and 1977 the proportion of new doctorate recipients in engineering, mathematics, and the physical sciences who secured academic positions dropped from 31 to 25 percent (see Table 9).

THE PRESENT AND THE FUTURE

Overall undergraduate enrollment in the nation's colleges and universities is expected to begin a decline within the next two or three years. The

TABLE 8 Doctorates Awarded in the United States

	Totals		Sciences and Engineering[a]	
Year	Number	Index	Number	Index
1960	9,732	100	4,674[b]	100
1965	16,340	168	8,307[b]	178
1970	29,500	303	13,637	292
1971	31,872	327	14,311	307
1972	33,044	340	13,984	299
1973	33,755	347	13,674	293
1974	33,046	340	12,950	277
1975	32,948	339	12,763	273
1976	32,936	338	12,222	261
1977	31,672	325	11,777	252

[a] Engineering, mathematics, physical sciences, and life sciences.
[b] From unpublished NRC tables.

SOURCES: National Research Council. *Summary Reports, 1977. Doctorate Recipients from U.S. Universities*. Washington.

TABLE 9 Employment Plans of New Doctorate Recipients (in percentages)

	Engineering Mathematics, and Physical Sciences		Nonsciences	
Fields of Employment	1969[a]	1977	1969[a]	1977
Postdoctoral study	20	28	2	4
Academic	31	25	80	70
Industry, government, and nonprofit	42	38	7	15
Other and unknown	6	9	11	11
TOTALS	100	100	100	100

[a] From unpublished NRC tables.

SOURCE: National Research Council. *Summary Report, 1977, Doctorate Recipients from U.S. Universities*. Washington.

reason is that the number of births in the United States has been declining since the end of the baby boom in 1964. As a result, we can expect a decline in the number of high school graduates seeking to enroll in universities during the early 1980's, returning to a situation that existed in the late 1950's.

This decline, however, will be more gradual than might be predicted on demographic grounds. There has been a significant increase in the number of adults—particularly women and minority-group members—attending college. More than half of all college students in 1977 were older than 21,[2] which was once the age at which the great majority of students received a first degree. Many of these older students, however, are found at community colleges and vocational institutions rather than at liberal arts colleges or universities.

A decline in the number of undergraduates, whatever its size, will probably not have a corresponding effect on the number of graduate students. Graduate enrollment in both the sciences and other fields is expected to continue to rise as more and more college graduates seek advanced degrees to bolster their chances on the job market. Many of these will be students who enter doctoral programs, only to drop out before they obtain their degree. Furthermore, foreign students comprise a substantial percentage of those enrolled in U.S. graduate schools, and that percentage is expected to continue growing for the next several years.[3]

Data for 1977–78 on first-year enrollments of graduate students suggest that earlier trends in individual disciplines may now be changing. First-year enrollments of graduate students in the physical sciences and engineering are up slightly, while first-year enrollments in the life sciences are slightly down.

Despite increased numbers of graduate students, total doctoral production—down by 25 percent in the physical sciences and engineering between 1971 and 1977[4]—seems likely to continue to decrease.

A smaller percentage of the new doctorate holders will obtain academic jobs than at any time in the recent past, partly because the number of undergraduates will decline and partly because the retirement rate of professors will be low. As of 1976, over half of those on college and university faculties had been hired during the preceding 15 years. Unless there is a marked movement among current faculty members to early retirement—an event that seems unlikely in a period of inflation and the recently legislated minimum mandatory retirement age of 70—the result will be relatively few openings for new faculty members until the 1990's.

Furthermore, as indicated earlier, a substantially larger proportion of individuals with recent doctorates in the sciences are seeking and obtaining jobs outside the universities than was true 10 years ago. This indicates that younger scientists are adjusting to the realities of the job market and the situation of the universities. Their ability to make this adjustment, however, is complicated by differences in demand. The demand for solid-state physicists, for example, has remained much higher than the demand for nuclear physicists. The demand for analytical chemists has stayed at a high level, even though the market for chemists generally has been relatively unfavorable. Predicting differences in future job markets is difficult, and it would seem to be increasingly desirable to search for modes of graduate and postdoctoral education that enable students in the sciences to adapt more easily to changes in demand.

MAINTAINING FACULTY BALANCE

The prospect of a period during which few younger scientists receive academic appointments brings with it the possibility of some impairment of the capacity of the universities to carry out basic research and sustain the vitality of their faculties. Younger faculty, as a group, bring to their work a high degree of enthusiasm, singleness of purpose, and freedom from administrative duties. As recent graduates, they also benefit from the currency of their knowledge and perspectives.

The existence of organized research units of the kind described in a subsequent section will enable universities to absorb some scientists and engineers without teaching responsibilities. The possibility that universities also will be engaged more extensively in applied research may provide additional employment opportunities for younger scientists and engineers. However, these positions are generally divorced from the central educational role of the universities.

Although the general outlook for faculty positions over the next five years is predictable, it will be important during this period to monitor and analyze such factors as the number of academic jobs available for younger faculty, trends in academic and nonacademic salaries, shifts of faculty out of academic life into other positions, retirement rates (including early retirement), and changes in tenure ratios, rules, and customs. Such studies should be helpful in assessing the need for remedial measures necessary to assure competent and balanced science faculties.

SUPPORT OF GRADUATE STUDENTS

During the past few years there has been a substantial shift in the sources of financial support for graduate students (see Table 10). For many years, most of the federal funds used to provide fellowships and traineeships for graduate students came from the G.I. bill. In the peak year of 1975, $370 million was provided for graduate students under this legislation,[5] but this

TABLE 10 Sources of Support for Full-Time Graduate Students in Sciences and Engineering in Doctorate-Granting Institutions (in percentages)

Sources of Major Support	1968	1971	1977
Federal	40	31	23
Institutional	32	37	37
Self	20	22	32
Other	8	10	8

SOURCES: National Science Foundation. *Graduate Student Support and Manpower Resources in Graduate Education, Fall 1970*, p. 73 and *Fall 1971*, p. 46; and *Graduate Science Education: Student Support and Postdoctorals, Fall 1977*, p. 77. Washington.

funding will fall to an estimated $207 million in 1979. Although substantially lesser in amount, the National Institutes of Health (NIH) training program in the biomedical sciences ($130 million in 1977)[15] and the National Science Foundation (NSF) graduate fellowship program ($11 million in 1978) have been especially helpful in providing assistance to graduate students in the sciences. However, these programs have declined in some areas or remained at the same level. The proportion of graduate students whose major source of support was the federal government has declined, from 40 percent in 1968 to 23 percent in 1977. Most of the difference has been made up by students themselves (by working, by the earnings of spouses, and by borrowing from private lenders) and by the universities, which have increased their funding for fellowships. One effect of this changed pattern has been to divert larger amounts of university funds to the support of graduate students, with a resulting reduction in the university funds available for other university functions.

WOMEN AND MINORITIES

The proportion of female graduate students in the universities increased from 38 percent in 1969 to 47 percent in 1978.[6] This proportion will probably increase very slowly over the decade ahead. The majority of women at the graduate level are still found in education, the humanities, and the social sciences.

Both the number of women receiving a doctorate in science or engineering and the proportion of all doctoral recipients in science or engineering who are women increased significantly between 1968 and 1977. The number of women receiving doctorates in science and engineering rose from 1,306 to 3,292; the proportion rose from 9.6 percent to 18.0 percent (see Table 11). These numbers and proportions, however,

TABLE 11 Women Securing Doctorates in Sciences and Engineering

	1968		1977	
Field	Number	Percent of Doctorates Awarded	Number	Percent of Doctorates Awarded
Physical sciences	232	5.0	431	9.9
Life sciences	510	13.8	957	20.1
Social sciences	552	15.8	1,830	28.1
Engineering	12	0.4	74	2.8
TOTALS	1,306	8.9	3,292	18.0

SOURCE: National Research Council. *Summary Report, 1977, Doctorate Recipients from U.S. Universities*, p. 8. Washington.

are both heavily weighted by the relatively large numbers of women in the life sciences, biology and medicine, and in the social sciences.

The number and proportion of women on university faculties will probably remain relatively small over the next decade, even with expected increases. As noted in the previous section of this chapter, there will be relatively few academic positions available during the 1980's, except in collegiate institutions not offering graduate degrees in the physical and life sciences and engineering.

Progress in enlarging the representation of nonoriental minorities in science has been slow and irregular. In 1977, for example, the proportion of doctorates granted to minority groups was quite small compared with that of whites, as shown in Table 12.

The low representation of blacks and other minorities among recipients of doctorates can be significantly increased only as efforts to erase the effects of generations of discrimination become effective. This will take time. Meanwhile, efforts must continue to strengthen the educational background of minority students who need help in order to deal successfully with graduate studies.

FUNDING OF ACADEMIC SCIENCE

Funds separately budgeted for the support of academic R&D (that is, funds designated specifically for research) come from a variety of sources: the general funds of universities, the federal government, state government, private foundations, industry, and individual donors. Virtually all the federal government's support comes to the universities as grant and

TABLE 12 Doctorate Recipient by Racial/Ethnic Group and Fields of Science (fiscal year 1977)

Racial-Ethnic Group	All Fields		Physical Sciences		Engineering		Life Sciences	
	No.	%	No.	%	No.	%	No.	%
American Indian	215	1	16	[b]	12	1	37	1
Asian	907	3	223	6	248	14	182	4
Black	1,186	4	44	1	15	1	68	2
Hispanic	471	2	56	2	22	1	35	1
White	23,411	86	3,051	85	1,412	79	3,506	88
Other and unknown	1,181	4	210	6	84	5	158	4
TOTAL[a]	27,371		3,600		1,793		3,986	

[a] Include U.S. citizens and non-U.S. with permanent visas.
[b] Less than 0.5 percent.
SOURCE: National Research Council. *Summary Report, 1977, Doctorate Recipients from U.S. Universities*, p. 16. Washington.

contract funds and, in 1977, constituted two-thirds of the separately budgeted funding. The universities themselves, the next most important source of funds for separately budgeted research, provided 22 percent. Private foundations and industry provided 8 percent and 3 percent, respectively.[7]

University contributions from general funds that are not separately budgeted are less clearly identifiable but they are large. Universities provide faculty and support staff salaries and benefits, as well as the general facilities and environment that make universities a particularly productive home for basic science.

In the case of public universities, these general funds are derived primarily from funds that the universities receive from state governments. In the case of private universities, the funds are taken primarily from tuition revenues, endowment income, and private gifts. The data presented in this section reflect only separately budgeted funds and therefore understate the actual level of university financial support for academic science.

LEVELS OF SUPPORT

Total support for academic R&D (that is, basic research, applied research, and development) has risen steadily since World War II in real terms except for the period 1968–74—when there was no increase in constant dollars, and the federal contribution actually declined by 8 percent, causing universities to increase their contribution in compensation. During the period 1974–79 there has been a 17 percent increase in support, which is helping to offset the adverse effects resulting from the years of level funding (see Table 13).

The funds over the last decade have had to be shared among a growing group of claimants. The number of scientists and engineers with doctorates employed by colleges and universities engaged primarily in research and development increased by 50 percent (from 33,600 to 50,400) from 1961 to 1969 as undergraduate enrollment and federal funds for research and development rose.[8] This group has now matured, and, as principal investigators, many now seek support. The combined effect of a slow increase in available funds and a rapid increase in claimants has been a general decline in the fraction of proposals funded by federal agencies. For example, in 1967, 82 percent of the proposals found worthy of support by the NIH were funded. By 1977, this percentage had fallen to 48.[9]

Funding for Basic Research

Basic research comprises about 68 percent of all research conducted in universities. In constant dollars, total support for basic academic research declined by 8 percent between 1968 and 1976, and federal support for basic academic science fell by 10 percent in constant dollars over that period.[10] To offset the cut in federal funds, universities increased their support of separately budgeted research from $334 million to $360 million in constant 1972 dollars, or 8 percent, over the 1968–76 period.[11] This increase in expenditures strained the resources of many of the universities.

Virtually level expenditures for all academic science, combined with the increased number of eligible investigators, had significant effects over this

TABLE 13 Recent Trends in Expenditures for Academic Research and Development (constant dollars in 1972 dollars)

	All Sources				Federal Government			
Year	Current Dollars	Percent Increase Over 1974	Constant Dollars	Percent Increase Over 1974	Current Dollars	Percent Increase Over 1974	Constant Dollars	Percent Increase Over 1974
1974	3,023		2,606		2,032		1,751	
1975	3,409	13	2,681	3	2,288	12	1,799	3
1976	3,730	23	2,789	7	2,501	23	1,870	7
1977	4,064	34	2,870	10	2,717	34	1,919	10
1978	4,585	52	3,018	16	3,075	51	2,024	16
1979	4,965	64	3,055	17	3,315	63	2,040	17

SOURCE: National Science Foundation. *National Patterns of R&D Resources: Funds and Personnel in the United States, 1953–1978/79*, pp. 29, 37. Washington.

1968–76 period, in addition to a reduction in the fraction of proposals accepted. As federal agencies sought to stretch their funds, budgets proposed by academic scientists were often cut, and a smaller proportion of proposals, as noted above, were funded for shorter time periods. Increasingly stringent economies became necessary in such things as shop and technical services, equipment purchases, graphics, and travel. Meanwhile, the rising competition for funds forced many researchers to devote much of their time to preparing lengthier and more detailed applications for research grants and contracts.

In recognition of the need, the federal government has recently raised its level of support. Between 1974 and 1977, federal funds for basic scientific research were increased 10 percent, in constant dollars. Acceptance by Congress of the President's proposed funding for basic scientific research in the universities during fiscal year 1980 would result in a 15 percent total increase, in constant dollars, between 1976 and 1980.

EFFECT OF CHANGES IN NATIONAL PRIORITIES

Changes in national priorities lead to changes in the research budgets of federal agencies. Between 1968 and 1978, for example, the portion of federal research funds accounted for by the Department of Health, Education, and Welfare (primarily NIH), NSF, and other agencies rose from 50 percent to 64 percent, while the portion accounted for by the Department of Defense, the Department of Energy, and the National Aeronautics and Space Administration (NASA) dropped from 50 percent to 33 percent.[16]

Such shifts affect both the universities and the agencies. Some individual researchers must change the direction of their research or seek new sponsors; opportunities for graduate students decline in some fields and rise in others; new research groups and research structures (such as centers and institutes) are formed, while older ones may be disbanded. Mission-oriented agencies seeking solutions for real and pressing problems tend to emphasize applied research and development at the expense of their support of basic research. NSF, which has sole responsibility among federal agencies for assuring adequate balance and general support of basic science, is faced with many demands as a result of decisions of other agencies. If these shifts continue, as is likely, it is possible that the coming years will see an increase in the proportion of university-based applied science and specialized institutes and nonteaching research centers.

CAPITAL EXPENDITURES

When funds are relatively plentiful (as they were in the 1960's), capital investments tend to rise. When funds are relatively scarce (as they have been in the 1970's), the tendency is to pay operating costs first and use what may be left over for capital investment. This is what accounts for the reduced federal investment in capital facilities for basic scientific research. Academic research in the natural sciences requires capital investment in buildings and large research installations. Laboratory equipment and special instrumentation are also capital costs, but their useful life is typically shorter than that of buildings.

Federal obligations for academic research facilities dropped from $212 million in 1966 to $28 million in 1975, in constant 1972 dollars.[12] Federal funds available to the universities for purchasing laboratory equipment also declined, although not as greatly. The proportion of funds allocated to NIH for laboratory equipment, for example, declined from 11.7 percent in 1966 to 5.7 percent in 1974.[13]

The federal government's decision to reduce allocations for capital investment and equipment was a rational response to budgetary problems. However, the period of low investment was so protracted that many research installations have become, or are becoming, obsolescent. Some experiments simply cannot be performed in existing facilities with an earlier generation of equipment. In short, the academic research system is consuming its capital, and the grace period during which the system could operate effectively on earlier capital investments is running out.

The need to lift the level of investment in research facilities and equipment has been recognized by the Administration and by Congress. Special funds—$93 million in 1978, $88 million in 1979, were provided, and $101 million is proposed by NSF for new capital investment and equipment in its 1980 appropriation request. Several years of funding at this general level will be required to refurbish the facilities and instruments necessary in modern academic science.

SUPPORT STRUCTURES

Effective research in scientific laboratories also requires an array of support services—shops, libraries, equipment, technicians. Some of the most urgent problems of the universities over the next five years will arise in funding these services.

PROJECT FUNDING

Although federal agencies support some broad areas of investigation through large grants or contracts that finance the work of many individuals and groups, the dominant mode is support of individual projects—that is, support for a defined research task spelled out in a proposal made by an investigator to a federal agency. This method of operation has many fundamental strengths:

- Recognition and encouragement of initiative on the part of individual investigators.
- Provision of an efficient mechanism for matching the legitimate needs of mission-oriented agencies for research in given areas with the legitimate desire of scientists to work on problems that interest them.
- Large-scale participation by good scientists in the decision process and assurance of quality through peer review of individual projects, while retaining for responsible government officials large-scale program decisions.
- Establishment of a means for shifting agency program emphasis without substantial disruption of specific research projects.
- Avoidance of the necessity, in distributing federal research funds, to judge the relative merits of universities.
- Disclosure to the Congress and the public of the content of research financed from public funds.

WEAKNESSES OF THE PROJECT SYSTEM

Yet the project system as currently administered is not without flaws. One problem, noted earlier, is a decreasing length of average periods of support for projects. The system is one that has been characterized as financing 10-year ideas with 3-year grants and 1-year appropriations. There is no short-term solution to this problem. If average periods of support are lengthened while total funds remain constant, fewer new starts in each year are possible or the average size of grants has to be cut. Worthwhile lines of investigation may be unfunded; cutting back the size of grants often impairs progress. Thus, the decisions of program managers are invariably compromises.

Federal agencies must obtain enough information about proposed research projects to enable them, or peer-review groups, to judge whether the work merits support. As funds have become tighter, applicants have taken to writing more detailed and documented proposals in the hope of gaining a competitive edge; and some agencies have encouraged, or even required, needlessly detailed applications. As periods of support and the average grant become smaller, the burden of more-frequent application writing rises. Many principal investigators now spend substantial portions of their time writing proposals—a fraction that could and should be substantially reduced without decreasing the rigor of the review process. NSF recently placed a 15-page limit on grant applications in order to reduce this burden on both reviewers and investigators.

Under the project system, expenditures under a given grant must be limited to activities specified under that grant, transfer of grant funds from one category of expenditures to another is limited, and funds must be spent within a specified period. These rules, which are intended to ensure prudent expenditure, often prevent research institutions from adjusting to unforeseen problems or responding to new developments or insights.

Finally, there is the ubiquitous question of indirect costs. When federally financed research is undertaken in academic institutions, two kinds of costs are generated. The first are direct costs, such as salaries and supplies required solely for the conduct of a specific project. The second are other costs arising from the use of common services provided for a number of projects. These include the provision of heat and light, library services, and use of accounting, contract, and administrative offices. University administrators generally hold that the federal government should pay the full direct and indirect costs of the research that it supports. They also argue that highly detailed accounting on such matters as the division of effort of investigators between research, teaching, and other functions is impracticable and unnecessary. At the same time, government auditors and contract administrators point to abuses that have occurred and argue for stricter accountability of investigators' time.

The debate over indirect costs has gone on for more than 20 years, but it intensified during the recent period of relatively scarce federal funds. Much more than an agreement over the technicalities of accounting is involved. Indirect costs have increased, in large part as a result of the increasing complexity of the project system and of the costs of complying with federal regulations and accounting procedures. Satisfactory resolution of the issue requires a consensus on the respective responsibilities and rights of the sponsoring agencies and universities and on what is properly allowable as indirect costs.

ACCOUNTABILITY, MANAGEMENT, AND REGULATION

Public concerns affecting both federal spending generally and research and development expenditures specifically are leading to closer supervision and management of academic science. The sheer size of the federal investment in basic and applied academic science—about $3 billion per year—directs that responsible attention be given to oversight of expenditures and the management of research activities.

Efforts to use federal funds more efficiently have the implicit support of both government and the academic community, but perspectives differ as to how and where these efforts should be directed.

Pressures for fiscal accountability will not decrease and may well increase. Grant and contract administration, therefore, is not likely to become less complex unless the special problems of the universities can be better recognized within government procurement procedures aimed at preserving public accountability. Distinct from the relatively small individual research grant or contract are the large research efforts carried out within the universities. These large projects, which often use big machines and the entire personnel of research institutes to attack broadly defined tasks, involve management techniques and resources quite different from those of the individual investigator. Improvements should be attainable in both areas through the joint efforts of university and agency administrators.

Finally, there is the area of social goals. In recent years, national efforts to erase racial discrimination have been joined by equally concerted efforts to eliminate discrimination on grounds of sex or age. Greater national attention is being paid to meeting the special needs of the physically and mentally handicapped, and to protecting the rights of those who serve as human subjects for research. Other goals include more astute regulation of pension plans and more adequate provision of health services in every community.

So far as the universities are concerned, the problems and costs of meeting these obligations and complying with federal and state laws and regulations are similar to those faced by other institutions. However, some regulations, such as those relating to laboratory safety and protection of human and animal research subjects, relate to problems that differ significantly between universities and other institutions and directly affect the cost and quality of scientific research.

Both the universities and federal agencies face complex obstacles in establishing reasonable regulations and administrative processes. Recent legislation directed toward social goals has demanded more in the way of new skills, attitudes, and resources, and more understanding by both parties, than could be generated in a short time. The consequence, too often, has been an adversarial relationship. Moreover, a number of existing regulations and their administration generate costs—in consumption of time and money—that are higher than necessary for both the federal agencies and the universities.

There needs to be a better forum in which the philosophical differences between the universities and the federal government can be considered, free from the atmosphere of adversarial negotiations and tight deadlines that have increasingly characterized discussion of these differences. A basic difficulty is that government funding of academic research is generally short term, while many of the university commitments are long term. More general is the question of whether the government is merely a purchaser of university services rather than a partner in a joint venture in research, sharing responsibility for the institutional health of the universities. If these problems and if differences are to be resolved, the next five years should see a major national effort on the part of both parties to achieve the purposes of each.

OTHER ISSUES

STRUCTURAL ADAPTATION

During the past three decades the growth in the size and complexity of scientific endeavor has required some modifications in the structure of academic science. Today, effective research often requires close collaboration among investigators from different disciplines, highly specialized equipment, and administrative arrangements that do not fall easily into the patterns of traditional university administration. As a result, the last 30 years have seen the rise of special research units—institutes, centers, laboratories, and others—within the universities.

These special units often function in a semi-autonomous way, with their own hierarchical structure, goals, and administrative arrangements. Each of the major research universities has a set of these special research units, operating within the university framework but with somewhat looser ties to the parent organization than the individual scientific departments, and differing in structure from one university to another. This adaptation to new directions in scientific exploration will continue into the indefinite future.

UNIVERSITIES AND INDUSTRY

Prior to World War II, industry provided the major portion of funding for academic science from external sources. After the war, however, industry's place as the chief outside supporter of academic science was assumed by the federal government. As of 1977, industry support for academic science amounted to $139 million in current dollars, about 3 percent of the total support from all sources.[14] This sum, provided for both general scientific research as well as work on specific projects of special interest to the industrial firms supporting them, does not take into account the large amount of work done by academic scientists working as occasional consultants within industry to help solve specific problems. Most of this consulting work in recent years has been done by members of chemistry, physics, and engineering departments.

It is clear that the nation would benefit from a closer and more extensive relationship between the universities and the industrial world. Basic scientific research often serves as a foundation for later technological advances that can be utilized by industry; industrial problems, in turn, often offer fertile ground for innovative scientific exploration. The need for a closer relationship between the universities and industry is now becoming more widely recognized. Administrators of federal agencies, university scientists, and corporate officials have all begun to realize the necessity for it and its potential benefits.

OUTLOOK

The following outlook section on academic science and graduate education is based on information extracted from the chapter and covers trends anticipated in the near future, approximately five years.

The major universities of the United States play a vital role in the progress of American science. In addition to educating young scientists, they are responsible for more than half of the nation's basic scientific research. Having had the benefit of strong national attention and a large commitment of federal funds, academic science during the past three decades has developed a momentum that will allow it to continue to make major contributions to the world's store of knowledge during the next five years and beyond. This period, however, will also be a time of adjustment for universities and academic science as they attempt to deal with such new circumstances as more stringent financial conditions, declining undergraduate enrollments, a sharp reduction in the number of new faculty positions, and the need to meet new regulatory and other administrative requirements. Adjusting to these new demands will require time and effort on the part of all those concerned with the vitality of academic science.

SCIENTIFIC RESEARCH AND GRADUATE EDUCATION

The close tie between scientific research and graduate education in the sciences can be expected to continue over the next five years. While most academic scientists will continue to perform their dual roles, teaching graduate and undergraduate students while supervising or conducting research, the number of specialized research institutes and centers may well increase.

ENROLLMENT

Total undergraduate enrollment in American colleges and universities will enter a period of decline during the next five years, chiefly because of the birthrate decline. Graduate school enrollment, however, will continue to grow, in both scientific and nonscientific fields. More college graduates will pursue master's degrees and doctorates in order to bolster their ability to find professional positions. In addition, the sizeable number of foreign students in graduate schools is expected to continue to rise.

DOCTORAL DEGREES

Somewhat paradoxically, the total number of doctorates to be awarded over the next five years will gradually decline, despite an increase in the number of those in graduate schools. As in the past, a substantial percentage of those seeking doctoral degrees will relinquish that goal. Furthermore, declining undergraduate enrollments will mean a reduction in the need for new faculty members with doctoral degrees. A substantial part of the decline in doctorates will be accounted for by a reduced number of doctoral awards in physics, chemistry, mathematics, and engineering.

JOBS

The relative unavailability of university faculty openings will cause a greater number of persons with doctoral degrees or other advanced education in the sciences to seek positions in industry, government, and nonprofit research organizations during the next five years. The small number of academic positions available during the next several years may also require special measures to assure the retention of the best young academics in science.

WOMEN AND MINORITIES

The large increase in the relative number of female graduate students and in the number of female holders of doctoral degrees that occurred in recent years will slow somewhat in the next five years. There will be slow but continued growth in the numbers of women in the sciences and engineering. The declining need for faculty members will also limit the job prospects of women with recent doctoral degrees. The outlook for minority students, particularly blacks and those of Latin American ancestry, is for continued slow improvement in the number of doctoral recipients over the next five years. Continued efforts to overcome the effects of disadvantaged background will be required.

FUNDING OF ACADEMIC SCIENCE

Funds for academic science come from a number of sources—the federal government, the universities themselves, private philanthropic foundations, and industry. The states, through their contributions to public universities, can also be considered a direct source of funding. Although federal funding for academic science has increased modestly in recent years, and may continue to do so, the universities will be under increasing pressure to increase their share of the funding, now more than 20 percent. The decline in federal support during the period 1968–74, combined with inflation and the rising number of scientific investigators within the universities, assures that the years immediately ahead will nonetheless be a time of emphasis on frugality within academic science. Universities and government will have to find ways to cope with the problems of aging buildings, obsolescent equipment, and rising support costs.

PROJECT FUNDING

During the next five years, most of the funds for academic research will continue to be distributed in the form of individual grants and contracts. Although this system has a number of strengths, attention should be given in the years immediately ahead to relieving the procedural rigidity that is now its chief weakness.

Continued attempts will also be made to resolve the differences between the universities and the federal government over the apportionment of indirect costs and over accounting procedures.

ACCOUNTABILITY, MANAGEMENT, AND REGULATION

Growing federal regulation in such areas as health, safety, financial security, and equal opportunity will also continue to affect the universities. In the recent past, the rapid proliferation of regulation has created strains between the federal government and academic sciences. During the next five years the efforts of both parties should be directed toward achieving the goals of regulation in a positive manner that enhances understanding and reduces total costs as well as the time diverted from scientific research.

OTHER ISSUES

The relationship between academic science and industry should become a closer one in the years immediately ahead.

REFERENCES

1. *Science Indicators 1976* (National Science Board. NSB-77-1). Washington, D.C.: Government Printing Office, 1977, p. 293.

2. *Current Population Reports* (Bureau of the Census). Series P-20, 1978, pp. 41 and 61.

3. *Counselor's Newsletter* (Institute of International Education) 11:1, 1978.

4. *Doctorate Recipients from U.S. Universities: Summary Report* (NRC Board on Human-Resource Data Analyses). Washington, D.C.: National Academy of Sciences, 1975 and 1977.

5. *Outlook and Opportunities for Graduate Education* (NAS National Board on Graduate Education). Washington, D.C.: National Academy of Sciences, 1975, p. 1.

6. Anderson, C. (ed.). *A Fact Book on Higher Education.* Washington, D.C.: American Council on Education, 1977, p. 77.114.

7. *National Patterns of R&D Resources, 1953–1977* (National Science Foundation, NSF 77-310). Washington, D.C.: Government Printing Office, 1977, pp. 37–38.

8. *Ibid.*, p. 34.

9. *Basic Data Relating to NIH* (National Institutes of Health). Washington, D.C.: Government Printing Office, 1977, p. 36.

10. *National Patterns of R&D Resources, 1953–1978-9* (National Science Foundation, in press). Washington, D.C., Table 12.

11. *National Patterns of R&D Resources, 1953-1977, op. cit.*, p. 30

12. *Science Indicators, 1976, op. cit.*, p. 217.

13. *Science Indicators, 1974* (National Science Board NSB 75-1). Washington, D.C.: Government Printing Office, p. 181.

14. *National Patterns of R&D Resources, 1953–1977, op. cit.*, p. 23.

15. National Research Council. *Personnel Needs for Biomedical and Behavioral Research, 1978 Report.* Washington, D.C., National Academy of Sciences, 1978.

16. *Federal Funds for Research and Development.* Vol. XXVI and XXVII (NSF 78-312). Washington, D.C., National Science Foundation, 1978.

(United Nations/Saw Lwin)

11 Institutions for International Cooperation

INTRODUCTION

The vigor and development of science and technology are intimately related to the interaction of ideas and activities of its many participants. This process takes place in several ways: personal contact, publications, symposia and workshops, coordinated but independent research, and closely coupled collaborative research. The process may be highly informal and unstructured, or it may be carried on within formal national or international institutions.

Here we emphasize formal international institutions. It is important to also recognize, however, the network of scientists and technologists that has operated effectively as an "invisible" international institution for more than a century.

Immediately after World War II, emphasis was placed on institutional arrangements that would help to rebuild the scientific infrastructure in Western Europe and Japan. Now, however, international institutions facilitate the traditional role of the scientific and technological enterprise in developing understanding and extending knowledge, and also in responding to global societal problems that have gained prominence since World War II. Since issues such as environmental quality, energy, population, health, food, water, urban settlements, poverty, and peace have global aspects, international institutions are important in coping with them. Clearly, we are still in an experimental phase of adapting existing institutions or fashioning new ones to address such problems, but their urgency requires a rare combination of imagination, wisdom, and discrimination in deploying the intellectual resources of increasingly interdependent nations.

This chapter attempts to put into perspective the rationale for international science at both the academic and intergovernmental levels and to do so in terms of the short-range prospects for international institutions of science and technology. Rather than attempting a catalog, we have concentrated on a selected few institutions of science and even fewer of those dealing with technology, and to draw attention to those that operate largely outside the established and well-studied framework of intergovernmental bodies. We have also tried to concentrate on institutions whose prime interest is in the fields of science, agriculture, and medicine, and have given comparatively little attention to institutions whose contribution to international cooperation is secondary to other missions. There are hundreds of ways in which scientists, engineers, physicians, agronomists, and social scientists share knowledge and cooperate with each other to discover new knowledge. Indeed, the diversity and richness of this interchange makes it virtually impossible to estimate the total scope of scientific cooperation taking place in the world at any given time.

This chapter, then, will briefly discuss international institutions concerned with science, engineering, agriculture, and medicine whose future seems to us to be particularly important in the near term. It will also try to show how they have evolved and the basis of their importance in today's world. The building, care, and maintenance of international institutions is a difficult and still evolving art. Our analysis has led us to the conclusion that much more needs to be done to make international institutions truly effective. We have tried to explain why this is so and to offer some glimpses of future possibilities.

THE RATIONALE FOR INTERNATIONAL COOPERATION

International cooperation in science is essential today, as it has been in the past. It contributes both to the vigor of U.S. science and to the health and well-being of our planet and its inhabitants. Through cooperation we can help to meet our global responsibilities, and we can address issues that transcend the concerns of any one nation.

In order to be shared, knowledge needs to be transmitted in some common language. Early observers of natural phenomena recognized that for others to see what they had seen required that they be able to describe their observations in words and numbers that meant the same thing to everyone. Thus one of the first, and even today one of the most important, aspects of international science is the development of common standards, units of measurement, and nomenclature, most of which is done by the scientific and professional societies and their international analogs.

Science is an international enterprise and communication is central to its existence. Today many hundreds of scientists move across national boundaries to conduct research in the laboratories of colleagues who are working on like problems.

Some idea of the dimension of the person-to-person contact is suggested by a 1978 survey[1] of a sample of 203 U.S. doctorate-granting institutions and medical schools that showed that three-fourths of these institutions had some faculty who collaborated on research with their foreign colleagues—principally in Western Europe, Latin America and the Caribbean, the Near and Middle East, and Japan. Moreover, during 1977–78 nearly 3,800 faculty members from doctorate-granting institutions participated in cooperative scientific and technical activity with developing countries.[2]

By contrast with basic research, institutionally directed and supported applied research, the totality of which also is difficult to measure, may well involve even more people and require more funds because of the scale of the projects. Institutionalized cooperation in technology is infinitely more complicated than it is in science. Technology embraces so many topics and has so many ancillary attributes, many of which are related to commercial profit-making and national security, that a meaningful examination of these institutions is not possible within the scope of this chapter. We would note here, however, that there are a number of areas in which the engineering and technical community of the United States has pioneered in cooperative ventures, space exploration being probably the most dramatic. Equally important, of course, are the intergovernmental relationships that regulate the movement of aircraft and ships and the agreements that permit our current level of telecommunications. Other vital technological cooperative ventures include the setting of industrial and commercial standards, agreements with respect to patents and their licensing, and the relatively free exchange of engineering information upon which our collective safety and progress depends. Primarily for simplicity, we will pursue a discussion of the history of scientific cooperation as illustrative, while recognizing that cooperation in engineering has had a parallel course. The two systems of international cooperation complement each other and are essential to the advancement and application of knowledge.

A scientific observation should be susceptible to duplication (excluding of course the observation of one-time or one-place phenomena). Thus if a scientific paper is properly written, the experiment or observation it describes can be reproduced in the laboratory of a distant colleague. This obviously provides for a desirable redundancy that produces independent verification or refutation of alleged advances of scientific knowledge. In recent years this facet of global science has been deliberately constrained by the scientists themselves (with some encouragement from legislators and budget-makers) when building the huge and very expensive machines of physics. Regular meetings of the directors of high-energy physics laboratories have enabled joint planning so that each major new machine has the capacity to do something the others cannot do and unnecessary duplication is avoided. Instead of building machines in many countries, the practice is to bring scientists from other countries to the machine, at obvious savings of resources.

Another element of the rationale for international science is the globally designed and executed experiment utilizing many observers throughout the world who follow common and agreed-upon instructions. Reaching back to the earliest astronomers who shared their observations about the course of the stars, we have come to such extraordinarily productive projects as the International Geophysical Year 20 years ago and today the Global Atmospheric Research Program (GARP) and the World Climate Program (WCP). In these programs highly sophisticated research is being planned by scientists with the understanding that major and minor powers of the world will contribute according to their resources of manpower and facilities, but they will all receive the full results of these efforts to understand the world's weather and climate.

The techniques of sharing our constantly expanding reservoir of knowledge have not changed very much in a long time except in scale. The basic medium for exchange is still the journal article, but because it often takes many months, sometimes years, to assure publication and because the number and specialization of journals has proliferated, international scientific meetings and symposia have become a complementary medium of great importance to scientists. Further, informal information acquired about a colleague's research in progress frequently leads to personal and group collaboration, avoidance of unintentional duplication, and modification of research plans, techniques, or instrumentation. Conducted today largely in English and drawing specialists from all over the world, the scientific conference serves several purposes. Here the common language of a discipline receives its regular infusion of new terms; and the process of agreeing upon names, values, and unit descriptions is begun. Also here, results and ideas are exchanged and new research strategies are often planned and sharpened. The significance of a recent experiment or theory can be examined by others similarly engaged, and plans for collaboration of mutual benefit can be worked out, whether the task be that of mapping the world's oceans or sharing the instrumentation of space facilities.

In recent years the scientific conference has been used as a model by the United Nations and its family of agencies for the organization of intergovernmental debates on population, the environment, food and nutrition, resources, and energy, albeit in a largely political rather than scientific atmosphere. These partially scientific, partially political conferences constitute public recognition of the need to apply science and technology to pressing public problems, most of which require international cooperation for their amelioration. The scientific community has felt their impact in terms of the redirection of research funding and priorities and in newly discovered interests and challenges.

SOME CHARACTERISTIC INTERNATIONAL INSTITUTIONS OF SCIENCE AND TECHNOLOGY

A very rough approximation of the division of labor among the international institutions of science and technology is that the nongovernmental entities tend to support the advances of science as an end, while the intergovernmental organizations tend to be devices to convey technology as a means. But it quickly becomes apparent that most nongovernmental and intergovernmental organizations are attempting both.

The broad sweep of scientific work of major international agencies such as the World Health Organization, the World Meteorological Organization, the International Atomic Energy Agency, the Food and Agriculture Organization (FAO) of the United Nations, the U.N. Development Program, or the U.N. Environmental Program, to list the most prominent, is generally well known. Each of these institutions has an objective for the furtherance of which the United States contributes in funds, ideas, people, and policy direction. We have chosen other lesser-known models of international cooperation where the work of individual scientists and scientific organizations has been and can be, we believe, particularly productive in the future. In the discussion that follows, we will, however allude to certain of the specific activities of some of the U.N. agencies as they are also illustrative of models to be pursued.

In the academic world the oldest, best-known, and most effective of the international nongovernmental organizations of science is the International Council of Scientific Unions (ICSU), a sort of umbrella organization for many specialized scientific bodies.[3] ICSU is composed of 18 autonomous scientific unions and more than 60 national members, such as academies of science, research councils, or like institutions. The main purpose of ICSU is "to encourage international scientific activity for the benefit of mankind."

In addition to fostering the efforts of the specialized unions and their component commissions serving the broad range of the physical and life sciences, ICSU initiates and coordinates multinational research programs in such areas as space, oceanographic, and polar research, and also serves as

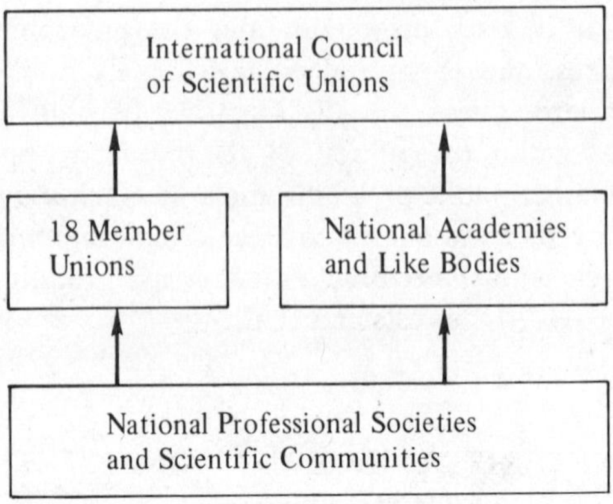

FIGURE 32 Organization of the International Council of Scientific Unions.

a vehicle to communicate scientific information and develop standards in methodology, nomenclature, and units.

Starting with the voluntary membership of individual scientists in their home country's professional societies, the institutionalization of international science flows through national committees for each discipline to national academies and similar bodies to the international unions of scientists (e.g., the International Union of Pure and Applied Chemistry), and then via the collaboration of these unions within the ICSU to form special project-oriented multidisciplinary committees and commissions that attempt to bring order and purpose to international science (Figure 32).

ICSU itself was formally created in 1931, but its origins can be traced back to the nineteenth-century associations of astronomers and earth scientists that were created, as are most institutions, to meet specific needs (in this case those whose work manifestly transcends national boundaries) and to the International Association of Academies (1900, London, Paris, St. Petersburg, Washington). Today it encompasses all of the natural sciences. Organizations comparable in terms of effectiveness or recognition do not yet exist in either the professional fields of medicine and engineering, or the disciplines of the social sciences, although there have been nominal analogs in existence for some time.

In the case of ICSU, the support and assistance of the major world science academies has been a decisive undergirding element. Neither medicine, engineering, nor the social sciences has equivalent national institutional bases, although strong and internationally minded engineering academies have been established in several countries in recent years.

The first international convocation of these academies was held in October 1978 under the auspices of the U.S. National Academy of Engineering.

The strengths of the international nongovernmental science unions are in a way also their weaknesses. They are sustained by the efforts of unpaid scientist volunteers, often supported by small and poorly paid secretariats. The work they do is subject to funding constraints that are always austere at best. Because of other priority calls upon the time of their leadership, projects move slowly and are largely dependent on the tenacity of a few very dedicated individuals.

A strength of ICSU, as distinct from the work of its member unions, lies in its eleven special and scientific committees, five interunion commissions, and three permanent services, which provide for shared disciplinary approaches to space, climate, water, science information, and other topics.

Three examples of ICSU scientific committees follow:

• The International Biological Program (IBP) was organized to coordinate research on the biological basis of productivity and human adaptability to environmental changes. American scientists studied ecosystems of major economic importance such as forests, grazing lands, and arid lands. International collaboration made possible the collection of comparable data from several sites around the world.

• There has been increasing recognition of the needs for interdisciplinary approaches to environmental problems. Accordingly, in 1970 ICSU established the Scientific Committee on Problems of the Environment (SCOPE). It has two main tasks: to advance knowledge about the influence of human activities on the environment and to serve as a nongovernmental source of advice on environmental problems.

SCOPE has seven activities at present: biogeochemical cycles, dynamic changes and evolution of ecosystems, environmental aspects of human settlements, ecotoxicology, simulation modeling of environmental systems, environmental monitoring, and communication of environmental information and societal assessment and response.

• More recently, the Committee on Genetic Experimentation (COGENE) was established by ICSU to serve as a nongovernmental, interdisciplinary, and international council of scientists and as a source of advice on recombinant DNA for governments, intergovernmental agencies, scientific groups, and individuals. Its functions are: to review, evaluate, and make available information on practical and scientific benefits, safeguards, containment facilities, and other technical matters; to consider environmental, health-related, and other consequences of any disposal of biological agents constructed by recombinant DNA techniques; to foster opportunities for training and international exchange; and to provide a forum through which interested national, regional, and other international bodies may communicate.

COGENE is currently concerned with: guidelines for research on recombinant DNA; risk assessment experiments; and benefits and applications of recombinant DNA.

In addition to these ICSU examples, there are joint activities with intergovernmental agencies.

In what has come to be viewed as a model in international institutionalized scientific cooperation, GARP began in 1961 when advances in understanding the theory of atmospheric processes and expanding computational capabilities led President Kennedy to propose to the United Nations that the ICSU and WMO mount an international program of atmospheric research. A joint organizing committee of 12 world-renowned meteorologists was appointed by the two organizations and accorded great autonomy in planning a comprehensive program of unprecedented scope.

By 1974, a large coordinated research effort was conducted in the eastern Atlantic Ocean with the objective, among others, of examining the relationship between convective activity and the large-scale wavelike disturbances that are frequently the progenitors of hurricanes. Several thousand scientists and technicians from 72 countries, aided by a thousand land stations, 40 ships, 12 aircraft, and 6 satellites, brought under observational surveillance 20 million square miles of land and ocean. The research was successful and mathematical models have now been developed that are capable of predicting the development of tropical atmospheric wave disturbances, their westward propagation, and of simulating their rainfall pattern.

This success led to an even larger experiment launched in December 1978.[4] Embracing the entire globe and involving 5,000 scientists and technicians from 150 countries, the Global Weather Experiment will last for 1 year and involve 5 geostationary satellites, up to 4 polar orbiters, more than 40 oceanographic research vessels, thousands of miles of specially instrumented aircraft flights each day, more than 300 instrumented balloons circulating freely about the equator at a height of 15 kilometers, and over 300 ocean buoys in the oceans of the Southern Hemisphere, each capable of transmitting observations for 9 months. In addition, thousands of commercial ships and aircraft have been instrumented to report data from wherever they are located.

Large-scale computers process the immense quantities of data received daily and put it in a form useful for the research effort that will extend over the next decade. Plans were laid at the World Climate Conference in

February 1979 to use this unique data base in a World Climate Program aimed at understanding natural climatic variability as well as change induced by human activity.

Testimony to ICSU's value is the continued donation of volunteer time and the continued use of ICSU to stimulate new scientific ventures. Within its formal institutional framework, it is estimated[5] that, on the average, each year about 40,000 scientists come into contact with colleagues from other countries through ICSU-sponsored conferences.

Among institutions where science is pursued as an end, there are portions of the U.N. family of specialized agencies—such as WMO andWHO—that support and sustain basic research (both directed, or applied, and nondirected). Almost invariably, however, the research now supported by U.N. agencies is linked to some social or economic goal and frequently it is tied directly to a technological means. Thus, the S in UNESCO, which used to be science *qua* science, is now labeled "natural sciences and their application to development."

Within the realm of the health sciences, three programs of the WHO deserve mention for the way they illustrate international cooperation. That smallpox has essentially been eradicated in the world is directly attributable to a program initiated in 1966 by the WHO. In this instance, international cooperation was required, first in the production of vaccine, second in its donation by producing countries and its eventual production in developing countries (which in a mere three years produced vaccine that was acceptable internationally for its potency and stability), and finally in an improved vaccination technique that required the development of a delivery system that was usable in all sorts of cultural environments by people of varying technical abilities. The program hinged on the cooperation of governmental health departments and private (in many cases profit-making) concerns who gave willingly and generously to the goals of the program.

A complementary WHO program, also begun in 1966, is international collaborative research on tropical diseases. In this instance, six diseases (malaria, schistosomiasis, filariasis, trypanosomiasis, leishmaniasis, and leprosy) have been selected for concentrated research by scientific working groups in a number of locations around the world under the general guidance of WHO steering committees. The heart of the program is the linkage, via formal and informal networks, of research being undertaken in these diseases in different centers in different countries and the agreement among the participants to exchange information and collaborate in the design and operation of the research projects.

WHO's third program, again more than 10 years old, is the special program of research development and training in human reproduction.

The program is designed to meet the needs of developing countries and has three main goals: developing contraceptive technologies, designing technologies that are acceptable and that can be easily implemented, and enhancing developing countries' increasing self-reliance in basic and applied research. The program, which is funded by both developing and industrialized countries, has met wide acceptance throughout the world.

The International Atomic Energy Agency (IAEA) is a well-known example of a multilateral intergovernmental organization devoted to the solution of problems generated by high technology. Coping with safe levels of irradiation of food, the disposal of nuclear waste, transport of hazardous nuclear materials, ionizing radiation, nonproliferation, and the world's long-term energy needs, this agency collaborates closely with WHO, the FAO, and with national nuclear regulatory agencies. The IAEA also supports research and training in nuclear sciences and engineering, as well as serving as the focal point for intergovernmental agreement on one of the most politically complicated technologies of our age.

Among the international institutions where science is precisely targeted is the Consultative Group on International Agricultural Research (CGIAR). This multinational and multi-institutional group stems from the agricultural research programs originally sponsored by the Rockefeller Foundation in the 1940's and later also by the Ford Foundation. CGIAR was started in 1971 to finance a revolution in the agricultural productivity of less-developed countries. It now supports research and training activities through 11 centers or programs located in the developing nations where almost 600 senior staff and a total of over 8,000 persons study the major food crops and animals in virtually all of the ecological zones of the developing world. The annual combined budget of the centers now exceeds \$100 million.[6] Jointly sponsored by the World Bank, the FAO, the U.N. Development Program, and the U.N. Environment Program, it has 18 donor governments, 4 private foundations, 3 regional development banks, the Commission of the European Communities, the International Development Research Centre (IDRC), and the Arab Fund for Social and Economic Development affiliated with its program.

The centers began with a strong orientation toward purely biological research, but they became the home of the green revolution. Centers are now developing plant types and practices to benefit small farmers who must operate without much help from controlled irrigation, fertilizers, and pesticides. The centers are also taking into account the social customs, consumer habits, and other societal habits affecting agricultural productivity.[7]

The CGIAR family consists of the following centers and programs with their locations and dates of founding:

CIAT	International Center for Tropical Agriculture (Colombia) 1967
CIMMYT	International Maize and Wheat Improvement Center (Mexico) 1943
CIP	International Potato Center (Peru) 1971
ICARDA	International Center for Agricultural Research in the Dry Areas (Egypt) 1976
ICRISAT	International Crops Research Institute for the Semi-Arid Tropics (India) 1972
ILCA	International Livestock Center for Africa (Ethiopia) 1974
ILRAD	International Laboratory for Research on Animal Diseases (Kenya) 1973
IITA	International Institute for Tropical Agriculture (Nigeria) 1968
IRRI	International Rice Research Institute (Philippines) 1960
WARDA	West Africa Rice Development Association (Liberia) 1971
GENES Board	International Board for Plant Genetic Resources (Italy) 1974

In September 1978 the CGIAR produced "An Integrative Report" on the Consultative Group and the International Agricultural Research System. Assessing the impact of high-yielding varieties of rice and wheat (the only two major innovations introduced so far by the CGIAR that have been in use for sufficient time to permit study) on agricultural production and in terms of social and economic impact, the CGIAR drew a number of positive conclusions. For example, in the 12 years since the high-yielding varieties were introduced, they have come to occupy over one-third of the total acreage being sown in these two grains.

One major conclusion of the review, concerning an issue in controversy, was that the principal research strategy of the international centers, that of developing yield-increasing technology, contributes both to more equitable distribution of the benefits of economic growth and greater agricultural productivity. A noncontroversial finding was that the returns on the dollars invested in agricultural research have been truly impressive.

Complementing this remarkable illustration of international cooperation in the application of science and technology to a specific global goal are other efforts more modest in scale and scope. For example, one of the most useful institutions for scientific cooperation with Western Europe is the Science Committee of NATO,which manages a $9 million program directed toward basic fundamental sciences rather than programs related to the military and defense objectives of the alliance. The bulk of the funding is used to provide grants or fellowships and for the support of specialized scientific meetings. In the past 20 years, some 75,000 individual scientists and engineers have participated in NATO's science activities.

Another NATO innovation was the Advisory Group for Aerospace Research and Development (AGARD), the history of which goes back almost to the founding of the NATO alliance. This group, which has a fundamental military orientation, is nonetheless a remarkable example of international scientific cooperation. AGARD has many of the characteristics of a professional society and has made contributions very like those made by the professional societies in other disciplines.

AGARD is but one example of very high level technical cooperation in practical fields that have received great impetus from the international military and security requirements of industrialized countries. The civilian applications of these technologies are in many cases as important as their military uses. Certainly NATO- and AGARD-initiated research on aircraft manufacture, safety, and space applications has equaled comparable contributions by other professional, technical, and scientific bodies.

The International Institute for Applied Systems Analysis (IIASA) near Vienna has been successfully launched by an *ad hoc* consortium of scientific bodies, supported by public funds, to explore new ways to understand and manage complex systems in areas such as energy, food, environment, water, and urban centers. It brings to bear the experience and insights of scholars and institutions from all over the industrialized world—from both market and planned economies.

IIASA, first organized in 1972, brings together scientists of different disciplines, cultures, and nationalities to work on problems of concern to mankind and to improve analytical techniques and their usefulness for decision making. IIASA's scientific staff consists of approximately 70 persons, supported by member dues, who spend varying periods of time, ranging from a few months to a few years at the institute, plus another 20–30 persons supported by grants and contracts from other sources. The institute's 1979 budget has been set by its 17-member international council, its governing body, at 130.6 million Austrian schillings (about $9.6 million).

During 1979 the institute expects to complete work on the final report of five years of research on global energy prospects. Approximately two years later work will be completed on its second major program, that in food and agriculture.

The Stockholm Institute for Peace Research (SIPRI) and the International Institute for Strategic Studies (IISS) in London are modest but scholarly and productive efforts addressing the problems of conflict. Despite the highly political and controversial nature of the issues with which they deal, they have acquired respect derived from their objective and scientific approach. SIPRI is an independent research institute with a particular focus on disarmament and arms control. Financed by the Swedish government, but with an international board, staff, and scientific council, it publishes an annual authoritative yearbook of world armament and disarmament, as well as reports and books on special topics in arms matters. Its funds (of approximately 4.5 million Swedish kroner, or about $1 million) come from the Swedish Parliament. By contrast, the IISS is an organization whose international membership is dedicated to scholarly analysis of all types of issues related to international security and specializes in the production of the *Adelphi Papers*—a series of essays. The IISS income, privately subscribed, is $240,000 per year.

The Norwegian government supports the International Peace Research Institute-Oslo (PRIO) which is a somewhat newer institution devoted to study of different types of violence. PRIO publishes studies and two journals, the *Journal of Peace Research* and the *Bulletin of Peace Proposals* under international auspices.

Another institutional innovation, modest in size, is the International Foundation for Science (IFS), which gives grants to young scientists for research relevant to development. Founded in 1972, IFS is a nongovernmental organization sponsored by scientific academies and research councils of 42 countries, two-thirds in developing and one-third in industrial parts of the world. At present 10 countries contribute to the foundation's budget, largely by government grants through academies or research councils. The budget for 1978 was $1.4 million. IFS supports the work of young grantees who must be native to, and carry out research in, a developing country, and where the institutions of the grantees contribute the salaries and basic support, often at amounts several times the foundation grant.

The IFS is *sui generis* in the sense that it is stimulating a network of individual researchers in several fields of applied biological research. The fields chosen have substantial economic and developmental impact potential and the research is of a sort especially appropriate to the needs of developing countries. The seven principal topical areas are: aquaculture; animal production; vegetables, oil seeds, and fruit; mycorrhiza studies, afforestation problems; fermentation, methods for food preparation; natural products; and rural construction.

Another institutional innovation is the International Development Research Centre (IDRC) established by the Parliament of Canada in 1970 to foster research in development problems and in the ways to apply knowledge to economic and social advancement in developing regions.[4] The IDRC enlists experts from Canada and other countries to help developing regions build research capabilities, innovative skills, and institutions to solve their own problems. It encourages coordination of research and fosters cooperation between developed and developing regions.

The goal of IDRC is to emphasize the role of the scientist and engineer in international development, and it encourages developing countries to use their own scientific communities. It is funded by and reports to the

Parliament of Canada, but it is administered by an international and autonomous board of governors.

It is a governmental aid organization created to support research projects originated and conducted by developing country researchers in their own countries and in terms of their own priorities. It has helped to create research networks that allow developing countries to share experiences and conduct studies with a common design in areas of mutual concern.

IDRC's areas of concentration are: agricultural and nutritional sciences, health sciences, information sciences, publications, and social sciences and human resources. Since its inception, over $100 million of IDRC support has been provided for 550 research projects, of which more than two-thirds were in science and technology.

Perhaps the best known of the basic scientific operating organizations financed by governments is CERN, the Conseil Européen pour la Recherche Nucléaire, formed by collaboration among 12 Western European nations. It is an outstanding example of international collaboration both on scientific and political levels. Its laboratories serve a community of more than 2,000 physicists, including many from the United States, some in permanent residence, but most of whom come for short working periods. CERN's present facilities represent an investment close to 1 billion current dollars by the European governments that support it. Its accelerators and support facilities are among the world's best. CERN has led major advances in our knowledge of the fundamental properties of matter and of the fundamental forces governing the behavior of matter in the universe. A proton–proton interacting storage ring is currently in operation, and a massive proton–antiproton interacting storage ring is proposed.

The important role of the large multipurpose funding institutions which support international scientific and technical activity deserves recognition. These institutions support research and applications in many ways, both directly and indirectly, as part of their broad programs. From the World Bank to the U.N. Development Program and the U.N. International Children's Education Fund (UNICEF), to the major private foundations of the world, there are many large and small patrons of science and the application of technology. Without the support these institutions provide to national and international research and its technological applications, many scientific and technological institutions would not survive.

SOME SPECULATION ABOUT FUTURE NEEDS AND PROSPECTS

Over the next five years the debate on how science and technology can contribute to a lessening of tensions between East and West, how science and technology can ameliorate conflict between North and South, and how they can solve the problems of the poorest of the poor will continue. How and to what extent can science and technology provide solutions to a wide range of essentially political problems? With accumulating experience we hope for increasing convergence between scientific and technical solutions and political problems, but the difficulties in achieving such convergence are manifest in the lengthy—and still not very fruitful—conferences on the Law of the Sea and in the thin margin of agreement and rather large residue of disagreement that remained after the 1977 U.N. Conference on Trade and Development. In these forums the consensus form of decision making and crosscurrents of national interest tend at best to reduce the crispness of definition of issues while simultaneously blurring the boundaries of acceptable compromise.

The international institutions that are now and will in the future be regarded as constructive and worth saving or encouraging are those that are perceived to reflect the interests of a number of nations, each of which, in fact, may have quite diverse motives and concerns.

Many of the major international intergovernmental organizations based on a one-country—one-vote principle appear to be coming under the effective control of developing country representatives. The result of this control—most notable in some agencies of the United Nations—may be a shift toward regional and bilateral arrangements where national interests are believed to be more precisely matched. In a sense, this return to free bilateralism may be likened to the rationale that underlay the initial arrangements for scientific cooperation in the early part of the last century.

We perceive three generic clusters of problems with which international institutions will be called upon to deal. These clusters of problems are perceived in the environment of East–West and North–South competition. In some respects they are manifestations of the differences that exist in natural resources and geographic limitations. In other respects they represent the differences in scientific and technological capacity and productivity that exist among countries.

- Interdependency problems such as the global effects of an increasing build-up of CO_2 (see p. 34), multinational management of water basins, and exploitation of the seabed under some yet-to-be-agreed-upon international arrangement.
- Distributive issues of equitable access to natural resources, including

assured food supply, reliable sources of energy at affordable prices under acceptable environmental constraints, and vastly expanded delivery of health care, adequate housing, and educational opportunity—all of which have science and technology components. These issues of distributive equity relate both to a nation's domestic concerns about meeting the needs of its people and to its international aspirations and the obstacles it faces in global competition.

- Vulnerability concerns such as those that have to do with radioactive wastes and the development of nuclear power, levels of acceptable environmental risk, natural catastrophies such as drought, and the worries of some that our technical capabilities are outpacing our political management skills.

THE INTERNATIONAL BACKGROUND

American interest in international institutions relates directly to our domestic experience. In the 1950's and 1960's America was the unchallenged leader of world science. Today, after monumental, postwar U.S. reconstruction assistance, science and technology have steadily expanded in Western Europe and Japan. America still contributes the single largest and most successful share of the total global scientific exercise—about one-third—but its role is changing.

One of the manifestations of the change in our scientific status is the decrease in postdoctoral exchanges between the United States and Western Europe. Reasons for this are complex and not fully understood, but it is clear that the young scientists of Europe no longer perceive a postdoctoral fellowship in the United States as critical to their career development.

At the same time, Americans are not going abroad for postdoctoral work to the extent they once did, because funding for this purpose is relatively scarce and because they may experience problems in reentering the U.S. job market. This is paradoxical because now, when our colleagues abroad have more to give, our postdoctoral community engaged in research at foreign institutions is at its lowest in years. We know, for example, that the number of those with new doctorates in science and engineering who had firm commitments for postdoctoral study abroad at the time of their degrees peaked in 1971 at 409. By 1977, that figure had dropped by 55 percent to 186. The proportion of this group with firm commitments for postdoctoral study abroad also decreased from 2.4 percent in 1971 to 1.2 percent in 1977, a 50 percent drop. This decrease occurred during a period when the proportion of doctors of science and engineering with firm postdoctoral study commitments increased slightly.[8]

The U.S. educational system is being utilized by increasing numbers of foreign students in our institutions of higher education. Their needs and the obvious importance they assume as the U.S. population of college-age youth declines have not escaped attention. The 235,000 foreign students currently enrolled in U.S. institutions of higher education now are expected to increase to a million within 10 years, reflecting the growing needs of the developing nations and their ability to provide funds. Today, 27 percent of all engineering doctorates and 45 percent of all engineering degrees go to foreign students.[9]

The United States is also still a favored country for advanced training by scholars from the Soviet Union, Eastern Europe, and now the People's Republic of China. The United States has some 16 bilateral scientific and technical agreements with the Soviet Union in addition to the long-standing bilateral exchange arrangements conducted by the National Academy of Sciences and International Research and Exchanges Board. Exchanges with the People's Republic of China have been conducted on a formal delegation-for-delegation basis since 1972, but are now entering into more conventional and productive exchanges between individual students and scholars, lecturers and researchers, which may become a truly significant means of reestablishing the much-desired personal relationships so important to sustained and fruitful international scholarly relations. The value of these exchanges is hard to measure, but the individual exchanges with the Soviet Union have gone on now for 20 years and surveys of participants indicate that in their judgment the exchanges are scientifically productive as well as politically useful.[10]

Competition

European strength in basic research and the impressive Japanese capacity for technological innovation have clear and obvious implications for American industrial concerns. In addition, the more advanced developing nations are determined to carve out a share of the technologically based Western consumer and capital goods markets. Thus, in the United States and in many other industrialized countries there are some fears about loss of jobs and business to foreign competition. While the magnitude of concern is difficult to gauge, there appears to be reasonably strong, albeit limited, protectionist sentiment developing with respect to the U.S. export of high technology.

In the developing world there is a strong pull toward technological self-reliance and a negative reaction to all forms of dependence on former colonial powers. There is also in developing nations a resurgence of interest in their traditional cultures and in the spiritual values embodied in those cultures. Congruent with this concern for their cultural heritage is a search for individual equity and international social justice.

Human Rights

Furthermore, a broad range of humanitarian concerns have assumed singular importance in the eyes of the scientific and technical community. These are the questions of human rights and repression, freedom of inquiry, travel, open immigration, dissemination of research results, and free circulation of scientists and ideas. None of these issues is new, but their enhanced visibility and the expectation of an accepted set of international standards of national behavior have opened new avenues for personal expression of concern by scientists. The revulsion against repression has caused some disruption of scientific exchange. For example, 2,400 U.S. scientists have recently announced their intention to restrict severely cooperation with the Soviet Union as a protest against prison terms meted out to certain Soviet scientists. To the extent that human rights are issues between individual members of national communities of scientists, they will affect and complicate the ways in which the scientific and technical community interacts in almost all of its global institutions.

Even in the most dire circumstances, scientists who are at bay in those countries where freedom of thought is proscribed insist that the exchange with their colleagues abroad is a more effective measure than the practice of boycotting. They make the point regularly and consistently that their professional lives as scientists depend on the ability to communicate with their colleagues abroad. They also take the long historical view and realize that scientific relations are fragile and take time to build, and, if they are upset by political concerns, these relations can wither quickly and at considerable cost.

WHAT DO WE NEED TO KNOW MORE ABOUT?

We have come to the realization that we still have much to learn about transferring technology to developing countries effectively. We are also concerned that these countries appear to be having substantial difficulty adapting, absorbing, and utilizing some of the technologies that have been or are in the process of being transferred.

Second, highly imaginative and creative efforts are needed to fashion new modes of collaborative research. With the resources for research concentrated in the industrialized world and pressing needs for research concentrated in the developing world, there is a gap that must be bridged. In part this issue is being addressed by some of the institutions described in the early part of this chapter and reference will be made later to other institutional innovations.

Third, far more knowledge is needed about the resources and availability of scientific and technical labor. We are witness to over- and undersupplies of scientists, engineers, managers, and technicians throughout the world. India ranks third among the countries of the world in terms of numbers of scientifically and technically trained people, but is first to admit it has yet to find effective ways to employ all or most of them.

Fourth, we also need to know more about how to induce change in international institutions. As a case in point, the major countries of the world for various reasons have long resisted the establishment of more agencies in the U.N. family. This resistance has been based on three considerations: First is the perception that the U.N. agencies now encompass just about anything one wants them to, so why create another agency if a new task can be undertaken in an existing agency? Second, there is no desire on the part of the funding nations to increase their annual dues. Third, as the *ad hoc* Working Group on Policy for Science and Technology within the U.N. system has said, the United Nations already contains a wide spectrum of activities in science and technology which are not at all well coordinated.[11] These issues must come to the surface and they will demand solution. Fourth, many of the global problems (e.g., climate) that require international cooperation and an appropriate institutional framework are interdisciplinary and we are still at a rather primitive stage in constructing institutions that are effective in crossing scientific/technical and social/political boundaries.

Finally, how far can the networking formula used so effectively by the CGIAR and some other highly directed institutions be applied? Networking is the organized and structured cooperative arrangements that are put in place between institutions to facilitate their communications and exchange of data to enable mutually reinforcing activities.

WHAT WE'VE LEARNED

Having alluded to the strengths and weaknesses of international institutions, it is important to identify the strengths upon which we can build more effective international scientific and technical cooperation.

Creativity and flexibility characterize the volunteer nongovernmental institutions. They can communicate very rapidly, move ideas quickly around the world, and develop unofficial communication arrangements of great sensitivity and usefulness. Theirs is a highly idiosyncratic and individualized activity based on people, not organizational momentum and mission. However, resources are usually insufficient for follow-through and they are dependent on external financial support.

In recent years the United States and partner nations have experimented with multinational technical assistance organizations and with bilateral science and technology agreements, joint commissions, and the like. At present the United States is engaged in 38 government-to-government bilateral agreements in areas that include agriculture, space, environment, energy, the oceans, natural resources, health, housing, defense, transportation, development assistance, and science and technology in general. A much larger number of agreements exists between U.S. governmental agencies and their counterparts abroad. There is no question in the minds of observers that, while both multilateral and bilateral approaches are needed, unless the goals sought are well designed and clearly articulated and unless adequate financial and human resources are dedicated, the experiments will be faulty. We have learned from sad experience that applying science and technology to political problems does not solve the problems.

As we have noted before, this chapter is not the place to analyze the bilateral and multilateral scientific and technological arrangements of the U.S. government. There are, however, a very few of these arrangements that might be cited for their unique characteristics with respect to scientific cooperation.

Bilateral support and networking of research institutions, the sharing of information, and the mutual recognition of priority scientific research targets can be extended into sectors of economic and developmental effort as well. A number of experiments in regional collaboration are getting under way and more should be expected, as should new forms of disciplinary collaboration and cooperative research.

For example, in 1970 the United States and South Korea agreed to collaborate in the establishment of a Korean Institute of Science and Technology (KIST). Starting from its incorporation in 1966, when little industrial R&D was being performed in Korea, KIST has become an organization that at the end of 1977 employed nearly 1,000 persons, (including 300 research scientists) with contracts worth about $13 million, 57 percent of which came from industry. The principal areas of R&D (by total contract value) were chemistry and chemical engineering, mechanical engineering, electricity and electronics, metallurgy and materials, and food science and feeds. KIST has supported Korean industrial development through its R&D and development of pilot-scale operations, analysis of advanced foreign technologies, development of new products and processes (from 1967–73, 131 patent applications were made), and repatriation of overseas Korean scientists and engineers. In addition, KIST has fostered the establishment of several promising new research organizations such as institutes for electronics and for ship-building and design.

The U.S. National Academy of Sciences and the Brazilian National Research Council undertook 10 years ago to modernize Brazilian chemical research resources. The result, after 17 young U.S. chemists had spent two to three years each teaching in the universities of Rio de Janiero and São Paulo and 15 senior U.S. chemists had made regular trips to Brazil, was the phased introduction of 196 highly trained Brazilian chemists into Brazilian education, industry, and research.

On the international scale, the Swedish International Development Agency has created a Swedish Agency for Research Cooperation with Developing Countries to promote research that helps developing countries achieve greater self-reliance and economic and social justice.

A consortium of 8 Sahelian countries and 12 industrialized donor nations has formed *Le Club des Amis du Sahel* to coordinate research to halt and reverse the problems of desertification and drought in the Sahel.

In the United States the prospective Institute for Scientific and Technological Cooperation (ISTC), within the government, has been designed as a grant-making and coordinating body to improve the availability and application of technology abroad and to expand knowledge and skills needed to meet developing-country problems in a framework of mutual benefit and partnership.

The key to ISTC operational style and technique will be direct collaboration of developing country and U.S. experts from the start. The purpose of this approach is to capitalize on the lessons learned in some 30 years of development assistance activity and simultaneously to recognize the presence in developing countries of a new generation of technically capable people who are determined to set their own domestic priorities. The specifics of ISTC's approach remain to be developed and will be central to its success.

Another U.S. innovation that we believe has promise was signaled by the adoption of Title V of the Foreign Relations Authorization Act of 1978, relating to the role of science and technology in foreign policy. This recognition by the Congress of the growing importance of science and technology in the conduct of diplomacy as well as in the management of foreign affairs is a positive step toward effective international cooperation.

PERSPECTIVES: THE NEAR-TERM ENVIRONMENT FOR INTERNATIONAL INSTITUTIONAL CHANGE

Science and technology have become increasingly visible as a part of foreign policy as well as of trade and academic exchange. The tradition has been that science is a part of world culture and technology is part of world commerce, but today these distinctions are blurred. Our views originate largely from the academic world and we have been concerned primarily with the scientific and technical cooperation initiated by scientists and engineers outside of government and carried out by a mixture of official agencies and nongovernmental voluntary organizations. Here, again, the distinctions are blurred because the scientific communities of individual nations cooperate with each other under a wide array of institutional mechanisms. Underlying these institutions, and fundamental to their existence, is the understanding of the need to extend and share knowledge. This is what scientific cooperation is all about.

In the international environment, most of the time institutional change appears to proceed at a very slow pace. Except for very occasional and sometimes dramatic political events that force rapid change, international institutions for science and technology move slowly.

We note that the growth of new international organizations more or less stabilized in the 1960's and, while the bulk of the new international institutions we have discussed (IDRC, IFS, CGIAR, etc.) were formed within the past decade, we do not predict a surge of new organizations in the next decade. Funds to start new organizations are more difficult to obtain and the realization of the immense effort required to make a new international institution work has permeated the thinking of both innovators and funding sources. We therefore expect a period of consolidation and shaking out as the more obviously successful and effective organizations lay claim to the available funds. How this will affect efforts to fill newly perceived and important gaps is not at all clear, but the existing institutions may not be capable of meeting these needs. Substantial renewal and adaptation will be required if international institutions are to meet the challenges facing them.

We foresee increased interest in process within institutions—a realization that our current ways and means might be modified to accelerate the transfer of information and of managerial technology. At the same time we foresee no diminution in the universal reliance on direct contact and the transfer of accumulated personal wisdom and experience.

As noted previously, for the past decade or so the major powers have sought to curb the institution-creating instincts of members of the United Nations. There are no signs that this mood is changing; therefore we suspect that most of the institutional risk-taking will be arranged outside the U.N. framework. However, with appropriate intervention, the U.N. agencies can work a significant improvement in the use of the large resources at their disposal. These agencies have the staff, funds, and technical means to coordinate their work in development and in other areas of scientific and technical support far more effectively than they are now doing.

A straw in the wind of change was the 1978 Buenos Aires U.N. Conference on Technical Cooperation among Developing Countries (or TCDC as it came to be known very quickly). Billed as a forerunner of the U.N. Conference on Science and Technology for Development (UNCSTD), TCDC may in fact have reached a decisive turning point with the decision of the developing countries to seek ways in which they can help support each other in the uses of science and technology. The meeting, which had little press coverage or participation from the industrialized countries, was highly regarded by the developing-world participants and is seen by them as a milestone in cooperative initiatives.

The August 1979 UNCSTD is a unique manifestation of the crosscurrents affecting the way science and technology will be used in the future. The attempt of the conference organizers to embrace the whole of science and technology in an economic, political, and diplomatic framework is proving to be very difficult. In the meantime, several privately sponsored preparatory meetings have developed constructive, albeit limited, options for the conference to consider. Some of these options are referred to in our text. Whether or not UNCSTD is declared a success is of little account, considering the groundswell of encouragement and the myriad solid technical proposals for human betterment that its planning and design phase has elicited. The conference is not an end in itself but part of a process that, we believe, will be destined to be both long term and productive.

While the trends of international institutions in science and technology are largely in the direction of applications to meet human needs, the institutions of science *qua* science will continue to be needed and to be fostered for the practical reasons of disciplinary order, as noted earlier, but, far more important, for the support they give to the continued search for new knowledge.

Two areas where institutional inventiveness might make a significant difference are the application of science and technology to economic development and the technological element of arms control—despite the argument that neither issue is centrally a science and technology problem and both are political, economic, and technical in nature.

In the case of economic development, we urge prompt and full exploration of the simultaneous creation of three forms of international cooperation linking national and international organizations: a consortium of financial donors, a network of knowledge-generating institutions in developing countries, and a mechanism to match specific scientific and technological skills of the industrialized nations to specific problems of developing countries. The donor consortium (which already exists in some respects) would be located predominantly in the developed countries. The knowledge-generating and implementing networks should be fostered mainly in the developing nations. The United States' proposed ISTC may approximate the beginning of a U.S. response to these needs, but it is too early to tell. The need for a well-articulated donor network of national foundations, interacting with national and international research centers and with the national agencies that actually create and maintain agricultural, health delivery, and other institutions, is apparent. This is a unique opportunity for exploration requiring innovative, vigorous, and effective leadership.

Arms control also presents a real challenge. Where there should be an international core of agencies funding research, virtually nothing exists. Where there should be a network of agencies generating and assembling knowledge at the international level, we find SIPRI, PRIO, and IISS, as well as a handful of more narrowly based institutions. We ask if this is sufficient.

The well-known Pugwash and Dartmouth conferences have been valuable for personal contact and wide-ranging exploration of issues. Pugwash is recognized as one of the earliest vehicles by which Soviet–American dialogues on disarmament were given high visibility by technical experts. Today the number of avenues for exploration of new arms control and disarmament initiatives is vastly more comprehensive than it was in the 1950's when Pugwash began. The question now is how to stimulate and sustain analysis in depth of the security issues that divide East and West and to grapple effectively with the arms races in the developing world. Pugwash interest in economic development has diluted its emphasis on disarmament.

The institutional politicization of the 1960's and early 1970's seems largely to have run its course. It is reasonable to anticipate an increasing degree of sophistication about the pressing needs of developing countries: technology transfer, the application of science and technology to development, and a greater recognition of the need for attention to our international institutional resources. The basic global infrastructure has not been built, but a substantial foundation has been laid.

It is clearly in the U.S. national interest to build and to sustain a community of cooperative institutions, working linkages, networks, and consortia that are sufficiently innovative to contribute to the quest for global accommodation as well as the health of U.S. science and its efforts to meet American technical responsibilities.

As we approach the many issues involved in building, renewing, and adapting institutions over the next five years, it is important that we do not become divided into the "uncritical lovers" and the "unloving critics" of institutions described so vividly by John Gardner. The former shield their institutions from life-giving criticism, literally smothering them in an embrace of death, loving their rigidities more than their promise. The latter are skilled in demolition, but almost willfully ignorant of the arts by which human institutions are nurtured, renewed, and made to flourish.[12]

OUTLOOK

The following outlook section on institutions for international cooperation is based on information extracted from the chapter and covers trends anticipated in the near future, approximately five years.

In the near future, a number of issues will draw increasing attention to the workings of international institutions of science and technology. One of these is the environment in which these institutions work, now beset by two types of polarization: conflicts between East and West and North and South. Both result from differences in scientific and technological capacity and productivity between and among nations. With the resources for research concentrated in the industrial world, and the pressing need for expanded research in the developing world, new methods of technology transfer and collaborative research must be found.

The issues we describe here are in many respects overlapping. Nonetheless, we foresee three major areas for concern:

- Interdependency problems such as the global effects of an increasing build-up of carbon dioxide, multinational management of water basins, and exploitation of the seabed under some yet-to-be-agreed-upon international arrangement.
- Distributive issues of equitable access to natural resources, including assured food supply, reliable sources of energy at affordable prices under acceptable environmental constraints, and vastly expanded delivery of health care, adequate housing, and educational opportunity. These issues have two facets. One is internal to each nation and is a direct measure of the well-being of individual citizens; the other relates to the nation's place in the world and its ability to secure the needs of its people in global competition.
- Vulnerability concerns such as those related to management of radioactive wastes and the development of nuclear power, levels of acceptable environmental risk, and natural catastrophes.

We do not see a surge of new organizations to deal with these problems in the next decade. We can expect a period of consolidation and possibly even thinning out as the more successful and effective organizations receive the available funds. We can anticipate changes in the character of a number of institutions that will require a complex mix of scientific and technical knowledge from more than a single discipline if they are to deal successfuly with global problems. Newly perceived institutional gaps remain to be filled, and it is important that a conservative stance toward the formation of new organizations not be used as an excuse to avoid institutional innovation.

The status of U.S. science and technology is changing and will continue to change as a result of international competition and as the more technologically advanced developing nations seek to acquire an increasing share of the technology market. This will cause increasing concern in the United States.

The developing countries are newly aware of the potential for cooperation among themselves. This, combined with their apparent influence in many international institutions, may result in a shift toward regional and bilateral arrangements that more precisely match national interests. Certainly over the next five years we can expect an expansion of the debate on how science and technology can be effectively used to help solve economic, social, and political problems.

Two specific areas, among many, where institutional inventiveness can make a significant difference are the application of science and technology to economic development and the technological aspects of arms control.

We predict increased governmental and academic interest in the process of change within institutions. How can organizations be better evaluated against clear criteria of effectiveness and how can organizational improvement be accelerated?

In applying science and technology to economic development, we believe it will be useful to explore carefully three kinds of cooperation linking national and international organizations: a consortium of financial donors, a network of knowledge-generating institutions, and a mechanism to match specific scientific and technological skills of the industrialized nations to specific problems of developing countries.

To bring the most effective intellectual forces to bear on matters of disarmament and arms control, the small group of existing organizations will have to be markedly strengthened and indeed here new institutions may be needed.

NOTES AND REFERENCES

1. *Scientific and Technical Cooperation with Developing Countries, 1977–78* (Higher Education Panel Reports, No. 40). American Council on Education, August 1978.

2. *International Scientific Activities of Selected Institutions, 1975–76 and 1976–77* (Higher Education Panel Reports, No. 37). American Council on Education, January 1978.

3. ICSU funding consists of $800–850,000 per year and the total expenditures of the whole ICSU family amount to approximately $3.8 million.

4. Gossett, B. First GARP Global Experiment. *WMO Bulletin* 28:5, January 1979.

5. From records maintained in the Office of the Foreign Secretary, National Academy of Sciences, Washington, D.C.

6. CGIAR funds amounted to $86.8 million in 1978 and $103.3 million in 1979 (projected).

7. *CGIAR—Consultative Group on International Agricultural Research.* CGIAR, New York, 1976.

8. *Highlights* (Commission on Human Resources, National Research Council 9/78). "Trends in Postdoctoral Appointments Abroad Scientists and Engineers."

9. Watkins, B.T. Foreigners' Enrollment Concerns Graduate Deans. *Chronicle of Higher Education*, December 11, 1978, p. 6.

10. *Review of U.S.–U.S.S.R. Interacademy Exchanges and Relations* (NRC Board on International Scientific Exchange). Washington, D.C.: National Academy of Sciences, 1977.

11. Berlinguet, L. "Views of the Transfer of Technology Based on Recent Programmes in the Third World," presented at a Workshop on Scientific and Technical Cooperation with Developing Countries, Organization for Economic Cooperation and Development, Paris, April 10–13, 1978.

12. 100th Commencement Address, Cornell University, Ithaca, N.Y., June 1968.

Reviewers and Additional Contributors

The assistance of the following persons is gratefully acknowledged. The contents of the report are the responsibility of the Steering Committee.

ENTIRE REPORT

Governing Board Members and Assembly/Commission Chairmen

JACOB BIGELEISEN, State University of New York, Stony Brook
R. H. BING, University of Texas, Austin
HARVEY BROOKS, Harvard University
ROBERT H. CANNON, JR., California Institute of Technology
W. KENNETH DAVIS, Bechtel Power Corporation
DAVID R. GODDARD, National Academy of Sciences
FREDERIC A. L. HOLLOWAY, National Academy of Engineering
R. DUNCAN LUCE, Harvard University
SAUNDERS MAC LANE, University of Chicago
COURTLAND D. PERKINS, National Academy of Engineering
E. R. PIORE, International Business Machines Corporation
FRANK W. PUTNAM, Indiana University
HARRISON SHULL, Indiana University
HERBERT A. SIMON, Carnegie-Mellon University
MITCHELL W. SPELLMAN, Harvard Medical School
H. GUYFORD STEVER, Assembly of Engineering, National Research Council

FRANK H. WESTHEIMER, Harvard University
GILBERT F. WHITE, University of Colorado

PLANET EARTH

SAMUEL S. ADAMS, Consulting Geologist
ARTHUR G. ANDERSON, International Business Machines Corporation
JAMES K. ANGELL, National Oceanographic and Atmospheric Administration
PAUL A. BAILLY, Occidental Minerals Corporation
ALBERT W. BALLY, Shell Oil Company
BRUCE A. BOLT, University of California, Berkeley
MICHAEL A. CHINNERY, Massachusetts Institute of Technology
PRESTON E. CLOUD, University of California, Santa Barbara
DANIEL J. FINK, General Electric Company
PETER T. FLAWN, University of Texas, Austin
ROBERT G. FLEAGLE, University of Washington
EDWARD A. FLINN, National Aeronautics and Space Administration
WILLIAM W. HAY, University of Miami
CLAUDE R. HOCOTT, Gulf Universities Research Consortium
JOHN D. ISAACS, University of California, La Jolla
DAVID S. JOHNSON, National Oceanographic and Atmospheric Administration
DON E. KASH, U.S. Geological Survey
WILLIAM M. KAULA, University of California, Los Angeles
CARL KISSLINGER, University of Colorado
JOHN KUTZBACH, University of Wisconsin
HELMUT LANDSBERG, University of Maryland
REUBEN LASKER, National Oceanographic and Atmospheric Administration
LESTER MACHTA, National Oceanographic and Atmospheric Administration
VINCENT E. MCKELVEY, U.S. Geological Survey
J. MURRAY MITCHELL, JR., National Oceanographic and Atmospheric Administration
JACK E. OLIVER, Cornell University
WILLIAM L. QUAIDE, National Aeronautics and Space Administration
ROGER REVELLE, University of California, San Diego
WALTER O. ROBERTS, Aspen Institute
DEREK W. SPENCER, Woods Hole Oceanographic Institution
JOHN STEELE, Woods Hole Oceanographic Institution
JAMES V. TARANIK, U.S. Geological Survey

WARREN S. WOOSTER, University of Washington

THE LIVING STATE

K. FRANK AUSTEN, Harvard Medical School
HARRY BEEVERS, University of California, Santa Cruz
FLOYD E. BLOOM, The Salk Institute
ALBERT C. ENGLAND, III, Center for Disease Control
LESLIE HICKS, Howard University
EDWARD KRAVITZ, Harvard Medical School
STEPHEN KUFFLER, Harvard Medical School
BEATRICE MINTZ, Institute for Cancer Reseach
HOWARD A. SCHNEIDERMAN, University of California, Irvine
GREGORY SISKIND, Cornell University Medical School
TERRY STROM, Harvard Medical School

THE STRUCTURE OF MATTER

PHILIP ANDERSON, Bell Laboratories
CHARLES P. BEAN, General Electric Company
JAMES BJORKEN, Stanford Linear Accelerator Center
WALTER BROWN, Bell Laboratories
CURT CALLAN, Princeton University
KENNETH CASE, Rockefeller University
SIDNEY DRELL, Stanford Linear Accelerator Center
GORDON H. DUNN, University of Colorado
E. NORVAL FORTSON, University of Washington
FRANK Y. FRADIN, Argonne National Laboratory
ARTHUR FREEMAN, Northwestern University
FREDERICK J. GILMAN, Stanford Linear Accelerator Center
MORTON HAMMERMESH, University of Minnesota
THEODORE W. HANSCH, Stanford University
DAVID S. HEESCHEN, National Radio Astronomy Observatory
VERNON W. HUGHES, Yale University
JOHN B. KETTERSON, Northwestern University
EUGEN MERZBACHER, University of North Carolina
JEREMIAH P. OSTRIKER, Princeton University Observatory
C. KUMAR N. PATEL, Bell Laboratories
P. JAMES E. PEEBLES, Princeton University
DAVID SCHRAMM, University of Chicago
GOPAL K. SHENOY, Argonne National Laboratory

MELVIN SHOCKETT, University of Chicago
RALPH O. SIMMONS, University of Illinois, Urbana
RONALD F. STEBBINGS, Rice University
VALENTINE LOUIS TELEGDI, University of Chicago

COMPUTERS AND COMMUNICATION

LEE L. DAVENPORT, General Telephone and Electronics Corporation
JERRIER A. HADDAD, International Business Machines Corporation
C. KUMAR N. PATEL, Bell Laboratories
ALAN J. PERLIS, Yale University
RICHARD J. ROYSTON, Argonne National Laboratory
THOMAS SCHELLING, Harvard University
MORRIS TANENBAUM, New Jersey Bell Telephone Company
VICTOR VYSSOTSKY, Bell Laboratories
FRANKLIN ZIMRING, University of Chicago

ENERGY

BETSY ANCKER-JOHNSON, General Motors Corporation
SOLOMON J. BUCHSBAUM, Bell Laboratories
ROBERT A. CHARPIE, Cabot Corporation
JAMES CROW, University of Wisconsin
LINCOLN GORDON, Resources for the Future
JOHN HOLDREN, University of California, Berkeley
STANFORD S. PENNER, University of California, San Diego
WILLIAM E. SHOUPP, Consultant

MATERIALS

RICHARD H. BOYD, University of Utah
WILLIAM DAUBEN, University of California, Berkeley
MERTON C. FLEMINGS, Massachusetts Institute of Technology
GEORGE R. HILL, University of Utah
ALLEN KNEESE, University of New Mexico
M. EUGENE MERCHANT, Cincinnati Milacron
GORDON MILLAR, Deere & Company
REINHARDT SCHUHMANN, JR., Purdue University
WILLIAM SLICHTER, Bell Laboratories
DAVID TURNBULL, Harvard University

NATHANIEL WOLLMAN, University of New Mexico

DEMOGRAPHY

CAROLYN S. BELL, Wellesley College
JUDITH BLAKE, University of California, Los Angeles
DEBBY BURGARD, Harvard University
ALBERT M. CLOGSTON, Bell Laboratories
ARTHUR A. CAMPBELL, National Institute of Child Health and Human Development
JOHN W. COLTMAN, Westinghouse Research Laboratories
KAREN CORCORAN, Harvard University
H. RICHARD CRANE, University of Michigan
LILLI S. HORNIG, Wellesley College
LLOYD G. HUMPHREYS, University of Illinois
DOROTHEA JAMESON, University of Pennsylvania
DIANE KENNY, Harvard University
JON A. KROSNICK, Harvard University
OLIVER H. LOWRY, Washington University
JACK E. MYERS, University of Texas, Austin
GLENN C. OAKLEY, Harvard University
JAMES F. PEDULLA, Harvard University
I. RICHARD SAVAGE, Yale University
LINDA UJIFUSA, Harvard University
CHEVES WALLING, University of Utah
SEYMOUR L. WOLFBEIN, Temple University

HEALTH OF THE AMERICAN PEOPLE

EDWARD H. AHRENS, Rockefeller University
JULIUS AXELROD, National Institute of Mental Health
THEODORE COOPER, Cornell University Medical College
WILLIAM H. DANFORTH, Washington University
JOHN R. EVANS, University of Toronto, Department of Medicine
JOHN W. FARQUHAR, Stanford University
WILLIAM H. FOEGE, Center for Disease Control
ROGER GUILLEMIN, The Salk Institute
PETER B. HUTT, Covington & Burling
ROBERT I. LEVY, National Institutes of Health
WALTER J. MCNERNEY, Blue Cross/Blue Shield
SHERMAN M. MELLINKOFF, University of California, Los Angeles

FREDERICK MOSTELLER, Harvard School of Public Health
ARNO G. MOTULSKY, University of Washington School of Medicine
CHARLES H. RAMMELKAMP, Case Western Reserve University, Department of Medicine
DOROTHY P. RICE, National Center for Health Statistics
DAVID E. ROGERS, The Robert Wood Johnson Foundation
DANIEL C. TOSTESON, Harvard Medical School
JAMES B. WYNGAARDEN, Duke University

TOXIC SUBSTANCES IN THE ENVIRONMENT

MYRON K. BRAKKE, University of Nebraska
JOHN J. BURNS, Hoffman-La Roche, Inc.
DOUGLAS GRAHN, Argonne National Laboratory
ALFRED G. KNUDSON, The Institute for Cancer Research
GENE E. LIKENS, Cornell University
BRIAN MACMAHON, Harvard School of Public Health
JOHN D. ROBERTS, California Institute of Technology
BERT VALLEE, Harvard Medical School
E. BRIGHT WILSON, Harvard University

ACADEMIC SCIENCE AND GRADUATE EDUCATION

ALBERT M. CLOGSTON, Bell Laboratories
JOHN W. COLTMAN, Westinghouse Research Laboratories
H. RICHARD CRANE, University of Michigan
ROBERT HILL, Duke University Medical Center
LILLI S. HORNIG, Wellesley College
LLOYD G. HUMPHREYS, University of Illinois
OLIVER H. LOWRY, Washington University
JAMES H. MULLIGAN, JR., University of California, Irvine
JACK E. MYERS, University of Texas, Austin
ROY RADNER, Harvard University
ISADORE M. SINGER, University of California, Berkeley
CHEVES WALLING, University of Utah
SEYMOUR L. WOLFBEIN, Temple University

INSTITUTIONS FOR INTERNATIONAL COOPERATION

LAWRENCE BOGORAD, Harvard University
JOSEPH FEINSTEIN, Varian Associates
DONALD FINK, Consultant
KEITH GLENNAN, Consultant
BRUCE HANNAY, Bell Laboratories
DIXON LONG, Case Western Reserve University
ROGER REVELLE, University of California, San Diego
EMIL SMITH, University of California School of Medicine
ARTHUR K. SOLOMON, Harvard Medical School
DOROTHY ZINBERG, Harvard University

Index

A

Academic Science, funding: 477–481
  basic research: 479–480
  expenditures: 481
  levels of support: 478–480
  national priorities: 480
  support structures: 481
accelerators: 183
  circular: 178, 201
  electron: 189, 201
  heavy-ion: 183
  high-energy: 189
  large particle: 201
  linear: 201
  proton: 178, 201
accidents: 382–383, 386
acid rains: 35, 275, 456
acoustic ranging: 29–30, 52
accretion disk: 154
ACTH: 414
adenosine triphosphate (ATP): 76, 103, 118, 119, 123, 124, 126, 129
Advanced X-ray Astronomy Facility (AXAF): 163
Advisory Group for Aerospace Research and Development (AGARD): 502
aerosol sprays: 34–35
aflatoxins: 450, 457
age structure: 347, 351–353
  change: 352–353
  future: 373–374
aging: 412–417
  cell: 83, 84, 413
  depression: 415
  health services: 416–417
  immunology: 413
  neurology: 413–414
  pharmacology: 414–415
  plant: 126
  research: 412–413
  sleep: 416–417
  see also elderly, the
agriculture: 37, 38, 40, 43, 47, 52, 116–133, 263, 273, 274, 446
  policy: 54
aircraft: 287
alcohol
  abuse: 411, 451
  effect: 76
algae, blue-green: 123, 127
algorithm: 227
  analysis: 233–234
  calculability: 234–236
  design and efficiency: 231–234
  probabilistic: 236
  proofs: 234
allergies: 92
alloys: 302–307

alpha particle: 181, 182, 186
  decay: 181
Alternating Gradient Synchrotron (AGS): 201
aluminum: 333, 334
aluminum gallium arsenide: 308
aluminum nitride: 308
aluminum phosphate: 310
amino acids: 64, 65, 72, 86, 98, 106
amorphous cell materials: 281
amorphous metals: 305
amorphous semiconductors: 168, 288
amphetamines: 410, 411
animals: 27–28, 30, 35
  and chemicals: 455–456
  models: 411–412
anisotropism: 303, 318
anorthosite: 334
antiatom: 192
antibodies: 67–68, 85, 86
  genes for: 88
  production: 88–89
antigens: 85, 88
antimatter: 192
antimolecule: 192
antineutrino: 199
antineutron: 192
  production: 200
antiproton: 189, 192
  production: 200
antiquark, charmed: 194
ARPANET: 236
arsenic: 451
arteriosclerosis: 385, 386
arthritis: 412
asbestos: 449, 458
astronomy, radio: 158–159
  infrared: 159–160
astrophysics, high-energy: 151–154
atherosclerosis: 84, 412
Atlantic Ocean: 11, 12
atmosphere: 263, 446
  circulation: 13
atoms
  alkali: 166
  configuration: 163, 165–166
  engineering: 168–169
  exotic: 176
  giant: 175
  muonic: 176
  structure: 170–171, 174–175
ATP: *see* adenosine triphosphate
auditory system: 110–111
auger spectroscopy: 307
autoimmunity: 92–93, 413
automatic gauge control: 317
automobiles: 257, 289, 304, 310, 316, 321, 327
auxins: 128
axon: 103
  conduction: 102–103

B

baby boom: 348, 375
bacteria, action on chemicals: 446
baryon: 192, 193, 196
  production: 200
BASIC: 243
basic research, funding for: 479–480
batteries: 166, 167
bauxite: 334
B cells: 86, 91
behavior: 99, 100
  intelligent: 248–249
  neuronal basis: 108–116
beryllium: 333, 334
beryllium oxide: 308
beta decay: 181, 191
Big Bang: 148, 149, 150, 157
binary systems: 153
biobehavioral science: 404–412
  treatment: 391
biogenic amines: 409
  hypothesis: 408–410
biomass: 261, 274, 275, 278, 282
birth defects: 455
birthrates: 346, 348–350
  projections: 367–368
bits: 219, 220, 221, 223
black holes: 151, 153–154, 157–158
black lung disease: 449
Blacks
  earnings: 364
  education: 367
  employment: 359, 362
  one-parent families: 357
  residence: 356
  teenage fertility: 349
BL-Lacertae objects: 156
bonding, chemical: 165
boron: 315

brain: 95–96, 97
evolution: 99
function: 407–408
hormones: 404
structure: 99–101, 112
see also neurotransmitters
Brazilian National Research Council: 510–511
breeder reactors: 269–270, 277, 283, 285
brines: 265
brown lung disease: 449
bubble memory: 221, 311
buildings: 257, 282, 287, 289
burst multiplexing: 237

C

cadmium: 451, 452
cadmium telluride: 310
cancer: 382, 383, 385, 386, 412
bladder: 401
breast: 399, 400
causes: 78–80, 84, 397
cervical: 399
detection: 399
epidemiology: 398, 454
immunology: 93
incidence: 453–454
lung: 399, 401, 448, 453–454
mortality: 399, 401–402
predisposition: 73, 79, 396
prevention: 400–401
risk factors: 396–398
skin: 36, 399
stomach: 454
treatment: 399–400
see also carcinogens, cigarettes
CANDU heavy water reactor: 284, 285
carbon cycle: 121–122
carbon dioxide: 30, 34, 37, 38, 39, 40, 45, 50, 116, 117, 118, 119, 121, 275, 505
atmospheric: 273–274
fixation: 119
carbon monoxide: 447
carcinogens: 78, 397–398, 441, 456–457
chemical: 449–450, 451–452
identification: 401
cardiovascular disease: 274, 382, 383, 386, 387–396, 412, 441, 455
modifying risk factors: 394–395
mortality: 388
prevention: 392–393, 395–396
risk factors: 393–396, 448
casting: 318–319
catalysts: 172, 307
cataracts: 110
cell biology: 75–85, 413
definition: 75
cells
adaptation: 75–76
aging: 83, 84
differentiation: 84
function: 75–76, 84–85
fusion: 88–89
membrane: 80–81, 97
motility: 82–83
reproduction: 77–78
structure: 75–76, 82–83, 84–85
transport: 80–81
centrifugal force: 189
ceramics: 308, 313–315
cerebral cortex: 110, 111, 115
CERN: 202, 504
chains: 166, 167
chemicals: 439–465
carcinogens: 78, 79, 398, 401
persistence: 446
production: 442–445
risks: 458
technology: 442–445
transformation: 446–447
chemotherapy: 400
CHESS 4.6: 249
childbirth risks: 350
children
daycare: 366
school: 366–367
chips, computer: 217–220, 221, 222–223
garnet: 221
silicon: 222
cholesterol: 393–394
Chlordane: 453
chloride cell: 327
chlorination: 448
chlorofluoromethanes: 34–35, 36
chlorophyll: 117–119, 173–174
photocenters: 117–119
chromosome: 67, 68, 82
structure: 69
viral: 79

cigarettes: 397–398, 401–404, 425, 441, 448, 451, 452, 453, 454, 457
adolescent smokers: 403–404
low tar and nicotine: 403
smokers: 403
circuitry: 311
digital: 311
integrated: 169, 216–217, 219, 320
solid state electric: 281
cirrhosis of the liver: 386
Civex process: 285
clays: 334
Clean Air Act: 457
climate: 28, 30, 31–40, 54, 275
fluctuations: 33, 37, 45
human impact: 33, 37, 263
prediction: 39
Climatic Impact Assessment Program: 36
cluster gas: 155–156
clusters: 166, 167
coal: 263, 266–267, 273, 277, 313
chemistry: 288
combustion: 261
conversion: 268–269
cobalt: 306, 317
alloys: 305
cobalt-base superalloys: 303, 314–315
cobalt-samarium alloys: 306
COBOL: 280
cogeneration: 289
collaboration, international: 460, 508–509
Coma: 155
combustion: 287
Committee on Genetic Experimentation (COGENE): 497–498
Communications Act of 1934: 244
communications and computers: 236–249
in-plant: 238
local: 238
competition, foreign: 326–327, 331, 333, 507
composite materials: 315–316
computed tomographic scanning: 399, 427
computer-aided design (CAD): 320–321, 322
computer-aided manufacturing (CAM): 320–321, 322
computer control, rolling mills: 317
computers: 215–251
climate: 43, 47–48
earthquakes: 17
games: 242, 243
home terminals: 242–244
materials: 311, 317, 318, 320
networks: 236
nuclear shell: 184–185
particle physics: 203
regulation: 244–245
reliability: 224–225
search: 232–234
single data-stream: 225–226
computer sciences
and mathematics: 231–236
theoretical: 231–236
conditioning: 114
conductors: 216, 217
congenital heart disease: 388–389
conservation
energy: 256–258, 290
materials: 330–331
consciousness: 115
Consultative Group on International Agricultural Research (CGIAR): 500–501, 509
continents: 13
continuous casting: 317
continuous rolling of strip: 317
conurbation: 354
converters, advanced: 284
copper: 308
coriolis force: 189
coronary artery disease: 391–392
corrosion: 305–306, 307
resistance: 306
Cosmic Background Explorer (COBE): 150–151
cosmology: 148–151
cost
academic science: 477–481
biomedicine: 62, 134
computers: 217–220, 222, 224, 226, 237, 243
earthquakes: 19
education: 470
energy: 257, 266, 269, 275, 278, 281
health: 419–420
international institutions: 512

materials: 302, 305–306, 308–309, 311
matter: 179
minerals: 21, 25
particle physics: 203–204
project funding: 482–485
space shuttle: 47
Crab: 156
crossed eyes: 110
credit card verification: 241
crown gall: 131
crust: 10–15
cryogenically cooled detectors and mirrors: 161–162
cryolite: 334
crystal: 217
electron: 167
growth: 168
metal: 302, 318–319, 325
semiconductor: 310
cyclotron: 189
cytochrome: 178
cytokinins: 128

D

dark reactions: 119, 121
Dartmouth conference: 514
data communications: 236
DATAPAC: 236
DDE: 453
DDN: 236
DDT: 448, 452, 453, 456
Deep Sea Drilling Project: 14, 53
defoliants: 446
Delaney Clause: 450, 451
demography: 343–378
data: 345–346
determinants: 343–344
method: 343
uses: 344–345
density of matter: 149
dependency ratio: 352–353
depression: 405, 408, 411
elderly: 415
desulfurization: 267
detergents: 264
deuterium: 276
deuterons: 182
developing nations
energy: 263, 265, 274, 333
diabetes mellitus: 386
diamond: 308
diarrhea: 386
dielectrics: 309
diffusion phenomena: 168
digital encoding: 237
dioxin (TCDD): 446, 453
diphtheria: 386
directional solidification: 303
disease causation
cellular: 84–85
environmental: 441, 453–455
genetic: 73, 91
immune system: 91
predisposition: 73
disease, plant: 131
divorce: 357, 358
DNA: 60, 64–70, 71, 72, 76, 82, 114, 133, 136, 137, 173
damage: 78, 131
duplication: 68–69, 77
recombinant: 66, 74, 124, 134, 497–498
doctorates: 471, 472, 474, 476–477, 493
dopamine: 106, 107, 116, 408, 414
hypothesis: 408, 410
double-alkali system: 267
double-blind experiments: 407
drilling: 14–15, 26, 52
drugs: 107, 442
abuse: 411, 451
and the aged: 414–415
development: 106, 411
effect: 76, 113, 115, 116
hypertension: 390, 394
mental illness: 407, 408
resistance: 77
side effects: 410–411
therapy: 404, 406
dual-phase steels: 305
dumand: 157

E

Earthquake Hazards Reduction Act of 1977: 20
earthquakes: 11, 13, 15–21, 45, 53
causes: 15–16
China: 16–17
cost: 15, 18, 19
hazards: 18–19
monitoring: 17

prediction: 16–19, 20–21
risk reduction: 19–21
economic development: 513
education: 322, 366–367, 469–489
and scientific research: 471
enrollment: 471–472
health: 395–396
hierarchy: 485
women and minorities: 357, 367, 474, 476–477
see also graduate education, schools, universities
elderly, the: 358, 412–417
electric furnace/scrap recycle: 327
electricity: 175, 276, 289–290
conduction: 166, 167
demand: 270
source: 165, 167, 260
electromagnetism
field: 176
radiation: 43, 146, 199, 309
spectrum: 164
see also interactions
electron beams: 311
electron-beam writing: 219
electronic displays: 310
electronic fast-connect circuit switching: 237
electronic inspection: 317
electronic mail: 239–240, 243, 244
electronics: 216
molecular: 177–178
electron microscopy: 179, 325
electron paramagnetic resonance (EPR): 173
electrons: 191, 199–200
arrangement: 163
heavy: 196
high-energy: 183
interaction: 166
low-energy: 165
scattering: 186
source: 216
surface: 166
electrostatic forces: 196
electrostatic generator: 189
elements, chemical: 148
emotion: 112–113
employment: 371–372
materials: 326–327
encryption
computers: 240
endocrine system: 414
endorphins: 113
energy: 252–298
conservation: 286–287
consumption: 256, 257–258, 372–373
economic growth: 286
farms: 263, 282
generation: 169
long-term options: 275–286
materials: 308–310, 323, 324, 327
mechanical: 15
mid-term options: 256–259
mid-term supply: 259–272
new sources: 75, 259–272
nuclear state: 184
policy: 54, 266
storage: 282
supply: 256
time perspective: 254–256, 265
see also geothermal, nuclear, solar
energy cycle
and climate: 32
changes: 33
environment: 265, 266, 323, 439–465
energy consumption: 288–289
energy supply: 272–275
health: 426
policy: 55
enzymes: 74–75, 173
epidemiology
cancer: 398, 402
epilepsy: 115
EPSS: 236
equilibrium, thermal: 169–170
EURONET: 236
evolution: 133–134
brain: 99
EXAFS: 172
exhaust-emission control: 310
extraction, materials: 323, 324
Exxon Donor Solvent (EDS): 269

F

fabrication, materials: 323, 324
faculty: 475, 477, 493
families: 375, 415
one-parent: 357–358
faults: 17, 18, 20, 43
see also San Andreas Fault

Fermi National Laboratory: 201, 202
FeMo-co: 123
ferredoxin: 123, 178
fertility: 347, 348–350, 351, 355, 368, 373–374, 375
  Blacks: 349
  current: 348
  teenage: 349–350
  Whites: 349
fiber-optic endoscope: 399
fiber optics: 310
Finnish hypertension program: 395–396
fire-flooding: 264
fires: 18–19, 43, 48
Fischer-Tropsch process: 269
Fishery Management and Conservation Act of 1976: 27–28
fission isomers: 186
flooding: 18–19
fluidized bed combustion: 267
forging: 321
FORTRAN: 230
fossil fuel: 39–40, 50, 54, 122–123, 263, 265, 266, 273, 274, 275, 279, 289, 313
frost resistance: 130–131
fuels, synthetic: 260, 263, 268–269
funds transfer, electronic: 241–242
fusion plasma: 276
future directions
  agriculture: 122, 124, 125, 126, 127, 129, 132 133
  astronomy: 156, 158
  atomic structure: 177–178
  biomedicine: 63
  cancer: 79–80, 82
  cell biology: 82, 137
  climate: 32, 36–37, 37–38, 38–39, 48, 49, 50
  computers: 215, 236, 237, 238, 240
  demography: 356, 358, 365, 367–368, 373–375
  DNA: 68, 69, 74
  earthquakes: 19–21
  education: 472, 474
  energy: 256, 258–259, 264, 265–266, 267, 268, 286–290
  health: 390, 398, 417–418, 423–428
  immunology: 90, 92, 137
  international cooperation: 505–514
  materials: 305, 320, 331, 333–335
  matter: 170
  minerals: 21, 24–26
  molecular genetics: 74–75, 135–136
  neuroscience: 96, 102, 109, 110–111, 112, 113, 115, 116
  nuclear structure: 186, 188
  oceans: 27, 29
  plate tectonics: 13, 14–15, 27

G

gadolinium–iron–garnet: 311
galaxies: 148
  hyperactive nuclei: 157–158
  see particular galaxy
gallium arsenide: 279, 308
Gamma Ray Observatory (GRO): 163
gamma rays: 154
  astronomy: 156–157
  instruments: 162–163
garnet: 220, 221
gas: 30–31, 54, 260, 263, 264–266, 275, 279
  cluster: 154–155
  infrared-absorbing: 35, 37
  intergalactic: 149
  thermal equilibrium: 169
gas-discharge devices: 311
gasification: 268–269
gas injection: 264
gas-turbine engines: 303, 315
genes: 64–66, 67–70, 82, 88, 136
  evolution: 133
  plant: 124
  recombination: 69–70
  regulation: 65–66, 78, 81, 84
  viral: 79
genetics
  abnormalities: 110
  behavior: 108–109
  disease: 73, 79, 88, 394, 417
  nitrogen fixation: 124
  plant research: 129, 132
Geodynamics Project: 14–15
geology: 44, 53
  see also minerals
geothermal energy: 43, 261, 275–276
germanium: 216, 217
giant dipole excitation: 186
G.I. bill: 475
gibberellins: 128
glass-reinforced plastics: 315

glassy metals: 305
Global Atmospheric Research Program (GARP): 44, 494, 498
Global Weather Experiment: 44, 51, 498
glucose: 117
gluons: 196
graduate education: 471, 474, 476
  support: 475–476
graphite: 315
graphite intercalation compounds: 166
GRAS: 450
gravity: 149, 154
greenhouse effect: 33
group practice: 418–419
growth control
  plants: 127–129
growth rate: 347
guayule: 316

H

Haber-Bosch process: 122
habituation: 113–114
hadron: 191, 192–193, 197
hafnium: 306, 307
halogenated hydrocarbons: 452–453
H-Coal: 269
health: 381–437
  energy consumption: 272–275, 289
  materials: 316
health care: 417–423
  access: 419
  goals: 426–427
  group practice: 418–419
  improvement: 425–426
  independent practice associations: 420
  payment: 419–420
  perspectives: 423–428
  practice: 418–419
  prepaid group practice: 419–420
  research: 423, 424
health maintenance organizations: 419–422
  assessments: 420–422
hearing: 110–111
heat pumps: 289
helper T cells: 88, 92
hepatitis: 89–90
herbicides: *see* pesticides
heuristics: 236
high-density energy storage: 308
High Energy Astronomical Observatory (HEAO): 156, 162
high-energy atom smashers: 325
high-powered pulsed lasers: 278
high-pressure blast furnace: 317
high-speed cutting: 321–322
high-strength steels: 304–305
high-temperature gas-cooled reactor: 285
hippocampus: 100, 101, 114
hormones: 98–99, 104, 112, 414
hot isostatic pressing: 319
households: 356–358
  number: 357–358
Hubble, Edwin: 148–149
human rights: 508
hunger: 112
hydrocarbons: 13, 289, 290
hydrodynamics: 169–170
hydroelectric power: 278, 281
hydrogenase: 126
hydrogen, recycling: 125–126
hydrometallurgical refining: 327
hypertension: 389–391
  treatment: 390–391
  unknowns: 391
hypothalamus: 112–113

I

illness
  burden of: 387, 426
  patterns: 385, 386
  tropical: 499
immigration: 347, 351
immune system diseases: 91
immunization: 89–90, 411, 425
immunology: 85–95, 413
  analytical technique: 94
  applications: 89–94
  cancer: 93, 399
implants: 316–317
imports
  materials: 326
income: 360
  city dwellers: 355–356
indium antimonide: 310
industry: 257, 259, 273, 287
  and universities: 485–486
  R&D: 510
infant morbidity: 350, 386–387, 403

infant mortality: 382, 383, 386
infection, response: 85
see specific disease, immunology
infectious disease: 385, 386
influenza: 385, 386
information dissemination: 306
Infrared Astronomical Satellite (IRAS): 162
infrared detectors: 310
innovation: 50–51
health care: 417–423
institutional: 513
materials: 302, 319, 331, 334
insect pests: 131
Institute for Scientific and Technological Cooperation (ISTC): 511, 513
institutional care: 407, 416–417
institutional factors
materials: 331, 333
institutions, international: 491–516
background: 506–508
change: 511–514
instruments, research: 146, 157, 160–163, 178–179, 189–190, 201–204
insulators: 216, 217, 308
intelligence, artificial: 246–249
INTELSAT: 237
interactions
electromagnetic: 183, 188, 191, 194, 197, 199
gravitational: 191
particle: 191
strong: 187–188, 191
weak: 188, 191, 194, 196–199
interferometer: 161
interferometry, very long baseline (VLBL): 159
interferon: 94
International Atomic Energy Agency (IAEA): 495, 500
International Biological Program (IBP): 497
international cooperation: 492–495
International Council of Scientific Unions (ICSU): 495–498, 499
International Decade of Ocean Exploration: 53
International Development Research Centre (IDRC): 500, 503–504
International Foundation for Science (IFS): 503
International Institute for Applied Systems Analysis (IIASA): 502
International Institute for Strategic Studies (IISS): 502–503
international issues
climate: 48–49
earth: 55–56
International Peace Research Institute —Oslo (PRIO): 503
Intersecting Storage Ring (ISR): 202
ion beams: 278
ion implantation: 219, 311
ions: 186
alkali: 167
heavy: 186
implantation: 169
light: 185
sources: 179
trapped: 175–176
iproniazid: 408–409
iron alloys: 305
isotopes, production: 183, 186

J

job market: 475–476
jojoba: 316
Josephson junction technology: 221
J particle: 194

K

Kepone: 448, 453
killer T cells: 86, 88, 93
Korean Institute of Science and Technology (KIST): 510

L

labor force: 357, 358–366
change: 365–366
women: 358, 360, 365–366
land: 372
Landsat: 43, 44, 55
landslides: 18
language:
assembly: 229
computer: 227–231, 243
neuronal basis: 115
robots: 247–248
large-scale integration (LSI): 217
lasers: 174–175, 179, 310, 311, 321
Law of the Sea: 27, 505

lead: 451, 452
learning: 113–115
  neuronal basis: 113
Le Club des Amis du Sahel: 511
legal aspects
  computers: 244–245
  demography: 351, 359, 366
  earth: 20, 23, 25, 27–28, 31
  education: 475–476
  energy: 266–267
  environment: 443, 450, 451, 457–458, 459
  health: 419
  materials: 317, 326, 327, 330
  project funding: 484
Leo: 150
leptons: 191, 196–199
life expectancy: 350–351, 383–385, 412
  by sex: 385
light
  plant response: 129–130
light-emitting diodes (LED): 310–311
lighting: 309
light water reactors (LWR): 272, 282–283, 284, 285
limbic system: 100–101, 114
Lindane: 453
liquefaction: 269
liquid crystals: 311
liquid metal fast breeder reactor (LMFBR): 283–284, 284–285
liquids: 163
  crystalline: 163
lithium: 409
lithium carbonate: 406
lithographic technique: 219, 311
Love Canal: 447
lymphocytes: 85, 86, 88

M

magnesium: 333, 334
magnesium oxide: 333
magnetic bubble technology: 220–221
magnetic mirror: 277
magnetism: 166, 167, 168, 172, 175, 187, 220–221, 311
  alloys: 306
  glassy metal: 305
  momentary: 183
  superconducting: 309
  ultra-thin film: 169
magnetohydrodynamic power generation: 309
mammography: 399
manic psychosis: 405, 408
marriage: 368, 375
  age: 356–357
mass, nuclear: 187
materials: 301–340
  biomedical: 316–317
  composite: 315–316
  consumption: 301
  crystalline: 164–165, 216, 220, 280
  cycle: 323–324
  development: 302–317
  energy-related: 308–310
  high-strength: 169
  information-related: 309–311
  inorganic: 313–315
  layering: 166–167
  manufacturing: 317–322
  metal-like: 167, 168
  near-term issues: 326–335
  noncrystalline: 279, 305
  novel: 166–167
  performance: 325
  processing: 317–322
  properties: 325
  R&D: 301–302, 323
  renewable: 316
materials processing: 317–322, 323, 324, 326
  barriers: 322
  primary: 317
  property improvement: 318
  secondary: 317, 318–322
materials science and engineering (MSE): 324–325
mathematics and theoretical computer sciences: 231–236
matter: 192
  condensed: 163–170
measles: 89, 90
measurement: 204
  particle physics: 203–204
meltdown: 271
membrane
  receptors: 81
  structure: 173
memory: 113–115, 414
  bubble: 221, 311
  computer: 311

neuronal basis: 113
silicon circuit: 221
meningitis: 89
mental illness: 404–412
aftercare: 407
treatment: 406
mercury: 446, 448, 451
mercury telluride: 310
meson: 192, 193, 194, 196
charmed: 194
D: 194
production: 200
metals: 12, 22, 24, 166, 168, 289, 302–307
processing: 317, 318
production facilities: 307
removal: 321–322
methane: 265
methyl mercury: 451, 455
microcomputer: 217, 220, 222–224, 226
microelectronics: 217, 320
microprocessors: 226
Mid-Atlantic Ridge: 12
Mid-Ocean Dynamics Experiment: 29
migration: 347, 353–356
implications: 368–373
Milky Way: 155, 156, 157, 160
mind: 95
minerals: 21–26, 43
exploration: 24–25
formation: 21–23
nutrition: 131–132
oceanic: 26–27
policy: 54
shortages: 21
minicomputers: 222–224
Minimata Bay: 451
mining: 25
oceanic: 26–27
Mirex: 448, 453
mirrors for solar power: 279
models
climate: 34, 39, 498
collective: 185
mineral economics: 24
nuclear: 184
nuclear shell: 184–185
plate tectonics: 13
surfaces: 165
molecular beam epitaxy: 168
molecules
biology: 60–62
electronics: 177–178
engineering: 168–169
genetics: 63–75
structure: 170–171, 174
molybdenum: 306
monoamine oxidase (MAO): 406, 409
mortality: 347, 350–351, 375
by race: 383
cancer: 399, 401–402
cardiovascular: 388
rates: 382–383
motivation: 112–113
motor system: 100
nerves: 109
neuronal basis: 111
Mount Hopkins: 161
Mount Palomar: 161
Mount Semirodnika: 161
muon: 176, 191, 196, 197, 200
decay: 200
muscle: *see* motor system
mutagens: 450, 456–457
mutations: 70–71, 78–80, 136, 455

N

National High Blood Pressure Education Program: 392
NATO: 501–502
natural gas: 263, 265
uses: 122
nebulae: 148
nephritis: 386
nerve cells
function: 102–107
nervous system: 96–102
neuroanatomy: 101
neurological disease: 452
neuron: 97–99, 100, 101, 102, 111, 113, 114
neuroscience: 95–116, 413–414
neurotransmitters: 81, 104, 106, 107, 111, 114, 408–410, 413–414
neutrino: 149, 154, 190, 191, 197, 199, 200, 203
astronomy: 156–157
neutron: 180–181, 182, 186, 189, 191
production: 200
nickel alloys: 305
nickel-base superalloys: 303, 314–315

niobium: 306
niobium-germanium: 309
niobium-tin: 309
nitrites: 450–451
nitrogenase: 126, 127
  structure: 123
nitrogenase enzyme system: 123, 124, 125–126
nitrogen fixation: 122–132
  genetic aspects: 124
  improvement: 124
nitrogen-fixing legumes: 124–125
nitrogen oxides: 36, 273, 274–275, 287, 289, 447
Nobel Prize
  computers: 216
  DNA: 68–69
Nonfuel Minerals Policy Study: 326
norepinephrine: 106, 112, 408
North Pacific Experiment: 28
nova: 160
nuclear fission: 260–261, 269–272, 275, 282–286, 290
nuclear fusion: 261, 275, 276–278, 290
  hazard: 277
  reactor: 277
nuclear magnetic resonance (NMR): 172–173, 179
nuclear plants: 270
nuclear power: 274, 506
nuclear reactors: 270
nuclear waste: *see* radioactivity, waste
nuclear weapons: 272, 284
nucleon: 180, 181, 184, 185, 186, 190, 191
nucleus: 179–181, 187
  artificial creation: 182–183
  characteristics: 183
  extreme states: 189
  forces: 180–181, 186
  interactions: 187–188
  particle decay: 181
  reactions: 185–186
  shell: 184–185
  structure: 179–190
nutrition: 124, 132–133
  plant: 131–132

O

Occupational Health and Safety Act: 457
oceans: 26–31, 54
  circulation: 13
  climate: 30, 32
  resources: 26–30
ocean thermal energy conversion: 281
office systems: 240–241
oil: 30–31, 39–40, 261, 263, 264–266, 275, 279, 290, 316
omega-minus: 193
optical fibers: 238–239
optical materials: 315
optical microscopy: 325
optic nerve: 108–109
optimization: 230
ore: 12
  deposits: 21–23, 24–25
  research: 23
  sources: 21–23
Orion: 162
outlook;
  agriculture: 54, 139–140
  age profile: 376
  age structure: 377
  aging: 432
  alloys: 336
  astrophysics: 206
  atomic structure: 207–208
  behavior and disease: 430–431
  burden of illness: 429
  cancer: 430
  cardiovascular disease: 429–430
  cell biology: 137
  ceramics: 337
  commerce: 377
  computer-aided design: 338–339
  computer-aided manufacturing: 338–339
  computers: 249–251
  condensed matter: 207
  conservation: 292, 339–340
  cosmology: 206
  demography: 376–377
  doctoral degrees: 487
  earth: 51–56
  earth, international aspects: 55–56
  education: 486–488
  education, women and minorities: 487
  employment: 377
  energy: 54, 291–297
  energy options: 294

energy-related materials: 336–337
energy research: 296–297
energy supply: 292–293
enrollment, science: 487
environment: 55, 293–294, 461–463
family: 376
federal regulation, education: 488
funding, academic science: 487–488
genes and environment: 432
geographic distribution, population: 376
health: 293–294, 428–433
health and basic science: 429
health care delivery: 432
health-services research: 432–433
immunology: 137–138
information-related materials: 336–337
international institutions: 515–516
jobs: 487
materials: 335–340
materials, composite: 337
materials development: 340–341
materials, near-term issues: 339
materials processing and manufacturing: 338
materials, renewable: 338
matter: 206–211
mental illness: 431
metals: 336
minerals: 54
molecular genetics: 135–136
molecular structure: 207–208
neuroscience: 138–139
nuclear structure: 208
nuclear fission: 295–296
nuclear fusion: 294
oceans: 54
particle physics: 208–209
physical sciences: 209–211
polymers: 337
population redistribution: 377
project funding: 488
recycling: 339–340
research and animals: 431
research and graduate education: 486
smoking and adolescence: 431
solar energy: 294–295
substitution: 339–340
overlapped instruction scheduling: 225–226
oxidation: 117, 121
oxygen steelmaking: 317
ozone: 35–36, 446, 447

P

Pacific Ocean: 10
mining: 27
packet switching: 237
pain: 113
Pangea: 12
Pap smear: 399
parasitic infections: 90
Parkinson's disease: 107, 116, 414
parity violation: 188
particles
annihilation: 199–200
creation: 199–200
decay: 193
laboratory production: 200–201
physics: 190–205
see also particle name
particulates: 289
PEP: 202
perception, neuronal basis: 109–111
permanent-magnet electric motors: 306
peroxyacetyl nitrate (PAN): 446
pertussis: 90
pesticides: 128–129, 442, 448, 452, 453
petroleum: 12, 43, 275, 313
phenothiazines: 406, 410
photocenters, chlorophyll: 117–119
photochemistry: 36, 446
photochrome: 129
photolithography: 217–218, 219
photons: 191, 192, 199, 203
photoperiod: 129–130
photorespiration: 121, 122
photosynthesis: 116–122, 127, 173, 174, 274
improvement: 119, 121
photovoltaics: 279, 281
piezoelectric materials: 310
pions: 187, 189–190, 191, 200, 203
decay: 200
placebo effect: 407
planets: 44–45
plants: 313, 316
growth: 127–129
stress: 127, 131
plasma etching: 219, 311
plastics: 311, 312

plate tectonics: 11–16, 45, 51
  implications: 14–15
  problems: 13
pleasure: 113
plutonium: 270, 277, 283, 284
pneumonia: 89, 385, 386
political factors
  materials: 331, 333
pollution: 327, 330, 443
  air: 35, 259, 273, 274, 275, 287, 288, 289, 443, 446, 447–448
  coal: 267
  water: 267, 446, 448–449
polybrominated biphenyls (PBB): 452, 453
polychlorinated biphenyls (PCB): 308, 443, 446, 447, 448, 452, 453, 455
polycrystalline cell materials: 281
polymers: 308, 311–315, 325
polypeptides
  brain: 106–107
polyvinylidene fluoride: 310
population: 347, 350, 368, 375
  aged: 412
  mental patient: 406–407
  redistribution: 368–373, 374
  urban: 355
population center: 346–347
positron: 181, 186, 191, 192, 199–200
positron emission techniques: 115
poverty: 357
powder forging: 319
powder metallurgy: 319–320
power devices: 308
power tower: 281
precision investment casting: 319
pressure-sensing materials: 310
primary health care: 422
primates: 411–412
programs
  analysis: 230
  development: 228–229
  languages: 227–231, 243
project funding: 482–485
  accountability: 483–485
  regulation: 484
  weakness of system: 482–483
projection masking: 219
prostaglandins: 389
prostheses: 316–317
protein
  structure: 72
  synthesis: 71–72
proton: 180–181, 182, 186, 191
  production: 200
psychopharmacology: 404–405
  neurotransmitters: 408–410
  research: 407–408
  therapy: 406
pulmonary dysfunction: 274
pulsars: 153, 156
pulse-code modulation: 237
psychoses: 115–116
Pugwash conference: 514

Q

quantum electrodynamics (QED): 175, 176
quantum number: 184
quark: 193–196
  charmed: 194
quartz: 217
quasars: 156, 157–158, 160

R

radar: 29
radiation: 150, 166
  damage: 70–71, 78, 277, 397
  electromagnetic: 43, 146, 199
  infrared: 159–160, 274
  relic: 150
  solar: 273
  sources: 164
  synchrotron: 151, 164, 171
  therapy: 400
  three-degree background: 150–151
radiative cooling: 176
radioactivity
  decay: 191
  wastes: 13, 54, 271, 277, 285, 286, 506
radio astronomy: 158–159
radioimmunoassay: 399, 404
railroads: 267
random-assignment experiments: 407
rare earths: 306
reactive metals: 306
reactor safety: 271–272
receptors, cell surface: 81
recycling: 323, 324, 330–331
red giant: 149
reflectors: 308

refractory metals: 306, 307
regulation: *see* legal aspects
REM: 111–112
renal disease: 452
reproduction
  human: 499–500
  plant: 129
research, current
  agriculture: 127, 128–129
  astronomy: 156, 157, 158, 160–163
  atomic structure: 175
  biochemical: 81
  biomedical: 62–63
  cell biology: 83, 84–85
  chromosome: 82
  climate: 36, 37–38, 41, 44, 47–48
  cosmology: 150–151
  earth: 51–52
  energy: 285
  environment: 459–460
  health: 427–428
  immunology: 88–89, 90, 91, 94
  matter: 163–164, 165, 166, 167, 168
  minerals: 23
  neuroscience: 96, 101–102, 103–104
  nuclear structure: 183, 185, 189
  oceans: 28–29
  particle physics: 194–195, 203–204
  photosynthesis: 117, 119, 121
  plate tectonics: 14–15
reserpine: 408–409
Resource Conservation and Recovery Act of 1976: 330
resources: 327, 334, 505–506
  and migration: 372–373
respiratory disease: 448
reticular formation: 100–101
retina: 108, 110
retirement: 353, 358–360, 415
rheocasting: 319
rheumatic heart disease: 389
Rh incompatibility: 90–91
rhizobia: 123, 124, 126
ribulose bisphosphate: 119, 121
ribulose bisphosphate carboxylase: 122
RNA: 64–65, 67–70, 71, 72, 114, 173
  messenger: 65–66, 67
robotry: 236, 246–249, 321
  language: 247–248
  sight: 248–249
rooftop water heaters: 278
rubbers: 316
  synthetic: 311–312
Rydberg constant: 175

S

saccharin: 450
Safe Drinking Water Act: 457
San Andreas Fault: 11, 17, 18, 19
satellites: 29, 38–39, 41, 43, 44, 49, 52, 55, 159, 163, 279
  communications: 237
  cosmology: 150–151
  digital communication technique: 238
  solar-power: 281
scanning microscopy: 307
schizophrenia: 115–116, 405, 410
schools: 366–367, 369–370, 375
Scientific Committee on Problems of the Environment (SCOPE): 497
scrubbing, exhaust gases: 267
security
  computers: 240
seismic shaking: 18, 19
selenium: 216, 217
self-employment: 358
semiconductors: 165, 166, 169, 216, 217, 310, 321
  amorphous: 168, 288
  technology: 222, 223
senile dementia: 412, 413–414
sensation, neuronal basis: 109–111
sensors: 309
sensory system: 99–100
serotonin: 106, 112, 408
Seveso: 453
sexual behavior: 112
Seyfert galaxies: 156, 157, 160
shale oil: 261, 263, 264, 313
SIALON: 313, 314
sight: 108–109, 110–111
  robots: 248–249
silicon: 217, 219, 279, 308, 310, 311, 325
  ceramics: 313–315
  chips: 222
  planar technology: 217
silicon carbide: 313, 314–315
silicones: 313
silicon nitride: 313, 314–315
silicosis: 449
single data-stream computers: 225–226

sleep: 111–112, 416
social services: 370–371
software: 227–231, 243
  coordinated parallel processes: 230
  problems: 226, 227–231
  shared data structures: 231
solar cells: 308–309
solar-electric systems: 288
solar energy: 32, 33, 75, 174, 260, 274, 275–276, 277, 278–282, 285, 290
solar radiation: 273–274, 278
solidification: 318–319
solids: 163
  amorphous: 164–165
  crystalline: 163–164
  magnetic: 163
  metal-like: 167
solid state
  physics: 216
  technology: 217
Solvent Refined Coal II (SRC-II): 269
Spacelab 2: 162
space shuttle: 40, 45–47, 160
space technology: 52, 493
SPEAR: 202
spectroscopy
  fluorescence: 171
  molecular: 171, 172–174
  nuclear: 188
  picosecond: 174
spin: 184
spin-labeling: 173
stability
  chemical: 446
  nuclear: 181, 182, 183, 189, 190
Stanford Electron Accelerator Center (SLAC): 201, 202
Stanford Heart Disease Prevention Program: 395
stars: 148, 159
  age: 149–150
  birth: 160
  neutron: 151, 153, 157
  white-dwarf: 151
  X-ray: 151
steels: 304–305
Stockholm Institute for Peace Research (SIPRI): 502–503
stomata: 130
storage, computer: 224
strange particles: 189
stratosphere: 35–37, 39
substitution: 331
suburbia: 353–356
suicide: 405, 410
sulfur oxides: 273, 274–275, 289, 447, 451
superalloys: 303
superconductors: 167, 168, 309–310
supernova: 160
superplasticity: 319–320
Super Protron Synchrotron: 202
suppressor T cells: 88, 92, 93
Surface Mining Control and Reclamation Act of 1977: 23
surfaces: 165–166, 321
  probes: 179
surgery, cancer: 399–400
symbiosis: 124–125
  associative: 126–127
symmetry: 204–205
synapse: 97–98, 102, 107, 109, 113, 114
  damage: 104, 106
  transmission: 103–104, 106
synchrotrons: 178, 201, 202
  see also radiation
synthesis gas: 268
synthetic gas: 268–269

T

tantalum: 306
tar sands: 265
tau: 197
T cells: 86, 91
technology transfer: 333
teenagers:
  Black: 362
  fertility: 349–350
  unemployment: 362, 364
telecommunications: 493
TELENET: 236
telescope
  ground-based: 161
  helium-cooled: 162
  infrared: 161–162
  radio: 158–159, 161
  space: 160
  X-ray: 162
temperature, and climate: 33, 34, 35, 36, 37, 38, 273–274

atmosphere: 263
Big Bang: 150
test standardization: 460
thermonuclear fusion: 169, 309
thin-film infrared devices: 310
thorium: 270, 283, 284
Three-Mile Island: 270, 271
threshold: 457
time-dilatation effect: 175
time-division networks: 237
tissue culture, plants: 129
titanium: 306, 307, 319, 333, 334
titanium dioxide: 310, 333
tokamak: 277
Toxic Substances Control Act (TSCA): 457
toxins, environmental: 439–465
transformers: 305
transistors: 216, 220, 311, 325
transmission electron microscopy: 303, 307
TRANSPAC: 236
transplantation: 91
transportation
electric: 289
mass: 373
tricyclic antidepressants: 406, 409, 410
Tris: 450
tritium: 276, 277
tsunamis: 18
tuberculosis: 385, 408
tungsten: 306
TYMNET: 236

U

ultrahigh vacuum technology: 165
U.N. Conference on Science and Technology for Development (UNCSTD): 512–513
U.N. Conference on Technical Cooperation among Developing Countries: 512–513
U.N. Conference on Trade and Development: 505
undergraduate education: 472, 473
unemployment: 357, 358, 360–365
by age: 362
by occupation: 364
by race: 364
by residence: 364
by sex: 364
duration: 364–365
teenage: 362, 364
United Nations: 495, 499, 500, 504, 505, 509, 512
unity, physical sciences: 205–206
universe
age: 148–149
expansion: 149, 150
universities
and industry: 485–486
funding for research: 470
U.N. Law of the Sea Conference: 26, 31
uranium: 269, 270, 277, 283, 284, 285, 286
urbanization
effect on climate: 33
energy consumption: 259–260
population: 353–356
U.S. Climate Program Act of 1978: 40

V

vaccines: 89–90
vapor-growth technique: 168
vector processors: 225, 226
Venus: 156
very large array (VLA): 161
Virgo: 155
viruses: 71
tumor: 79
volcanoes: 12, 13, 21, 39, 43, 45

W

wastes
chemical: 447
for energy production: 263, 282
gases: 33
recycling: 330–331
see also radioactivity
water
consumption: 289
fluoridation: 425
injection: 264
plants: 130
use: 263
weather: 28, 43, 498
forecasting: 42
see also climate
wind power: 278, 281

women
  daycare: 366
  education: 357, 367, 474, 476–477
  employment: 360, 365–366
  heads of household: 357–358
  income: 364
  taxes and benefits: 366
  unemployment: 364
World Bank: 500, 504
World Climate Program (WCP): 494, 498
World Health Organization: 495, 498, 499–500
World Meteorological Organization: 495
W particle: 199, 202

X

X-ray astronomy: 149
X-rays: 171, 178, 311
  bursters: 151
  diffraction: 325
  EXAFS: 172
  instruments: 162–163
  lithography: 219
  pulsars: 153
  sources: 154–156

Z

Zero Gradient Synchrotron (ZGS): 201
zero population growth: 373
zirconium: 306, 307
zirconium dioxide: 310